Molecular-Based Study of Fluids

Molecular-Based Study of Fluids

J. M. Haile, EDITOR
Clemson University

G. A. Mansoori, EDITOR
University of Illinois at Chicago

Based on a symposium
jointly sponsored by the
ACS Divisions of Industrial
and Engineering Chemistry
and Physical Chemistry
at the 182nd Meeting
of the American Chemical Society,
New York, New York
August 23–28, 1981

ADVANCES IN CHEMISTRY SERIES **204**

AMERICAN CHEMICAL SOCIETY
WASHINGTON, D.C. 1983

Library of Congress Cataloging in Publication Data

Molecular-based study of fluids.
(Advances in chemistry series; 204)

"Based on a symposium jointly sponsored by the ACS Divisions of Industrial and Engineering Chemistry, and Physical Chemistry at the 182nd ACS National Meeting, New York, New York, August 26, 1981."

Bibliography: p.
Includes index.

1. Molecular theory—Congresses. 2. Fluids—Congresses.
I. Haile, J. M. II. Mansoori, G. A. III. American Chemical Society. Division of Industrial and Engineering Chemistry. IV. American Chemical Society. Division of Physical Chemistry. V. American Chemical Society. National Meeting (182nd: 1981: New York, N.Y.) VI. Series.

QD1.A355 no. 204 [QD461] 540s [541.2'2]
ISBN 0-8412-0720-8 82-24373

PRINTED IN THE UNITED STATES OF AMERICA

FOREWORD

ADVANCES IN CHEMISTRY SERIES was founded in 1949 by the American Chemical Society as an outlet for symposia and collections of data in special areas of topical interest that could not be accommodated in the Society's journals. It provides a medium for symposia that would otherwise be fragmented, their papers distributed among several journals or not published at all. Papers are reviewed critically according to ACS editorial standards and receive the careful attention and processing characteristic of ACS publications. Volumes in the ADVANCES IN CHEMISTRY SERIES maintain the integrity of the symposia on which they are based; however, verbatim reproductions of previously published papers are not accepted. Papers may include reports of research as well as reviews since symposia may embrace both types of presentation.

ABOUT THE EDITORS

J. M. HAILE is an Associate Professor of Chemical Engineering at Clemson University. He received his Ph.D. and M.E. degrees in Chemical Engineering from the University of Florida and his B.S. in Chemical Engineering from Vanderbilt University. He has been a visiting research scientist in the Physics Department at the University of Guelph, Guelph, Ontario; at the Science Research Laboratory of the U.S. Military Academy at West Point; and in the Neutron and Solid State Physics Branch at Chalk River National Laboratories, Chalk River, Ontario. During the 1982–83 academic year, he spent a sabbatical leave as a Visiting Associate Professor in the School of Chemical Engineering at Cornell University. His primary research interests are associated with the study of dense fluids by computer simulation; however, he has also published theoretical work on the microscopic structure of gas–liquid interfaces and has performed neutron diffraction experiments on mixtures of dense gases. Professor Haile is the 1980 recipient of the Clemson University, McQueen Quattlebaum Faculty Achievement Award.

G. ALI MANSOORI is a Professor of Chemical Engineering at the University of Illinois at Chicago. He received his Ph.D. from the University of Oklahoma, M.Sc. from the University of Minnesota, and B.Sc. (Summa Cum Laude) from the University of Tehran, all in Chemical Engineering. Before joining the University of Illinois, he spent a year as a postdoctoral fellow at Rice University. He has been a consultant with Argonne National Laboratory and Science Applications, Inc. Dr. Mansoori's primary field of research is statistical mechanics and thermodynamics for prediction of fluid properties and phase equilibria for chemical and engineering process design.

CONTENTS

PREFACE

THE MOLECULAR THEORY OF LIQUIDS AND DENSE GASES is currently in the midst of a healthy period of growth and expansion. Much of the activity in the area has been instigated by the wide-spread availability of high-speed digital computers coupled with significant advances in a number of experimental methods for measuring fluid properties and exploring fluid-phase behavior. The computer has not only provided the means for generating quantitative results for problems defying analytic solution, but it also has enabled direct simulation of molecular behavior in fluids via techniques known as Monte Carlo, molecular dynamics, and Brownian dynamics. Important experimental advances include high-flux nuclear reactors and pulsed-neutron sources for determining a variety of static and dynamic fluid properties; lasers for extracting information on dynamic relaxation processes; improved molecular beams for ascertaining details of intermolecular pair potential functions; and ellipsometry for probing fluid interfaces. These various computer simulation and experimental methods are providing molecular theorists, as never before, with a wealth of data to be digested, organized, interpreted, and made predictable.

Numerous theoretical tools have been developed in attempts to cope with the profusion of simulation and experimental data. The more successful theoretical developments include integral equations for molecular distribution functions, perturbation and variational theories, analytic expressions for the thermodynamic properties of the hard-sphere and Lennard-Jones fluids, and improved forms for intermolecular potential energy functions. The success of the molecular approach to the study of fluid behavior is indicated by the fact that many of these theoretical advances are replacing empiricisms in engineering design and process analysis computations. Furthermore, molecular-based corresponding states and conformal solution theories are now widely used by the engineering community.

Thus, the molecular-based study of fluids is a multidisciplinary endeavor that involves chemists, physicists, and engineers. This volume reflects the breadth of the endeavor as indicated by the variety of phenomena under investigation, the diversity of scientists and engineers involved in the research, and the internationally recognized importance of the problems to be solved. In this collection of papers, we have emphasized, with some exceptions, static properties at the expense of dynamic properties, because substantially more progress has been made in resolving difficulties in the theory of static properties. The only constraints we have placed on the authors are to insist that results take priority over methodology and that the papers present a juxtaposition of two from the following triad: theory, experiment, and computer simulation. We hope this collection of papers communicates to the research specialist, the curious nonspecialist, and the practicing engineer the recent progress made towards a more complete explanation of fluid-phase behavior.

J. M. HAILE
Cornell University
Ithaca, NY 14853

G. A. MANSOORI
University of Illinois at Chicago
Chicago, IL 60680

November 29, 1982

1

Molecular Study of Fluids: A Historical Survey

G. A. MANSOORI

University of Illinois at Chicago, Department of Chemical Engineering, Chicago, IL 60680

J. M. HAILE

Clemson University, Department of Chemical Engineering, Clemson, SC 29631

This introductory chapter traces the development of the molecular theory of fluids as it has evolved over roughly the last 200 years. Many of the modern variations of molecular theory applied to fluids originated in the last quarter of the 1800s with the contributions of van der Waals; this chapter is organized to reflect that fact. The overview of present day techniques presented here includes brief discussions of theoretical, experimental, and computer simulation methods. The intent is to provide some historical perspective for the remainder of this volume.

THE OBJECTIVE OF THE STUDY OF FLUIDS from a molecular basis is to develop means for accurately predicting thermophysical properties and local structure in fluid systems. The thermophysical properties of interest include thermodynamic properties, transport properties, and phase equilibrium behavior. Local structure in fluids is measured by spatial and temporal distribution functions; in general, these distribution functions are proportional to the probability of finding molecules at particular points in the fluid at particular times. To attain the desired predictive capability, molecular theories usually start from a few well-defined characteristics of the constituent molecules. These characteristics typically include the geometric structure of individual molecules; the nature of forces acting among different molecules (i.e., intermolecular potential energy functions); and the nature of forces acting among sites on individual molecules (i.e., intramolecular potential energy functions).

In recent years, the molecular-based study of fluids has been motivated not only by scientific demands to improve on existing knowledge, but also by practical demands from increasingly sophisticated industry. Hence, developments in molecular theory are serving as a foundation for engineering design calculations in a growing number of industrial

0065-2393/83/0204-0001/$07.75/0

situations in which fluids are the primary media for transporting matter and energy and for supporting chemical reactions.

Throughout this book, *fluid* refers collectively to the liquid and gaseous states of matter, including those states in the region of the phase diagram above the critical temperature and critical pressure. The distinctions between gases and liquids and between liquids and solids are not so easy to put into words as one might think. It should suffice to say that by *gas* we mean a substance whose volume increases continuously and indefinitely as the system pressure is reduced isothermally. In contrast, a *liquid* is a substance whose volume does not change continuously, without limit, if the system pressure is either increased or decreased isothermally. Further, a liquid is an equilibrium state of matter and, therefore, is distinct from an amorphous solid. Figure 1 gives a representation of the three phases.

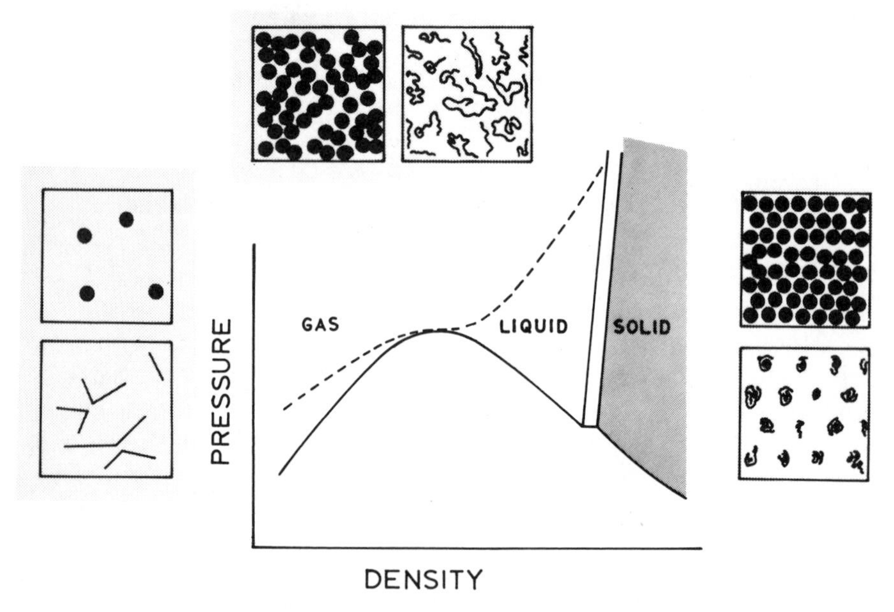

Figure 1. Macroscopic and microscopic descriptions of solid, liquid, and gas phases.

The center is a schematic pressure–density diagram for a pure substance. The solid lines represent two-phase equilibria between pairs of the three basic states: solid, liquid, and gas. The broken line represents the critical isotherm, which goes through a point of inflection at the gas–liquid critical point. The inserts are schematic representations of computer simulation results for each of the three basic states. Each insert shows a typical packing configuration for the spherical molecules at one instant of time and typical molecular trajectories for a period of time. The left insert represents the gas phase, which is characterized by low density and long, straight-line trajectories occasionally interrupted by binary collisions. The solid phase, represented in the right insert, exhibits high density and close packing of the molecules, which are essentially confined to fixed sites. The central insert represents the dense fluid and liquid states, which have features, on the molecular level, between the extremes represented by solid and gas.
(Reproduced with permission from Ref. 149. Copyright 1981, Scientific American.*)*

In this chapter we survey many of the significant developments in the evolution of molecular theory applied to fluids. The goal here is to provide background and perspective that will, to some extent, give context to the various chapters of this volume. Because of the central importance of the work of van der Waals, we have organized the presentation into three major sections:

1. The era before van der Waals
2. The contributions of van der Waals
3. The era after van der Waals

In the hundred years or so since van der Waals, progress toward a comprehensive molecular theory of fluids has been propelled by the interaction of theory, laboratory experiment, and, more recently, computer simulation. Each of these three modi operandi is addressed in the discussion of the era after van der Waals.

The scope of this survey chapter is necessarily restricted. Although intramolecular and intermolecular potential energy functions are usually taken as known in molecular-based studies of fluids, the measurement of such potentials and development of realistic model potentials is a complex undertaking. We do not discuss studies of potential functions, per se, other than to describe a basic pair potential in Figure 2. The chapter by Murthy et al. in this volume and other literature (*1–3*) deal with that subject in more detail. There are approaches toward prediction of physical properties in which detailed knowledge of potential functions is avoided; Kerley's contribution to this book is one such approach.

More generally, the scope of this introductory chapter is limited to prominent developments in the study of static properties of fluids. The study of dynamic properties—transport properties, relaxation processes—is an important and active area of current research. In this volume, the chapter by Kiefer and Visscher presents an original attack on one of the many problems in that area. Other publications focus on the description of dynamic properties of fluids (*4–6*).

We have, in this introductory chapter, sacrificed thoroughness in documentation in favor of an educational tone directed toward those readers who are uninitiated in the mysteries of fluid state physics. We hope those experts who do not find their work directly referenced here will be able to find such citations in other chapters of this volume. In any event, a thorough discussion of the historical development and recent progress toward a complete molecular theory of fluids would be a veritable feast compared to the meager apéritif presented here.

The Era Before van der Waals

Aside from a fairly superficial discussion of forces in fluids by Isaac Newton in his *Principia Mathematica* near the end of the 17th Century

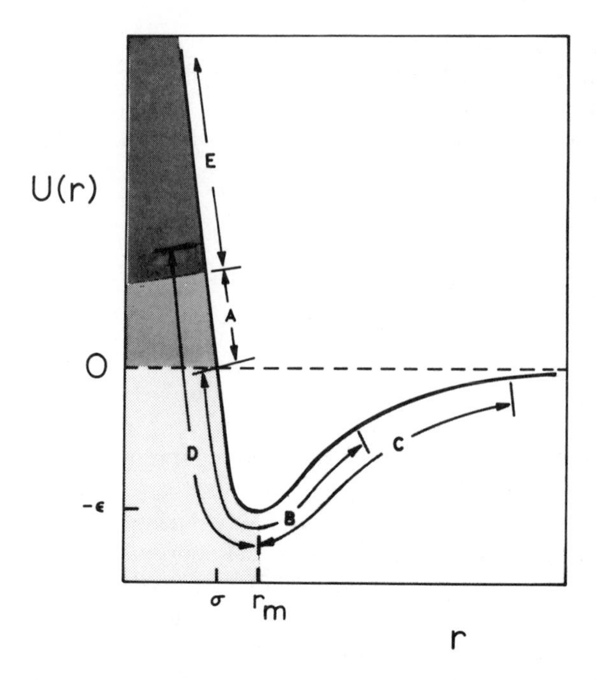

Figure 2. Intermolecular potential functions.

A schematic intermolecular potential energy function is shown for two spherically symmetric molecules, separated by a distance r. *For separations* $r < r_m$, *the net force acting on the molecules is repulsive and arises primarily from overlap of electron clouds. For* $r > r_m$, *the net force is attractive and is primarily due to induction of temporary dipoles via momentary distortion of the electron clouds. In general, a number of experimental techniques are required to determine the full potential curve shown here. These techniques include: high* (A) *and low* (B) *temperature measurements of second virial coefficients, high* (D) *and low* (C) *temperature measurements of dilute gas viscosities, and molecular beam experiments* (B *and* E). *The letters* (A–E) *indicate the approximate portion of the potential curve that is most sensitive to the corresponding experiment. In addition, solid-state properties may be used to check values of the potential parameters* ε *and* σ *determined from other experiments and to estimate the magnitude of multi-body interactions, as opposed to the purely two-body potentials described in this figure* (3).

(7), there was little important experimental or theoretical work done on dense fluids until the early 1800s. In 1808 Pierre Simon Laplace published the first book of his multivolume treatise, *Mécanique Céleste*, which contained a description of fluid equilibrium and fluid motion (8). In a supplement to the treatise, Laplace developed a theory of capillary phenomena that would influence many later workers, including van der Waals.

Early Motivation for the Study of Fluids. There were several reasons for the early research on fluids at moderate to high densities. A primary motivation was to demonstrate that any substance would, under the proper conditions of temperature and pressure, exist in any of the three states of matter—solid, liquid, and gas. This had been conjectured much earlier than 1800; John Dalton was merely repeating the conjecture when in 1807 he included the following statements in his formulation of the atomic theory (9):

> *There are three distinctions in the kinds of bodies, or three states . . . ; namely, those which are marked by the terms elastic fluids, liquids, and solids. A very familiar instance is exhibited to us in water, . . . which, in certain circumstances, is capable of assuming all the three states.*

A second motivation was the study of deviations from ideal gas behavior. The combined laws of Charles, Boyle, and Gay-Lussac produced an ideal gas equation of state of the form

$$PV/T = R \tag{1}$$

in which P is the absolute pressure exerted by a gas of specific volume V at absolute temperature T, and R is a constant. There was considerable interest in determining whether Equation 1 applied to all gases and in discovering the range of conditions of state over which Equation 1 was valid.

A third motivation was the accumulation of indirect evidence for the existence of intermolecular forces in matter. Again, Dalton had reached very clear ideas on this subject, as indicated by his conclusion (*10*) that

> *The constitution of a liquid, as water, must then be conceived to be that of an aggregate of particles, exercising in a most powerful manner the forces of attraction and repulsion, but nearly in an equal degree.*

Early Work to 1850. The experimental challenge in research on fluids in the nineteenth century was to liquefy all substances that were known to be gaseous near ambient temperature and atmospheric pressure. It is not known who first discovered that gases could be liquefied by a combination of cooling and compression, but the Dutch chemist van Marum was probably the first to knowingly liquefy a substance—it happened to be ammonia—liquefied in the latter part of the eighteenth century (*11, 12*). By 1800, several gases had been liquefied; Michael Faraday has given a historical review of much of that early work up until 1823 (*13*).

In the years 1822–23 Charles Cagniard de la Tour reported (*14–16*) his observations of what we now call the gas–liquid critical point. Cagniard de la Tour was not attempting to liquefy gases; rather, he was approaching the problem from the opposite direction by vaporizing liquids in a sealed tube. He found that above a certain temperature, a liquid could be completely vaporized in an enclosed container. Although largely neglected, this discovery was of crucial importance to both the practical problem of gas liquefaction and the more abstract problem of developing a molecular theory of fluids.

In 1823 Sir Humphrey Davy and Faraday reported the temperatures and pressures at which several gases could be liquefied (*17*). Their studies included chlorine, ammonia, hydrogen sulfide, hydrochloric acid, and a number of other substances. Subsequently, Faraday undertook his well-known work on electricity and electrolyte solutions, but in the early 1840s he sought relief from that intense labor by returning to the problem of gas liquefaction. He reported successful liquefaction of a number of gases (*18*), reaching temperatures of about −90 °C and pressures of 50 atm. However, he was unable to liquefy hydrogen, oxygen, nitrogen, carbon monoxide, methane, and nitric oxide. These became known as "permanent" gases and, over the next decade or so, many researchers made serious, though unsuccessful, attempts to liquefy them. Prominent among those attempts was the work of Natterer, who, it is reported (*19–22*), applied pressures of as much as 3000 atm., but at room temperature.

Faraday himself did no further work on the problem, but he was familiar with Cagniard de la Tour's discovery and understood its implications. Thus, in 1845 we find Faraday writing (*18*):

> *Again, that beautiful condition which Cagniard de la Tour has made known, and which comes on with liquids at a certain heat, may have its point of temperature for some of the bodies to be experimented with, as oxygen, hydrogen, nitrogen, etc., below that belonging to the bath of carbonic acid and ether; and in that case, no pressure which any apparatus could bear would be able to bring them into the liquid or solid state.*

Faraday's published papers on gas liquefaction have been assembled into a little volume by the Alembic Club of Edinburgh (*23*).

In 1850 Pierre Eugene Marcelin Berthelot performed an experiment that demonstrated the presence of cohesive forces in liquids. The experiment involved sealing a liquid in a glass tube and heating the system until the liquid expanded, filling the tube. The tube was then cooled, producing tensions of as much as 50 atm. before the liquid collapsed. Two decades later, van der Waals incorporated the idea of cohesive forces into his equation of state.

Heidelberg, 1851–61. During the 1850s, while fruitless efforts were being made to liquefy the "permanent" gases, a collaboration between Robert Bunsen and Gustav Kirchhoff at Heidelberg culminated in 1859 with the development of spectroscopy—an analytical technique that was to have far-reaching impact on much of science, including the study of fluids. Almost immediately, Kirchhoff turned the spectroscope towards the sun and identified a number of elements in the solar spectrum. In the mid-1860s Janssen (*24*) and Lockyer (*25*) performed spectral analyses of the solar atmosphere and identified a new element that Lockyer named helium. A popularized account of the Bunsen–Kirchhoff discoveries has been given by Gingerich (*26*).

During the period 1859–60 the Russian chemist Dmitri Ivanovich Mendeleev was visiting Bunsen and Kirchhoff at Heidelberg. Mendeleev established a small laboratory of his own there and studied capillary phenomena of fluids, deviations from ideal gas behavior, and thermal expansion of liquids (*27–30*). In 1860, Mendeleev was studying surface tensions of liquids when he rediscovered the gas–liquid critical point—which he called the "absolute boiling point" (*31*). This discovery confirmed the earlier work of Cagniard de la Tour and was repeated a year later by Andrews. While in residence at Heidelberg, Mendeleev attended a conference in Karlsruhe that directed his thinking along the path to the periodic table.

The Era of Andrews and van der Waals, 1861–73

Thomas Andrews was trained as a medical doctor and had practiced medicine before assuming a position as professor of chemistry in Belfast. Andrews had begun his research on fluids by attempting to liquefy the permanent gases. Failing at this, he turned, in 1861, to carbon dioxide. He found that at temperatures below 31 °C, carbon dioxide could be liquefied by applying sufficient pressure. However, above 31 °C liquefaction would not occur at any pressure. Andrews hypothesized that such a state existed for all fluids and called it the critical point. His experiments were first made public in the 1863 edition of W. A. Miller's textbook (*32*). In 1869 Andrews gave the Royal Society's Bakerian lecture, which he entitled "On the continuity of the gaseous and liquid states of matter" (*33–35*).

During the years 1862–72 a young Dutchman, Johannes Diderik van der Waals, was engaged in his doctoral work at Leiden. His research on fluids was a theoretical study based on Maxwell's kinetic theory of gases and Laplace's studies of capillary phenomena. On learning of Andrews's identification of the critical point, van der Waals resolved that his theory should account for the behavior of fluids both above and below the critical point. By accepting the molecular hypothesis, including the ideas that molecules are of finite size and exert forces on one another, van der Waals arrived at his celebrated equation of state (*36*)

$$P = RT/(V-b) - a/V^2 \qquad (2)$$

Here, a and b are parameters characteristic of a particular fluid. Parameter a measures the attractive forces among the molecules, and parameter b measures the molecular volume. This equation of state unified the experimental knowledge of 1875, for it not only accounted for deviations from the ideal gas law, Equation 1, but also predicted gas–liquid equilibrium and the existence of a critical point. Further, van der Waals concluded that critical phenomena result from a balance of contributions from short-range repulsive forces and long-range attractive forces acting between molecules. This conclusion is perfectly valid today, and we now know that the a/V^2 term in Equation 2 is the rigorous consequence of assuming weak attractive forces acting at long range (*37*, *38*).

Surprisingly enough, the qualitative accuracy of Equation 2 extends far beyond the experimental knowledge of the 1870s, for by extending it to mixtures, van der Waals and others (*39*, *40*) have predicted a wealth of phase equilibrium behavior. For example, van der Waals used Equation 2 to predict the possibility of phase separations in binary mixtures above the critical point. This is gas–gas equilibrium, so-called, and its existence was experimentally verified in 1940 (*41*). Gas–gas equilibrium is discussed in the chapter by Deiters in this book. A general classification scheme for fluid phase equilibria, originally based on solutions of Equation 2 for mixtures, is discussed in the chapter by Shing and Gubbins. Though qualitatively correct, the van der Waals equation is quantitatively inaccurate in the high density regions of the phase diagram. In industrial situations, it has been supplanted by more reliable, albeit more complicated, equations of state.

In addition to the equation of state, Equation 2, van der Waals developed the principle of corresponding states. This principle hypothesizes that the functional relations among pressure, temperature, and volume are the same for all fluids, and hence the phase diagrams for all fluids can be made to coincide by a proper scaling of P, T, and V. Physically, the critical point is the "corresponding" state among all fluids, so the scaling can be accomplished by using the critical values P_c, T_c, and V_c. A graphic interpretation of the idea of corresponding states is presented in Figure 3.

The thesis of J. D. van der Waals "On the Continuity of the Liquid and Gaseous States," was published in 1873. In 1910 he was awarded the Nobel prize for his work on fluids. In his Nobel lecture, van der Waals referred to the importance of a molecular approach to fluids in the following way (*42*):

> *It will be perfectly clear that in all my studies I was quite convinced of the real existence of molecules, that I never regarded them as a figment of my imagination, nor even as mere centres of force effects. I considered them to be the actual bodies . . . . When I began my studies I had the feeling that I was almost alone in holding that view. And when, as occurred*

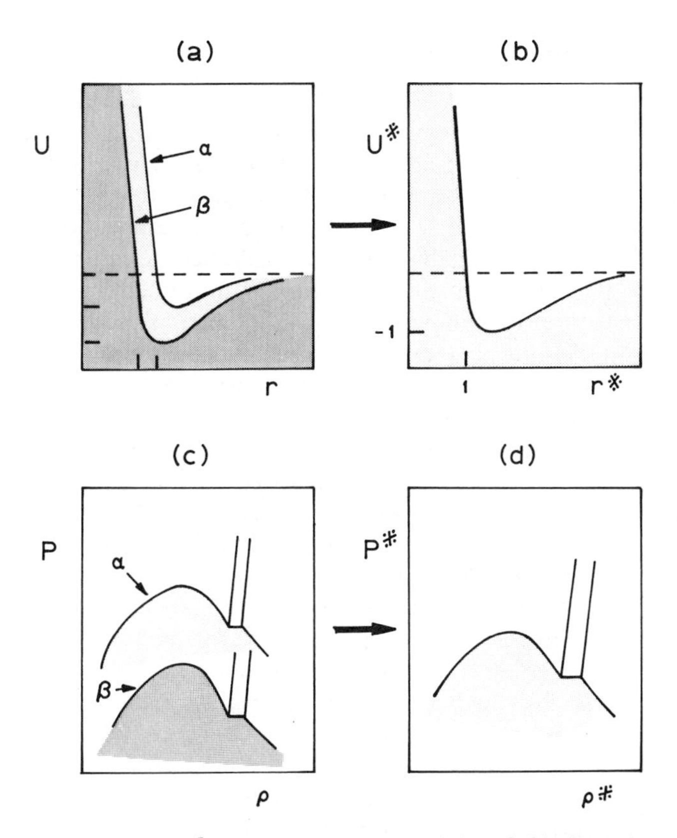

Figure 3. Corresponding states. Top, parts (a) and (b) show the microscopic explanation of corresponding states. Bottom, parts (c) and (d) show use of this concept.

In (a), the intermolecular potential energy functions of two pure substances α and β have the same functional form but different values of the parameters ε and σ. Thus, these two potential curves may be made to coincide by plotting in reduced quantities, $U^* = U/\varepsilon$ and $r^* = r/\sigma$, as is done in (b). The utility of the corresponding states idea is illustrated in (c) and (d). In (c), the phase diagrams of substances α and β occupy different regions of pressure–density space. However, the diagrams can be made to coincide by plotting in reduced quantities, as in (d). The reduction may be done either in terms of potential parameters, as shown here, or in terms of critical properties.

already in my 1873 treatise, I determined their number in one gram-mol, their size and the nature of their action, I was strengthened in my opinion, yet still there often arose within me the question whether in the final analysis a molecule is a figment of the imagination, and the entire molecular theory too. And now I do not think it any exaggeration to state that the real existence of molecules is universally assumed by physicists. Many of those who opposed it most have ultimately been won over, and my theory may have been a contributory factor. And precisely this, I feel, is a step forward.

The Era After van der Waals

Molecular Theories. BACKGROUND WORK, 1873–1905. Even though the van der Waals equation of state, Equation 2, does not tell us much about microscopic structure in dense fluids, it represents the first successful interpretation of macroscopic fluid properties in terms of molecular quantities. Thus, by 1873 a reasonably accurate theoretical description of fluids was available before the practical requirements of science and engineering demanded it. During the remaining years of the nineteenth century, two gifted theorists—Boltzmann and Gibbs—established the foundations of statistical mechanics, the bridge that connects molecular behavior with macroscopic fluid properties.

Ludwig Boltzmann set out to obtain a molecular interpretation of the second law of thermodynamics. That work culminated in a demonstration that the second law has a statistical character (personified as Maxwell's demon). During the course of his work, Boltzmann accomplished a number of other objectives. These included a rigorous proof of Clerk Maxwell's kinetic theory of gases; development of many of the fundamental concepts of statistical mechanics, such as phase space, ergodic systems, and the H-theorem; and a derivation of the Boltzmann transport equation. His work on fluids is summarized in *Vorlesungen über Gastheorie*, for which there is an English translation by Brush (*43*).

Boltzmann was as much interested in the philosophy of science, metaphysics, as he was in ordinary physics, which he called orthophysics. Thus, he was involved in arguments raging at the time as to whether molecules actually exist or are merely models of nature. By 1895, Boltzmann was satisfied that the kinetic theory of gases verifies the atomic theory (*44*):

> *But this theory (the kinetic theory) agrees in so many respects with the facts, that we can hardly doubt that in gases certain entities, the number and size of which can roughly be determined, fly about pell-mell.*

J. Willard Gibbs was the preeminent American scientist of the nineteenth century, though this fact was not immediately recognized. Gibbs earned a Ph.D. in 1863 from the Sheffield Scientific School of Yale University for his thesis, "On the Form of the Teeth of Wheels in Spur Gearing," a decidedly practical problem. He also invented and patented a railway car brake and invented a governor for steam engines (*45*). It is interesting that two of the theoreticians with the most fertile minds of the last hundred years—Gibbs and Einstein—should both have served apprenticeships in such practical, mechanical environments as provided by gears, brakes, and patent offices. From 1866 to 1869 Gibbs traveled in Europe, absorbing the leading scientific ideas of the time. During

1868–69 he visited Heidelberg, where he was influenced by Helmholtz and Kirchhoff.

From 1871 until his death in 1903 Gibbs served as professor of mathematical physics at Yale. From 1876 to 1878 he published his magnum opus, "On the Equilibrium of Heterogeneous Substances," which appeared in the obscure journal, *Transactions of the Connecticut Academy* (*46, 47*). With this one work, Gibbs single-handedly established the field of fluid phase equilibrium and solved many of its important problems. Among other things, the work included the criteria for phase equilibrium, the Gibbs adsorption equation for concentration in fluid interfaces, and the celebrated phase rule. The work is a monumental achievement in classical thermodynamics; all its results are deduced solely from the first and second laws of thermodynamics. Following publication of that work, Gibbs turned to a study of the molecular explanations underlying his classical thermodynamic results. The product of this further study was the first textbook in statistical mechanics, published in 1902 (*48*).

By the turn of the century, several empirical and semiempirical equations of state, in addition to Equations 1 and 2, were in use. Prominent among them were various forms of the virial equation of state. The virial equation is a Taylor's expansion of the compressibility factor $Z = PV/RT$ in density, ρ, about that for an ideal gas ($Z = 1$),

$$Z = 1 + B\rho + C\rho^2 + D\rho^3 + \ldots \tag{3}$$

where B, C, etc., are the virial coefficients, e.g.,

$$B = \left[\frac{\partial Z}{\partial \rho}\right]_{\rho=0} \tag{4}$$

Kamerlingh Onnes (*49, 50*) in 1901 was one of the first to write down a form of Equation 3, and it was Onnes who suggested the name *virial coefficients*. A summary of the historical development of the virial equation has been given by Spurling and Mason (*51*).

MODERN APPROACHES. For a number of years, the virial equation of state was treated as an empirical expression; values of the coefficients were estimated by fitting to experimental data. In the 1930s, Joseph Mayer (*52, 53*) showed that the virial equation has a rigorous derivation in statistical mechanics and, further, that the coefficients are related to intermolecular forces in an appealing way. Thus, the second virial coefficient, B, is related only to two-body interactions, the third, C, only to two-body and three-body interactions, and so forth. The virial equation has, therefore, assumed importance as both a practical tool and an enlightening statement on the theory of macroscopic and microscopic property relations. Consequently, the problems of, at least, thermodynamic properties of imperfect gases can be considered to be solved.

Unfortunately, the virial equation fails to converge at liquid densities, so new attacks have had to be formulated for a statistical mechanical description of the liquid state. A survey of this work is given in Reference 152. In the brief summary here, we divide these approaches into (1) interpretive techniques, (2) predictive techniques, and (3) perturbation and variational techniques.

Interpretive Approaches. Interpretive approaches to the molecular theory of fluids begin with an approximate description of microscopic structure in fluids, as shown in Figure 4. These approaches are called *lattice* theories because the liquid structure is customarily assumed to resemble the regular lattice structure of crystalline solids (*54, 55*). Assumptions regarding structure must be guided both by physical reality and by the ability to calculate thermodynamic properties of the substance under consideration. For crystalline solids, these requirements are harmonious because solids are known to have regular structures that are disturbed only slightly by thermal motion. Such regular and more or less static structures lead to a statistical mechanical partition function that can be used for calculations of thermodynamic properties. Liquids, however, offer a serious challenge to the viability of interpretive approaches because liquid structure is continually changing and can be visualized only on an instantaneous basis. To account for this physical reality in liquids, a static average over the instantaneous structure may be used. Such an approach may lead to a satisfactory lattice theory for liquids. Advanced lattice theories of the liquid state require complicated combinatorial mathematics to achieve realistic models of liquid structure (*55*).

Predictive Approaches. Predictive statistical mechanical techniques place initial emphasis on the process by which the intermolecular forces determine the structure, in the hope that a correct mathematical description of this process will lead to equations whose solutions describe the actual liquid structure (*37, 38, 56–58*). Theories of this class are often called *distribution function* theories, because the resulting equations involve molecular distribution functions that specify the probability of finding sets of molecules in particular statistical mechanical configurations. The measurement of local structure in fluids via molecular distribution functions was introduced by John Kirkwood in the mid-1930s (*59*). The definition of such functions was motivated by the discovery in the 1920s that X-ray diffraction could be used to obtain the radial distribution function $g(r)$ in atomic fluids. The function $g(r)$ is a measure of the probability of finding two atomic centers separated by a distance r. One of the earlier and best experimental papers that describes the measurement of $g(r)$ is an extensive one on argon (*60*).

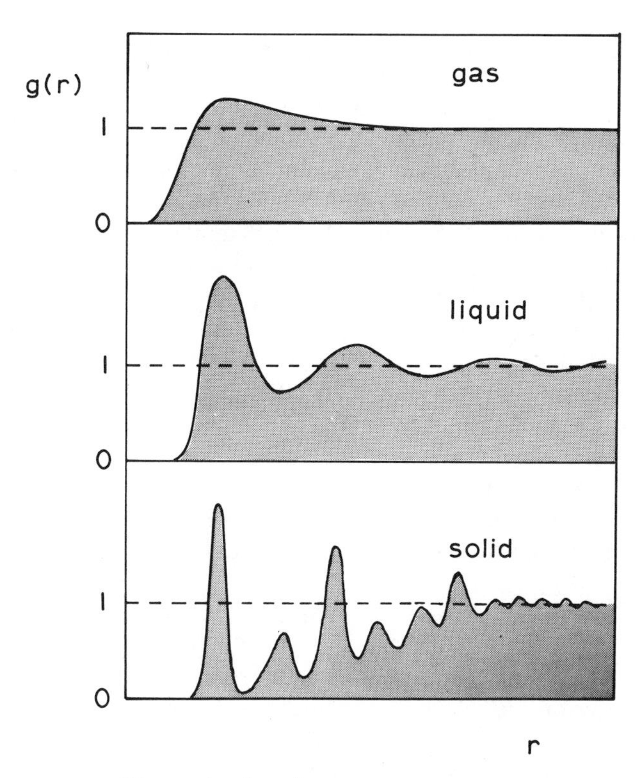

Figure 4. Local structure in matter.

Local structure may be characterized by a pair distribution function g(r) *that is related to the probability of finding pairs of molecules separated by distances* r. *The limiting values of* g(r) *with distance and density are illustrated here. At short range,* g(r) *vanishes because of excluded volume effects; at long range,* g(r) *goes to unity because* g(r) *is normalized by the bulk density. At low densities there is essentially no local structure in the fluid and* g(r) *quickly assumes its long range limit. In crystalline solids, structure persists to larger* r, *and* g(r) *is a series of Gaussian functions, corresponding to molecular vibrations about lattice sites. At liquid densities, the fluid exhibits short range structure similar to the solid and long range disorder characteristic of the gas phase.*

Three distribution function theories, the Yvon–Bogoliubov–Born–Green–Kirkwood (YBBGK), Percus–Yevick (PY), and hypernetted chain (HNC) theories, are the basis for the various theories now available in the predictive class. These three theories have rather different origins, and each requires specific initial assumptions to produce tractable results. It is generally conceded that the PY theory is capable of predicting thermodynamic properties of liquids and vapors more accurately than the other two theories. The YBBGK theory does, however, possess unique features, such as the ability to predict qualitatively a freezing transition. Overall, these theories are deemed sufficient for predicting properties of atomic fluids such as argon.

For molecular liquids, mathematical solution of the distribution function theories is more complicated than for atomic fluids. In the HNC theory for molecular fluids, simplifications can be obtained by expanding a logarithmic term; this leads to the linearized (LHNC) and the quadratic (QHNC) versions of the theory (*61, 62*). The LHNC theory turns out to be equivalent to another theory of molecular fluids called the generalized mean field theory (*63*). In general, LHNC and QHNC are capable of predicting properties of dipolar and quadrupolar fluids. Such developments in the predictive approach have produced analytic results that are of both scientific and industrial importance (*58, 64*).

Perturbation and Variational Approaches. These theories arise from expansions of the free energy, or partition function, of a substance about a relatively simple reference substance for which analytic molecular thermodynamic relations are in hand (*58, 65–71*). The reference system could be a fluid whose molecules interact with a simple potential function and whose properties have been obtained from interpretive or prediction theories or from computer simulation. This is precisely the motivation for the work on hard convex bodies presented by Boublik later in this book.

The basis of perturbation and variational theories is an idea that can be traced back to the van der Waals equation of state, Equation 2. According to this idea, repulsive intermolecular forces are the dominant effect in determining the structure in dense fluids, while attractive forces play only a minor role in determining structure. Attractive forces are largely responsible for maintaining the stability of liquids at high densities.

The successful perturbation and variational theories usually expand about reference fluids whose intermolecular forces are repulsive only. However, there are several exceptions to this general observation. One is the case of associating molecular fluids (those with hydrogen bonding) for which attractive forces among the molecules are so large that they compete with repulsive forces in determining structure (*72*). Another exception is the case of fused salts. In fused salts the attractive charge–charge interactions produce a local structure known as *charge layering* (*73*).

If we consider Equation 2, the van der Waals equation of state, as the first perturbation theory for fluids, then the next advance occurred in the early 1950s when Zwanzig (*74*) introduced a perturbation theory for atomic fluids and Pople (*75*) made an analogous development for molecular fluids. Further progress was not made until the 1960s, when computer simulation data became available for simple model fluids that could serve as references in the theoretical expansions. Perturbation and variational theories have been quite successful in predicting thermodynamic properties of gases, liquids, and solids, and there remain broad prospects for extending these approaches to various kinds of fluid systems.

In this book, the chapter by Henderson summarizes the development of perturbation theories for atomic fluids. In the chapters by Kohler and Quirke and by Smith and Nezbeda perturbation theories are applied to problems posed by molecular fluids. In the chapter by Singh and Shukla these ideas are extended to mixtures of molecular fluids.

Variational theories offer at least one unique feature among molecular theories for fluids. This feature is the prediction of the melting transition or melting curve (*65, 66*). Applications of the variational approach are discussed in detail in the chapter by Kerley.

FLUID INTERFACIAL PHENOMENA. The interfacial regions between bulk phases of matter (e.g., liquid–vapor, solid–liquid, and liquid–liquid regions) have posed fascinating and challenging problems for both theorists and experimentalists. A meaningful description of such interfaces is much more difficult to attain than description of bulk phases because of the nonhomogeneous, anisotropic nature of the interfacial region. Study of interfacial phenomena is of practical importance, for example, in attempting to promote mass and energy transfer across phase boundaries, in catalysis, in lubrication, and in fuel cell technologies. The primary problems to be solved include (1) descriptions of relations between interfacial properties, such as surface tension, and molecular distribution functions; and (2) prediction of species adsorption in the interfacial region.

The historical development of theories for interfacial properties largely parallels that for bulk fluids. Thus, as mentioned earlier, Laplace worked on a theory of capillary phenomena in the early 1800s (*8*). J. D. van der Waals developed a theory of fluid interfaces that was published in 1893 but that has only recently been translated into English, by Rowlinson (*76*). Gibbs also studied interfacial phenomena; he developed much of the classical thermodynamics of interfaces and showed how the interfacial tension is affected by adsorption of species in the interface (*46, 47*).

Modern theories for describing fluid interfaces include an updating of the van der Waals theory (*77*), distribution function theories initiated by Kirkwood and Buff (*78, 79*), and perturbation theories for atomic (*80*) and molecular (*81–83*) fluids. Recently, ellipsometry has been used to measure experimentally the microscopic adsorption of material in a liquid–liquid interface (*84*). Extensive reviews of work on interfacial problems are available (*85–88*) and a new work on fluid interfaces (*153*) is a valuable addition to the literature. The chapter by Fischer in this book deals with vapor–liquid interfaces while the chapter by Henderson addresses the wall–ionic fluid problem. Figure 5 provides an example of the application of computer simulation to problems of interfacial phenomena.

Experimental Techniques. In the 19th Century, conjectural relations between microscopic and macroscopic phenomena in fluids were largely based on inferences from measurements of macroscopic properties. With the development of sophisticated experimental methods has

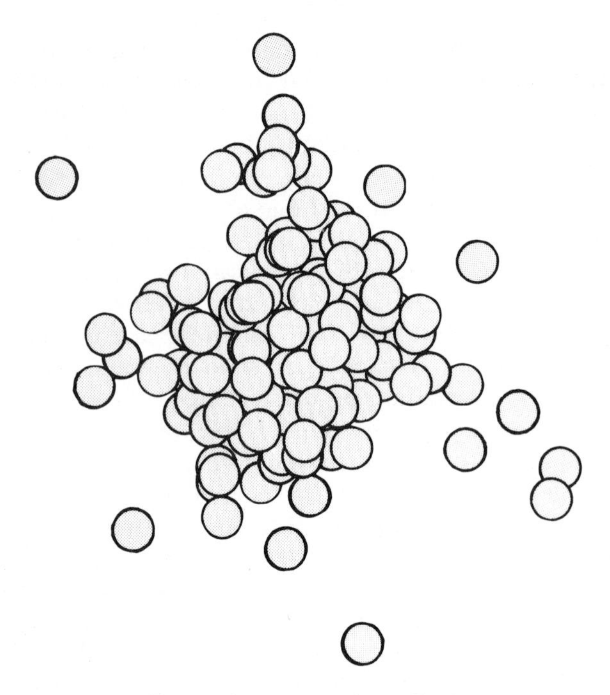

Figure 5. Nonuniform fluids.

This figure shows one configuration from a molecular dynamics simulation of a liquid droplet, in vacuo, initially formed from 138 spherical molecules. The intermolecular interactions in this simulation were modeled by a Lennard–Jones 6-12 potential function. The temperature of the droplet, in reduced units, was $kT/\varepsilon = 0.71$ (150).

come the ability to probe more directly the molecular-scale nature of matter. In this superficial review, we discuss available experimental methods in the following three categories:

1. Thermodynamic property measurements, such as pressure–volume–temperature (*PVT*) experiments, calorimetric methods, and phase equilibrium studies
2. Radiation scattering experiments, including Raman and Rayleigh light scattering, X-ray diffraction, and neutron scattering
3. Molecular relaxation processes, such as transport, dielectric relaxation, ionic diffusion, and migration

Through the van der Waals era, measurement of bulk fluid thermodynamic properties (particularly *PVT* experiments, calorimetry, and phase equilibrium studies) was the primary means of gleaning information on the molecular nature of fluids. The prominent experimentalists of the last hundred years include Dewar, who invented the vacuum insulated bottle and first liquefied air and hydrogen in large quantities (*89*); Ramsay, who, in collaboration with Lord Rayleigh, discovered argon and went on to discover the other noble gases—krypton, neon, and xenon (*90–92*); and Kamerlingh Onnes, who first liquefied helium and explored the behavior of matter at temperatures near absolute zero (*93–96*). The tradition of high quality experimental work at Leiden under Kamerlingh Onnes was continued by Michels. Giauque performed a number of accurate experiments to test the validity of the third law of thermodynamics (*97*). In the first half of the 20th Century several prominent experimentalists resided in the U.S.; these include: Beattie at Massachusetts Institute of Technology, Bridgman (*98*) at Harvard, Dodge at Yale, Kurata at Kansas State, and Sage and Lacey at California Institute of Technology. For an abbreviated historical summary, *see* Reference 99.

Today, the study of thermodynamic properties and phase diagrams under conditions of extreme pressures and temperatures remains of importance both for satisfying industrial needs and for investigating intermolecular forces (*100, 101*). In this book, the chapter by Deiters reports on recent experimental *PVT* and phase equilibrium results and illustrates how such data help support theoretical developments. Often the infinitely dilute region of mixture phase diagrams is neglected in the molecular study of fluids; in this volume, Jonah assesses the importance of infinitely dilute solutions.

Traditional exploration of intermolecular forces in terms of thermophysical properties was carried out only indirectly; that is, assumed forms for intermolecular pair potentials were fitted to experimental data such as second virial coefficients and viscosities. Recently, Smith and coworkers (*3, 102, 103*) have devised methods for directly inverting macroscopic property data to obtain the pair potential in atomic fluids. Further, recent studies of short-range intermolecular forces have been carried out under extreme pressures by imparting shock waves to the fluid (*104*).

Direct experimental study of molecular phenomena in fluids now primarily relies upon radiation scattering methods, particularly light, X-ray, and neutron diffraction. The basic theoretical analysis of light scattered from a sample was first performed by Lord Rayleigh in 1871 (*105*). Since the wavelength of visible light is two to three orders of magnitude larger than the intermolecular spacing at liquid densities, light scattering is unable to provide details of local structure in liquids (*106*). However, in the region of the critical point of both pure fluids and mixtures, spatial correlations become long range, and therefore light scattering is widely used in the study of critical phenomena (*106*). Further, methods are under development for using depolarized light scattering to probe local orientational structure in fluids (*107–109*).

From the time of Roentgen in the 1890s (*110*), there was interest in developing a method for using X-ray diffraction to analyze the molecular structure of matter. By 1915 Debye (*111*) and Ehrenfest (*112*) were aware of how this could be done, and in the 1930s Menke and Debye (*113*) introduced X-ray diffraction as a viable quantitative technique for measuring local structure in fluids. The method may also be used to deduce forms for the intermolecular pair potential in simple fluids (*114*). Today, X-ray diffraction is widely used to study atomic and molecular fluids, fluid mixtures, and liquid metals (*115, 116*). A simplified description of the method is given in Figure 6. For a recent discussion of local structure in fluids, *see* Reference 151.

Neutron scattering is a more recent method, which probes both the static and the dynamic structure in fluids. It is, therefore, a more powerful experiment than X-ray diffraction, which is limited to determination of static structure. Results from neutron scattering experiments are more expensive to obtain and more difficult to analyze than are X-ray diffraction results. The classic book on the neutron diffraction method is that by Bacon (*117*); several more up-to-date reviews on the methods (*118–120*)

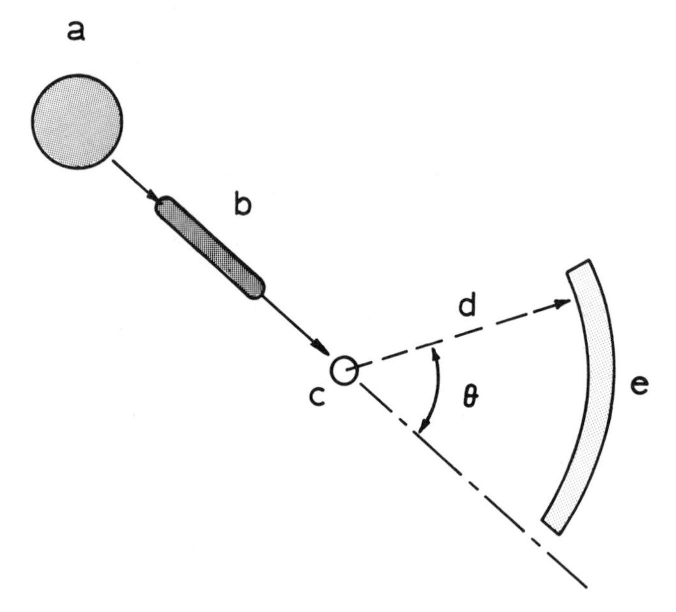

Figure 6. Measurement of local structure.

A simplified schematic diagram is shown of an X-ray or neutron diffraction experiment for measuring local structure in fluids. An X-ray or neutron source, (a) provides an incident beam of radiation that is collimated and made monochromatic (b). The beam is then directed onto the fluid sample (c). One measures the intensity of the scattered radiation (d) with detectors (e) as a function of the diffraction angle θ. The measured intensity I(θ) is normalized with respect to an appropriate reference and corrected for a variety of secondary effects, such as background radiation, absorption, multiple scattering, and inelastic scattering. The normalized, corrected intensity produces the static structure factor S(q), where $q = (4\pi/\lambda) \sin(\theta/2)$, and λ is the wavelength of the incident beam. The Fourier transform of S(q) yields the pair distribution function g(r) (151).

and applications (*121, 122*) are available. Egelstaff and coworkers have recently constructed a sensitive neutron diffractometer and used it in studies of structure and intermolecular potentials in gases (*123–125*).

The general method in molecular relaxation studies is to apply a stress to the fluid and measure the time required for the fluid to come into equilibrium with the stress. Alternatively, the stress is applied and then removed, and the time needed for the fluid to relax back to equilibrium is measured. In dielectric relaxation studies, the stress is an applied electric field; in measurements of diffusion coefficients, it is a gradient in concentration; for viscosity, it is a shear stress, and so forth. Such relaxation studies have not proven as fruitful as the static property measurements mentioned above. It seems likely that further advances in the fundamental theory of nonequilibrium processes will be necessary before a more detailed description of molecular motions can be extracted from experiment.

Computer Simulation. IMPORTANCE. In some ways, molecular theory attains its highest degree of conceptualization in the simulation of matter on digital computers. These simulations use individual particles as the basic entity under study, and a collection of several hundred or a few thousand particles comprises the *system* to be simulated. To define the system properly, one must specify four items:

1. The structure of the individual particles
2. The state of the system; e.g., the number of particles per unit volume of system, the number of particles present of a particular species, the average kinetic energy of the particles
3. The nature of forces acting among the particles
4. The nature of the interaction of the system with its surroundings

The simulation then determines how the system behaves under these four basic constraints. Typically, the system behavior is measured by monitoring the particle positions and various properties that depend on those positions. In some simulations, various time derivatives of the positions and related quantities are also evaluated. Depending on the particular nature of the particles and their relation with the environment, as embodied in the four attributes cited above, the simulation may evoke an interpretation as any of a variety of physical situations. Examples include such interpretations as molecules in a beaker of water, atoms on a polyethylene molecule, molecules on a two-dimensional surface, ions in a high energy plasma, or stellar material forming a galaxy. The book by Hockney and Eastwood gives an overview of many of these applications of computer simulation (*126*).

Much of the importance of computer simulation stems from its interaction with and stimulation of theoretical and experimental work. In comparisons of theoretical and simulation results, simulation plays the role of experiment with the object of ascertaining the adequacy of a proposed theory under conditions that can be met in both the theory and the simulation. In this way certain ambiguities that may arise in comparisons of theory and experiment can be avoided. An example is the study of bulk fluids, in which the same intermolecular force law can be adopted in both theory and experiment, removing the force law as an unknown in testing the theory. Much work has been performed using simulation in this way and one often sees simulation referred to as a *computer experiment* (*127–131*).

An alternative use of simulation is the comparison of experiment and simulation in which one attempts to mimic nature as closely as possible. Using bulk fluid studies again as an example, consider the problem of determining the form of an intermolecular potential function for a particular fluid. A form for the potential is chosen, the simulation is performed, and properties obtained from the simulation are compared with laboratory results for the real fluid. At this point, a trial-and-error approach is adopted in which one alternately modifies the potential function and performs further simulation. In this case, the simulation becomes an extension of theoretical modeling and probably should not be interpreted as an experiment (*72*). Because of the importance of water, extensive efforts have been made to mimic water accurately by computer simulation. In this volume, the chapters by Rossky and Hirata and by Beveridge et al. report on recent advances in the study of water via simulation–experiment comparisons.

This modeling aspect of simulation assumes a larger role than merely an attempt to approximate nature. As in most modeling studies, simulations in this vein enable one to probe how changes in particular aspects of the model affect the overall system behavior. For example, studies may be performed to discover the effects of modifying interparticle and intraparticle forces; system temperature, pressure, or composition; or the strength or character of the system's interaction with its surroundings. The chapter by Szczepanski and Maitland in this book is an illustration of this type of work. Such studies have suggested fruitful lines of inquiry for further work by both theorists and experimentalists.

METHODS AND APPLICATIONS. The application of computer simulation to the study of fluids has its origins in the mid-1950s at certain U.S. national laboratories. The first simulation method to be applied to fluids was the Monte Carlo technique, which appeared in 1953 (*132*), followed by the molecular dynamics technique in 1959 (*133*). Monte Carlo embraces the ensemble averaging concepts of Gibbs (*48*) and is an adaptation of the standard Monte Carlo method for evaluating multidimensional integrals. The basis of the method is the realization of ensemble averages for properties of interest from a Markov chain of particle positions generated for a single system. Wood has given a description of the basic Monte Carlo method (*134*); a more recent collection of applications of the method is also available (*135*).

The molecular dynamics method is based on the kinetic theory of Maxwell (*136*) and Boltzmann (*43*) and involves the determination of particle positions by numerically solving the coupled equations of motion given by Newton's laws. The paper by Hoover et al. in this book summarizes the historical development and current applications of molecular dynamics. More detailed descriptions of the method are also available (*127, 131, 137–139*).

Since the early 1960s, Monte Carlo and molecular dynamics have been applied to systems of increasing complexity. From fluids composed of hard spheres, methods have evolved to deal with soft spheres, rigid linear molecules, rigid nonlinear polyatomics, and flexible polyatomic molecules.

In addition to the ability to simulate complicated molecules, methods have been devised to simulate various system boundary conditions. Examples include gas–liquid and solid–fluid interfaces, as well as moving boundaries that give rise to shear or compression of the fluid. The growing number of problems that can be attacked with simulation has been accompanied by increasing availability of inexpensive, powerful computers. The result is that a large number of researchers are engaged in a variety of very different simulation studies. A catalog of applications of simulation to 1976 has been given by Wood and Erpenbeck (*140*).

In addition to the evolution of methods for simulating particular physical systems, efforts have been made to improve the computing efficiency of the methods. The results of such efforts usually involve some hybrid of Monte Carlo and molecular dynamics that is particularly useful for a specific class of problems. These hybrid methods encompass a spectrum from the purely random motion occurring in Monte Carlo to the purely deterministic motion in molecular dynamics. Such a spectrum is depicted in Figure 7 (*141*). There is not space here to discuss all of these variations on the computer simulation theme; however, the Brownian dynamics method should be mentioned as of growing importance in the study of large molecules. In this book, the chapter by Evans gives a description of the method, and in the chapter by Weber et al. the method of Brownian dynamics is applied to polyethylene.

Review papers are available describing the application of computer simulation to the following problems: fluids composed of hard-core molecules (*138*), dynamic properties of monatomic liquids (*142*), equilibrium properties of fluids composed of linear molecules (*143*), water (*72*), polar fluids (*144*), supercooled liquids and solid–liquid phase transitions (*145, 146*), dynamics of protein molecules (*147*), and intermolecular spectros-

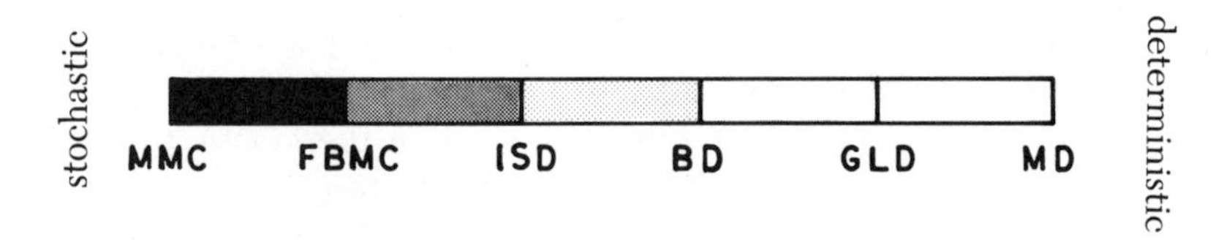

Figure 7. Relative degree of determinism in various simulation methods. *Key: MMC, Metropolis Monte Carlo; FBMC, force-biased Monte Carlo; ISD, impulsive stochastic dynamics; BD, Brownian dynamics; GLD, general Langevin dynamics; and MD, molecular dynamics* (141).

copy (*148*). In the present volume, Shing and Gubbins review the computer simulation methods that have been devised for determining the free energy in pure fluids and mixtures.

Conclusion

The study of liquids and dense gases has a long history; as with most such endeavors, this history is one of problems recognized, encountered, and often overcome. Progress in the study of fluids met an impasse in the first half of the twentieth century when formal theoretical advances were far ahead of the calculational and experimental methods of the day. However, the power and availability of the digital computer has rectified that situation.

In addition to the important advances made possible by the simulation methods discussed in the previous section, the computer makes other contributions to fluid research. For example, the computer has made possible the numerical solution of complicated, nonlinear integro-differential equations and evaluation of the multidimensional integrals that seem to be mainstays of molecular theory. Further, the computer has assumed an important role in controlling experimental devices, as well as in gathering and analyzing experimental data. There is some concern that the impasse of the first half of the 1900s may be reversed in the last quarter of the century, with the amount and detail of data provided by computer simulation and computer controlled laboratory experiments far outstripping the immediate ability of theory to organize and interpret that data.

In any event, the field of fluid state physics is today flowering in richness and diversity. As the seeds of recent successes spread to related fields, we expect to see continued growth in the fundamental study of fluids and in the application of the newly acquired knowledge to problems of both industrial and academic concern.

Literature Cited

1. "Intermolecular Interactions: From Diatomics to Biopolymers"; Pullman, B., Ed., Wiley: New York, 1978.
2. "Intermolecular Forces"; *J. Chem. Soc., Faraday Disc.* **1965,** *40.*
3. Maitland, G. C.; Rigby, M.; Smith, E. B.; Wakeham, W. A. "Intermolecular Forces"; Oxford University Press, Oxford, 1981.
4. Mazo, R. M. "Statistical Mechanical Theories of Transport Properties"; Pergamon Press: Oxford, 1967.
5. Chapman, S.; Cowling, T. G. "The Mathematical Theory of Non-uniform Gases"; Cambridge University Press: Cambridge, 1939.
6. Boon, J. P.; Yip, S. "Molecular Hydrodynamics"; McGraw-Hill: New York, 1980.
7. Newton, I. "Philosophiae Naturalis Principia Mathematica"; Koyré, A.; Cohen, I. B., Eds., Harvard University Press: Cambridge, MA, 3rd ed., 1972.
8. Marquis de la Place, P. S. "Celestial Mechanics"; Bowditch, N. Trans., Chelsea: Bronx, NY, 1966.
9. Dalton, J. "A New System of Chemical Philosophy" Philosophical Library: New York, 1940; Part I, p. 141.
10. *Ibid.*, p. 150.
11. Goodstein, D. L. "States of Matter"; Prentice-Hall: Englewood Cliffs, NJ, 1975; p. 266.
12. *Ibid.*, p. 450.
13. Faraday, M. *Quart. J. Sci.* **1824,** *16*, 229.
14. Cagniard de la Tour, C. *Ann. Chim.* **1822,** *21*, 127.
15. *Ibid.*, **1823,** *22*, 410.
16. *Ibid.*, **1823,** *23*, 267.
17. Faraday, M., *Phil. Trans.* **1823,** *113*, 189.
18. Faraday, M., *Phil. Trans.* **1845,** *135*, 155.
19. Davies, M. "The Physical Principles of Gas Liquefaction and Low Temperature Rectification"; Longmans, Green, and Co.: London, 1949; p. 1.
20. Travers, M. W. "The Experimental Study of Gases"; Macmillan: London, 1901; p. 162–3.
21. *Ibid.*, p. 172.
22. Travers, M. W. "Encyclopedia Britannica"; London, 1929; 14th ed., p. 173.
23. Faraday, M. "The Liquefaction of Gases"; Alembic Club reprint no. 12, University of Chicago Press, 1912.
24. Janssen, P. *Compt. Rend.* **1868,** *67*, 838.
25. Lockyer, J. N. "Contributions to Solar Physics"; Macmillan: London, 1874.
26. Gingerich, O. *Sky and Telescope* **1981,** *62*, 13.
27. Mendeleev, D. I. *Compt. Rend.* **1860,** *50*, 52.
28. *Ibid.* **1860,** *51*, 97.
29. Mendeleev, D.I., *Liebigs Ann.* **1860,** *114*, 165.
30. *Ibid.* **1861,** *119*, 1.
31. Partington, J. R. "A History of Chemistry"; Macmillan: London, 1972; Vol. 4, p. 893.
32. Miller, W. A. "Elements of Chemistry, Theoretical and Practical"; London, 1863; 3rd ed., Vol. I, p. 328.
33. Andrews, T., *Phil. Trans.* **1869,** *159*, 575.
34. *Ibid.* **1876,** *166*, 421.
35. *Ibid.* **1887,** *178*, 45.
36. van der Waals, J. D., Ph.D. Thesis, Leiden, 1873; Threlfall, R.; Adair, J. F., Trans., *Phys. Memoirs* **1890,** *1*, 333.

37. Temperley, H. N. V.; Trevena, D. H. "Liquids and Their Properties"; Halsted Press: New York, 1978.
38. Watts, R. O.; McGee, I. J. "Liquid State Chemical Physics"; Wiley: New York, 1976.
39. Scott, R. L.; Van Konynenburg, P. H. *J. Chem. Soc., Faraday Disc.* **1970,** *49*, 87.
40. Leland, T. W. *Adv. Cryogenic Eng.* **1976,** *21*, 466.
41. Krichevskii, I. R. *Acta Phys. Chem., USSR* **1940,** *12*, 480.
42. van der Waals, J. D. In "Nobel Lectures in Physics"; Elsevier: Amsterdam, 1967; Vol. 1, p. 264–5.
43. Boltzmann, L. "Lectures on Gas Theory"; Brush, S. G., Trans.; Univ. of CA Press: Berkeley, 1964.
44. Boltzmann, L. *Nature* **1895,** *51*, 414.
45. Gibbs, J. W. "The Early Work of Willard Gibbs"; Henry Schuman, Inc.: New York, 1947.
46. Gibbs, J. W. *Trans. Conn. Acad.* **1876,** *3*, 108.
47. *Ibid.* **1878,** *3*, 343.
48. Gibbs, J. W. "Elementary Principles in Statistical Mechanics"; Yale Univ. Press: New Haven, 1902.
49. Kamerlingh Onnes, H. *Comm. Phys. Lab Leiden* **1901,** *71*.
50. *Ibid.* **1901,** *74*.
51. Mason, E. A.; Spurling, T. H. "The Virial Equation of State"; Pergamon: Oxford, 1969.
52. Mayer, J. E. *J. Chem. Phys.* **1942,** *10*, 629.
53. Mayer, J. E.; Mayer, M. G. "Statistical Mechanics"; Wiley: New York, 1977; 2nd ed.
54. Hirschfelder, J. O.; Curtiss, C. F.; Bird, R. D. "Molecular Theory of Gases and Liquids"; Wiley: New York, 1954.
55. Barker, J. A. "Lattice Theories of the Liquid State"; Pergamon: Oxford, 1963.
56. Hill, T. L. "Statistical Mechanics"; McGraw-Hill: New York, 1956.
57. Boublik, T.; Nezbeda, I.; Hlavaty, K. "Statistical Thermodynamics of Simple Liquids and their Mixtures"; Elsevier: New York, 1980.
58. Barker, J. A.; Henderson, D. *Rev. Mod. Phys.* **1976,** *48*, 587.
59. Kirkwood, J. G., "Theory of Liquids"; Alder, B. J., Ed., Gordon and Breach: New York, 1968.
60. Prins, J. A. *Z. Phys.* **1927,** *41*, 184.
61. Patey, G. N. *Mol. Phys.* **1977,** *34*, 427.
62. *Ibid.* **1978,** *35*, 1413.
63. Henderson, R. L.; Gray, C. G. *Can. J. Phys.* **1978,** *56*, 571.
64. Stell, G.; Patey, G. N.; Hoye, J. S. *Adv. Chem. Phys.* **1981,** *48*, 183.
65. Mansoori, G. A.; Canfield, F. B. *Ind. Eng. Chem.* **1970,** *62*, 12.
66. Mansoori, G. A.; Ali, I. *Chem. Eng. J.* **1974,** *7*, 173.
67. Girardeau, M. D.; Mazo, R. M. *Adv. Chem. Phys.* **1973,** *24*, 187.
68. Rasaiah, J. C.; Stell, G. *Mol. Phys.* **1970,** *18*, 249.
69. Rasaiah, J. C.; Stell, G. *Chem. Phys. Lett.* **1970,** *4*, 651.
70. Weeks, J. D.; Chandler, D.; Andersen, H. C. *J. Chem. Phys.* **1971,** *54*, 5237.
71. *Ibid.* **1971,** *55*, 5422.
72. Stillinger, F. H. *Adv. Chem. Phys.* **1975,** *31*, 1.
73. Hansen, J. P.; McDonald, I. R. *Phys. Rev.* **1975,** *11A*, 2111.
74. Zwanzig, R. W. *J. Chem. Phys.* **1954,** *22*, 1420.
75. Pople, J. A. *Proc. Roy. Soc.* **1954,** *221A*, 498, 508.
76. Rowlinson, J. S. *J. Stat. Phys.* **1979,** *20*, 197.

77. Widom, B. In "Phase Transitions and Critical Phenomena"; Domb, C.; Green, M. S., Eds., Academic Press: New York, 1972; Vol. 2, Ch. 3.
78. Kirkwood, J. G.; Buff, F. P. *J. Chem. Phys.* **1949,** *17*, 338.
79. Buff, F. P. Z. *Elektrochem.* **1952,** *56*, 311.
80. Toxvaerd, S. *J. Chem. Phys.* **1971,** *55*, 3116.
81. Haile, J. M.; Gubbins, K. E.; Gray, C. G. *J. Chem. Phys.* **1976,** *64*, 1852.
82. *Ibid.* **1976,** *64*, 2569.
83. Thompson, S. M.; Gubbins, K. E.; Haile, J. M. *J. Chem. Phys.* **1981,** *75*, 1325.
84. Beaglehole, D. *J. Chem. Phys.* **1981,** *75*, 1544.
85. Ono, S.; Kondo, S. *Handbuch der Phys.* **1960,** *10*, 134.
86. Toxvaerd, S. In "Statistical Mechanics"; Singer, K., Ed. Chemical Society: London, 1975, Vol. 2.
87. Croxton, C. A. "Statistical Mechanics of the Liquid Surface"; Wiley: New York, 1980.
88. Evans, R. *Adv. Phys.* **1979,** *28*, 143.
89. Dewar, J. "Collected Papers of Sir James Dewar"; Lady Dewar, Dickson, J. D. H.; Ross, H. M.; Dickson, H. C. S., Eds., Cambridge Univ. Press: London, 1927; 2 Vols.
90. Strutt, J. W.; Ramsay, W. *Phil. Trans.* **1895,** *186A*, 187.
91. Strutt, J. W. "Scientific Papers"; Dover: New York, 1964; 6 Vols.
92. Ramsay, W. "The Gases of the Atmosphere: The History of their Discovery"; Macmillan: London, 1896; Horton, G. K. In "Rare Gas Solids"; Klein, M. L.; Venables, J. A., Eds., Academic Press: London, Chap. 1, 1976.
93. Kamerlingh Onnes, H. *Leiden Comm.* **1908,** 108.
94. *Ibid.* **1911,** 122b.
95. *Ibid.* **1911,** 124c.
96. Burton, E. F.; Smith, H. G.; Wilhelm, J. O. "Phenomena at the Temperature of Liquid Helium"; Reinhold: New York, 1940.
97. Giauque, W. F. In "Nobel Lectures in Chemistry, 1942–1962"; Elsevier: Amsterdam, 1964.
98. Bridgman, P. W. "Collected Experimental Papers"; Harvard Univ. Press: Cambridge, MA, 1964.
99. Bradley, R. S. In "High Pressure Physics and Chemistry"; Bradley, R. S., Ed., Academic Press: New York, 1963; Vol. 1.
100. "Physics of High Pressures and the Condensed Phase"; van Itterbeek, A., Ed., North-Holland: Amsterdam, 1965.
101. Schneider, G. M. *Adv. Chem. Phys.* **1970,** *17*, 1.
102. Cox, H. E.; Crawford, F. W.; Smith, E. B.; Tindall, A. R. *Mol. Phys.* **1980,** *40*, 705.
103. Smith, E. B.; Tildesley, D. J.; Tindell, A. R.; Price, S. L. *Chem. Phys. Lett.* **1980,** *74*, 193.
104. Ross, M. *J. Chem. Phys.* **1980,** *73*, 4445.
105. Strutt, J. W. *Phil. Mag.* **1871,** *41*, 107, 274, 447.
106. McIntyre, D.; Sengers, J. V. In "Physics of Simple Liquids"; Temperley, H. N. V.; Rowlinson, J. S.; Rushbrooke, G. S., Eds., North-Holland: Amsterdam, 1968; p. 467.
107. Berne, B. J.; Pecora, R. "Dynamic Light Scattering"; Wiley: New York, 1976.
108. Hoye, J. S.; Stell, G. *J. Chem. Phys.* **1977,** *66*, 795.
109. Bertucci, S. J.; Burnham, A. K.; Alms, G. R.; Flygare, W. H. *J. Chem. Phys.* **1977,** *66*, 605.
110. Roentgen, W. K. *Ann. Phys. Chem., Wied.* **1892,** *45*, 91.

111. Debye, P. *Ann. Phys.* **1915,** *46*, 803.
112. Ehrenfest, P. *Proc. Acad. Sci., Amsterdam* **1915,** *17*, 1184.
113. Debye, P.; Menke, H. *Phys. Z.* **1930,** *31*, 797.
114. Karnicky, J. F.; Reamer, H. H.; Pings, C. J. *J. Chem. Phys.* **1976,** *64*, 4592.
115. Karnicky, J. F.; Pings, C. J. *Adv. Chem. Phys.* **1976,** *34*, 157.
116. Blum, L.; Narten, A. H. *Adv. Chem. Phys.* **1976,** *34*, 203.
117. Bacon, G. E. "Neutron Diffraction"; Clarendon Press: Oxford, 1975; 3rd ed.
118. Marshall, W.; Lovesey, S. W. "Theory of Thermal Neutron Scattering"; Clarendon Press: Oxford, 1971.
119. Enderby, J. E. In "Physics of Simple Liquids"; Temperley, H. N. V.; Rowlinson, J. S.; Rushbrooke, G. S., Eds., North-Holland: Amsterdam, 1968; Chap. 14.
120. Howe, R. A. In "Progress in Liquid Physics"; Croxton, C. A., Ed., Wiley: New York, 1978; Chap. 13.
121. "Chemical Applications of Thermal Neutron Scattering"; Willis, B. T. M., Ed., Oxford Univ. Press: London, 1973.
122. Boutin, H.; Yip, S. "Molecular Spectroscopy with Neutrons"; MIT Press: Cambridge, MA, 1968.
123. Teitsma, A.; Egelstaff, P. A. *Phys. Rev.* **1980,** *21A*, 367.
124. Egelstaff, P. A.; Teitsma, A.; Wang, S. S. *Phys. Rev.* **1980,** *22A*, 1702.
125. Egelstaff, P. A. *J. Chem. Soc. Faraday Disc., London* **1978,** *66*, 1.
126. Hockney, R. W.; Eastwood, J. W. "Computer Simulation Using Particles"; McGraw-Hill: New York, 1981.
127. Verlet, L. *Phys. Rev.* **1967,** *159*, 98.
128. Barojas, J.; Levesque, D.; Quentrec, B. *Phys. Rev.* **1973,** *7A*, 1092.
129. Granchetti, S. *Nuovo Cimento* **1977,** *42B*, 85.
130. Hansen, J. P. *Phys. Rev.* **1973,** *8A*, 3096.
131. Rahman, A. *Phys. Rev.* **1964,** *136*, 405.
132. Metropolis, N. A.; Rosenbluth, A. W.; Rosenbluth, M. N.; Teller, A. H.; Teller, E. *J. Chem. Phys.* **1953,** *21*, 1087.
133. Alder, B. J.; Wainwright, T. E. *J. Chem. Phys.* **1959,** *31*, 459.
134. Wood, W. W. In "Physics of Simple Liquids"; Temperley, H. N. V.; Rushbrooke, G. S.; Rowlinson, J. S., Eds., North-Holland: Amsterdam, 1968; Chap. 5.
135. "Monte Carlo Methods in Statistical Physics"; Binder, K.; Ed., Springer-Verlag: New York, 1979.
136. Maxwell, J. C. "Scientific Papers"; Niven, W. D., Ed., Cambridge Univ. Press: Cambridge, 1890.
137. Vesely, F. "Computerexperimente an Flüssigkeitsmodellen"; Physik Verlag: Weinheim, 1978.
138. Erpenbeck, J. J.; Wood, W. W. In "Statistical Mechanics"; Berne, B. J., Ed., Plenum: New York, 1977; Part B, p. 1.
139. Kushick, J.; Berne, B. J. In "Statistical Mechanics"; Berne, B. J., Ed., Plenum: New York, 1977; p. 41.
140. Wood, W. W.; Erpenbeck, J. J. *Ann. Rev. Phys. Chem.* **1976,** *27*, 319.
141. Ceperley, D.; Tully, J., Eds., *Nat. Res. Comp. Chem. Proc. 6* **1979**.
142. Copley, J. R. D.; Lovesey, S. W. *Rep. Prog. Phys.* **1975,** *38*, 461.
143. Streett, W. B.; Gubbins, K. E. *Ann. Rev. Phys. Chem.* **1977,** *28*, 373.
144. Alder, B. J.; Pollock, E. L. *Ann. Rev. Phys. Chem.* **1981,** *32*, 311.
145. Frenkel, D.; McTague, J. P. *Ann. Rev. Phys. Chem.* **1980,** *31*, 491.
146. Angell, C. A.; Clarke, J. H. R.; Woodcock, L. V. *Adv. Chem. Phys.* **1981,** *48*, 397.

147. McCammon, J. A.; Karplus, M. *Ann. Rev. Phys. Chem.* **1980,** *31*, 29.
148. Frenkel, D. In "Intermolecular Spectroscopy and Dynamical Properties of Dense Systems"; Van Kranendonk, J., Ed., North-Holland: Amsterdam, 1980; p. 156.
149. Barker, J. A.; Henderson, D. *Sci. Am.* **1981,** *245*, 130.
150. Thompson, S. M., unpublished data.
151. Egelstaff, P. A. *Adv. Chem. Phys.* **1983,** in press.
152. Rowlinson, J. S.; Swinton, F. "Liquids and Liquid Mixtures"; Butterworths: London, 1982; 3rd ed.
153. Rowlinson, J. S.; Widom, B. "Molecular Theory of Capillarity"; Clarendon Press: Oxford, 1982.

RECEIVED for review March 9, 1982. ACCEPTED for publication June 7, 1982.

2

Historical Development and Recent Applications of Molecular Dynamics Simulation

WILLIAM G. HOOVER, A. J. C. LADD, and V. N. HOOVER
University of California, Davis, CA 95616

The development of molecular dynamics is traced from Galileo's day to present day computation. Several applications are described. These indicate the broad scope of present day molecular dynamics: location of phase equilibria, characterization of both linear and nonlinear transport problems, simulation of solid-phase plastic flow, and simulation of fluid-phase shock waves.

MOLECULAR DYNAMICS IS THE STUDY of molecules in motion under the influence of intermolecular forces. The first studies of molecular motion were applied mainly to gases, because gases, in which particles move about freely, were easiest to investigate. Although it was realized even before 1900 that the same treatment could, in principle, be applied to liquids and solids, these did not become important subjects of molecular dynamics until the advent of fast computers. A complete historical review can be found in Reference 1.

As a separate field, molecular dynamics is barely 100 years old, dating from Maxwell's and Boltzmann's introduction of statistical methods to study large numbers of particles. But its origins go back to the beginning of true scientific endeavor by Galileo nearly 400 years ago. Galileo Galilei (1564–1642) was the first to experiment systematically with moving objects, finding laws for velocity and acceleration. Around the same time, Johannes Kepler (1571–1630) labored to formulate the laws of planetary motion. Isaac Newton (1642–1727) combined and generalized the discoveries of Galileo and Kepler to show that the force acting on falling objects on earth and on celestial objects was the same, that of gravity. Newton also developed calculus—the mathematical machinery needed to describe, through his laws of motion, a complete mechanical view of the universe. His precise treatment of mechanical phenomena has had an overwhelming impact and a validity unchallenged until quantum mechanics and relativity theory arrived in this century.

0065-2393/83/0204-0029$06.00/0

Eighteenth century scientists generalized and applied Newton's laws. Two Swiss colleagues, Leonhard Euler (1707–80) and Daniel Bernoulli (1700–82), fruitfully combined mathematics with mechanics. Euler conceived the principal formulas of fluid dynamics. He formulated the equations of motion for simplified macroscopic fluid models. Bernoulli developed macroscopic models for fluids and solids that included wave motion.

Near mid-eighteenth century, Euler's protégé, Joseph Lagrange (1736–1813), produced a general variational description of Newtonian mechanics which became known as Lagrangian or analytical mechanics. A more general formulation of mechanics, which was later seen to underlie quantum mechanics, was embodied in 1834 in Hamilton's "principle of least action." William Rowan Hamilton (1805–65), child prodigy in languages as well as mathematics, generalized Newton's equations into a form in which particle paths can be represented as minimal paths, and from which Lagrangian and Newtonian mechanics follow logically. Hamilton's principle grew out of an analogy with his main research in optics, which is related to modern wave mechanics.

The dynamical studies that resulted from Newton's work emphasized both celestial motion and that of tangible earthly matter. Eighteenth century experimenters formulated the gas laws to describe their observations on the relations between pressure and volume, and later temperature, of gases. Extension of the macroscopic laws of motion to the molecular level came much later, in spite of the fact that particle theories of matter go back to suggestions by Leucippus, Democritus, and Epicurus around 400 B.C. (Greek *atomos* means indivisible). But Aristotle, for whom metaphysics, not the objective world, was basic reality, rejected the atomic notion. His prestige caused the particle idea to be suppressed during long centuries of Aristotelian supremacy. By the seventeenth century, the idea hesitantly reappeared. Newton cautiously assumed a corpuscular view of matter, but avoided detailing it; his ideas of inertia, momentum and gravity did not depend on the ultimate division of matter.

The first to relate experimental gas law results to a dynamical theory involving motion of gas particles was Daniel Bernoulli. Bernoulli showed mathematically that gas pressure comes from the impact of minute gas particles against a surface. At the time, this original kinetic theory had astonishingly little effect on scientific thought. Bernoulli's theory was too advanced for his time and could not be accepted until more was learned about the nature of heat and the nature of particles themselves.

Heat was a puzzling phenomenon to early scientists. Was it a substance or was it motion? Orthodox opinion dating from the Greeks held it to be a distinct material. But Francis Bacon (1561–1626) claimed, "Heat itself, its essence and quiddity, is Motion and nothing else." In Newton's time, Robert Hooke (1635–1703) concluded that "heat is nothing but a brisk agitation of the insensible parts of an object." But the eighteenth century, dominated by the concept of heat as a measurable quantity, rejected the vague idea of heat as motion, even after Bernoulli gave it mathematical precision in 1738 in his kinetic theory. The continuing official viewpoint into the nineteenth century regarded heat as a tangible fluid substance transferred from hot to cold objects, to which the name *caloric* was given in 1787 by Antoine Lavoisier (1743–94). Doubts were cast over the caloric theory by Benjamin Thompson, Count Rumford (1753–1814), an ingenious American turned European, whose observations on heat appearing in the process of boring cannons convinced him by 1804 that heat is vibratory particle motion. His ideas were taken further by the German physician Julius Mayer (1814–78), who in 1842 suggested the general principle of conservation and equivalence of all forms of energy. Within a year, Mayer's radical proposal was verified by careful experiments on the mechanical equivalent of heat performed by James Prescott Joule (1818–89) in his Scottish brewery laboratory.

The concept that heat and work were equivalent manifestations of energy formed the basis for the science of thermodynamics, and is stated in its first law. The principle underlying the second law appeared in the 1824 memoir of Sadi Carnot describing his work on efficiency of steam engines. But not until 1852 did William Thomson, later Lord Kelvin (1824–1907), formally proclaim the "universal tendency in nature to the dissipation of mechanical energy." Thomson's dissipation principle was given its modern focus in 1865 when Rudolf Clausius (1822–79) devised the word *entropy* for describing the irretrievable degradation of all forms of energy into heat.

The law of increasing entropy, by introducing a one-way direction to the workings of nature, was a major jolt to the mechanistic Newtonian system, which apparently could run just as well backwards as forwards. The reversibility of Newton's equations is only apparent, not real, because the equations are mathematically unstable for strongly coupled degrees of freedom. This means that a small change in initial conditions leads to catastrophic changes in subsequent particle trajectories; the numerical precision required to reverse trajectories grows exponentially with elapsed time. Any tiny fluctuation, as is always found in real systems, suffices to introduce mathematically irreversible behavior.

While macroscopic thermodynamics studied heat and energy, microscopic particle motion was clarified early in the nineteenth century through chemistry. Direct contact with then-hypothetical particles being impossible, it was left to chemists to establish atoms by examining chemical combinations of various substances. John Dalton's studies of combining ratios in compounds resulted in the law of multiple proportions. Gay-Lussac in France also investigated chemical reactions. Neither made a distinction between atoms and molecules. It was the task of Amadeo

Avogadro (1776–1856) to show, in 1811, that the ultimate *atoms* combine into divisible *molecules* (*molecule* means little masses in Latin). The discovery in 1827—not then understood—by biologist Robert Brown (1773–1858) of the continual agitated motion of particles viewed through a microscope later gave strong support to the atomic–molecular theory, by explaining Brownian motion as a result of molecular bombardment. At the end of the nineteenth century, Paul Langevin (1872–1946), in his work on molecular structure, helped link Brownian motion to kinetic theory.

The chemical concept of molecular structure joined with the thermodynamic notion of heat to advance kinetic theory. Joule went on from his heat–work measurements to calculate in 1848 the average velocity that molecules must have to produce an observed pressure by impact on a container—Bernoulli's work was being vindicated. About 10 years later, Clausius described a model of elastic spheres colliding and studied gas diffusion. Lord Kelvin, early a supporter of Carnot, Joule, and Clausius, used his prestige to establish kinetic theory. James Clerk Maxwell (1831–79), best known for electromagnetic discoveries, had an equally great influence on kinetic theory by his idea of average velocity of gas molecules within a sample, with actual velocities being distributed probabilistically.

If to Maxwell goes the credit for first applying probability to kinetic theory, the development of Maxwell's idea and its relation to thermodynamics was the achievement of Ludwig Boltzmann (1844–1906). Boltzmann stated the law of increasing entropy in terms of the tendency for molecular motion to become more random or disordered. Boltzmann attempted to justify Maxwell's hypothesis by relating statistics and entropy by means of his H-theorem. His work was the real start of statistical mechanics, which, by applying probability to molecular motion, avoids the need to follow the time development of particle trajectories.

In 1873 J. D. van der Waals (1837–1923) in Holland included in kinetic theory actual sizes of molecules and introduced intermolecular forces. His work showed that kinetic theory could explain not only properties of gases, but also the transition between gas and liquid. By the turn of the century, J. Willard Gibbs (1839–1903) had constructed a general statistical mechanical method applicable to all three states of matter. In 1916, solutions to Maxwell's transfer equations were given by Sydney Chapman (1888–1970). In the following year, David Enskog (1884–1947) similarly solved the Boltzmann equation describing the dynamical evolution of gases. This double solution made it possible to compare kinetic theory with viscous flow and heat conduction experiments and also predicted thermal diffusion, later found experimentally.

Just after the turn of the century, Max Planck (1858–1947) introduced his revolutionary quantum hypothesis, showing that energy levels (in electric oscillators) were quantized, or limited to discrete values that are multiples of a definite quantum of energy. That natural phenomena do actually proceed by jumps and not continuously as envisioned by Newton's mechanics and its tool, calculus, was a blow even to Planck. His discontinuity hypothesis was initially viewed with suspicion. Einstein's explanation of the photoelectric effect finally helped quantum theory gain acceptance as an abstract system explaining discrepancies between Newton's laws and observed reality. Although conceptually closer to reality than classical mechanics, quantum mechanics, through the uncertainty principle, adds enormous calculational difficulties to treating real materials. Consequently, in statistical mechanics and in molecular dynamics, classical Newtonian mechanics remains a functional tool, actively used to this day.

Equations of Motion and Forces

Kinetic theory, armed with statistical averaging techniques that make it feasible to treat large numbers of particles, provides the theoretical basis for the actual calculations of molecular motion undertaken by molecular dynamics. These molecular dynamics calculations consist of series of "snapshots" of particle coordinates and momenta that closely satisfy microscopic equations of motion. For many years such intricate studies involved too much calculation to permit meaningful results, but nearly 40 years ago computing technology became sufficiently advanced to be applied fruitfully to many-body systems. At about this time, progress changed from the sort of individual endeavor of previous centuries to organized team work, resulting from the changeover to computer aided scientific activity.

The early molecular dynamics calculations were carried out at the University of California's Los Alamos and Livermore laboratories, where computers became available as a fringe benefit of weapons work. Modelled on celestial mechanics, with molecules represented by mass points interacting with central forces, these calculations led to rapid advances in both equilibrium and nonequilibrium systems (*2, 3*). Computational teamwork tested the validity of the equilibrium statistical mechanics of Gibbs, and the kinetic theory of Boltzmann and Maxwell. The computer results showed that Boltzmann's equation does correctly describe the approach to equilibrium and that the equation of state derived from statistical Monte Carlo averaging agrees with that found by dynamical time-averaging (*4, 5*).

The more recent proliferation of molecular dynamics calculations to dozens of institutions makes it impossible and even undesirable to present a comprehensive review of developments. The enormous increase produced some welcome duplications and verifications of results as well as less welcome computations of questionable value. The growth of low-cost computing has so facilitated calculation that it has become simpler to calculate than to understand the theory underlying the numbers. Even a very slow machine can readily produce too much output for a competent investigator to explain. Thus the most relevant advances in software are those that speed assimilation of computed information. Particularly valuable are stereoscopic plotting routines, contour plotters, and automatic movie-making devices. These features greatly reduce the amount of the researcher's time necessary for interpretation.

Definite accomplishments of recent calculations include a complete description of the equilibrium fluid and solid phases for particles interacting with the argonlike Lennard–Jones interparticle potential (inverse 6th power attraction and inverse 12th power repulsion) (*6–8*), the development of increasingly accurate liquid-phase perturbation theories (*9, 10*), based on hard-sphere, computer generated properties that closely reproduce these equilibrium properties, and new methods for measuring thermodynamic and transport properties as functions of volume and energy for a wide range of force laws.

The simplest force-law models of Boltzmann and van der Waals viewed particles as hard spheres or billiard balls with mutual attractions added to explain gas–liquid coexistence. Empirical "force laws" describe the mutual interaction of molecules as a function of their relative orientation and separation. Solid-phase calculations emphasize force-law derivatives and were instrumental in developing the many analytic "potentials" (integrated forces) used in the last 30 years.

Two distinct kinds of extensions have been made from the early mass-point calculations (*11–14*). First, bigger polyatomic molecules have been treated, although such calculations take one or two orders of magnitude longer than atomic ones. Second, the microscopic effects of macroscopic thermodynamic heat and work have been included by incorporating temperature and strain-rate constraints in the equations of motion (*15, 16*).

The most straightforward approach to polyatomic problems, treating each molecule as an aggregate of mass points interacting with its neighbors through central forces, is not physically realistic. Intramolecular angle-dependent and multipolar forces are required to study even relatively simple dynamical problems. Evans simulated the dynamics of benzene molecule collisions (*17*), while Helfand and Weber (*18, 19*) studied the torsional motions of long aliphatic carbon chains (*see* Figures 1 and 2). The successful treatment of polyatomic molecules such as benzene as rigid bodies by Evans resurrected interest in Hamilton's quaternions, angular analogs of vectors which are dynamically better behaved than Euler's angles.

Following the motion of large molecules made up of dozens of atoms

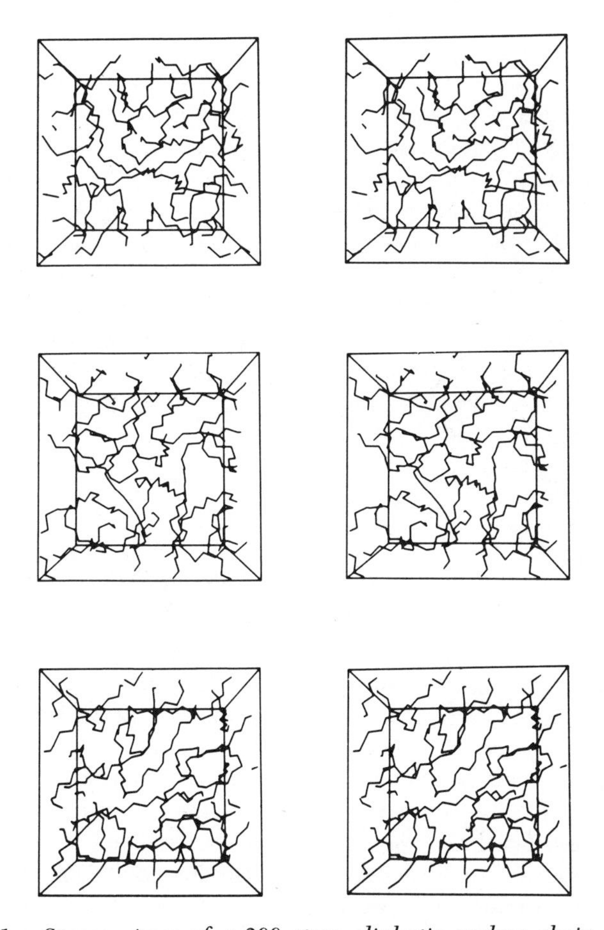

Figure 1. Stereo views of a 200-atom aliphatic carbon chain. (Reproduced with permission from Ref. 18. Copyright 1980, American Institute of Physics.)

taxes even large computers and has led to the use of approximate statistical models, based on Langevin's ideas, for simulating the interaction of such molecules with the surrounding medium. Langevin originally used statistical interactions to explain Brown's observations on moving pollen grains. The postulated and largely unknown random forces can be assigned in many ways—producing either the velocities or the accelerations characteristic of a certain temperature, for instance. Because the choice influences final nonequilibrium results, complete calculations are essential to validate these ad hoc models.

Validation is becoming more difficult. Polyatomic simulations are today moving rapidly toward increased realism (*see* Figure 3) at the cost of complexity and kinematic indeterminacy. The latter loss, inherent in

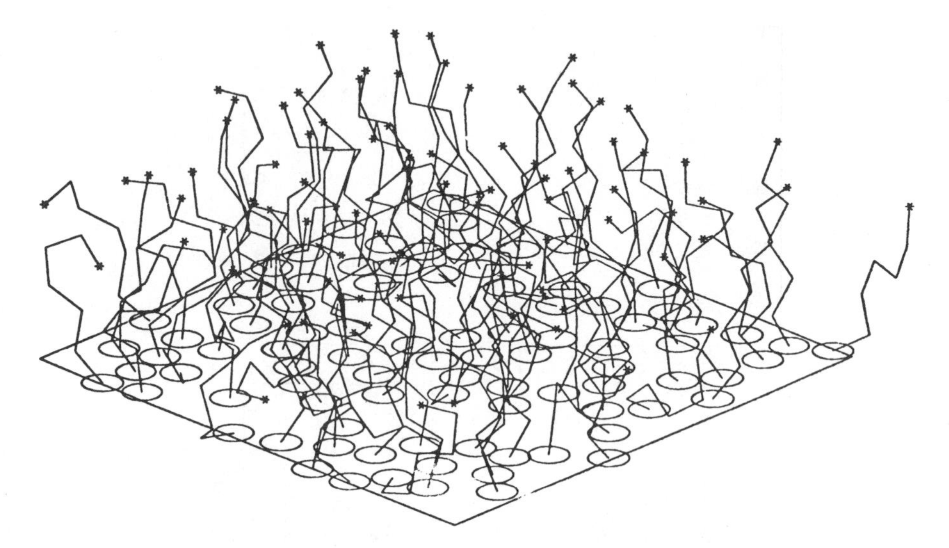

Figure 2. Conformation of a lipid monolayer. (Reproduced with permission from Ref. 14. Copyright 1980, Nature.*)*

random forces, complicates numerical verification because reversibility of the equations of motion and conservation of energy and momentum can no longer be used to test solutions.

The complexity introduced into polyatomic deterministic simulation by the wide range of time scales between slow conformational degrees of freedom and fast bond oscillations may be reduced if a new method suggested by Pechukas proves feasible. Because details of the bond oscillations are ordinarily of little interest, Pechukas has treated these as sources and sinks of energy to be added to a rigid-bond Hamiltonian. This added energy varies with molecular conformation to conserve the action of the oscillating modes. Including the extra energy leads to exact equations of motion for the conformational degrees of freedom in the adiabatic (high-frequency) limit. The obstacle to practical use of this method has so far been the difficulty of separating the conformational and vibrational degrees of freedom.

On a microscopic scale, molecular dynamics measures temperature by averaging kinetic energy. Gradual temperature changes can be imposed by continuously scaling the momenta of the particles during dynamical calculations, thereby adding or subtracting heat energy from the simulation. Gradual adiabatic changes can similarly vary the energy by performing pressure–volume work in a way consistent with the first law of thermodynamics. Both momentum scaling and adiabatic coordinate scaling have been successfully incorporated in microscopic equations of motion.

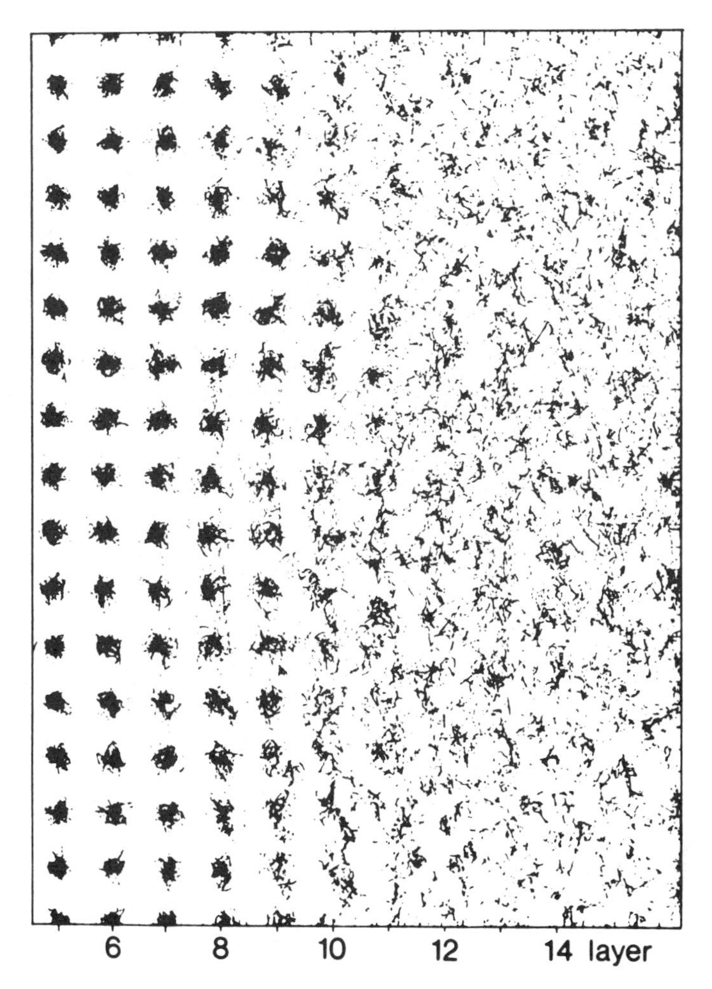

Figure 3. Time exposure of a solid–fluid interface. (Reproduced with permission from Ref. 22. Copyright 1980, American Institute of Physics.)

Application to Phase Equilibrium

The rough corresponding states similarity among phase diagrams of widely varying substances suggests that even very simple interparticle-force models can explain the qualitative properties of real matter. The classical calculations of Alder, Wainwright, and Wood, based on hard-sphere, square-well, and Lennard–Jones force laws, justified this expectation by reproducing, qualitatively, the solid–fluid melting line and gas–liquid–solid triple-point equilibria found in real systems (*20*).

A few phase diagrams that include quantum corrections have been calculated. Hansen's plasma calculations in France (*21*) and the ongoing calculations of Ceperley and Alder at Livermore on the absolute zero phase diagrams of boson and fermion systems represent the present limit of numerical quantum statistical mechanics. These equilibrium quantum calculations are much more time consuming and intricate than the corresponding classical ones. They can be carried out only at low temperature, where the ground state is important, or at temperatures high enough for perturbation theory to be applied to the classical theory. Rigorous quantum calculations cannot yet deal with the complications involved in intermediate-temperature or time-dependent systems.

Some early Lennard–Jones and square-well calculations encountered two-phase liquid–vapor states. These states were qualitatively interesting to see, but were quantitatively difficult to analyze, simply because interfacial boundaries are relatively thick on an atomic scale. Cape, Ladd, and Woodcock (*22, 23*) have used simulations of equilibrating phases, both at the triple point and along the melting line, as primary means of locating equilibrium pressures and temperatures and determining interfacial properties. Such calculations require many particles (as many as 7680 were used) and care in choosing initial conditions. Now, approximate equilibria are first obtained using smaller systems. Several similar small systems are then grouped to make a large compound system for further examination. The "time exposure" of a solid–fluid interface shown in Figure 3 indicates the detail obtainable in surface morphology.

The coexisting-phase properties obtained by these direct equilibrations are consistent with earlier triple-point thermodynamic predictions based on single-phase free-energy simulations with far fewer particles. This is only one of many examples in which self-consistency between two or more approaches has confirmed the accuracy of computer generated data in regions where rigorous theory gives little a priori guidance. Nonequilibrium effects are important to the direct simulation of coexisting phases because the equilibration of large phases is controlled by heat diffusion. The computational difficulty due to heat diffusion can be sidestepped by carrying out the molecular dynamics isothermally (*15, 24*). If Newton's equation of motion, $\dot{p} = F$, has added to the right side a momentum dependent force $-\zeta p$, then the constant-temperature constraint $d/dt \Sigma p^2 = 0$ can be identically satisfied by choosing $\zeta = \Sigma F \cdot p / \Sigma\, p \cdot p$. The resulting trajectories conserve kinetic energy and provide an example of what we call *nonequilibrium* molecular dynamics, in which the equations of motion are modified to satisfy desirable constraints, at the expense of energy conservation.

Application to Fluid Transport

The conceptually simplest nonequilibrium situations involve linear flows of mass, momentum, and energy proportional to the corresponding gradients of chemical potential, velocity, and temperature. These simple prototype flows form a convenient bridge between the well understood statistical mechanics, which can describe linear transport by dynamical perturbation theory, and the largely undeveloped theory of nonequilibrium nonlinear flows.

Viscosity, principally shear viscosity (the response of stress to changes in shape), dominates nonequilibrium flows, determining whether these are turbulent or laminar. Three different molecular dynamics methods have been used to compute the coefficient of shear viscosity. To demonstrate the simplest type of shear deformation, suppose that the fluid's x velocity component is proportional to the y coordinate, $\dot{\varepsilon} = du_x/dy$. Such a deformation can be described using Hamiltonian mechanics. The so-called "Doll's Tensor" Hamiltonian,

$$H = H_{\text{eq}} + \dot{\varepsilon}\sum yp_x$$

was inferred from the corresponding equations of motion (*25*),

$$\dot{q} = (p/m) + q \cdot \nabla u \quad \text{and} \quad \dot{p} = F - \nabla u \cdot p$$

which reproduced exactly the desired macroscopic flow field and also led to the macroscopic energy conservation relation between P_{xy} and the strain rate $\dot{\varepsilon}$. The shear viscosity η can also be obtained by applying Green–Kubo linear response theory to the nonequilibrium Hamiltonian in the limit of vanishing strain rate $\dot{\varepsilon}$, with the result that $P_{xy} = -\eta\dot{\varepsilon}$ where η is the shear viscosity

$$P_{xy} = -\dot{\varepsilon}(V/kT)\int_0^\infty \langle P_{xy}(0)P_{xy}(t)\rangle_{\text{eq}}dt$$

Thus, the time-averaged decay of equilibrium pressure fluctuations can be used to give estimates of transport coefficients (*26*). Holian has recently shown (*27*) that for finite systems the two viscosities just described can differ. Computer simulations suggest that the number dependence is reduced by using the Doll's Tensor approach.

The linear-response approach has been followed more literally and less formally by Jacucci and coworkers (*28*) who actually applied a finite but still very small perturbation. Then the difference between the two slightly different dynamical many-body trajectories—one perturbed and the other unperturbed—was followed in time, and the resulting stress differences used to estimate the viscosity coefficient. The nonlinear response to the same form of perturbation has been studied too, through the steady state that develops with a large and continual isothermal rate of shear (*29*).

These three methods for determining viscosity agree fairly well with each other and with real viscosity measurements. They agree also in predicting a shear-thinning decrease in viscosity with increasing frequency or strain rate (*30*). The viscosity decrease is not well understood, exceeding, by orders of magnitude, predictions based on the corresponding mode-uncoupling theories.

Analogous calculations for bulk viscosity (the irreversible response of stress to changes in volume) require the periodic adiabatic dilation and compression of space simultaneously with the molecular dynamics calculations. These calculations reveal a variation of viscosity with dilation frequency stronger than theoretical predictions and evidently quite unrelated to the experimental frequency dependence (*31*, *32*)—which apparently diverges as $\omega^{-5/2}$ at low frequency. The computer results have pointed out the need to revise the 1926 Chapman–Enskog bulk viscosity theory, which overpredicts bulk viscosity by nearly an order of magnitude under some conditions and which also fails to explain either of the low-frequency bulk viscosities observed in laboratory or computer experiments. Ultrasonic data suggest a very strong frequency dependence of the moderate-density bulk viscosity, but Hickman and Hoover, applying nonequilibrium molecular dynamics to that problem, found considerably smaller values for frequencies large enough and system sizes small enough for computer simulation.

Most computer flow simulations are necessarily nonlinear, so that the pressure-tensor perturbations caused by the deformations can be distinguished from background thermal fluctuations. The nonlinearity has interesting consequences. A system undergoing adiabatic compression, for instance, deviates in its pressure by a bulk viscous term proportional to the strain rate. The virial theorem has been used (*16*) to show that along with this pressure shift there is a corresponding temperature shift, so that the strain-rate-caused deviations of $P(T, V)$ and $P(E, V)$ from the equilibrium pressure are not the same.

Nonlinear effects are sometimes controversial. The coupling of heat flow with rotation is an example. According to Boltzmann's low-density kinetic theory, Coriolis's accelerations in rotating systems can prevent heat flow from paralleling the temperature gradient. On the other hand, certain formal approaches to macroscopic continuum mechanics rule out such violations of Fourier's law (*33*). The direction of the heat flow was studied using nonequilibrium molecular dynamics (*34*). A dense, two-dimensional fluid, constrained to rotate at constant angular velocity in the presence of a temperature gradient, developed a heat flux in good agreement with the predictions of Boltzmann's kinetic theory.

The same adiabatic perturbation to the Hamiltonian used to shear fluids is being used to study dislocation motion in solids (*see* Figure 4). Imperfect solids are plastically strained at relatively high amplitudes and

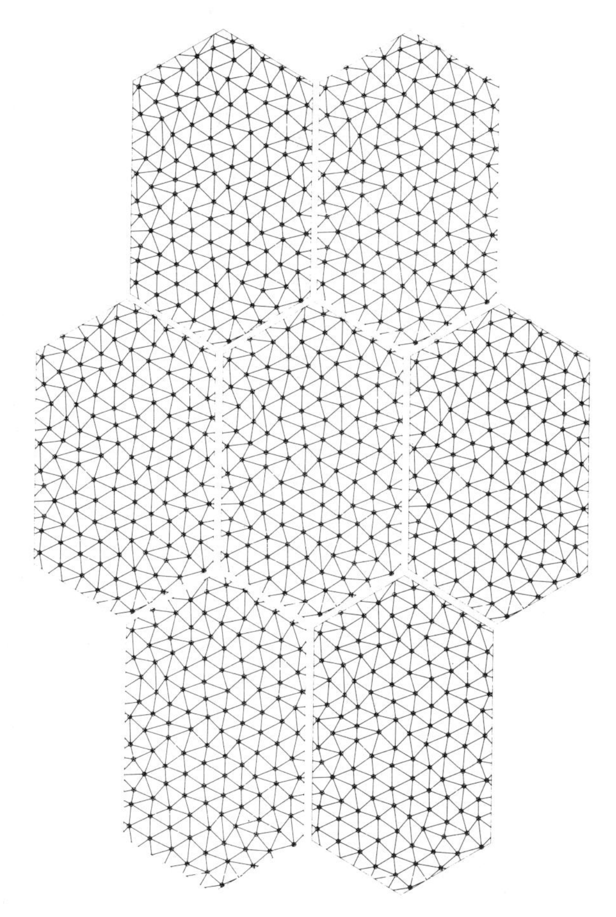

Figure 4. Periodic plastic flow, an application of adiabatic nonequilibrium molecular dynamics. (Reproduced with permission from Ref. 35. Copyright 1980, Metallurgical Society of the American Institute of Mechanical Engineers.)

gigahertz frequencies (*35*). By including the constant-temperature restriction, these solid-phase studies conform to fluid studies, showing an increase of shear stress with density and strain rate, and a decrease with temperature and system size (*30*). Results of such calculations can be compared with corresponding continuum mechanics calculations and used in macroscopic plasticity and fracture simulations. In these material failure simulations, dislocations act as point particles obeying equations of motion deduced from atomistic simulations. This work will eventually lead to improved constitutive descriptions of plastic flow in solids. The Doll's Tensor (Σqp) Hamiltonian has been applied to crystal structure stability studies too, by treating the pressure tensor as the independent variable, which governs the time-varying strain-rate tensor (*36*).

Today there is a need for critical evaluation of different possible definitions of nonequilibrium nonlinear coefficients. Work now in progress, both on the theory of nonlinear flows and on their simulation, will lead to major advances in understanding rheological problems.

Application to Fluid Shock Wave Structure

Slow heating and deformation could be described by equilibrium molecular dynamics, but in a case involving extremely rapid heating and deformation, such as a shock or detonation wave (*37*), when large changes occur in the time of only one atomic vibration, equilibrium simulations are inappropriate. Macroscopic heating usually occurs by conduction or convection from the boundary, whereas microscopic systems can easily be "heated" homogeneously throughout. Likewise, the homogeneous microscopic deformations associated with the Doll's Tensor Hamiltonian $H = H_{eq} + \Sigma qp{:}\nabla u$ are more naturally replaced by shock deformation on a macroscopic scale.

Fast shock wave compression can be simulated by inhomogeneously shrinking one space dimension in a microscopic molecular dynamics simulation (*see* Figure 5). Laboratory shockwave studies have been undertaken in liquids, solids, and gases for years. These experiments, plus additional recent work on the structure of gas-phase shockwaves, have been particularly valuable in obtaining equation of state information under extreme conditions at pressures up to tens of megabars. The structure of weak—and therefore broad—shockwaves in solids has also been studied experimentally and used to refine constitutive flow models. Through computer simulations, fluid shock waves are fairly well understood, and some progress has been made in simulating the much more complex solid phase shock waves.

The computer shock wave, in which cold material is suddenly compressed adiabatically and in the absence of nearby boundaries to high pressure, is an ideal nonequilibrium problem because the walls that complicate both simulation and analysis are absent. Theoretical treatment of even the low-density Boltzmann limit is incomplete, so that computer simulations of dense fluid shock waves very far from equilibrium are challenging tests for macroscopic theories.

A 4800-particle molecular dynamics simulation was used to generate shock wave profiles corresponding to shock compression of liquid argon to nearly twice its normal density (*37*). The resulting stress and temperature profiles, shown in Figure 6, agreed surprisingly well with Navier–Stokes continuum theory, a linear theory in which the transport coefficients are assumed to be independent of the velocity or temperature

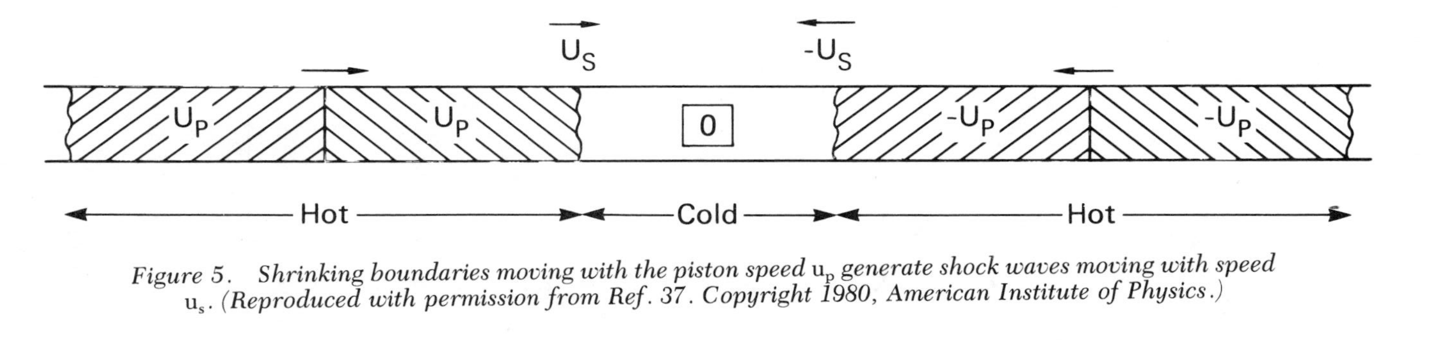

Figure 5. *Shrinking boundaries moving with the piston speed* u_p *generate shock waves moving with speed* u_s. *(Reproduced with permission from Ref. 37. Copyright 1980, American Institute of Physics.)*

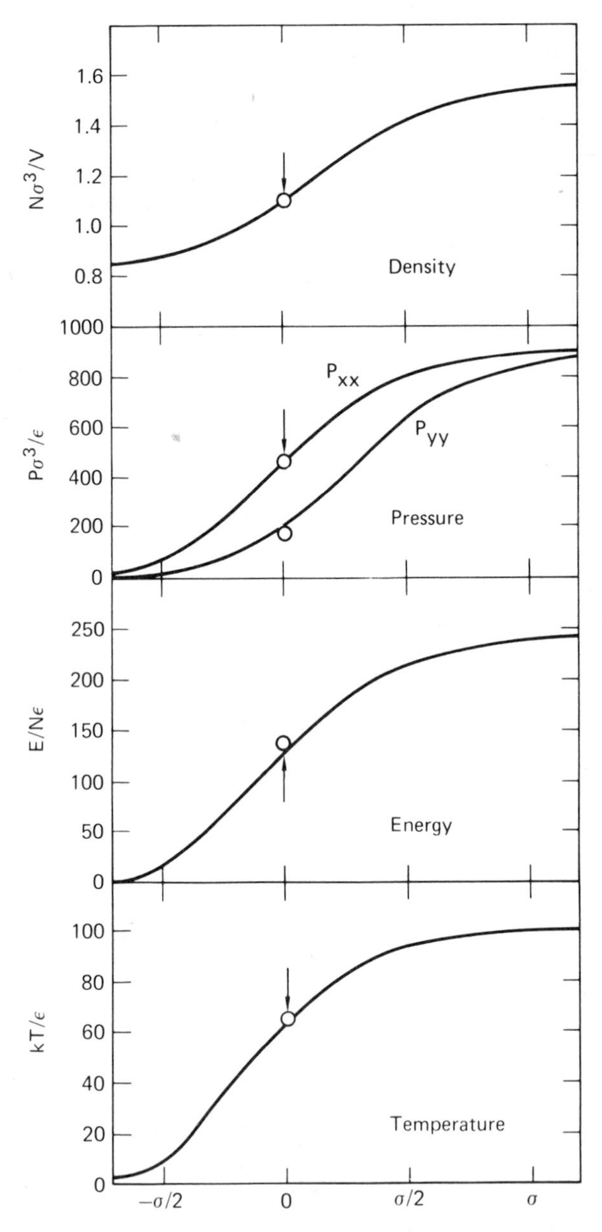

Figure 6. Profile of a strong, dense-fluid shock wave. For liquid argon, such a shock wave heats the fluid from the triple point to 1 eV in one atomic vibration time. For argon, the distance scale shown covers about 6 Å. The points were generated with nonequilibrium molecular dynamics. The smooth curves are solutions of the Navier–Stokes equations. (Reproduced with permission from Ref. 37. Copyright 1980, American Institute of Physics.)

gradients. This good agreement suggests that shockwave experiments could be used to define slowly varying nonlinear transport coefficients.

By appending chemical reactions to simulations including viscosity and conduction, the related problem of detonation wave structure can be studied. It is difficult for molecular dynamics to deliver the realism required in applications, because most real detonations are dominated by the effects of impurities. Nevertheless, models of simple liquid-phase detonations should be useful for exploring the region where chemistry is coupled with thermal and viscous effects. Except in the cases of rare-gas excitation reactions, simulations including chemistry require the development of potential surfaces for polyatomic molecules.

The natural high-pressure periodic boundary conditions have seldom been used in potential-surface calculations, but there is presently a tremendous effort devoted to representing zero-pressure polyatomic potential surfaces and incorporating these surfaces into molecular calculations. The success of these efforts should lead to an understanding of polyatomic systems on a par with today's quantitative understanding of simple fluids and solids.

Acknowledgment

The authors are pleased to acknowledge support of this work by the Army Research Office, Research Triangle Park, NC.

Literature Cited

1. Brush, S. G. "Kinetic Theory"; Pergamon: New York, Vols. 1–3, pp. 1956–1972.
2. Fermi, E.; Pasta, J. G.; Ulam, S. M. "Studies of Nonlinear Problems", LASL Report LA-1940, 1955.
3. Tuck, J. L.; Menzel, M. T. *Advan. Math.* **1972,** *9*, 399.
4. Alder, B. J.; Wainwright, T. E. *Proc. IUPAP Symp. Transp. Processes Statist. Mech.* **1956,** 97.
5. Metropolis, M.; Rosenbluth, A. W.; Rosenbluth, M. N.; Teller, A. H.; Teller, E. *J. Chem. Phys.* **1953,** *21*, 1087.
6. Ree, F. H. *J. Chem. Phys.* **1980,** *73*, 5401.
7. Levesque, D.; Verlet, L. *Phys. Rev.* **1969,** *182*, 307.
8. Hansen, J. P. *Phys. Rev.* **1970,** *2A*, 221.
9. Mansoori, G. A.; Canfield, F. B. *J. Chem. Phys.* **1969,** *51*, 4958.
10. Barker, J. A.; Henderson, D. *Rev. Mod. Phys.* **1976,** *48*, 587.
11. Helfand, E. *J. Chem. Phys.* **1978,** *69*, 1010.
12. Fixman, M. *Proc. Nat. Acad. Sci.* **1974,** *71*, 3050.
13. Gottlieb, M.; Bird, R. B. *J. Chem. Phys.* **1976,** *65*, 1467.
14. Kox, A. J.; Michels, J. P. J.; Wiegel, F. W. *Nature* **1980,** *287*, 317.
15. Evans, D. J. *Mol. Phys.* **1979,** *37*, 1745.
16. Hoover, W. G.; Evans, D. J.; Hickman, R. B.; Ladd, A. J. C.; Ashurst, W. T.; Moran, B. *Phys. Rev. A* **1980,** *22*, 1690.
17. Evans, D. J. *Mol. Phys.* **1977,** *34*, 317.

18. Weber, T. A.; Helfand, E. *J. Chem. Phys.* **1979,** *71*, 4760.
19. Weber, T. A.; Helfand, E. *J. Chem. Phys.* **1980,** *72*, 4014.
20. Wood, W. W. In "The Physics of Simple Liquids", Temperley, H. N. V.; Rowlinson, J. R.; Rushbrooke, G. S. Eds.; North Holland: Amsterdam, 1968; p. 115.
21. Hansen, J. P.; McDonald, I. R. *Phys. Rev.* **1981,** *23A*, 2041.
22. Cape, J. N.; Woodcock, L. V. *J. Chem. Phys.* **1980,** *73*, 2420.
23. Ladd, A. J. C.; Woodcock, L. V. *Chem. Phys. Lett.* **1977,** *51*, 155.
24. Ashurst, W. T., Ph.D. Dissertation, University of California at Davis-Livermore, 1974.
25. Hoover, W. G.; Ladd, A. J. C.; Hickman, R. B.; Holian, B. L. *Phys. Rev.* **1980,** *21A*, 1756.
26. Levesque, D.; Verlet, L.; Kürkijarvi, J. *Phys. Rev.* **1973,** *7A*, 1690.
27. Holian, B. L., unpublished data.
28. Ciccotti, G.; Jacucci, G. *Phys. Rev. Lett.* **1975,** *35*, 789.
29. Evans, D. J. *J. Stat. Phys.* **1980,** *22*, 81.
30. Heyes, D. M.; Kim, J. J.; Montrose, C. J.; Litovitz, T. A. *J. Chem. Phys.* **1980,** *73*, 3987.
31. Cowan, J. A.; Ward, P. W. *Can. J. Phys.* **1973,** *51*, 2219.
32. Cowan, J. A.; Ball, R. N. *Can. J. Phys.* **1980,** *58*, 74.
33. Jou, D.; Rubi, J. M. *J. Non-Equilib. Thermodyn.* **1980,** *5*, 125.
34. Hoover, W. G.; Moran, B.; More, R. M.; Ladd, A. J. C. *Phys. Rev.* **1981,** *24A*, 2109.
35. Hoover, W. G.; Ladd, A. J. C.; Hoover, N. E. *Proc. AIMETMS*; Fall Meeting, Pittsburgh, 1980.
36. Parrinello, M.; Rahman, A. *Phys. Rev. Lett.* **1980,** *45*, 1196.
37. Holian, B. L.; Hoover, W. G.; Moran, B.; Straub, G. K. *Phys. Rev.* **1980,** *22A*, 2798.

RECEIVED for review January 27, 1982. ACCEPTED for publication September 9, 1982.

3

Perturbation Theory, Ionic Fluids, and the Electric Double Layer

DOUGLAS HENDERSON
IBM Research Laboratory, San Jose, CA 95193

The properties of many fluids can be regarded as those of a simpler fluid (usually a hard-sphere fluid) plus some corrections. Perturbation theory, which is based on this idea, is reviewed briefly. Many earlier approaches, such as the virial series and the van der Waals theory, can be regarded as special cases of perturbation theory. Perturbation theory is applied to ionic fluids and is found to be useful provided that the coulomb potential is resummed. It is useful to restructure the perturbation expansion so that the mean spherical approximation is the leading term in the series. Finally, perturbation theory is applied to electrified interfaces, where results similar to those of the mean spherical approximation are obtained using simple arguments.

PERTURBATION THEORY HAS BEEN, at the very least, one of the most significant developments in the theory of liquids during the past two decades. Perturbation theory combines accurate results for the thermodynamic properties with a pleasing physical picture and relatively straightforward numerical calculations. In particular, perturbation theory avoids the often frustrating convergence problems characteristic of the iterative procedures used in the numerical solution of integral equations arising from, for example, the hypernetted chain equation.

Although it is only recently that the power of perturbation theory has been fully appreciated, perturbation theories have a venerable history. The van der Waals theory of dense gases and liquids is an early form of perturbation theory. The van der Waals theory is not surveyed here because the connection between it and perturbation has been pointed out previously (*1*). The virial expansion of a dense gas is another early form of perturbation theory.

In this chapter, perturbation theory is briefly reviewed with an emphasis on pointing out its generality. Its application to electrolytes is considered. The chapter concludes with an application of perturbation theory to an interfacial problem, the electric double layer.

0065-2393/83/0204-0047/$07.25/0

Perturbation Theory

Our starting point is the free energy,

$$A = -kT \ln \int \exp\{-\beta\Phi\} d\boldsymbol{r}_1 \ldots d\boldsymbol{r}_N + \text{terms independent of the density} \quad (1)$$

where $\beta = 1/kT$, T is the temperature, and $\Phi(\boldsymbol{r}_1 \ldots \boldsymbol{r}_N)$ is the potential energy of the N molecules whose centers of mass are at $\boldsymbol{r}_1 \ldots \boldsymbol{r}_N$.

For simplicity, assume that the potential energy is pairwise additive

$$\Phi = \sum_{i<j} u(r_{ij}) \quad (2)$$

where $r_{ij} = |\boldsymbol{r}_i - \boldsymbol{r}_j|$.

Hence, the free energy becomes

$$A = -kT \int \prod_{i<j} e(r_{ij})\, d\boldsymbol{r}_1 \ldots d\boldsymbol{r}_N + \ldots \quad (3)$$

where

$$e(r_{ij}) = \exp\{-\beta u(r_{ij})\} \quad (4)$$

Further, assume that $e(r_{ij})$ depends upon a parameter γ, i.e.,

$$e(r) = e(\gamma, r) \quad (5)$$

which is small enough so that the free energy can be expanded as a series in γ

$$A = A_0 + \gamma\left(\frac{\partial A}{\partial \gamma}\right)_{\gamma=0} + \frac{1}{2}\gamma^2\left[\frac{\partial^2 A}{\partial \gamma^2}\right]_{\gamma=0} + \ldots \quad (6)$$

where A_0 is the known free energy of the system when $\gamma = 0$ (the unperturbed or reference system).

Thus,

$$\beta \frac{\partial A}{\partial \gamma} = -\frac{1}{2}\rho^2 \int g(12)\, e_\gamma(12)\, d\boldsymbol{r}_1 d\boldsymbol{r}_2 \quad (7)$$

and

$$\beta \frac{\partial^2 A}{\partial \gamma^2} = -\frac{1}{2} \rho^2 \int g(12)\, e_{\gamma\gamma}(12)\, d\boldsymbol{r}_1 d\boldsymbol{r}_2$$
$$- \rho^3 \int g(123)\, e_\gamma(12)\, e_\gamma(12)\, d\boldsymbol{r}_1 d\boldsymbol{r}_2 d\boldsymbol{r}_3$$
$$- \frac{1}{4} \rho^4 \int [g(1234) - g(12)\, g(34)] e_\gamma(12)\, e_\gamma(34)$$
$$\times\, d\boldsymbol{r}_1 d\boldsymbol{r}_2 d\boldsymbol{r}_3 d\boldsymbol{r}_4 \tag{8}$$

where $\rho = N/V$ (V is the volume)

$$e_\gamma = e^{-1}\, \partial e/\partial \gamma \tag{9}$$

and

$$e_{\gamma\gamma} = e^{-1}\, \partial^2 e/\partial \gamma^2 \tag{10}$$

The functions g(1 . . . h) are the h-body distribution functions. In the limit $\gamma = 0$, they are the distribution functions of the reference fluid.

Equation 8 is valid only in the canonical ensemble. An extra term must be added to obtain results that are valid for an infinite system. However, this correction term is not relevant for our discussion here. We refer to the review of Barker and Henderson (2) for details.

Equation 8 is often difficult to use. An approximation, due to Barker and Henderson (2), which is often useful, is

$$\beta \frac{\partial^2 A}{\partial \gamma^2} = -\frac{1}{2} \rho^2 \int g(12)\, e_{\gamma\gamma}(12)\, d\boldsymbol{r}_1 d\boldsymbol{r}_2$$
$$- \rho^3 \int g(12)\, g(23)\, e_\gamma(12)\, e_\gamma(23)\, h(13)\, d\boldsymbol{r}_1 d\boldsymbol{r}_2 d\boldsymbol{r}_3$$
$$- \frac{1}{4} \rho^4 \int g(12)\, g(34)\, e_\gamma(12)\, e_\gamma(34)$$
$$\times [2h(13)h(24) + 4h(13)h(14)h(24)$$
$$+ h(13)h(14)h(23)h(24)]\, d\boldsymbol{r}_2 d\boldsymbol{r}_3 d\boldsymbol{r}_4 \tag{11}$$

where $h(12) = g(12) - 1$. Although approximate, Equation 11 is applicable to an infinite system.

The higher order terms involve many integrals. For some applications the ring diagrams are the most important. If only the ring diagrams for the third- and fourth-order terms are displayed,

$$\beta \frac{\partial^3 A}{\partial \gamma^3} = -\rho^3 \int e_\gamma(12)\, e_\gamma(13)\, e_\gamma(23)\, g(123)\, dr_1 dr_2 dr_3 + \ldots \quad (12)$$

and

$$\beta \frac{\partial^4 A}{\partial \gamma^4} = -\rho^4 \int e_\gamma(12)\, e_\gamma(13)\, e_\gamma(24)\, e_\gamma(34)\, g(1234) \times dr_1 dr_2 dr_3 dr_4 + \ldots \quad (13)$$

In general, there are three functional dependences of $e(\gamma;r)$ on γ that have been considered. For references see Barker and Henderson (2). The first is

$$e(\gamma;r) = \exp\{-\beta[u_0(r) + \gamma u_1(r)]\} \quad (14)$$

where $u_0(r)$ is the pair potential of the reference system. Hence

$$e_\gamma(r) = -\beta u_1(r) \quad (15)$$

and

$$e_{\gamma\gamma}(r) = [\beta u(r)]^2 \quad (16)$$

This case is useful when the perturbation energy, $u_1(r) = u(r) - u_0(r)$, is small.

In some applications, $u_1(r)$ is large and positive. For such applications $e_\gamma(r)$, given by Equation 15, is not small. Then it may be more appropriate to use

$$e(\gamma;r) = e_0(r) + \gamma e_0(r) f_1(r) \quad (17)$$

where

$$e_0(r) = \exp\{-\beta u_0(r)\} \quad (18)$$

and

$$f_1(r) = \exp\{-\beta u_1(r)\} - 1 \quad (19)$$

In Equations 18 and 19, $u_0(r)$ is the reference pair potential and $u_1(r) = u(r) - u_0(r)$. For this case

$$e_\gamma(r) = f_1(r) \tag{20}$$

and

$$e_{\gamma\gamma}(r) = 0 \tag{21}$$

Even if $u_1(r)$ is large and positive, $e_\gamma(r)$ is bounded. In principle, this approach could be used with large and negative perturbations. However, for this situation $e_\gamma(r)$ would then be very large and this approach would be of limited value.

A third procedure, also applicable to potentials that are large and positive, is based upon

$$e(\gamma;r) = \exp\left\{-\beta u\left(d + \frac{r-d}{\gamma}\right)\right\} \tag{22}$$

An expansion based upon Equation 22 is an expansion in an inverse steepness parameter about a hard-sphere reference fluid, where d is the hard-sphere diameter. Expansions based upon Equation 22 are useful when $u(r)$ is large, negative, and steep. For this case,

$$e_\gamma(r) = \beta u'\left(d + \frac{r-d}{\gamma}\right)e(\gamma;r)\,\frac{r-d}{\gamma^2} \tag{23}$$

A perturbation theory for a given system is developed by making choices as to what is an appropriate reference fluid and which of the above three procedures is to be used. Other choices besides the three above are possible. These are just the ones that have been used.

Virial Expansion

The simplest reference fluid is the perfect gas, where $g(1\ldots h) = 1$. If we use the perfect gas as a reference fluid, then the perturbation is the entire potential. Obviously, the u-expansion of Equations 14–16 is inappropriate. It is better to use the f-expansion of Equations 19–21. The first-order term is easily evaluated. For a perfect-gas reference fluid, Equation 11 is free of approximation. The first term in Equation 11 vanishes because $e_{\gamma\gamma}(r) = 0$. The other two terms vanish because, in the limit $\gamma = 0$, $h(r) = 0$. For similar reasons only the ring diagrams contribute to the third- and fourth-order terms. Hence,

$$\frac{(A - A_0)}{NkT} = -\frac{1}{2}\rho \int f_{12}\, dr_2 - \frac{1}{6}\rho^2 \int f_{12} f_{13} f_{23}\, dr_2 dr_3 - \frac{1}{8}\rho^3 \int f_{12} f_{13} f_{24} f_{34}\, dr_2 dr_3 dr_4 + \ldots \qquad (24)$$

where $f_{12} = f(r_{12}) = \exp\{-\beta u(r_{12})\} - 1$.

If the terms in Equation 24 are grouped by the power of ρ rather than by the number of f-functions, Equation 24 is the virial expansion of A. We will not use Equation 24. Our purpose is to point out that the virial expansion can be regarded as the simplest form of a perturbation theory. Another reason for writing down Equation 24 is that it bears a striking similarity to the perturbation expansions which we will use for ionic solutions.

Lattice Gas

Another simple application of perturbation theory is gained by considering the lattice gas in which the N molecules are restricted to L lattice sites. For this system

$$u(r) = \begin{cases} \infty & r = 0 \\ -\varepsilon & r = \text{nnd} \\ 0 & \text{otherwise} \end{cases} \qquad (25)$$

where nnd means nearest neighbor distance.

The unperturbed system is a lattice gas of noninteracting molecules, subject only to the restriction that only one molecule can occupy a lattice site. Thus

$$A_0/NkT = \ln x + \frac{1 - x}{x} \ln (1 - x) \qquad (26)$$

where $x = N/L$ plays the role of the density.

Because the perturbation potential is small and negative, the u-expansion of Equations 14–16 can be used. The $g(1 \ldots h)$ terms of the unperturbed system are equal to unity when all the molecules occupy different sites, and are zero otherwise. For such a reference fluid, Equation 11 is valid.

To evaluate the terms in the perturbation expansion we note that, in the limit $\gamma = \beta\varepsilon = 0$, $h(12)$ is zero unless molecules 1 and 2 are on the same site and $u_1(12)\, g(12)$ is zero unless molecules 1 and 2 are nearest neighbors. If z is the number of nearest neighbors of the lattice

$$\left.\frac{\partial(A/NkT)}{\partial(\beta\varepsilon)}\right|_{\beta\varepsilon=0} = -\frac{1}{2}\,xz \tag{27}$$

and

$$\left.\frac{1}{2}\,\frac{\partial^2(A/NkT)}{\partial(\beta\varepsilon)^2}\right|_{\beta\varepsilon=0} = -\frac{1}{4}\,x(1 - x)^2 z \tag{28}$$

In contrast to many of the other examples we will consider here, diagrams in addition to the ring diagram given in Equation 12 contribute to the third-order term. Nonetheless, it is of interest to display the result. It is

$$\left.\frac{1}{6}\,\frac{\partial^3(A/NkT)}{\partial(\beta\varepsilon)^3}\right|_{\beta\varepsilon=0} = -\frac{1}{12}\,x(1 - x)^2(1 - 2x)^2 z - \frac{1}{6}\,x^2(1 - x)^3\,\xi \tag{29}$$

where ξ is the total number of triangles of nearest neighbors that can be formed on the lattice, divided by N.

The first-order term is the result of the van der Waals theory. To this order, the perturbation contribution changes the energy of the lattice gas without changes in entropy or structure. The higher order terms give the effects on the free energy of changes in structure resulting from the perturbation.

These higher order terms become small at high densities where $L \sim N$. This is because the lattice is nearly fully occupied, and rearrangements in structure are difficult since only one molecule can occupy a lattice site. This means that at high densities the perturbation expansion will converge rapidly even if $\beta\varepsilon$ is not small. This is a very important observation. It is true for many other systems and is one of the main reasons why perturbation theory is so useful.

At lower densities, the perturbation expansion converges slowly. Thus, if the expansion is to be used in the neighborhood of the critical point, many terms are needed. For the lattice gas, these terms can be obtained fairly easily. For other systems this is not true, and so it is only for the lattice gas that critical point properties can be examined. This is one reason why the lattice gas is of such great interest.

Simple Liquids

For a simple liquid consisting of spherical molecules with a steep repulsion curve, an appropriate reference potential is the positive part of the potential. Thus, using the u-expansion

$$\frac{A - A_0}{NkT} = 2\pi\beta\rho \int_{\sigma}^{\infty} u(r)\, g_0(r)\, r^2 dr + \ldots \tag{30}$$

where A_0 and $g_0(r)$ are the free energy and radial distribution function of the reference fluid, and σ is the value of r for which $u(r) = 0$.

The integral in Equation 30 is very nearly independent of density and temperature. Thus, to a good approximation, the first-order term has the lattice gas form given in Equation 27. The second-order term is also similar to the lattice gas result, Equation 28; in particular, it is small at high densities.

Because the reference potential is steep, the higher order terms will be small at high densities, just as was the case for the lattice gas. Even with just the first-order term, the perturbation series gives good results at high densities. With two terms, excellent results are obtained at high densities. Even at lower densities the results are quite good.

Despite these results, the perturbation theory outlined above is not very practical because, in the above form, A_0, $g_0(r)$, and the higher order distribution functions must be determined by computer simulations for every state that is considered. One might as well perform the computer simulations directly for the actual system.

The step that makes perturbation theory practical for simple liquids is the replacement of A_0 and $g_0(r)$ by the hard-sphere A_{HS} and $g_{HS}(r)$. Using Equation 23 for $0 \leq r \leq \sigma$, Barker and Henderson (*3*) showed that

$$A_0 \simeq A_{HS} \tag{31}$$

and

$$g_0(r) \simeq g_{HS}(r) \tag{32}$$

if the hard sphere diameter (d) is chosen by

$$d = \int_0^{\sigma} [1 - \exp\{-\beta u_0(z)\}]dz \tag{33}$$

Since the thermodynamic properties and distribution functions of hard spheres are well known, perturbation theory becomes a simple and accurate theory of liquids.

Perturbation theory also leads to a simple picture of a liquid. At high densities, where the molecules are packed close together, the liquid molecules behave much as gas molecules at the same density. The main contribution of the perturbation is to provide the potential well in which the molecules move.

Results for a fluid whose potential is given by the Lennard–Jones interaction

$$u(r) = 4\varepsilon\left[\left(\frac{\sigma}{r}\right)^{12} - \left(\frac{\sigma}{r}\right)^{6}\right] \tag{34}$$

have been given earlier (*1, 2*). Perturbation theory results for mixtures of liquids are also available (*2*).

Charged Hard Spheres

A system of charged hard spheres, where

$$u_{ij}(r) = \begin{cases} \infty & r < \sigma \\ z_i z_j\, e^2/\varepsilon r & r > \sigma \end{cases} \tag{35}$$

is a useful model ionic fluid. In Equation 35, $z_i e$ is the charge of an ion of species i, ε is the dielectric constant of the solvent, which is taken to be a dielectric continuum, and σ is the diameter of the hard spheres.

Let us apply perturbation theory to this system. Using the u-expansion with a hard-sphere reference fluid

$$\frac{A - A_0}{NkT} = -\frac{1}{2}\beta\rho \sum_{ij} x_i x_j \int u_{ij}(12)\, g_0(12)\, dr_2 + \ldots \tag{36}$$

where A_0 and $g_0(r)$ are the free energy and radial distribution function for hard spheres of diameter σ. Because of charge neutrality

$$\sum_i z_i x_i = 0 \tag{37}$$

where $x_i = N_i/N$ (N_i is the number of hard spheres with charge $z_i e$), the first-order term in Equation 36 vanishes.

To the second-order term, only the first term in Equation 8 contributes. The other terms vanish because of charge neutrality. Likewise, in the third-order term only the ring diagram is nonzero. Thus

$$\frac{A - A_0}{NkT} = -\frac{1}{4}\beta^2\rho \sum_{ij} x_i x_j \int u_{ij}^2(12)\, g_0(12)\, d\mathbf{r}_2$$
$$+ \frac{1}{6}\beta^3\rho^2 \sum_{ijk} x_i x_j x_k \int u_{ij}(12)\, u_{ik}(13)\, u_{jk}(23) \times g_0(123)\, d\mathbf{r}_2 d\mathbf{r}_3 + \ldots \tag{38}$$

We see that the cancellation that leads to small values of the higher order perturbation terms at high densities is not present for this system. The perturbation series will converge more slowly.

Let us restrict our attention to the case of a two-component system where $z = |z_1| = |z_2|$. Equation 38 becomes

$$\frac{A - A_0}{NkT} = -\frac{\kappa^4}{64\pi^2\rho}\int \frac{g_0(12)}{r_{12}^2}\, d\mathbf{r}_2$$
$$+ \frac{\kappa^6}{384\pi^3\rho}\int \frac{g_0(123)}{r_{12}r_{13}r_{23}}\, d\mathbf{r}_2 d\mathbf{r}_3 + \ldots \tag{39}$$

where κ is the Debye screening length and is defined by

$$\kappa^2 = \frac{4\pi\beta z^2 e^2 \rho}{\varepsilon} \tag{40}$$

Equation 39 is our starting point. A quick inspection shows that each integral is divergent because of the long range of the coulomb potential. To get anything useful we must sum the divergent terms.

It is convenient to rewrite Equation 39 as

$$\frac{A - A_0}{NkT} = -\frac{\kappa^4}{64\pi^2\rho}\int_{r_{12}>\sigma}\left\{\frac{1}{r_{12}} - \frac{\kappa^2}{6\pi}\int_{r_{ij}>\sigma}\frac{d\mathbf{r}_3}{r_{13}r_{23}} + \ldots\right\}\frac{d\mathbf{r}_2}{r_{12}}$$
$$-\frac{\kappa^4}{64\pi^2\rho}\int_{r_{12}>\sigma}\frac{g_0(12) - 1}{r_{12}^2}\, d\mathbf{r}_2$$
$$+\frac{\kappa^6}{384\pi^3\rho}\int_{r_{ij}>\sigma}\frac{g_0(123) - 1}{r_{12}r_{13}r_{23}}\, d\mathbf{r}_2 d\mathbf{r}_3 + \ldots \tag{41}$$

Each of the terms in the first integral is divergent. All integrations are outside the cores (i.e., $r_{ij} > \sigma$). However, to evaluate the first integral, let us extend the range of the integrations to include all $r_{ij} > 0$. No error is introduced since we have merely added and subtracted the regions $0 < r_{ij} < \sigma$. The first integral is called a ring or chain sum since the terms are simple ring diagrams consisting of repeated convolutions of $1/r_{ij}$.

In principle, we could sum the integrand of the first term in Equation 41. However, the sum is complex. It is much easier to rewrite this term as

$$\int_{r_{12}>0} \left\{ \frac{\kappa^4}{r_{12}} - \frac{\kappa^6}{6\pi} \int_{r_{ij}>0} \frac{d\mathbf{r}_3}{r_{13}r_{23}} + \ldots \right\} \frac{d\mathbf{r}_2}{r_{12}}$$

$$= \int_0^{\kappa} \frac{\partial}{\partial \kappa'^2} \left[\int_{r_{12}>0} \left\{ \frac{\kappa'^4}{r_{12}} - \frac{\kappa'^6}{6\pi} \int_{r_{ij}>0} \frac{d\mathbf{r}_3}{r_{13}r_{23}} + \ldots \right\} \frac{d\mathbf{r}_2}{r_{12}} \right] d\kappa'^2$$

$$= 2 \int_0^{\kappa} \left[\int_{r_{12}>0} \left\{ \frac{\kappa'^2}{r_{12}} - \frac{\kappa'^4}{4\pi} \int_{r_{ij}>0} \frac{d\mathbf{r}_3}{r_{13}r_{23}} + \ldots \right\} \frac{d\mathbf{r}_2}{r_{12}} \right] d\kappa'^2 \tag{42}$$

Thus, defining $\mathscr{C}(r)$ as

$$\mathscr{C}(r_{12}) = \frac{\kappa^2}{r_{12}} - \frac{\kappa^4}{4\pi} \int_{r_{ij}>0} \frac{d\mathbf{r}_3}{r_{13}r_{23}} + \ldots \tag{43}$$

and taking the Fourier transform

$$\hat{\mathscr{C}}(k) = \frac{4\pi}{k} \int_0^{\infty} r\mathscr{C}(r) \sin kr \, dr \tag{44}$$

we have

$$\hat{\mathscr{C}}(k) = \frac{4\pi\kappa^2}{k^2} \left\{ 1 - \frac{\kappa^2}{k^2} + \frac{\kappa^4}{k^4} - \ldots \right\} = \frac{4\pi\kappa^2}{k^2 + \kappa^2} \tag{45}$$

Hence

$$\mathscr{C}(r_{12}) = \kappa^2 \frac{e^{-\kappa r_{12}}}{r_{12}} \tag{46}$$

Therefore,

$$\int_{r_{12}>0} \left\{ \frac{\kappa^4}{r_{12}} - \frac{\kappa^6}{6\pi} \int_{r_{ij}>0} \frac{d\mathbf{r}_3}{r_{13}r_{23}} + \ldots \right\} \frac{d\mathbf{r}_2}{r_{12}}$$

$$= 8\pi \int_0^{\kappa} \kappa'^2 \left[\int_0^{\infty} e^{-\kappa' r} \, dr \right] d\kappa'^2$$

$$= 16\pi \int_0^{\kappa} \kappa'^2 \, d\kappa' = \frac{16\pi}{3} \kappa^3 \tag{47}$$

Substitution of Equation 47 into 41 gives the Stell–Lebowitz (SL) series (*4*)

$$\frac{A - A_0}{NkT} = -\frac{\kappa^3}{12\pi\rho} - \frac{\kappa^4}{6\pi\rho}\int_0^\infty h_0(r)dr + \ldots \quad (48)$$

Retaining only the first term gives the Debye–Huckel (DH) theory (*5*). Of course, Debye and Huckel did not obtain their theory in this manner. Except for exceedingly small κ, the DH theory gives poor results.

The SL corrections to the DH approximation converge very slowly. The corrections can only be evaluated using approximate Padé summation methods. Rather than use their procedure, we shall use a more powerful scheme. We have derived the SL expansion from perturbation theory. Stell and Lebowitz did not use this method but obtained their series in a more direct manner. Although the method given here is less direct, it indicates how improvements may be made. The integrations in the SL expansion are over all space ($r_{ij} > 0$). This is natural in their series. However, in the original perturbation expansion, the integrations are over $r_{ij} > \sigma$. This suggests that improved results might be obtained by taking the integrals for the region $0 < r_{ij} < \sigma$ and combining them with the κ^3 term. Thus

$$\frac{A - A_0}{NkT} = -\frac{\kappa^3}{12\pi\rho}\left[1 - \frac{3}{4}\kappa\sigma + \frac{3}{4}\kappa^2\sigma^2 - \frac{7}{8}\kappa^3\sigma^3 + \ldots\right] - \frac{\kappa^4}{16\pi\rho}\int_\sigma^\infty h_0(r)dr + \ldots \quad (49)$$

At first sight, $\kappa^3(1 - 3\kappa\sigma/4 + 3\kappa^2\sigma^2/4 - 7\kappa^3\sigma^3/8 + \ldots)$ seems to be an unpromising combination; it is, in fact, $(2\Gamma)^3\,(1 + 3\Gamma\sigma/2)$ where κ and Γ are related by

$$\kappa = 2\Gamma(1 + \Gamma\sigma) \quad (50)$$

We note that $2\Gamma \leqslant \kappa$. Expanding

$$\Gamma = \frac{\kappa}{2} - \frac{\kappa^2\sigma}{4} + \ldots \quad (51)$$

Thus,

$$\frac{A - A_0}{NkT} = -\frac{(2\Gamma)^3(1 + 3\Gamma\sigma/2)}{12\pi\rho} - \frac{\kappa^4}{16\pi\rho}\int_\sigma^\infty h_0(r)dr + \ldots \quad (52)$$

Retaining only the first term gives the mean spherical approximation (MSA) (*6, 7*).

As is seen in Figure 1, the first (MSA) term gives fairly good agreement with computer simulations (*8–10*). In principle, Equation 52 gives a series of corrections to the MSA. At low concentrations, where κ is small, the corrections are negligible. At higher concentrations the convergence is fairly poor and a Padé summation is required.

Henderson and Blum (*11*) have suggested changing the expansion parameter from κ to 2Γ. They obtain

$$\frac{A - A_0}{NkT} = -\frac{(2\Gamma)^3(1 + 3\Gamma\sigma/2)}{12\pi\rho} - \frac{(2\Gamma)^4}{16\pi\rho}\int_\sigma^\infty h_0(r)dr + \ldots \qquad (53)$$

The higher order terms in this expansion are given by Henderson and Blum. Since $2\Gamma \leqslant \kappa$, this series of corrections to the MSA is better behaved. In fact, the corrections are negligible at normal ionic concen-

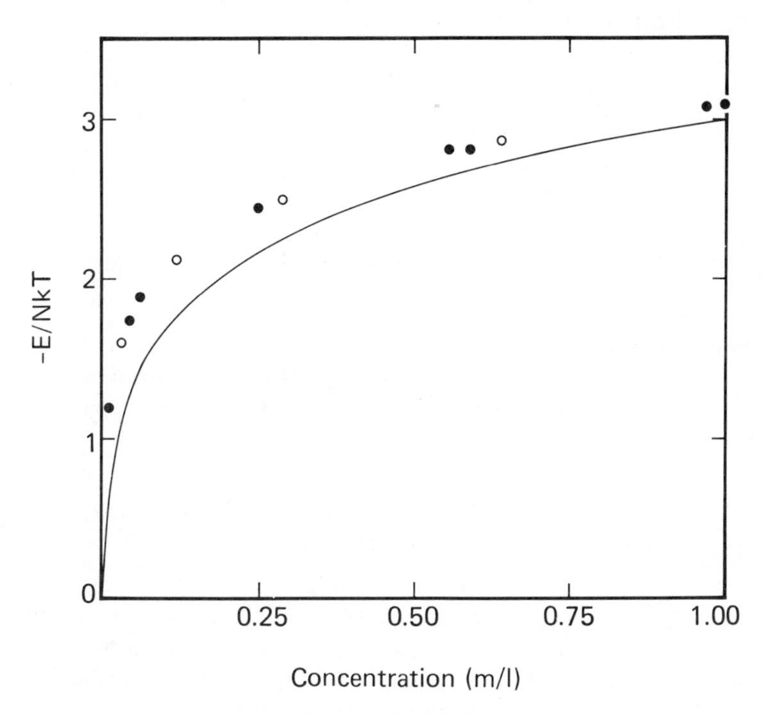

Figure 1. Internal energy of a 2:2 model ionic solution. Conditions, σ = 4.2 Å; ε = 78.358; and T = 298.16 K. Key: ○, computer simulation values of van Megen and Snook (10); ●, computer simulation values of Valleau et al. (8, 9); and —, MSA results.

trations. At the densities characteristic of fused salts, the correction terms would make an appreciable contribution.

At low concentrations, the correction terms, given in Equations 52 and 53, are not the most important corrections to the MSA result. This is because there is, in the fourth-order term, the contribution

$$\frac{\Delta A}{NkT} = -\frac{1}{48}\beta^4\rho\sum_{ij} x_i x_j \int u_{ij}^4(r) g_0(r) dr \tag{54}$$

Expressions analogous to this appear in every even-order perturbation term.

The above integral converges because r^{-4} goes to zero sufficiently quickly to prevent a divergence. However, since the series in Equation 53 works well at higher densities, the contribution of this β^4 term must disappear at these higher densities due to some cancellation with terms which are higher order in the density. Hence, even though a resummation is not forced upon us to prevent a divergence, a resummation is desirable as it approximates this cancellation. The effect of the resummation in Equations 42 to 46 is to replace

$$u_1(r) = \frac{z_i z_j e^2}{\varepsilon r} \tag{55}$$

by

$$\bar{u}_1(r) = \frac{z_i z_j e^2}{\varepsilon}\frac{e^{-\kappa r}}{r} \tag{56}$$

Further, the effect of the combination of the core terms is to replace $e^{-\kappa r}/r$ by the appropriate MSA expression. To a good approximation this is $e^{-\kappa(r-\sigma)}/r$. Thus, we can add the approximate correction term

$$\frac{\Delta A}{NkT} = -\frac{1}{2}\rho\sum_{ij} x_i x_j \int \sum_{n=2}^{\infty}\frac{1}{(2n)!}\left[\frac{\beta z_i z_j e^2}{\varepsilon r} e^{-\kappa(r-\sigma)}\right]^{2n} g_0(12)\, dr_2 \tag{57}$$

Equation 57 is correct at low concentrations where $\kappa \to 0$. At higher concentrations where κ is appreciable, this correction term is ad hoc, but does become small. Correction terms similar to Equation 57 have been considered by many authors (*12*).

Perturbation theory provides a reasonably good theory of the model ionic fluid given by Equation 35. The main deficiency is the primitive model of the solvent. The solvent appears only through the dielectric constant ε. What is needed is a more realistic treatment of the solvent.

Dipolar Hard Spheres

A model of a solvent that is an interesting system to which to apply perturbation theory is the dipolar hard-sphere fluid, for which the pair potential (which is a function of orientation as well as position) is

$$u(r_{12}, \Omega_1, \Omega_2) = \begin{cases} \infty & r < \sigma \\ -(\mu^2/r_{12}^3)D(1,2) & r > \sigma \end{cases} \tag{58}$$

where Ω_1 and Ω_2 are variables specifying the orientation of molecules 1 and 2, μ is the dipole moment

$$D(12) = 3(\hat{\mu}_1 \cdot \hat{r}_{12})(\hat{\mu}_2 \cdot \hat{r}_{12}) - \hat{\mu}_1 \cdot \hat{\mu}_2, \tag{59}$$

and $\hat{\mu}_1$ and $\hat{r}_{12}$ are unit vectors.

Before considering this system of hard spheres with embedded point dipoles, a few general comments about the application of perturbation theory to molecules with nonspherical pair potentials are in order. Nonspherical molecules are a more complex system than the simple spherical systems discussed so far. The dipolar hard spheres considered here are an especially simple system because the hard core is spherical. The u-expansion is appropriate and everything proceeds in a reasonably straightforward manner. For more complex systems, where the core is nonspherical, the situation can be more complex. If a spherical reference system is used then some of the perturbation energy may be large and positive. If so, a u-expansion is inappropriate and an f-expansion is preferable, at least for the regions where the perturbation is large and positive. However, the penalty we pay is that in the region where $u_1(r,\Omega)$ is negative, $f(r,\Omega)$ can be very large, which might result in convergence problems. Ideally one would like a nonspherical reference fluid. Should such a reference fluid be available, perturbation theory might be more widely useful for nonspherical potentials. Expressions for the free energy of many fluids consisting of nonspherical molecules are available (2). However, there are no expressions presently available for $g(r,\Omega)$ for such systems.

Fortunately, such problems need not concern us when considering dipolar hard spheres. Using the u-expansion

$$A = A_0 + \sum_{n=1}^{\infty} (\beta\mu^2)^n A_n \tag{60}$$

where A_0 is the free energy of hard spheres of diameter σ

$$\frac{A_1}{NkT} = \frac{1}{2}\rho \int g_0(r_{12})dr_2 \int u_1^*(12)\, d\Omega_1 d\Omega_2 \tag{61}$$

and $u_1^*(12)$ is the pair potential divided by μ^2. The angular integral in Equation 61 is zero. Hence $A_1 = 0$. After performing the angular integrations

$$A_2/NkT = -\frac{1}{6}\rho \int r^{-6} g_0(r) dr \tag{62}$$

and

$$A_3/NkT = \frac{1}{54}\rho^2 \int u_{123} g_0(123)\, d\mathbf{r}_2 d\mathbf{r}_3 \tag{63}$$

where

$$u_{123} = \frac{1 + 3\cos\theta_1 \cos\theta_2 \cos\theta_3}{(r_{12} r_{13} r_{23})^3} \tag{64}$$

and r_{ij} and θ_i are the sides and interior angles of a triangle formed by molecules 1, 2, and 3.

We see that these expressions for dipolar hard spheres are quite similar to the virial expansion and to the expressions for charged hard spheres. Only the ring diagram survives in the third-order term. In fact, if we took only the diagrams that contribute to the virial expansion and then expanded in powers of β we would obtain Equations 38 and 60–63.

The integral in Equation 62 is easily evaluated. Barker et al. (*13*) have calculated the integral in Equation 63. As is seen in Figure 2, the free energy series obtained from Equation 60 converges very slowly. The perturbation terms seem to alternate in sign. Rushbrooke et al. (*14*) have employed the Padé sum

$$A = A_0 + \beta^2 \mu^4 \frac{A_2}{1 - \beta\mu^2 A_3/A_2} \tag{65}$$

and found it to be in good agreement with computer simulations (*15*). This is shown in Figure 2.

There is some indirect evidence (*15*) from computer simulations that the terms in the perturbation sum, Equation 60, are negligible for $n > 4$. Hecht et al. (*16*) have obtained formal expressions for A_4 and A_5, and Tani et al. (*17*) have made some progress towards calculating A_4. It will be interesting to see if a truncated series agrees with the simulation results.

In the case of the charged hard spheres, it was found helpful to

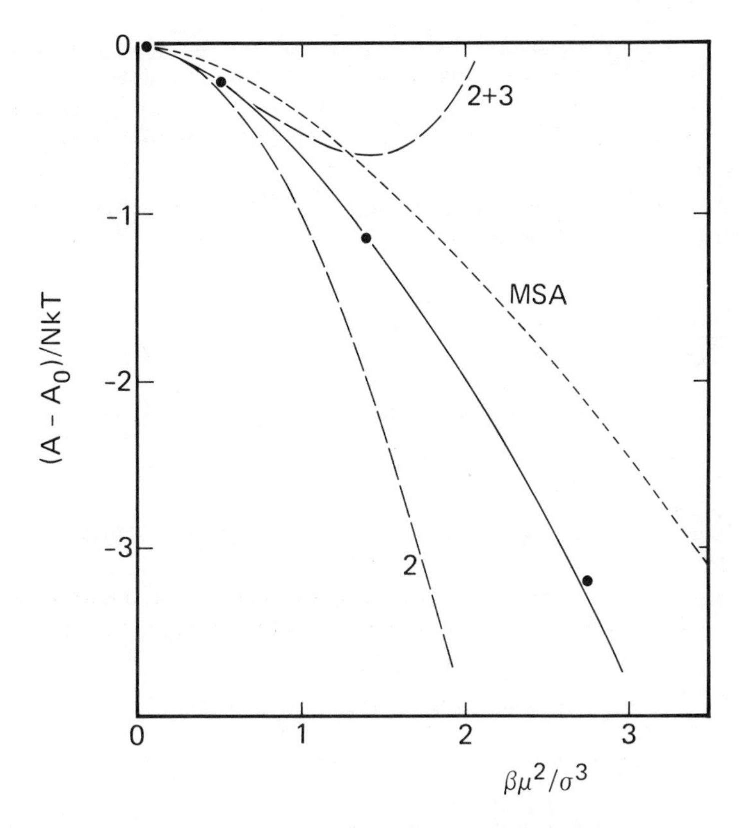

Figure 2. Free energy of a dipolar hard-sphere fluid ($\rho\sigma^3 = 0.8344$) as a function of reduced dipole moment. Key: ●, computer simulation values of Valleau and Patey (15); ---, 2 and 2 + 3, results of Equation 60 when truncated after 2 and 3 terms, respectively; ---, MSA results; solid curve, results of Equation 65.

remove the MSA results from the perturbation series and write the series as a correction to the MSA. This could also be done for the dipolar hard spheres. If this is done

$$A/NkT = A_{\mathrm{MSA}}/NkT - \frac{27y^2}{8\pi\rho}\int_{\sigma}^{\infty} r^{-4}h_0(r)dr + \ldots \tag{66}$$

where $y = 4\pi\rho\beta\mu^2/9$. The similarity to Equation 52 is striking. Terms important at low densities can be constructed in a manner analogous to Equation 57. We should not expect this series of correction terms to the MSA result to converge quickly. A Padé summation may still be required. It may or may not be preferable to the original series, Equation 60.

However, in view of the fact that A_{MSA} is not too bad an approximation to A, as is seen in Figure 2, using A_{MSA} as a starting point, rather than A_0, does not seem too bad an idea. At low densities, terms analogous to those in Equation 57 would have to be included in the perturbation expansion.

Perturbation theory can also be applied to the calculation of the dielectric constant. The result is

$$\frac{(\varepsilon - 1)(2\varepsilon + 1)}{9\varepsilon} = y\left[1 + \frac{\beta^2\mu^4}{9}\rho^2 \int \frac{3\cos^2\theta_3 - 1}{(r_{13}r_{23})^3} g_0(123)d\boldsymbol{r}_2 d\boldsymbol{r}_3 + \ldots\right] \tag{67}$$

Including only the first term in the above series gives the Onsager result for ε (*18*).

This series for ε can be rewritten by expressing it as a correction to the MSA for ε rather than as a correction to the Onsager result. Hence

$$\frac{(\varepsilon - 1)(2\varepsilon + 1)}{9\varepsilon} = \left[\frac{(\varepsilon - 1)(2\varepsilon + 1)}{9\varepsilon}\right]_{MSA} + \frac{9}{16\pi^2} y^3 \int \frac{3\cos^2\theta_3 - 1}{(r_{13}r_{23})^3}\{g_0(123) - 1\}\, d\boldsymbol{r}_2 d\boldsymbol{r}_3 + \ldots \tag{68}$$

The integrals in Equations 67 and 68 can be calculated by the same techniques as those used in the calculation of the integral in Equation 63. However, the numerical problems are more difficult because the integral is long ranged. Despite this, Tani et al. (*17*) have been able to calculate this integral and thereby obtain ε. Their results are promising but still preliminary and so are not presented here.

The perturbation theory presented here is based upon the *u*-expansion. Dipolar hard spheres have been treated by *f*-expansion techniques also (*19*). The method has some advantages but is more complex since the angle averages cannot be performed analytically.

In the past two sections, perturbation theory has been applied to the ions and the solvent separately. What is needed is a treatment of a mixture of ions and dipoles. This has not yet been done. The formulation of perturbation theory for this system would be more difficult as dipoles as well as ionic terms will have to be resummed to avoid divergences.

Electric Double Layer

Let us consider first ions near a charged electrode. As before, we consider the ions to be charged hard spheres of diameter σ. The electrode is approximated as a uniform, hard, charged wall. First let us consider the case where the solvent is a uniform dielectric medium.

If the electrode is charged, there will be an accumulation near the electrode of ions whose charge is opposite to that of the electrode. We can then speak of a double layer of charge. To apply perturbation theory to the system, we consider the electrode to be a large ion whose diameter is $R >> \sigma$ and whose charge is Q. Eventually, we will take the limit $R \to \infty$.

Since the concentration of this large ion is $1/N$

$$A = A_0 - \frac{N\beta e^4\rho}{4\varepsilon^2} \sum_{ij} x_i x_j z_i^2 z_j^2 \int \frac{g_0(r)}{r^2}\, dr - \frac{\beta e^2 Q^2 \rho}{2\varepsilon^2} \sum_i x_i z_i^2 \int \frac{g_0^w(r)}{r^2}\, dr + \ldots \quad (69)$$

where the sums are over the bulk charged hard spheres and $g_0^w(r)$ is the radial distribution between the large hard sphere and the bulk hard spheres.

Hence, the excess free energy due to the presence of this large ion is

$$\begin{aligned} A &= -\frac{\kappa^2 Q^2}{2\varepsilon} \int_{(R+\sigma)/2}^{\infty} g_0^w(r)\, dr \\ &= -\frac{\kappa^2 Q^2}{2\varepsilon} \int_0^{\infty} dr - \frac{\kappa^2 Q^2}{2\varepsilon} \int_0^{\infty} h_0^w(r) dr + \ldots \end{aligned} \quad (70)$$

Replacing the divergent integral in Equation 70 by the ring sum of which it is the first member gives

$$\begin{aligned} \Delta A &= -\frac{\kappa^2 Q^2}{2\varepsilon} \int_0^{\infty} e^{-\kappa r} dr + \frac{\kappa^2 Q^2}{2\varepsilon} \int_0^{(R+\sigma)/2} dr \\ &\quad - \frac{\kappa^2 Q^2}{2\varepsilon} \int_{(R+\sigma)/2}^{\infty} h_0^w(r) dr + \ldots \\ &= -\frac{\kappa Q^2}{2\varepsilon} + \frac{\kappa^2 Q^2}{2\varepsilon} \left(\frac{R+\sigma}{2} \right) + \ldots \end{aligned} \quad (71)$$

The last integral has been neglected since $h_0^w(r) \simeq 0$ for normal ionic concentrations.

Introducing Γ, defined by Equation 50, gives

$$\Delta A = -\frac{Q^2}{\varepsilon}\Gamma(1 - \Gamma R) + \ldots$$

$$= -\frac{Q^2}{\varepsilon}\frac{\Gamma}{1 + \Gamma R} \tag{72}$$

For large R, Equation 72 becomes

$$\Delta A = -\frac{Q^2}{\varepsilon R} + \frac{Q^2}{\varepsilon \Gamma R^2} \tag{73}$$

The first term in Equation 73 is the free energy of a sphere. The second term is the free energy of the double layer,

$$\Delta A_{DL} = \frac{Q^2}{\varepsilon \Gamma R^2} \tag{74}$$

The potential difference across the double layer is

$$V = \frac{\partial(\Delta A_{DL})}{\partial Q}$$

$$= \frac{2Q}{\varepsilon \Gamma R^2} \tag{75}$$

If the surface charge density is $E/4\pi$ then $Q = ER^2/4$. Thus,

$$V = \frac{E}{\varepsilon(2\Gamma)} \tag{76}$$

which is the MSA result (*20*).

The MSA potential is linear in the charge density on the electrode. However, as is seen from Figure 3, where Equation 76 is compared with computer simulation (*21*), this is true only for a small charge density. For larger charge densities additional perturbation terms must be included in Equation 69. Something more sophisticated than a generalization of Equation 57 seems to be needed, since in such an approximation all of the higher order terms in Q (or E) have the same sign as the linear term in Equation 76. As is seen from Figure 3, the simulation results lie below the linear term. This means some of the higher-order terms in Q must be negative.

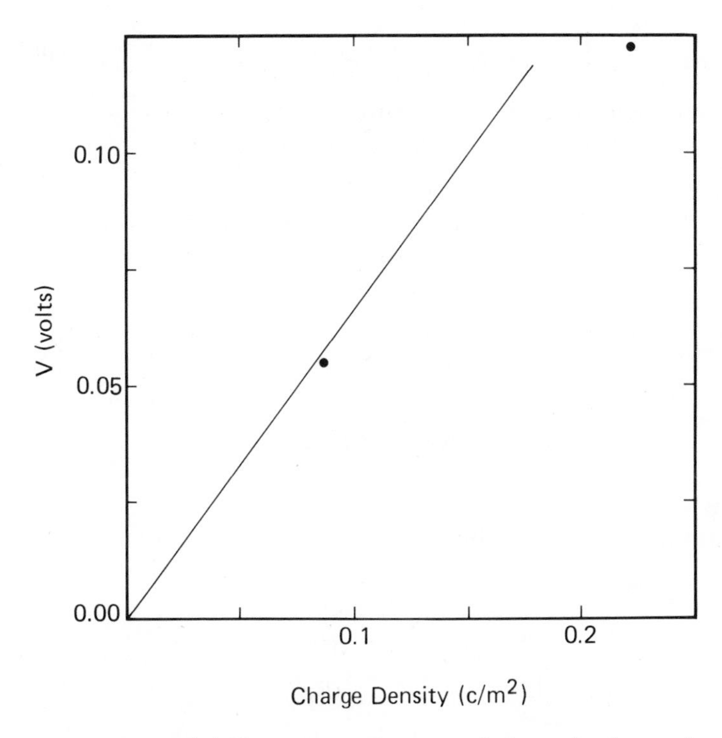

Figure 3. Potential difference as a function of electrode charge density for an electrified interface. Key: ●, computer simulation of Torrie and Valleau (21) for a 1:1 electrolyte (σ = 4.25 Å, ε = 78.5, T = 298 K); —, MSA result for this system.

A qualitatively correct expression for the dipolar hard-sphere solvent's contribution to the double layer potential can also be obtained from perturbation theory. The contribution to the surface free energy is

$$\Delta A = -\frac{1}{2}\beta\rho_s \int u_{ws}^2\,(r,\Omega)g_0^{ws}(r)drd\Omega \tag{77}$$

The first-order contribution vanishes because the integral of $u_{ws}\,(r,\Omega)$ over orientation is zero. Now

$$u_{ws}(r,\Omega) = \begin{cases} \infty & r < \sigma_s \\ \dfrac{Q\mu}{r^2}\,\hat{\mu}\cdot\hat{r} & r > \sigma_s \end{cases} \tag{78}$$

Hence

$$\Delta A = -\frac{2\pi}{3}\beta\rho_s Q^2\mu^2 \int_{(R+\sigma_s)/2}^{\infty} r^{-2}g_0^{ws}(r)dr \tag{79}$$

since $\langle \cos^2 \theta \rangle = \frac{1}{3}$. To evaluate this contribution, the integral in Equation 79 must be evaluated. However, a qualitative expression can be obtained by assuming that $g_0^{ws}(r) = 1$. To this approximation

$$\Delta A = -\frac{3yQ^2}{R + \sigma_s} \tag{80}$$

where $y = 4\pi\rho\beta\mu^2/9$.

The surface part of Equation 80 is

$$\Delta A_{\mathrm{DL}} = \frac{3yQ^2\sigma_s}{R^2} \tag{81}$$

so that the solvent contribution to the potential difference is

$$\begin{aligned} V &= \frac{6yQ\sigma_s}{R^2} \\ &= 3y\,\frac{E\sigma_s}{2} \end{aligned} \tag{82}$$

Equation 82 can be made more similar to the MSA result by recalling that for small y, $3y = \varepsilon - 1$ (cf. Equation 67). At large values of y we could write $3y \simeq (\varepsilon - 1)/\lambda\varepsilon$ where λ is some unspecified parameter reflecting the relation between y and ε and the fact that $g_0^{ws}(r)$ is not identically unity. Substituting for $3y$

$$V = \frac{\varepsilon - 1}{\lambda}\left(\frac{E\sigma_s}{2\varepsilon}\right) \tag{83}$$

Equation 83 is the same as the MSA result (22, 23). However, the perturbation theory value for λ will differ from the MSA result. In the MSA, λ is a weak function of ε since if $\varepsilon = 1$, $\lambda = 1$ and if $\varepsilon = 78$, $\lambda = 2.76$. Presumably, this is true in perturbation theory also. If so, the MSA and perturbation theory values of λ will be similar.

Hence, including both the ionic and solvent contributions

$$V = \frac{E}{\varepsilon(2\Gamma)} + \frac{(\varepsilon - 1)E\sigma_s}{2\varepsilon\lambda} \tag{84}$$

which becomes, in the limit of small κ,

$$V = \frac{E}{\varepsilon\kappa} + \frac{E\sigma}{2\varepsilon} + \frac{(\varepsilon - 1)E\sigma_s}{2\varepsilon\lambda} \tag{85}$$

Equation 85 is the same as the MSA result (*22, 23*).

Summary

Perturbation theory is a powerful tool in the theory of fluids. In this chapter we have briefly reviewed the general formalisms of perturbation theory and its application to simple fluids. Much of this is well known. However, the observation that the virial series is a form of perturbation theory may not be well known.

For ionic fluids, we have shown that perturbation theory gives some interesting results. In particular, it can be reformulated so that the zeroth-order approximation is the mean spherical approximation. This is fairly promising since the MSA often gives a fairly reasonable approximation to the free energy. Presumably, a perturbation theory for the corrections to the mean-spherical-approximation free energy will be better behaved than one for the corrections to the hard-fluid free energy. However, this conjecture will have to be tested. The work of Tani et al. (*17*) is of interest in this regard.

Finally, perturbation theory has been applied to the electric double layer. With little effort, the MSA results for the double layer are recovered. Unfortunately, the double layer problem is highly nonlinear. Hence, approximations which are more sophisticated than those presently available will have to be developed before perturbation theory can be applied at high charge densities on the electrode. Nonetheless, perturbation theory as a theory of the double layer is promising because of its simplicity.

Glossary of Symbols

A Helmholtz free energy
A_0 Helmholtz free energy of the unperturbed fluid
A_{HS} hard-sphere free energy
A_n nth order perturbation term in the expansion of the free energy
$\mathscr{C}(r)$ ring or chain sum
$\hat{\mathscr{C}}(k)$ Fourier transform of chain sum
$D(12)$ angular part of dipolar interaction
d hard sphere diameter

Continued on next page

Glossary of Symbols—Continued

E electric field near a charged electrode; also 4π times the charge density on the electrode
e magnitude of the electronic charge
$e_\gamma(r)$ $e^{-1}\partial e/\partial\gamma$
$e_{\gamma\gamma}(r)$ $e^{-1}\partial^2 e/\partial\gamma^2$
$f(r)$ $\exp\{-\beta u(r)\} - 1$
$f_1(r)$ $\exp\{-\beta u_1(r)\} - 1$
$g(r)$ radial distribution function
$g_0(r)$ radial distribution function of the unperturbed fluid
$g_0^w(r)$ radial distribution function of an unperturbed fluid molecule near large ion
$g_{HS}(r)$ hard-sphere radial distribution function
$g(1 \ldots h)$ h-body distribution function
$h(r)$ $g(r) - 1$
k Boltzmann constant
N total number of molecules and ions
Q total charge on electrode
R diameter of large ion before wall limit is taken
r_i position vector of ith molecule or ion
T temperature
$u(r)$ intermolecular pair potential
$u_0(r)$ reference fluid pair potential
$u_1(r)$ $u(r) - u_0(r)$
$u^*(r)$ reduced pair potential
V volume
x density of a lattice gas
x_i concentration of molecules or ions of species i
y $4\pi\rho\beta\mu^2/9$
z number of nearest neighbors in a lattice
z_i valence of ion of species i

Greek Letters

β $1/kT$
Γ renormalized screening parameter in MSA
γ expansion parameter in perturbation series
ε parameter describing strength of attractive part of pair potential; also dielectric constant
κ $(4\pi\beta z^2 e^2\rho/\varepsilon)^{1/2}$
μ dipole moment
ξ number of triangles of nearest neighbors that can be formed on a lattice, divided by N
ρ N/V

ρ_s solvent density, N_s/V
σ hard-sphere diameter, or the value of r for which $u(r) = 0$
$\Phi(\boldsymbol{r}_1 \ldots \boldsymbol{r}_N)$ potential energy of a collection of N molecules
Ω variable specifying the orientation of a molecule

Abbreviations

DH Debye–Huckel
HS hard sphere
MSA mean spherical approximation
nnd nearest neighbor distance
SL Stell–Lebowitz

Acknowledgment

This work was supported in part by National Science Foundation grant number CHE 80–01969.

Literature Cited

1. Henderson, D. ACS ADVANCES IN CHEMISTRY SERIES No. 182; ACS: Washington, D.C., **1979**, p. 1.
2. Barker, J. A.; Henderson, D. *Rev. Mod. Phys.* **1976,** *48*, 587.
3. Barker, J. A.; Henderson, D. *J. Chem. Phys.* **1967,** *47*, 4714.
4. Stell, G.; Lebowitz, J. L. *J. Chem. Phys.* **1968,** *49*, 3706.
5. Debye, P.; Huckel, E. *Z. Phys.* **1923,** *24*, 133, 305.
6. Waisman, E.; Lebowitz, J. L. *J. Chem. Phys.* **1972,** *56*, 3086, 3093.
7. Blum, L. *Theor. Chem.* **1980,** *5*, 1.
8. Valleau, J. P.; Cohen, L. K. *J. Chem. Phys.* **1980,** *72*, 5935.
9. Valleau, J. P.; Card, D. N. *J. Chem. Phys.* **1980,** *72*, 5942.
10. Van Megen, W.; Snook, I. K. *Mol. Phys.* **1980,** *39*, 1043.
11. Larsen, B.; Rasaiah, J. C.; Stell, G. *Mol. Phys.* **1977,** *33*, 987.
12. Andersen, H. C.; Chandler, D.; Weeks, J. D. *Adv. Chem. Phys.* **1976,** *34*, 105.
13. Barker, J. A.; Henderson, D.; Smith, W. R. *Mol. Phys.* **1969,** *17*, 579.
14. Rushbrooke, G. S.; Stell, G.; Hoye, J. S. *Mol. Phys.* **1973,** *26*, 1199.
15. Patey, G. N.; Valleau, J. P. *Chem. Phys. Lett.* **1973,** *21*, 297; *J. Chem. Phys.* **1976,** *64*, 170.
16. Hecht, C.; Henderson, D., unpublished data.
17. Tani, A.; Henderson, D., unpublished data.
18. Onsager, L. *J. Am. Chem. Soc.* **1936,** *58*, 1486.
19. Perram, J. W.; White, L. R. *Mol. Phys.* **1972,** *24*, 1133.
20. Blum, L. *J. Phys. Chem.* **1977,** *81*, 136.
21. Torrie, G.; Valleau, J. P. *J. Chem. Phys.* **1980,** *73*, 5807.
22. Carrie, S. L.; Chen, D. Y. C. *J. Chem. Phys.* **1980,** *73*, 2949.
23. Blum, L.; Henderson, D. *J. Chem. Phys.* **1981,** *74*, 1902.

RECEIVED for review March 2, 1982. ACCEPTED for publication August 25, 1982.

4

A Review of Methods for Predicting Fluid Phase Equilibria: Theory and Computer Simulation

KATHERINE S. SHING[1] and KEITH E. GUBBINS
Cornell University, School of Chemical Engineering, Ithaca, NY 14853

This chapter first reviews computer simulation methods for calculating the free energy or chemical potential in a mixture. Particular attention is given to methods suitable for dense gas or liquid mixtures, including umbrella sampling and test particle methods. This is followed by a review of mixture theories based in statistical mechanics. We focus on theories developed since 1967, and include perturbation theory for spherical and nonspherical molecules as well as the fluctuation formulas of Kirkwood and Buff.

AMONG THE MOST SIGNIFICANT ADVANCES for future work on phase equilibria has been the development of perturbation theories and computer simulation methods. Computer simulation studies provide data on precisely defined model fluids that can be used to test theoretical approximations. Such tests are of great value in discriminating among theories and are a desirable prelude to comparisons with experimental data on real fluids. For applications to phase equilibria it is particularly useful to calculate the chemical potentials of the mixture components by simulation. Such calculations require specialized techniques, and it is these techniques that we review in the first section of this chapter. We place special emphasis on those methods (umbrella sampling and test particle methods) that have been developed recently and that are useful at liquid densities.

We next review some of the most useful statistical mechanical theories that have been developed since 1967. These include perturbation theories for both spherical and nonspherical molecules, and theories based on the fluctuation formulas of Kirkwood and Buff. For spherical molecules, the theoretical situation is relatively satisfactory, except for mixtures where the molecules differ greatly in size (e.g., dilute solutions,

[1] Current address: University of Southern California, Department of Chemical Engineering, Los Angeles, CA 90007

0065-2393/83/0204-0073$09.50/0

supercritical extraction systems). For nonspherical molecules, there have been substantial improvements in methods in the last six years, and calculations based on perturbation theory are now quite accurate for mixtures in which the liquids are completely miscible. However, these methods are still rather poor for highly nonideal mixtures (e.g., those where hydrogen bonding is important), mainly because the intermolecular potential functions are poorly known.

The calculations carried out so far usually assume rigid molecules and neglect effects due to quantum corrections or multibody forces. Quantum corrections are important for mixtures containing hydrogen or helium, but are usually small otherwise. Some calculations have been made that include three-body dispersion and induction forces. The rigid molecule approximation precludes the use of the theoretical methods reviewed here for long-chain molecules.

Free Energy and Chemical Potential by Computer Simulation

The usual Monte Carlo and molecular dynamics techniques used to simulate fluids can yield the internal energy and pressure with reasonable accuracy, but do not give good results for the Helmholtz free energy, A, or chemical potential, μ. The conventional Monte Carlo and molecular dynamics methods are reviewed elsewhere (*1, 2*). (For a more detailed discussion of the older methods used to calculate A or μ, *see* References 3–5.)

For pure fluids, a knowledge of A is equivalent to knowing μ, but this is not the case for mixtures. The first few methods described below give the free energy rather than the chemical potential. The grand canonical Monte Carlo and test particle methods then described give the chemical potential directly; this is to be preferred where mixture phase equilibria are to be studied.

Methods for Calculating Helmholtz Free Energy. THERMODYNAMIC INTEGRATION. In thermodynamic integration, the difference in the free energy between two states is calculated by numerically integrating over states between the initial and final state, using some thermodynamic relationship. For example

$$A(\rho_1, T) = A(\rho_0, T) + N \int_{\rho_0}^{\rho_1} \frac{P}{\rho^2}\, d\rho \tag{1}$$

or

$$\frac{A(\rho, T_1)}{T_1} = \frac{A(\rho, T_0)}{T_0} + \int_{1/T_0}^{1/T_1} U d\left(\frac{1}{T}\right) \tag{2}$$

where $\rho = N/V$ is the number density, P is pressure, and U is internal energy. Values for P or U from computer simulations for 10 or so points along an isotherm or isochore are usually fitted to polynomials and then integrated according to Equation 1 or 2. The free energy for the initial state is usually known (usually one takes the high-temperature or low-density limit).

The integration variable is not restricted to thermodynamic variables. The initial and final states could represent two systems at the same thermodynamic state, but having different intermolecular forces. Thus we can write

$$A(\lambda_1) - A(\lambda_0) = \int_{\lambda_0}^{\lambda_1} \left\langle \frac{\partial \mathcal{U}(\lambda)}{\partial \lambda} \right\rangle_{\lambda} d\lambda \qquad (3)$$

where, for example, the state λ may represent a quadrupolar Lennard–Jones fluid, whereas λ_0 represents a Lennard–Jones fluid. Here $\mathcal{U}$ is the intermolecular potential energy. In this particular example, then $\langle \partial \mathcal{U}(\lambda)/\partial\lambda\rangle_\lambda$ is given by $\langle \mathcal{U}_p \rangle_\lambda$ where $\mathcal{U}(\lambda) = \mathcal{U}_{LJ} + \lambda \mathcal{U}_p$. Therefore the integration variable λ is a measure of the quadrupole strength.

The thermodynamic integration method has been used by various authors to study model fluids, including hard spheres (*6*), soft repulsive spheres (*7, 8*), Lennard–Jones atoms (*9–13*), the dipolar–quadrupolar Lennard–Jones fluid (*14*), one-component plasma (*15*), and hard dumbbells (*16*). Thermodynamic integration is tedious and poses problems when phase transitions occur along the path of integration. In the two-phase region, simulation gives large uncertainties in P and U used in Equations 1 and 2, so that the resulting values of A are rather unreliable. To overcome these problems, artificial constraints have to be imposed on the system to reduce the fluctuations. In the case of the melting transition, Hoover and Ree (*17*) used a single occupancy model where the center of each particle is confined to a cell centered at the lattice site. In the case of the vapor–liquid transition, Hansen and Verlet (*18*) divided the Monte Carlo box into subcells and restricted the density fluctuations in the box by imposing upper and lower limits on the density of each subcell. When such artificial constraints are imposed, one is essentially performing a series of simulations on artificial systems, the sole purpose of which is to allow integration to obtain A and which is otherwise of no particular interest or utility.

METHOD OF MCDONALD AND SINGER. Several more direct methods of finding A have been proposed. McDonald and Singer (*19–21*) write the free energy for the two states T_1 and T_0 as

$$\frac{A(T_1)}{T_1} - \frac{A(T_0)}{T_0} = -k \ln \int_{-\infty}^{\infty} f_{T_0}(\mathcal{U}) \exp\left[-\frac{\mathcal{U}}{k}\left(\frac{1}{T_1} - \frac{1}{T_0}\right)\right] d\mathcal{U} \qquad (4)$$

Again, the two states need not be restricted to thermodynamic states and (with obvious changes to the integrand) could equally well represent two systems having different intermolecular forces at the same thermodynamic state. $A(T_1) - A(T_0)$ is easily obtained from Equation 4. In this method, a Monte Carlo simulation at the state T_0 is used to obtain the distribution $f_{T_0}(\mathcal{U})$ which is then integrated according to Equation 4 to obtain $\Delta A \equiv A(T_1) - A(T_0)$. Typically, $f_{T_0}(\mathcal{U})$ is a rather narrow Boltzmann distribution and covers only the range of $\mathcal{U}$ important to the state T_0. Therefore, $f_{T_0}(\mathcal{U})$ obtained from the simulation at T_0 will allow calculation of $A(T_1)$ only if state T_1 is close to state T_0; that is, when the range of $\mathcal{U}$ important to T_1 overlaps sufficiently the range important to T_0. For Lennard–Jones fluids, McDonald and Singer (*19, 21*) found that this method works if the states T_1 and T_0 differ in temperature by less than 15%.

MULTISTAGE SAMPLING. As described in the last subsection, the free energy difference between two systems 1 and 0 having intermolecular potential energies $\mathcal{U}_1$ and $\mathcal{U}_0$ can be written as

$$\Delta A = A_1 - A_0 = -kT \ln \int_{-\infty}^{\infty} f_0(\mathcal{U}_p) \exp(-\mathcal{U}_p/kT)\, d\mathcal{U}_p \tag{5}$$

where $f_0(\mathcal{U}_p)$ is the probability density for observing the difference $\mathcal{U}_p \equiv \mathcal{U}_1 - \mathcal{U}_0$ in the reference (0) system, where $\mathcal{U}_1$ and $\mathcal{U}_0$ are the potential energies that would be observed in the 1 and the 0 systems for the molecular configuration in question. Valleau and coworkers (*22, 23*) noted that when the systems 1 and 0 are rather different, $f_0(\mathcal{U}_p)$ as obtained from a single simulation at the state 0 will not overlap the distribution $f_0(\mathcal{U}_p) \exp(-\mathcal{U}_p/kT)$, which is the integrand in Equation 5. The integral in Equation 5 will then be underestimated. They suggested the use of bridging distributions that bridge the gap between $f_0(\mathcal{U}_p)$ and $f_0(\mathcal{U}_p) \exp(-\mathcal{U}_p/kT)$, and used Boltzmann distributions corresponding to physically realistic states between 0 and 1 as the bridging functions. Unless states 0 and 1 are quite close together, several distributions (or stages) are needed to bridge the gap—hence the name multistage sampling. This method has been used to study Coulombic hard spheres (*22*), dipolar hard spheres (*23*), and diatomic Lennard–Jones molecules using hard diatomics as reference (*23, 24*). The method is superior to thermodynamic integration, since a knowledge of $f_0(\mathcal{U}_p)$ over the range of $\mathcal{U}_p$ relevant to states 0 and 1 allows accurate interpolation for the continuous spectrum of states between 0 and 1. (Jacucci and Quirke (*24, 25*) introduced a method called marquee sampling, in which an analytic form for the potential function of the intermediate ensemble is given.)

For many systems, multistage sampling requires fewer simulations than thermodynamic integration. Exceptions occur at low temperature and for large systems, since for those cases the Boltzmann bridging distributions become very narrow, and overlap is usually poor. These shortcomings prompted Valleau and Torrie to seek other more optimum choices for bridging functions. This resulted in the development of the method of umbrella sampling, described later in this chapter.

BENNETT'S METHOD. Bennett (*26*) derived expressions for the optimal estimation of the free energy difference between two systems with temperature scaled potentials $\mathcal{U}_1' \equiv \mathcal{U}_1/kT_1$ and $\mathcal{U}_0' \equiv \mathcal{U}_0/kT_0$ using data from simulations of finite length.

$$\Delta A_{\mathrm{est}} = \ln \frac{\sum_{n_1} \{f(\mathcal{U}_0' - \mathcal{U}_1' + C)\}}{\sum_{n_0} \{f(\mathcal{U}_1' - \mathcal{U}_0' - C)\}} + C - \ln\left(\frac{n_1}{n_0}\right) \qquad (6)$$

For a run of sufficient length the first ln term on the right-hand side of Equation 6 will converge to zero, and we have

$$\Delta A_{\mathrm{est}} = C - \ln\left(\frac{n_1}{n_0}\right) \qquad (7)$$

Here ΔA_{est} is the estimated free energy difference, n_1 and n_0 are the number of configurations generated in systems 1 and 0, respectively, C is a shift constant, and $f(x) = 1/[1 + \exp(x)]$ is the Fermi function.

Two separate simulations, one for the 0 system and one for the 1 system, are made, and histograms of the energy distribution functions $h_0(\mathcal{U}_p' \equiv \mathcal{U}_1' - \mathcal{U}_0')$ and $h_1(\mathcal{U}_p')$ are constructed. Equal computer time should be devoted to each simulation, and this determines the optimal ratio n_1/n_0. Using the distributions $h_0(\mathcal{U}_p')$ and $h_1(\mathcal{U}_p')$, ΔA is calculated by making guesses for C and iterating using Equations 6 and 7 until convergence is achieved. The success of this method depends on the overlap between the distributions $h_1(\mathcal{U}_p')$ and $h_0(\mathcal{U}_p')$. When these do not overlap, ΔA cannot be estimated reliably. A comparison of Bennett's method and multistage sampling has been made by Quirke and Jacucci (*24*, *25*) in a study of Lennard–Jones diatomics. They also suggested a new method for correcting the results of short Monte Carlo simulations.

UMBRELLA SAMPLING. Umbrella sampling (*27*, *28*) allows calculation of the difference $\Delta A = A_1 - A_0$ of two states that are not necessarily close to each other, using only one (or at most a few) simulations. It is convenient to write Equations 4 and 5 in a more general form

$$\left(\frac{A}{T}\right)_1 - \left(\frac{A}{T}\right)_0 = -k \ln \int_{-\infty}^{\infty} f_0(\Delta) \exp\left[-\mathcal{U}_p'(\Delta)\right] d\Delta \qquad (8)$$

where $\mathcal{U}_p' = (\mathcal{U}/kT)_1 - (\mathcal{U}/kT)_0$ and Δ is an appropriate variable. For example, Δ equals $\mathcal{U}$ in Equation 4 and $\mathcal{U}_p$ in Equation 5.

In umbrella sampling (*27, 28*) $f_0(\Delta)$ in Equation 8 is not determined from Monte Carlo simulation at the reference state 0 alone (as was the case in the work of McDonald and Singer), nor is it determined from a range of physically realistic states in between and including the states 0 and 1 [as was the case in the multistage sampling work of Valleau, Card, and Patey (*14, 22*), as well as in Bennett's method (*26*)]. Instead, f_0 is determined from one or a few simulations for artificial systems that give rise to configurations typical of many states between 0 and 1. An artificial system is generated by replacing the Boltzmann distribution $\exp(-\beta\mathcal{U})$ with a new distribution $W(\Delta)\exp(-\beta\mathcal{U})$, where $W(\Delta)$ is a suitable weighting function, and Δ is a suitable integration variable. With a judicious choice of $W(\Delta)$ the simulation for this artificial system gives a weighted distribution $f_w(\Delta)$ that is much broader than the Boltzmann distribution f_0 (*see* Figure 1). The original Boltzmann distribution $f_0(\Delta)$ can be recovered from $f_w(\Delta)$ by reweighting according to

$$f_0(\Delta) = \frac{f_w(\Delta)}{W(\Delta)} \Big/ \left\langle \frac{1}{W(\Delta)} \right\rangle_w \tag{9}$$

where $\langle\ \rangle_w$ indicates an ensemble average over the weighted chain of configurations. In this way f_0 is obtained over a much wider range of Δ.

Umbrella sampling has been used by various authors to study several model fluids, including pure Lennard–Jones fluids (*28*), Lennard–Jones mixtures (*29*), dipolar hard spheres (*30*), and quadrupolar Lennard–Jones

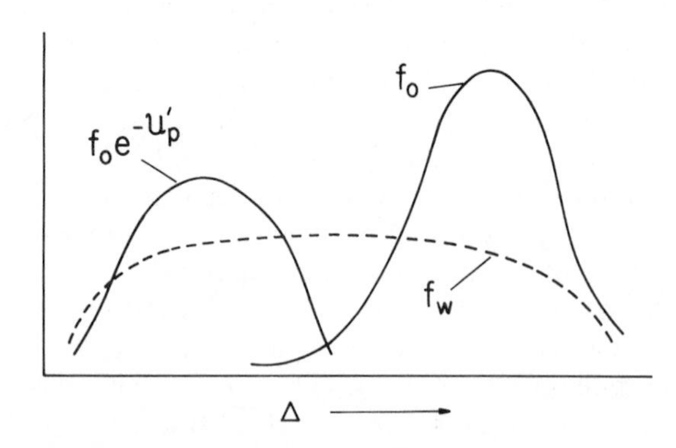

Figure 1. Probability density distribution functions for a typical liquid condition for the unbiased system (f_0) *and the weighted system* (f_w). *For small* Δ *values,* f_0 *and the desired curve* $f_0 \exp(-\mathcal{U}_p')$ *are calculated from* f_w *using Equation 9.*

fluids (*31*). It has also been used to study the surface tension of water (*32*).

Umbrella sampling has several advantages. It has been shown that ΔA is less dependent on system size than A itself, and good results for free energies have been obtained using only 32 atoms (*27, 30; see also 24, 25*). Since umbrella sampling simulations are carried out for artificial systems representing states between the system of interest (1) and the reference state (0), in cases where the system of interest phase-separates, the umbrella sampling simulation may show no phase transition and thus allow more accurate results to be obtained. Disadvantages of this method are that the reference free energy must be known; also, the weighted sampling makes it uneconomical to obtain some of the properties other than A, such as correlation functions. The distribution functions f_0 and f_w narrow as the size of the system increases or when temperature decreases; therefore the number of stages or separate simulations required to cover a specified range in Δ also increases. The selection of a suitable weighting function requires some trial and error. Finally, the method is restricted to Monte Carlo calculations and cannot be used in molecular dynamics.

Methods for Calculating Chemical Potential. GRAND CANONICAL MONTE CARLO. The partition function for the grand canonical ensemble is given by

$$\Xi = \sum_{N=0}^{\infty} \frac{1}{N!} \lambda^N \int \exp\left[-\mathcal{U}(x^N)/kT\right] dx^N \tag{10}$$

where $dx^N = dx_1 dx_2 \ldots dx_N$ and x_i represents the coordinates that specify the configuration of molecule i (e.g., $x_i = \mathbf{r}_i$ for spherical molecules, $x_i = \mathbf{r}_i\omega_i$ for nonspherical rigid molecules), and $\lambda = (q_{qu}/\Lambda_t^3\Lambda_r) \exp(\mu/kT)$; here q_{qu}, Λ_t and Λ_r are the usual quantal, translational, and rotational partition functions, respectively, for a single molecule.

In this ensemble, μ, T and V are fixed and the density fluctuates. The mean density and $\langle N \rangle$ are found as ensemble averages at the end of the simulation. This simulation method has been implemented in somewhat different ways by Norman and Filinov (*33*), Adams (*34–37*), and Rowley, Nicholson, and Parsonage (*38*). The method involves essentially two steps. The first step is the displacement of the particles in the system and is identical to the procedure used in the canonical ensemble simulation. The second step involves adding or removing particles according to the requirements of the grand canonical ensemble weighting function, so that the number density in the system fluctuates. In their studies of the Lennard–Jones fluid, both Norman and Filinov, and Adams observed abrupt changes in density at a particular value of μ corresponding to the vapor–liquid transition; however, such jumps were very infrequent.

Adams also used this simulation method to locate the coexistence curve of the Lennard–Jones fluid. He observed that close to the phase transition, if the simulation was started from the wrong phase for the prescribed μ, the system would ultimately converge to the correct phase. No flipping back and forth between the two phases was observed, however. Grand canonical Monte Carlo works best at high temperatures and low densities, since the addition of molecules is then allowed with sufficient frequency for adequate sampling of the density fluctuations relevant to this ensemble. Recently, Mezei (39) has modified the sampling procedure so that states of higher densities can be studied. In his cavity biased (μTV) Monte Carlo method, a network of uniformly distributed test points was generated in the fluid and the fraction of these points in a suitable cavity was found. Insertion of a new particle was attempted at this cavity instead of at randomly selected points. The fraction of test points in the chosen cavity also allows the proper normalization of the ensemble averages. For the Lennard–Jones fluid, Mezei found that the efficiency of the insertion process was increased by a factor of 8 while the required central processing unit (CPU) increased by a factor of 2.5.

Grand canonical Monte Carlo has the advantage of giving the chemical potential directly. Since the number of particles in the system is allowed to fluctuate, the ensemble permits density fluctuations and also permits concentration fluctuations in the case of mixtures. Therefore, this ensemble should be more suitable for studying systems close to phase transitions and for systems close to the critical region. Grand canonical Monte Carlo simulations are more complex and more time consuming than canonical ensemble Monte Carlo simulations. Adams noted that the results are sensitive to errors in the random-number generators used. At low temperatures and high densities, it is very difficult and time consuming to sample the density fluctuations adequately, even with Mezei's improved cavity biased version. This problem will become more severe if a more complex, angle-dependent potential is used, because the success of the insertion attempts will then also be angle dependent. Furthermore, since it is μ that is specified, the density (and also the composition in the case of mixtures) is not known until the end of the simulation; this is inconvenient in practice. This method is restricted to Monte Carlo calculations, and does not appear to be useful for molecular dynamics.

TEST PARTICLE METHOD. This method is based on an expression for the chemical potential derived by Widom (*40*)

$$\begin{aligned} \mu_{\alpha r} &= -kT \ln \langle \exp(-\mathcal{U}_{t\alpha}/kT) \rangle_{N-1} \\ &= -kT \ln \int_{-\infty}^{\infty} f_{N-1}(\mathcal{U}_{t\alpha}) \exp(-\mathcal{U}_{t\alpha}/kT)\, d\mathcal{U}_{t\alpha} \end{aligned} \tag{11}$$

where $\mu_{\alpha r} = \mu_\alpha - \mu_\alpha^{id}$ is the residual chemical potential for the species α in a fluid composed of $N-1$ real molecules of which N_A are molecules of species A, N_B are molecules of species B, and so forth, in a volume V at temperature T; $\mathcal{U}_{t\alpha}$ is the potential energy felt by an invisible test particle (molecule 1 of species α); and $\langle . . . \rangle_{N-1}$ denotes an ensemble average in a system of $N-1$ real molecules. It should be emphasized that in Equation 11 the test particle has no influence on the $N-1$ real molecules.

Adams (*34*) used this method to calculate the free energy of hard spheres in order to test the scaled particle theory. Romano and Singer (*41*) later implemented this method in both Monte Carlo and molecular dynamics simulations to calculate the chemical potential of bromine and chlorine using a two-center Lennard–Jones model. The sampling procedure used by Romano and Singer was as follows: after every molecular dynamics time step or after every few hundred Monte Carlo configurations, the value of $\mathcal{U}_{t\alpha}$ for three mutually perpendicular test particles at each of a few hundred uniformly distributed lattice sites was calculated. The ensemble average $\langle \exp(-\mathcal{U}_{t\alpha}/kT)\rangle_{N-1}$ was found by an unweighted average over all values of $\mathcal{U}_{t\alpha}$ obtained over the course of the simulation. Powles (*42*) further studied the method and used the results of Romano and Singer to calculate the vapor–liquid coexistence properties of bromine and chlorine.

The advantages of the test particle method are that it is much simpler than the grand canonical Monte Carlo algorithm, and requires only the addition of a simple subroutine to perform the test particle sampling. Also, Romano and Singer have shown that it can be implemented in both canonical Monte Carlo and molecular dynamics simulations, although they found that the convergence was poorer in molecular dynamics. It is conceivable that this method can also be used in other ensembles, for example, the isothermal–isobaric ensemble. Since the test particle sampling does not affect the real particle configurations, the internal energy, pressure, and correlation functions can be obtained in the usual way. The disadvantage of the method is that it fails at normal liquid densities because of the predominance of configurations in which the test particle overlaps one or more of the real molecules, causing the Boltzmann factor of the test particle to be negligibly small. Such failure is closely related to the failure of the grand canonical Monte Carlo method at high densities. In mixtures, this difficulty increases whenever one considers a solute that interacts strongly (through large size or large attractive force) with the other solvent molecules. There are two sampling problems associated with the test particle method at liquid densities:

1. Given a fixed configuration of the real particles, how can one efficiently sample the relevant range of $\mathcal{U}_{t\alpha}$ by placing

the test particle in locations that overlap real particles as little as possible?

2. How can the sampling be concentrated on those real particle configurations that exhibit "holes" and thus make a major contribution to the average in Equation 11?

These problems can be overcome by one or more recently developed techniques which are now briefly described.

Restricted Umbrella Sampling. The first problem mentioned above can be solved by using a procedure called restricted umbrella sampling (*43*), in which a weighting function is used to force the test particle to move mainly in regions with holes. The weighted distribution is normalized by comparing it to the unweighted distribution over the range of $\mathcal{U}_{t\alpha}$ where the two overlap. The weighting function acts only on the motion of the test particle, which is invisible to the real particles. In other words, through the weighting function the test particle can see the real particles, but the reverse is not true. For a Lennard–Jones fluid at a reduced density of $\rho^* = \rho\sigma^3 = 0.7$, restricted umbrella sampling is able to increase the number of configurations that contribute significantly to the integral in Equation 11 by a factor of 30.

Combined f-g Sampling. This method is designed to solve the second problem, i.e., that of adequately sampling real molecule configurations that exhibit holes. It is based on a combination of Widom's expression, Equation 11, and its inverse (*43, 44*)

$$\mu_{\alpha r} = kT \ln \langle \exp (\mathcal{U}_{t\alpha}/kT)\rangle_N$$

$$= kT \ln \int_{-\infty}^{\infty} g_N(\mathcal{U}_{t\alpha}) \exp (\mathcal{U}_{t\alpha}/kT)\, d\mathcal{U}_{t\alpha} \tag{12}$$

where $\langle . . . \rangle_N$ is now an ensemble average over the system of N molecules in which the test particle is one of the N real molecules. Equation 12 has been derived independently by Oliviera (*44*) and by Shing (*43, 45, 46*).

From Equations 11 and 12, it can be shown that (*43, 46*)

$$g_N(\mathcal{U}_{t\alpha}) = \exp (\mu_{\alpha r}/kT)\, f_{N-1}(\mathcal{U}_{t\alpha}) \exp (-\mathcal{U}_{t\alpha}/kT) \tag{13}$$

This means that when there is a range of $\mathcal{U}_{t\alpha}$ over which g_N and f_{N-1} overlap, the chemical potential $\mu_{\alpha r}$ can be calculated. The functions f_{N-1} and g_N are shown for a typical liquid density in Figure 2. It should be noted (*see* Equations 11 and 13) that g_N is proportional to the integrand needed to calculate the chemical potential. This is as expected, since g_N gives the distribution over $\mathcal{U}_{t\alpha}$ for a real molecule and thus samples the hole region adequately.

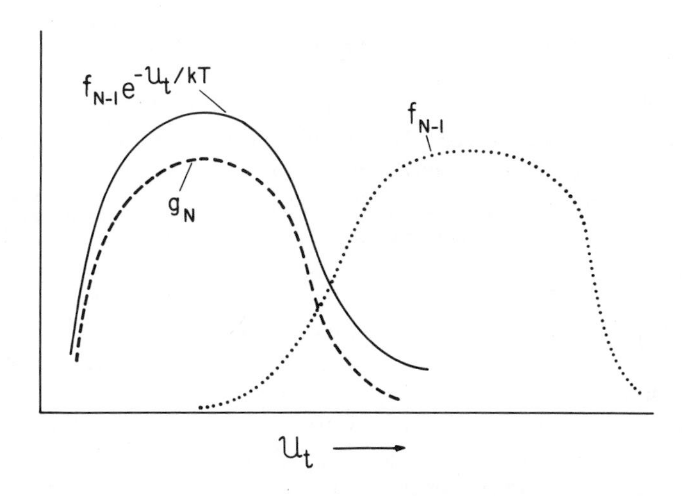

Figure 2. Probability density distribution functions needed in the test particle method for a typical liquid state condition. The function f_{N-1} *is a very broad distribution and extends to very large* $\mathcal{U}_t$ *values; the width of this distribution is compressed here for convenience of plotting. For small* $\mathcal{U}_t$ *values the desired integrand,* $f_{N-1} \exp(-\mathcal{U}_t/kT)$, *is obtained by using* g_N *in Equation 13 to calculate* f_{N-1}.

Using restricted umbrella sampling and *f-g* sampling, Shing and Gubbins (*43*) have shown that for the Lennard–Jones fluid, the chemical potential up to the triple-point density can be found. They also used this method to calculate the chemical potential of highly nonideal Lennard–Jones mixtures (*43*, *46*). Results for the Henry constant in Lennard–Jones mixtures using this technique are shown in Figures 6 and 7.

Test Particle Method with Full Umbrella Sampling. If the density is very high, or if the test particle interacts strongly (for example, for large or strongly interacting solutes, in the case of mixtures) the distributions g_N and f_{N-1} may no longer overlap. In such cases the spontaneous generation of holes in the fluid occurs so rarely that it is almost never sampled in a simulation of normal length. Therefore it is necessary to bias the sampling artificially in such a way as to emphasize configurations with suitable holes. This is done by coupling the motion of the test particle to that of the real particles through a weighting function (*47*). The test particle and the real particles are then mutually visible.

The advantage of this procedure is that it works at very high densities and for very nonideal mixtures. However, since the motion of the test particle is now felt by the real molecules, the structure of the fluid is altered and it is no longer practical to calculate some of the usual properties (for example, the correlation functions). The fact that the weighting function now affects the motion of both the test particle and the real particles also means that this full umbrella sampling test particle method cannot be used in molecular dynamics simulations.

SUMMARY OF METHODS FOR CALCULATING CHEMICAL POTENTIAL. The methods for calculating the chemical potential will usually be the most useful for testing theories of mixtures, particularly when the ability of the theory to predict phase equilibria is important. Of the two methods described, the test particle method or one of its modifications works under most conditions and is relatively simple to use. The grand canonical Monte Carlo method will offer advantages over the test particle procedure when large density or concentration fluctuations are likely in the real system, i.e., near phase transitions or critical points. At present, this method is more complex to program and is restricted to moderate densities unless special techniques are used. Mezei's method (*39*) removes this last restriction to some extent, but does not solve the problem completely for very dense liquids.

RECENT WORK. Several articles have appeared recently on computer simulation methods for determining the chemical potential. An interesting application of the grand canonical Monte Carlo method to chemical equilibria in mixtures of bromine and chlorine has been described (*48*), and a new method of implementing the grand canonical Monte Carlo method has been proposed and applied to the Lennard–Jones fluid (*49*). The test particle method has been further studied using molecular dynamics in place of the Monte Carlo technique (*50*). These authors study a shifted-force Lennard–Jones fluid using both the direct Widom method and the combined *f*-*g* sampling technique. They find little to choose between these two variants of the test particle method, in contrast to the results found in the Monte Carlo studies of the Lennard–Jones fluid (*43*). The molecular dynamics runs of Powles et al. (*50*) are considerably longer than the Monte Carlo calculations of Shing and Gubbins (*43*), are at different state conditions for a larger system, and are for a different potential model, so that the calculations cannot be directly compared. It is possible that the molecular dynamics method samples phase space more efficiently, so that the direct Widom method is adequate for dense fluids with spherical potentials, at least when they are pure. Careful tests comparing the Monte Carlo and molecular dynamics methods for the same potential, state conditions, and length of runs are needed to clarify these points.

Theory

In this section we concentrate on the more successful theoretical approaches that have been developed since 1967. The most promising approach at present is perturbation theory, and various forms of perturbation theory are described below. Most recent work has proceeded by evaluating the partition function, but an important alternative approach is to start from the formulas of Gibbs and Kirkwood and Buff, discussed below, which relate the mixture thermodynamics to concentration fluctuations. We omit any discussion of many of the older theories (cell and lattice theories, regular solution theory, random mixture theory, and so forth) and theories that apply only to specialized states (for example, the critical region) or particular sorts of mixtures (polymer solutions, aqueous mixtures, fused salts, and so forth).

It is convenient to classify binary phase behavior on the basis of the types of critical and three-phase lines present and on the way these intersect. For fluid phase equilibria, the classification scheme shown in Figure 3 is convenient, and includes all the known binary types. Class I systems are often fairly ideal in a thermodynamic sense, and do not exhibit liquid–liquid immiscibility. The remaining five classes display liquid–liquid separation of various kinds. The most common types of behavior are Classes I, II, and III. These classes of phase behavior are further complicated by the presence of solid phases, and many subclasses occur. Detailed discussions of these various types of phase behavior are given elsewhere (*51–55*).

Mixtures of Hard Molecules. The simplest nonideal mixtures are those composed of hard particles. Their study provides valuable insight into the effects of molecular size and shape on the thermodynamic properties, in the absence of complications from attractive forces. In addition to integral equation and perturbation theory, it is possible to study such mixtures using scaled particle theory. In scaled particle theory, one calculates the work required to add a hard molecule to the fluid by first adding a point molecule and then scaling this molecule up to its full size (*56*). The derivation of scaled particle theory is valid only for mixtures of molecules of the same shape, although the final expression obtained seems to work quite well even when the molecules have different shapes. Studies of mixtures of hard bodies have been made by Percus-Yevick theory (*57*), scaled particle theory (*56*), various modified forms of scaled particle theory (*58–61*), and computer simulation (*62–66*). These studies are in general agreement, and lead to the following three conclusions concerning the thermodynamics of mixtures of hard molecules:

1. The excess volume, V^E, and hence G^E, is always negative. (For such mixtures $U^E = 0$ and G^E is simply the integral of V^E over pressure.) There is therefore no fluid–fluid phase transition in such mixtures, though it is possible that isotropic–nematic phase transitions may occur for very elongated molecules.
2. Molecular shape has virtually no effect on the shape of the V^E curve, and only a minor effect on its magnitude.

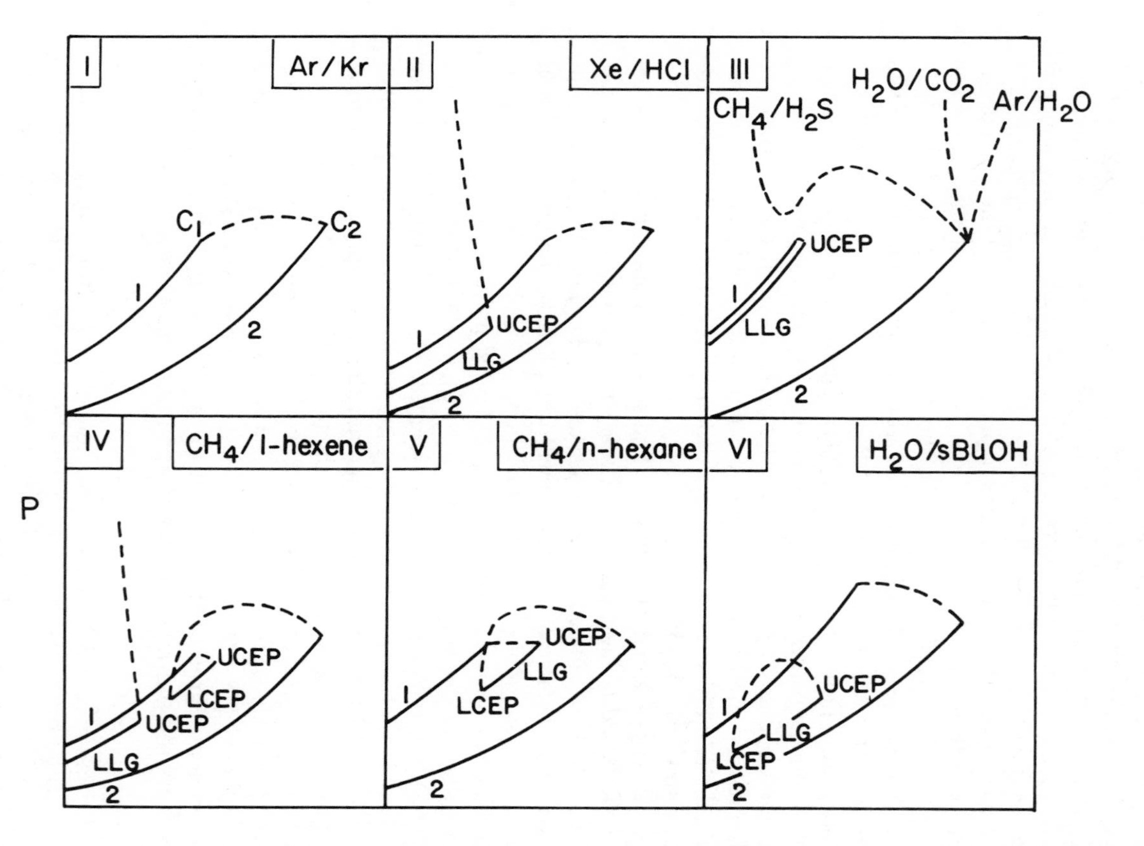

Figure 3. Classification of binary phase diagrams. Curves 1 and 2 are pure component vapor pressures, C_1 and C_2 are critical points of pure components 1 and 2, dashed lines are critical loci, and LLG is the liquid–liquid–gas line.

3. The size ratio of the two molecules and the pressure both have a large effect on the shape and magnitude of the V^E curves.

The small influence of molecular shape on the compressibility factor is shown in Figure 4 for a mixture of hard spheres and spherocylinders of equal volumes. It is seen that the values of $P/\rho kT$ for such a mixture are similar to those for pure hard spheres and pure spherocylinders, provided these molecules all have the same volume. The values of V^E for this mixture are very small, because the nonideality arises entirely from the difference in shape of the two species, their volumes being the same.

Perturbation Theory for Spherical Molecules. Until 1971, work on the theory of liquid mixtures focused almost exclusively on simple,

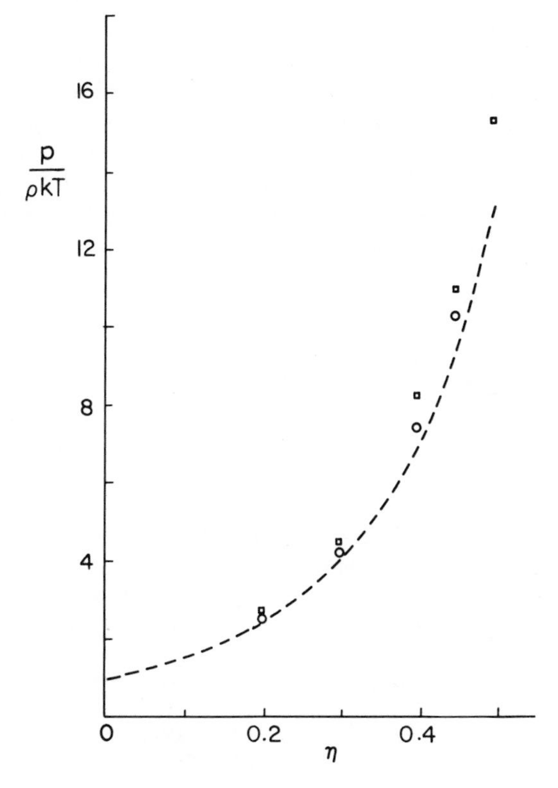

Figure 4. The compressibility factor for pure hard spheres (---), an equimolar mixture of spheres and spherocylinders of equal volumes (○), and pure spherocylinders (□). Here $\eta = \rho\Sigma_\alpha x_\alpha v_\alpha$, where ρ = N/V is number density, x_α is mole fraction, and v_α is molecular volume of component α. (Reproduced with permission from Ref. 66. Copyright 1980, Taylor and Francis, Ltd.)

spherical molecules. Perturbation and conformal solution theories, which relate the free energy of the real mixture to a hard-sphere mixture and to an ideal mixture, respectively, are successful for spherical molecules (neon, argon, krypton, xenon) and are quite good even for weakly non-spherical molecules (e.g., nitrogen, methane). These theories give a good account of systems with phase diagrams of Classes I and III of Figure 3, provided that the molecules are spherical and not too different in size. They can predict qualitative behavior of Classes II, IV and V, but this generally requires the use of potential parameters that are physically unrealistic. They cannot predict behavior of Class VI. We give only a brief account of these theories here; for details, reviews are available (*67–71*).

CONFORMAL SOLUTION THEORY. In this approach it is assumed (1) that the molecules are conformal; i.e., they obey the same intermolecular force law, differing only in the values of the potential parameters $\varepsilon_{\alpha\beta}$ and $\sigma_{\alpha\beta}$, and (2) the values of $\varepsilon_{\alpha\beta}$ and $\sigma_{\alpha\beta}$ for the various molecular pairs are not too different from each other. Because of the second assumption, it is possible to expand (*72*) the Helmholtz free energy A about that of an ideal solution of molecules, all of which have the same parameters ε_x and σ_x, and to terminate the series at the first-order term. The most successful of these theories is the van der Waals 1 fluid (vdW1) theory. The expansion parameters in vdW1 theory are the combinations $\varepsilon\sigma^3$ and σ^3, which appear to give more rapid convergence than other choices that have been tried (*67–69, 72, 73*). This leads to

$$A = A_x + R_\varepsilon \sum_\alpha \sum_\beta x_\alpha x_\beta (\varepsilon_{\alpha\beta}\sigma^3_{\alpha\beta} - \varepsilon_x \sigma^3_x) + R_\sigma \sum_\alpha \sum_\beta x_\alpha x_\beta (\sigma^3_{\alpha\beta} - \sigma^3_x) + \cdots \quad (14)$$

Here R_ε and R_σ are pure fluid integrals for the reference fluid, the precise form of which need not concern us here; A_x is the free energy of the reference mixture; and σ_x and ε_x are the size and energy parameters for the reference fluid. If ε_x and σ_x are now chosen according to the vdW1 mixing rules,

$$\varepsilon_x \sigma^3_x = \sum_\alpha \sum_\beta x_\alpha x_\beta \, \varepsilon_{\alpha\beta}\sigma^3_{\alpha\beta} \quad (15)$$

$$\sigma^3_x = \sum_\alpha \sum_\beta x_\alpha x_\beta \, \sigma^3_{\alpha\beta} \quad (16)$$

then the first order term in Equation 14 vanishes, and Equation 14 reduces to the simple result $A = A_x$. This is the basis of corresponding states treatments for mixtures, and for the shape factor methods.

Clearly Equations 14–16 can be expected to apply only for mixtures in which the molecules are not too different, i.e., in which σ_{AA}/σ_{BB} and $\varepsilon_{AA}/\varepsilon_{BB}$ are not too different from unity for binary *A–B* mixtures. If the series in Equation 14 is extended to second order, it is found that these terms involve triple summations over the mole fractions (*74, 75*). Thus for mixtures of components of very different critical volumes or critical temperatures, Equations 15 and 16 are likely to be unsatisfactory.

A test of this theory is shown in Figure 5 for argon–krypton mixtures, and in Figures 6 and 7 for infinitely dilute solutions of *A* in *B*, where *A* and *B* are spherical Lennard–Jones molecules. The points in these last two figures are exact computer simulation results for such mixtures (*43, 46*), obtained by the modified test particle method described earlier, while the dashed lines give the results for the theory. The usual Lorentz–Berthelot rules

$$\varepsilon_{AB} = (\varepsilon_{AA}\varepsilon_{BB})^{1/2}$$

$$\sigma_{AB} = {}^1/_2(\sigma_{AA} + \sigma_{BB}) \tag{17}$$

are used in these calculations. It is seen from these figures that the vdW1 treatment gives quite a good description of the chemical potential for $1/4 \leq \varepsilon_{AA}/\varepsilon_{BB} \leq 4$ (corresponding to critical temperatures that vary by as much as a factor of four) when the molecular sizes are the same; these limits correspond to $1/2 \leq \varepsilon_{AB}/\varepsilon_{BB} \leq 2$. However, the theory does not describe well the effect of molecular size differences (Figure 7), particularly when the solute is much larger than the solvent molecules.

The vdW1 theory has been extended in two ways. In the first (*74, 75*), the expansion of Equation 14 is extended to second order. The second-order terms must be calculated explicitly and involve three-body integrals, so that the simplicity of the first-order theory is lost. Detailed numerical calculations do not seem to have been reported, except for hard-sphere mixtures (*72, 75*). For that case, the second-order theory seems to give quite good results even for ratios σ_{BB}/σ_{AA} as large as three, where perturbation theory based on a pure hard sphere reference system fails (See Figure 8). The second extension is the so-called two-fluid theory, or vdW2 theory (*67–70, 76*), in which the properties of the real mixture are equated to those of an ideal mixture of two pure pseudo-components; i.e., components *A* and *B* (in the case of binary mixtures) are referred to different pure reference fluids. The vdW1 mixing rules of Equations 15 and 16 are now replaced by

$$\varepsilon_{x\alpha}\sigma_{x\alpha}^3 = \sum_{\beta} x_\beta \varepsilon_{\alpha\beta}\sigma_{\alpha\beta}^3 \tag{18}$$

$$\sigma_{x\alpha}^3 = \sum_{\beta} x_\beta \sigma_{\alpha\beta}^3 \tag{19}$$

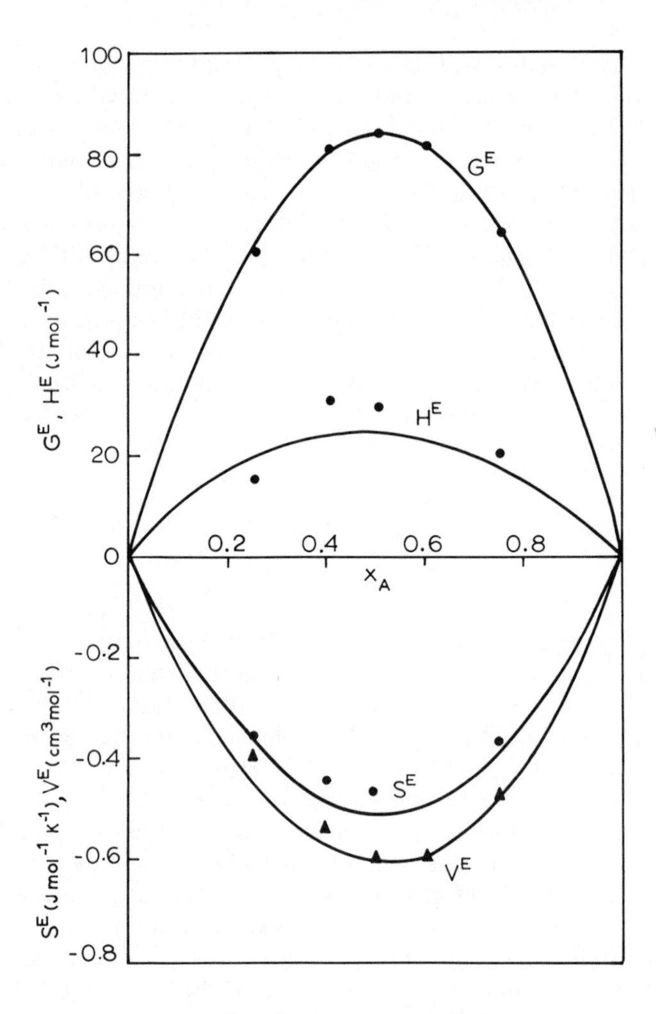

Figure 5. Test of vdW1 theory (lines) against Monte Carlo data (points) (67–70, 77) *for Lennard–Jones mixtures at 115.8 K. The Lennard–Jones parameters are chosen to simulate argon–krypton mixtures.*

where $\alpha = A,B$, and so forth. The vdW2 theory has been tested against simulation data for Lennard–Jones mixtures in which the molecules are of nearly the same size, and is poorer than the vdW1 theory. It does not seem to have been tested for highly nonideal Lennard–Jones mixtures, where the size or energy parameters of the two molecular species are very different.

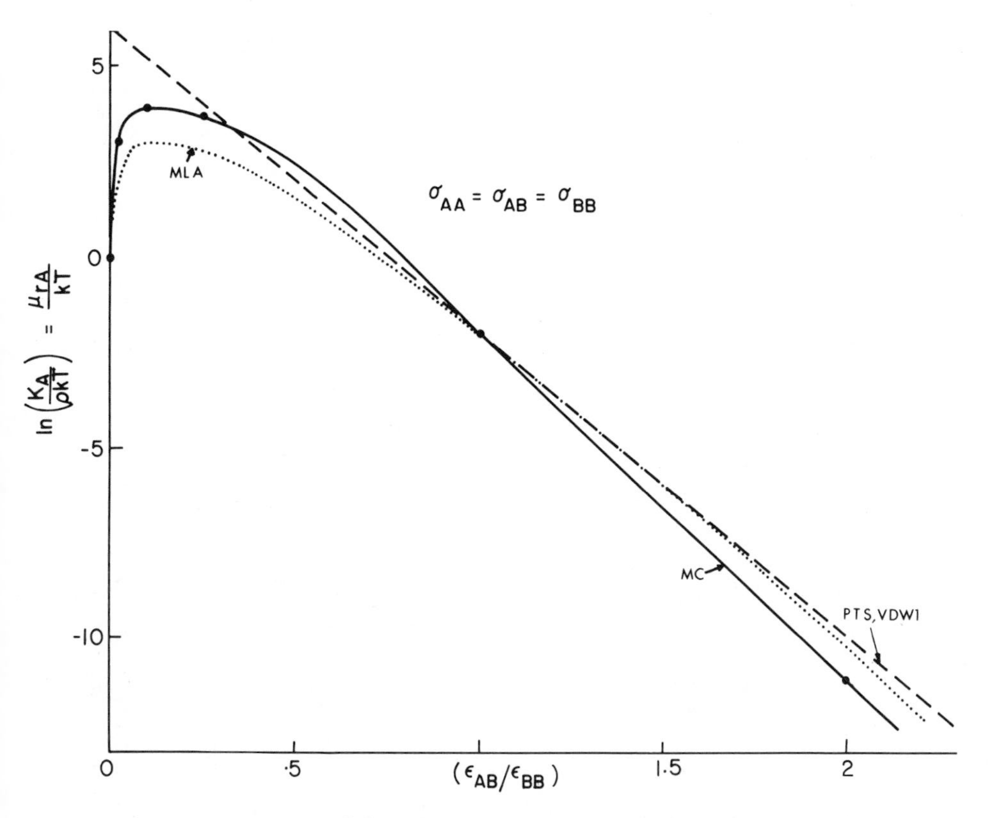

Figure 6. Variation of the Henry constant K_A, *with the ratio* $\varepsilon_{AB}/\varepsilon_{BB}$ *for Lennard–Jones mixtures at* $kT/\varepsilon_{BB} = 1.2$, $\rho\sigma_{BB}^3 = 0.7$. *The molecules are of the same size,* $\sigma_{AA} = \sigma_{BB}$. *In this case, the results of the vdW1 theory are the same as those for a perturbation expansion (46) about the pure solvent (PTS). Key: ·····, result of the Mansoori-Leland approximation; ——, result of Monte Carlo technique; and ---, result of vdW1 theory. (Reproduced with permission from Ref. 46. Copyright 1982, Institute for Physical Science and Technology.)*

PERTURBATION ABOUT A HARD-SPHERE FLUID. In this approach the intermolecular potential energy $\mathcal{U}$ is usually written in the form

$$\mathcal{U}_\lambda = \mathcal{U}_0 + \lambda \mathcal{U}_p \tag{20}$$

where $\mathcal{U}_0$ is the reference system potential energy, $\mathcal{U}_p$ is the perturbing energy, and λ is a perturbation parameter. Choosing $\lambda = 0$ gives the reference potential, while $\lambda = 1$ gives the potential for the full system. If we expand the Helmholtz free energy A in powers of λ and subse-

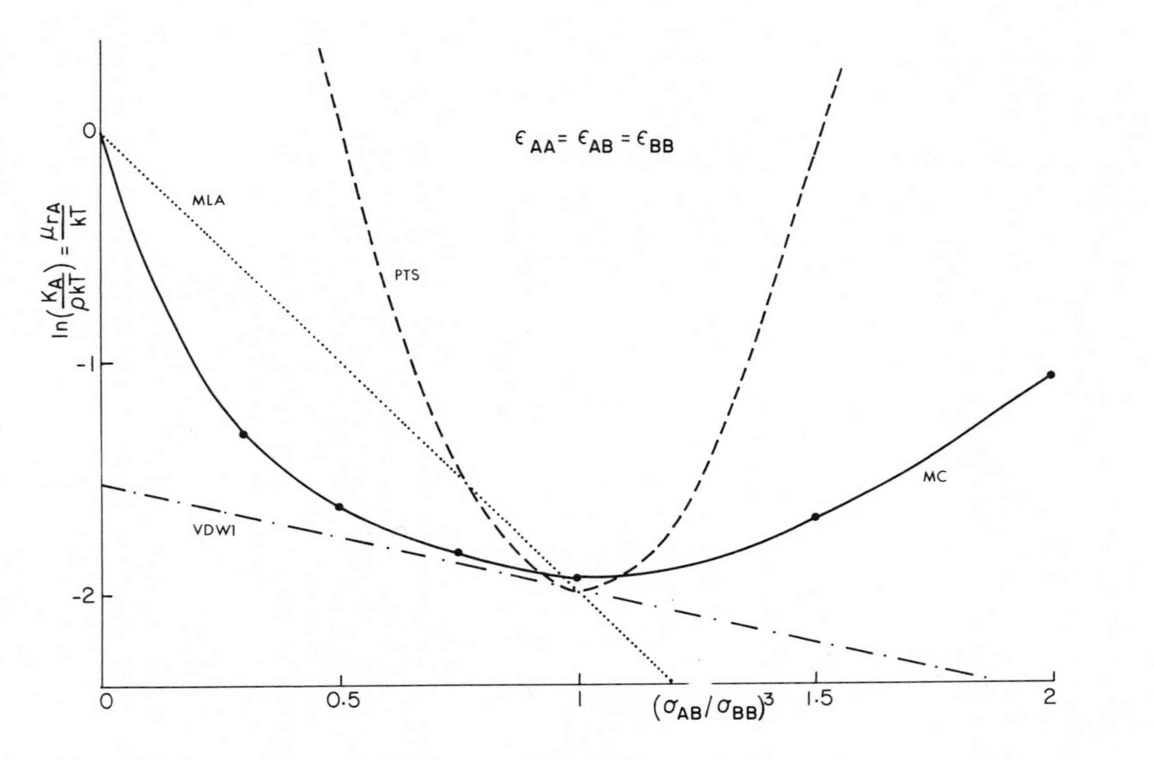

Figure 7. Variation of Henry constant with $(\sigma_{AB}/\sigma_{BB})^3$ for Lennard–Jones mixtures, $kT/\varepsilon_{BB} = 1.2$, $\rho\sigma_{BB}^3 = 0.7$. *Key as in Figure 6.*

quently set $\lambda = 1$ we obtain the perturbation expansion

$$A = A_0 + A_1 + A_2 + A_3 + \ldots \tag{21}$$

where A_0 is the free energy for the reference fluid, A_1 is the first-order perturbation term, etc. The reference fluid is chosen to be one in which the molecules interact with a purely repulsive potential, so that the attractive forces are included in $\mathcal{U}_p$. The reference system properties are subsequently related to those of a fluid of hard spheres through a second expansion, usually in some form of inverse steepness parameter (the hard-sphere potential being infinitely steep). The expansion can be carried out either about a pure hard-sphere fluid, or about a hard-sphere mixture. The latter gives the better results, particularly if the molecules are much different in size (Figure 8 shows results of using a pure hard-sphere reference fluid).

Three variations on this approach have been proposed—due to Leonard et al. (*79, 80*), Lee and Levesque (*81*), and Mansoori and Leland (*82, 83*). The theories of Leonard et al. and Lee and Levesque differ mainly in the definition of the reference potential, the former using the part of the potential for $r < \sigma$, and the latter that part for $r < r_m$, where r_m is the separation corresponding to the potential minimum. The Mansoori–Leland theory is an extension of the variational approach to mixtures. These three theories give similar results (*67–70*).

These theories have been extensively reviewed (*67–71*), so that we do not dwell on them here. They are more complicated to use than vdW1 theory and its extensions, since the evaluation of A_1 and higher order terms requires the evaluation of integrals over the hard-sphere correlation functions. They give better results for highly nonideal mixtures, however, particularly if the molecules are much different in size, and are not restricted to conformal mixtures. They can be used as a starting point for the derivation of empirical equations of state, such as the van der Waals and Redlich–Kwong equations (*67–71, 84*). Such an approach makes clear the approximations in such equations, and can be used to suggest new equations.

THE MANSOORI–LELAND APPROXIMATION. In the Mansoori–Leland approximation (*85*), the true mixture radial distribution function, $g_{\alpha\beta}(r)$, is replaced by the corresponding function for a pure fluid evaluated at a reduced temperature $kT/\varepsilon_{\alpha\beta}$, a reduced distance $r/\sigma_{\alpha\beta}$, and a reduced density $\rho\sigma_x^3$, with σ_x^3 given by Equation 16. Results for the Mansoori–Leland approximation are included in Figures 6 and 7. It gives quite a good account of the effects of varying $\varepsilon_{AA}/\varepsilon_{BB}$ (and hence T_{cA}/T_{cB}, where T_c is critical temperature) when the molecules are similar in size, but does not work well when the molecules are much different in size.

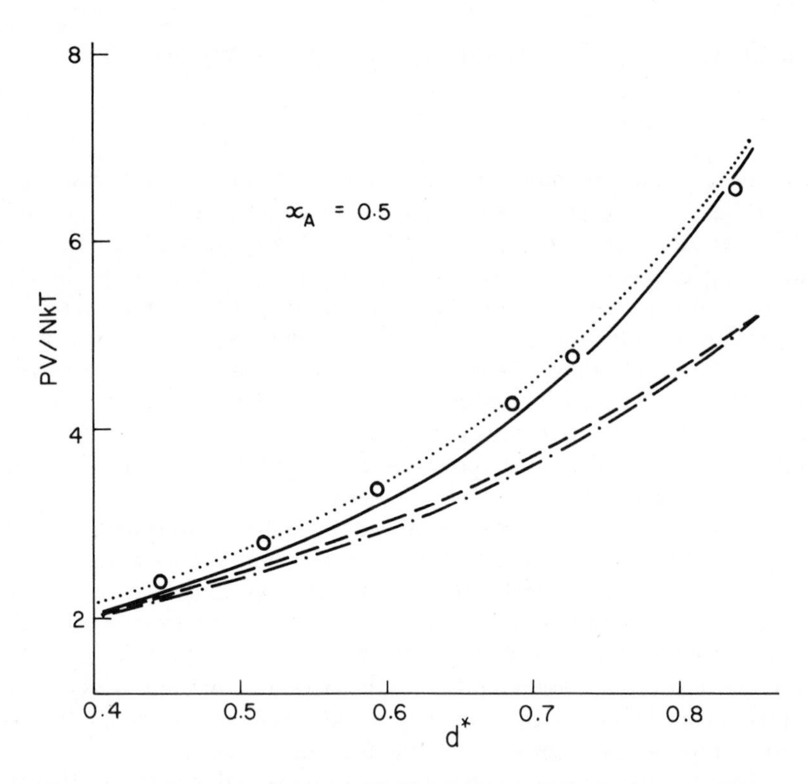

Figure 8. The compressibility factor for an equimolar mixture of hard spheres with $\sigma_{BB}/\sigma_{AA} = 3$*. The points are molecular dynamics results of Alder (62–66) and the curves are theoretical results. Key: ····, Percus–Yevick (compressibility) equation; -----, vdW1; ———, second-order vdW1 theory; and — · — · —, Henderson–Barker perturbation expansion about a pure hard sphere fluid (79). Here* $d^* = (x_A\sigma_{AA}^3 + x_B\sigma_{BB}^3)\rho$*. (Reproduced with permission from Ref. 75. Copyright 1971, Taylor and Francis, Ltd.)*

Perturbation Theory for Nonspherical Molecules. For nonspherical molecules perturbation theory calculations have been made using reference fluids of both spherical and nonspherical molecules. The most extensive comparisons with experiment have been made for spherical reference molecules using the Padé approximation of the Pople expansion (*86*) suggested by Stell et al. (*87*).

$$A = A_0 + A_2(1 - A_3/A_2)^{-1} \tag{22}$$

The term A_1 vanishes in this series. In this expansion the reference fluid pair potential $u^0_{\alpha\beta}(r)$ is defined by

$$u^0_{\alpha\beta}(r) \equiv \langle u_{\alpha\beta}(r\omega_1\omega_2)\rangle_{\omega_1\omega_2} \tag{23}$$

where α and β are mixture components and $\langle . . . \rangle$ indicates an unweighted average over molecular orientations. Usually some simple model, such as the Lennard–Jones 6-12 or 6-n is chosen for u^0. For the 6-12 model both the equation of state and the pair correlation function are known from computer simulation studies (*88*), and it is a simple matter to relate the properties of the 6-n fluid to those for the 6-12 case (*89*). General expressions have been given (*90*) for A_1 and A_2; they involve two- and three-body integrals over the correlation functions for the reference system. These integrals have been evaluated for a variety of potential forms and fitted to simple functions of temperature and density (*88*). Equation 22 is found to agree well with computer simulation results for fluids in which the molecules have spherical or near-spherical cores, even when strong electrostatic forces are present. It is less satisfactory for molecules with highly nonspherical shapes (*90*).

In calculations based on Equation 22, the pair potential is written as a sum of terms

$$u(12) = u_0(r) + u_{elec}(12) + u_{ind}(12) + u_{dis}(12) + u_{ov}(12) \qquad (24)$$

where u_{elec}, u_{ind}, u_{dis}, and u_{ov} are the electrostatic, induction, anisotropic dispersion, and anisotropic overlap terms, respectively. The anisotropic dispersion term is usually approximated by the London expression, and the remaining anisotropic potential contributions are represented by the first few terms in an expansion in generalized spherical harmonics. Explicit expressions for the terms in Equation 24 are given in Reference 91. Equation 22 has been used to explore the relationship between intermolecular forces and the resulting phase diagram (*90–93*). Thus, if one or both of the components interacts with a Lennard–Jones plus a dipole–dipole term, it is possible to obtain any of Classes I to V of Figure 3 by suitable adjustment of the parameters in the potential (*see* Figure 9). Similar results are obtained if a quadrupole–quadrupole term is used in place of the dipole–dipole one. If, instead, the anisotropic part of the potential consists only of an overlap term designed to simulate the shape of a linear molecule, then only Classes I and III are obtained (*90*). The anisotropic overlap and dispersion parts of the potential seem to have a relatively small effect on the phase diagram, whereas the effect of electrostatic forces is large. A detailed study of the effect of various potential terms on systems showing gas–gas immiscibility has been made by Gibbs (*93*). Jonah et al. (*84*) have recently carried out a study of the influence of various types of intermolecular potential terms on the dissolving power of solvents used in supercritical extraction of liquids and solids. They found that quadrupolar forces are particularly effective in increasing the dissolving power.

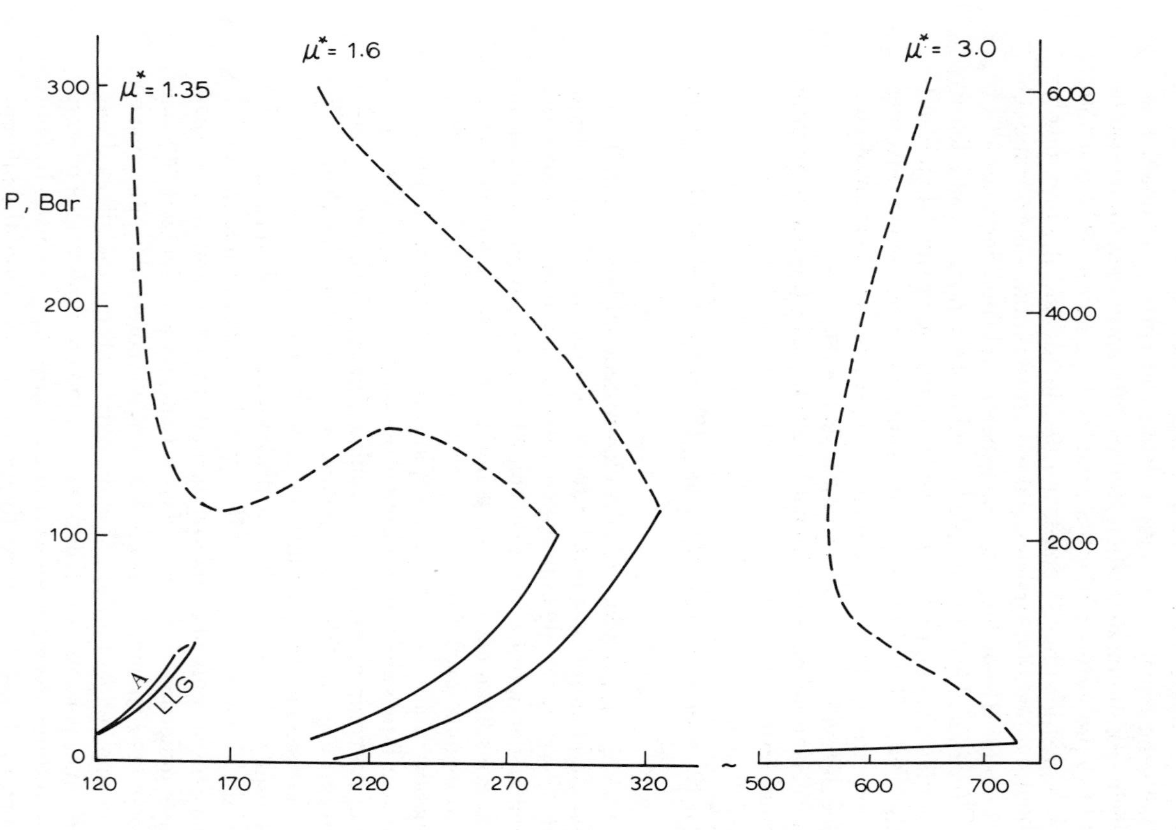

Figure 9. *Pressure–temperature diagram for mixtures of Lennard–Jones molecules (component A) with molecules (component B) interacting with Lennard–Jones molecules plus dipole–dipole potential;* $\varepsilon_{BB}/\varepsilon_{AA} = 1.5$; $\sigma_{BB}/\sigma_{AA} = 1$; ε_{AB} *and* σ_{AB} *are given by the Lorentz/Berthelot rules. Solid lines are vapor pressure curves for the pure components; dashed lines are critical loci; and* $\mu^* \equiv \mu_B^* = \mu_B/(\varepsilon_{BB}\sigma_{BB}{}^3)^{1/2}$. *The dipole is added to the less volatile component leading to Class III behavior. Gas–gas immiscibility occurs for large* μ^*.

Comparisons of Equation 22 with experiment have been carried out for liquid mixtures involving inorganic fluids (nitrogen, oxygen, carbon monoxide, carbon dioxide, carbon tetrafluoride, nitrous oxide, hydrogen chloride, and hydrogen bromide), hydrocarbons (acetylene, ethylene, and *n*-alkanes up to C_8), and the first four primary alcohols. Work up to 1978 has been reviewed (*89*). More recent studies include those by Shukla et al. (*94, 95*) for simple inorganics (nitrogen, oxygen, carbon monoxide, and carbon dioxide) and methane; by Clancy et al. (*96–99*) on mixtures involving hydrogen chloride, hydrogen bromide, ethylene carbon tetrafluoride, hydrogen, methane, and xenon; by Machado et al. (*101, 102*) on the mixtures carbon dioxide–ethane, ethane–ethylene, xenon–nitrous oxide, and nitrous oxide–ethylene; by Gibbs (*93*) on mixtures involving the *n*-alkanes up to C_8 and the first four primary alcohols; and by Moser et al. (*102*) for mixtures carbon monoxide–methane, xenon–hydrogen bromide and xenon–hydrogen chloride.

For mixtures having Class I phase behavior (no liquid–liquid equilibria) agreement between Equation 22 and experiment is usually good. Results for the system carbon dioxide–ethane are shown in Figure 10. In the theoretical calculations the pair potential model is that of Equation 24, with u_0 being the 6-*n* model, and the electrostatic potential being represented by a quadrupole–quadrupole term. The three 6-*n* parameters (n, ε, σ) were adjusted to best fit the vapor pressure and saturated liquid density of the pure components, and ε_{12} was adjusted to best fit the data at 263.15 K. Equation 22 gives an excellent fit to the data over the temperature range of 60 K. The dashed lines are the best fit obtainable using the van der Waals 1 theory, using the same fitting procedure and number of parameters, but with the 6-*n* potential alone. The vdW1 theory gives good results at the temperature where ε_{12} is fitted, but fails to reproduce the variations of *Pxy* with temperature. This conclusion is borne out by calculations for other mixtures (*100, 101*). Equation 22 usually gives good results for mixtures in which the molecules are appreciably different in size and for polar mixtures (*102*), provided the mixture is still of Class I.

As we pass from the Class I systems to systems having greater nonideality (Classes II, III, and so forth) the agreement between theory and experiment becomes poorer. Of particular importance is the recent work of Gibbs (*93*), who has studied the systems methanol–methane and methanol–ethane, both of which are Class III in the classification scheme of Figure 3. Gibbs uses a potential model of the type given in Equation 24 with the electrostatic potential approximated by dipolar and quadrupolar terms. The nonaxial quadrupole of methanol is approximated by an effective axial quadrupole. Good results are obtained for the phase equilibria and three-phase (L_1L_2G) line at low temperatures for these methanol systems (Figure 11), and the theory predicts the correct qual-

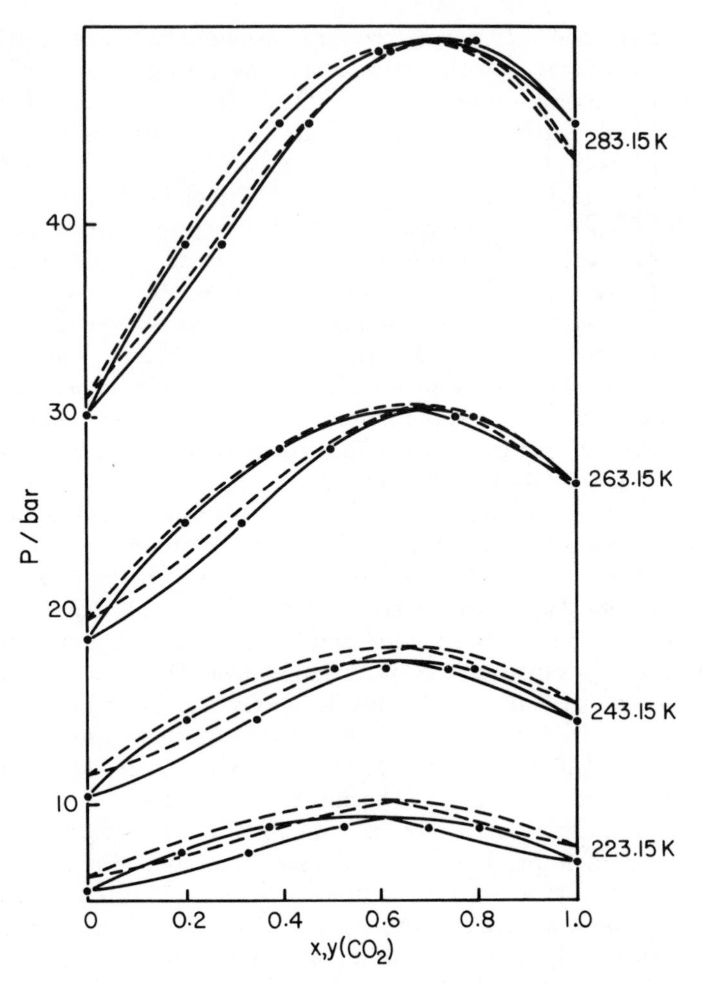

Figure 10. Theory versus experiment for carbon dioxide–ethane. Solid lines are from Equation 22, dashed lines are vdW1 theory, and points are experimental data.

itative Class III behavior. However, the quantitative agreement is poor at higher temperatures (Figure 12). This is principally because the methanol model predicts a critical temperature and a critical pressure that are too high. Thus the critical locus is of the correct shape, but lies at pressures that are too high. When the quadrupolar terms are omitted from the methanol potential model, the results are much worse, and the theory predicts Class I behavior.

Machado et al. (*100, 101, 103*) have recently carried out calculations for nonaxial molecules, with particular attention being paid to nonaxial quadrupole effects. Such nonaxial effects are important in general. The

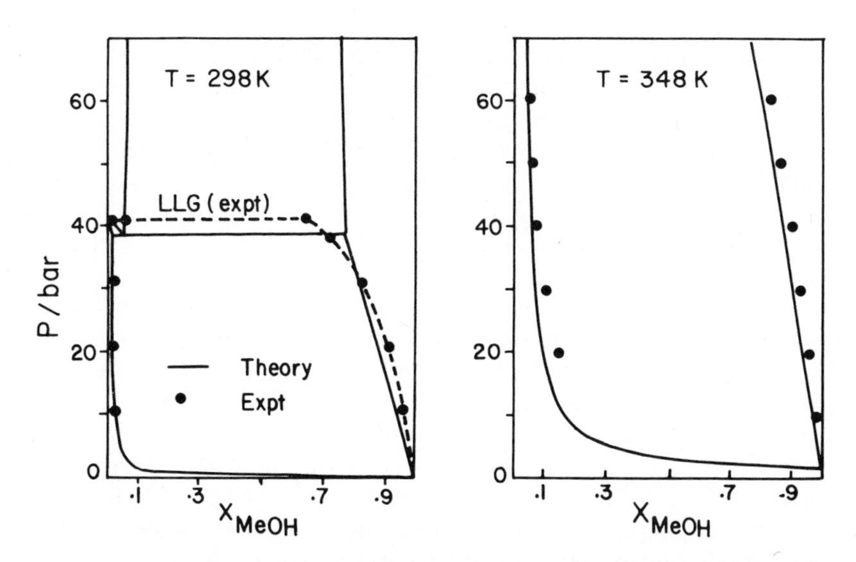

Figure 11. Theory versus experiment for methanol–ethane. Solid lines are from Equation 22, dashed lines and points are experimental data. (Reproduced with permission from Ref. 93.)

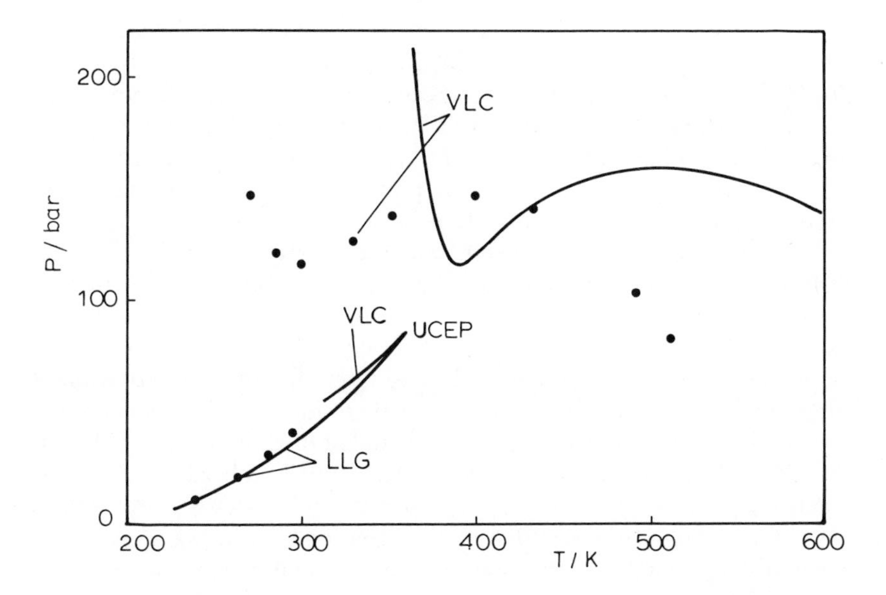

Figure 12. Theory versus experiment for methanol–ethane. Solid lines are theoretical and points are experimental data. (Reproduced with permission from Ref. 93.)

simplest example of a nonaxial molecule is ethylene (D_{2h} point group). An approximate calculation of the excess enthalpy for a nitrous oxide–ethylene equimolar mixture at 183.2 K and zero pressure gives 357 J/mol by the correct nonaxial treatment, and 84 J/mol by the effective axial method (*103*). The experimental value is 271 J/mol. Molecules such as water and methanol are more nonaxial than ethylene (*103*), and the errors in using an effective axial treatment would probably be substantially greater.

The neglect of quantum effects usually leads to small errors unless the system contains hydrogen as one of the constituents. Clancy and Gubbins (*98*) have studied hydrogen and hydrogen–xenon mixtures using Equation 22, but accounting for quantum corrections using the first correction to the partition function of order $\hbar^2$. It is found that the simple $0(\hbar^2)$ treatment gives good results for hydrogen for temperatures down to about 100 K, the precise temperature depending on the density (quantum effects being more important at higher densities). Below this temperature a full quantal treatment becomes necessary. Figure 13 shows a comparison of theory and experiment at 100 K for the hydrogen–methane system. In this case the anisotropic intermolecular forces make only a small contribution, but the quantum effects are substantial. While the theory gives quite good results, the empirical equations of state fail to converge over much of the range of composition, presumably because of the neglect of quantum corrections.

The effect of three-body dispersion forces on thermodynamic properties have been studied recently by Shukla et al. (*94, 95*) and by Clancy (*98–99*) and were found to be significant. Preliminary calculations (*104–107*) also suggest that multibody induction forces are important, but there do not seem to have been any comparisons with experiment so far.

Mixture calculations based on perturbation theory using a nonspherical molecule reference system have been made by Boublik (*108–112*). A Kihara potential is used, and the properties of this fluid are expanded about those for a fluid of hard convex molecules. Calculations have been reported for the excess properties of argon–nitrogen, argon–oxygen, nitrogen–oxygen, nitrogen–methane and carbon disulfide–cyclopentane, and are in moderately good agreement with experiment. This approach accounts in a more realistic fashion for the effects of molecular shape than does the Padé approximant of Equation 22, but omits the effects from electrostatic forces. Enciso and Lombardero (*113*) have used nonspherical reference perturbation theory to calculate excess properties for argon–nitrogen and argon–oxygen mixtures using a two-center Lennard-Jones model. Agreement with experiment is poor. Several other authors have proposed perturbation theories based on a nonspherical reference potential, but have not compared them with experiment.

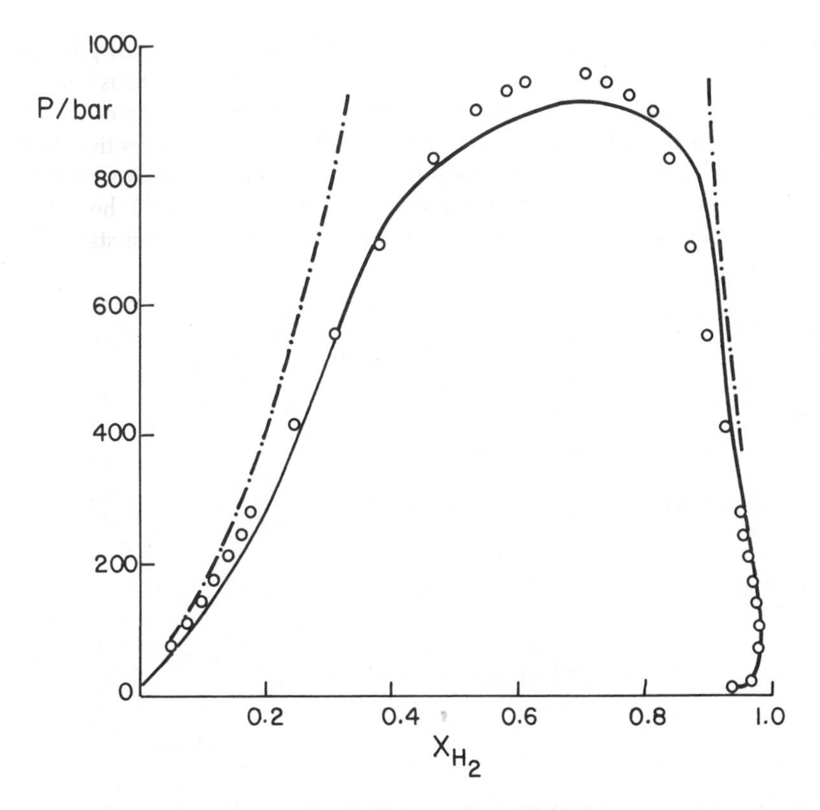

Figure 13. Vapor–liquid equilibrium data for hydrogen–methane from experiment (○), from theory including quantum correction (——), and from the Redlich–Kwong equation of state (– · –), at 100 K. Parameters in the equation of state were fitted to the data at 130 K in each case. (Reproduced with permission from Ref. 98. Copyright 1981, Taylor and Francis, Ltd.)

Finally we mention the work of Starling et al. (*114*) who have developed a form of conformal solution theory based on the Pople perturbation theory. This involves a third parameter in the expansion that accounts for anisotropy of the intermolecular forces, and gives good results in mixtures that are not too nonideal. It has been applied to hydrocarbon mixtures.

Kirkwood–Buff Theory. It is possible to relate mixture thermodynamic properties to composition fluctuations in a particularly direct and simple way. Such equations were first derived by Gibbs (*115*), but were put in a convenient form much later by Kirkwood and Buff (*116*), who showed that the composition fluctuations can be written in terms of integrals over the pair correlation functions $g_{\alpha\beta}(r)$ between molecular centers. The resulting equations are quite general, and make no assumption concerning the type of molecule (spherical or nonspherical, rigid or nonrigid), or additivity of the intermolecular potentials.

The Kirkwood–Buff theory gives expressions for the composition derivatives of the chemical potentials, the partial molecular volumes, $\overline{V}_\alpha = (\partial V/\partial N_\alpha)_{TPN'}$, and the isothermal compressibility, $\chi = -V^{-1}(\partial V/\partial P)_{TN}$. For a binary mixture of A and B, some of these relations are (*116–117*)

$$\left(\frac{\partial \mu_A}{\partial x_A}\right)_{TP} = kT\left[\frac{1}{x_A} + \frac{\rho_B(2H_{AB} - H_{AA} - H_{BB})}{1 - \rho_B x_A(2H_{AB} - H_{AA} - H_{BB})}\right] \quad (25)$$

$$\overline{V}_A = [1 + \rho_B(H_{BB} - H_{AA})]/\eta \quad (26)$$

$$\overline{V}_B = [1 + \rho_A(H_{AA} - H_{AB})]/\eta \quad (27)$$

$$\chi = [1 + \rho_A H_{AA} + \rho_B H_{BB} + \rho_A\rho_B(H_{AA}H_{BB} - H_{AB}^2)]/kT\eta \quad (28)$$

where

$$H_{\alpha\beta} = 4\pi\int_0^\infty [g_{\alpha\beta}(r) - 1]r^2\,dr \quad (29)$$

$$\eta = \rho_A + \rho_B + \rho_A\rho_B(H_{AA} + H_{BB} - 2H_{AB}) \quad (30)$$

Expressions similar to Equation 25 can be derived for the other composition derivatives of μ_A and μ_B (*116, 117*). It is seen from Equations 25–30 that all of the mixture thermodynamic properties can be obtained if it is possible to calculate the integrals $H_{\alpha\beta}$. Also, from Equation 25 we see that if

$$H_{AB} = \frac{1}{2}(H_{AA} + H_{BB}) \quad (31)$$

the last term on the right-hand side of Equation 25 vanishes, and we have the ideal solution (Lewis rule) form for μ_A; i.e., $d\mu_A = kTd\ln x_A$. Equation 31 provides the molecular definition of such an ideal solution.

The corresponding expressions for a multicomponent mixture have been given by Kirkwood and Buff (*116*) and in somewhat more general form by O'Connell (*118*). O'Connell has also shown that the Kirkwood–Buff theory expressions can be rewritten in terms of integrals $C_{\alpha\beta}$ in place of $H_{\alpha\beta}$, by using the relation (in matrix form)

$$H = C + \rho CHX \quad (32)$$

where H and C have elements $H_{\alpha\beta}$ and $C_{\alpha\beta}$, and X is a diagonal matrix whose nonzero elements are x_A, x_B, ... Here $C_{\alpha\beta}$ is given by

$$C_{\alpha\beta} \equiv 4\pi \int_0^\infty c_{\alpha\beta}(r) r^2 \, dr \tag{33}$$

where $c_{\alpha\beta}(r) \equiv \langle c_{\alpha\beta}(r\omega_1\omega_2)\rangle_{\omega_1\omega_2}$ is the centers direct correlation function. The advantage of the C-form of Kirkwood–Buff theory is that $c_{\alpha\beta}(r)$ is a simpler and more short-ranged function than $g_{\alpha\beta}(r)$, so that it should be possible to develop simpler approximations to the C integrals.

O'Connell and coworkers (*119–123*) have used the C-form of Kirkwood–Buff theory to make numerical calculations for highly nonideal mixtures containing supercritical components. It is found that the C integrals, when put in reduced form, are rather insensitive to the type of intermolecular forces involved, so that simple corresponding states correlations can be developed.

Conclusion. For mixtures of spherical or near-spherical molecules the existing forms of perturbation theory give good results unless the molecules are very different in size. For small, nonspherical molecules, including strongly polar or quadrupolar molecules, the Padé theory of Equation 22 gives good results and is a significant improvement over vdW1 theory or empirical equations of state, particularly for nonpolar–polar mixtures. Further improvements in perturbation theory calculations will come with the development of perturbation theory methods based on a nonspherical reference system. The Kirkwood–Buff theory provides an alternative starting point because of its simplicity and generality, and seems to have been relatively little used by engineers.

Acknowledgments

This work was supported by National Science Foundation Grant No. CPE-7909168 and Gas Research Institute Grant No. 5014-363-0119.

Literature Cited

1. Valleau, J. P.; Whittington, S. G. In "Modern Theoretical Chemistry"; Berne, B. J., Ed.; Plenum: New York, 1977; Vol. 5, Part A, Chap. 4.
2. Hoover, W. G.; Ladd, A. J. C.; this volume, 1983; Chap. 5.
3. Valleau, J. P.; Torrie, G. M. In "Modern Theoretical Chemistry"; Berne, B. J., Ed., Plenum: New York, 1977; Vol. 5, Part A, Chap. 5.
4. Barker, J. A.; Henderson, D. *Rev. Mod. Phys.* **1976,** *48*, 596.
5. Quirke, N. "Proc. NATO Summer School on Superionic Conductors, Odense, Denmark"; Plenum: New York, 1980.
6. Hoover, W. G.; Ree, F. H. *J. Chem. Phys* **1968,** *49*, 3609.
7. Hoover, W. G.; Ross, M.; Johnson, K. W.; Henderson, D.; Barker, J. A.; Brown, B. C. *J. Chem. Phys.* **1971,** *52*, 4931.

8. Hoover, W. G.; Gray, S. G.; Johnson, K. W. *J. Chem. Phys.* **1971,** *55*, 1128.
9. Levesque, D.; Verlet, L. *Phys. Rev.* **1969,** *182*, 307.
10. Hansen, J.-P. *Phys. Rev.* **1970,** *A2*, 221.
11. McDonald, I. R.; Singer, K. *Molec. Phys.* **1972,** *23*, 29.
12. *Ibid.* **1972,** *24*, 464.
13. Nicolas, J. J.; Gubbins, K. E.; Streett, W. B.; Tildesley, D. J. *Mol. Phys.* **1979,** *37*, 1429.
14. Patey, G. N.; Valleau, J. P. *J. Chem. Phys.* **1976,** *64*, 170.
15. Hansen, J.-P. *Phys. Rev A.* **1973,** *8A*, 3096.
16. Tildesley, D. J.; Streett, W. B. *Mol. Phys.* **1980,** *41*, 85.
17. Hoover, W. G.; Ree, F. H. *J. Chem. Phys.* **1967,** *47*, 4873.
18. Hansen, J.-P.; Verlet, L. *Phys. Rev.* **1969,** *184*, 151.
19. McDonald, I. R.; Singer, K. *J. Chem. Soc., Discuss. Faraday,* **1967,** *43*, 40.
20. McDonald, I. R.; Singer, K. *J. Chem. Phys.* **1967,** *47*, 4766.
21. McDonald, I. R.; Singer, K. *J. Chem. Phys.* **1969,** *50*, 2308.
22. Valleau, J. P.; Card, D. N. *J. Chem. Phys.* **1972,** *57*, 5457.
23. Patey, G. N.; Valleau, J. P. *Chem. Phys. Lett.* **1973,** *21*, 297.
24. Jacucci, G.; Quirke, N. *Mol. Phys.* **1980,** *40*, 1005.
25. Quirke, N.; Jacucci, G. *Mol. Phys.* **1982,** *45*, 823.
26. Bennett, C. H. *J. Comput. Phys.* **1974,** 22, 245.
27. Torrie, G. M.; Valleau, J. P. *Chem. Phys. Lett.* **1974,** *28*, 578.
28. Torrie, G. M.; Valleau, J. P. *J. Comput. Phys.* **1977,** *23*, 187.
29. Torrie, G. M.; Valleau, J. P. *J. Chem. Phys.* **1977,** *66*, 1402.
30. Ng, K.-C.; Valleau, J. P.; Torrie, G. M.; Patey, G. N. *Mol. Phys.* **1979,** *38*, 781.
31. Shing, K. S.; Gubbins, K. E. *Mol. Phys.* **1982,** *45*, 129.
32. Lee, C. Y.; Scott, H. L. *J. Chem. Phys.* **1980,** *73*, 4591.
33. Norman, G. S.; Filinov, V. S. *High Temp. Res. USSR,* **1969,** *7*, 216.
34. Adams, D. J. *Mol. Phys.* **1974,** *28*, 1241.
35. *Ibid.* **1975,** *29*, 307.
36. *Ibid.* **1976,** *32*, 647.
37. *Ibid.* **1979,** *37*, 211.
38. Rowley, L. A.; Nicholson, D.; Parsonage, N. G. *J. Comput. Phys.* **1975,** *17*, 401.
39. Mezei, M. *Mol. Phys.* **1980,** *40*, 901.
40. Widom, B. *J. Chem. Phys.* **1963,** *39*, 2808.
41. Romano, S.; Singer, K. *Mol. Phys.* **1979,** *37*, 1765.
42. Powles, J. G. *Mol. Phys.* **1980,** *41*, 715.
43. Shing, K. S.; Gubbins, K. E. *Mol. Phys.* **1982,** *46*, 1109.
44. M. J. de Oliviera, private communication.
45. Shing, K. S., Ph.D. Thesis, Cornell University, 1982.
46. Jonah, D. A.; Shing, K. S.; Gubbins, K. E. "Proc. 8th Symposium on Thermophysical Properties"; American Society of Mechanical Engineers: New York, 1982; p. 335.
47. Shing, K. S.; Gubbins, K. E. *Mol. Phys.* **1981,** *43*, 717.
48. Coker, D. F.; Watts, R. O. *Mol. Phys.* **1981,** *44*, 1303.
49. Yao, J.; Greenkorn, R. A.; Chao, K. C. *Mol. Phys.* **1982,** *46*, 587.
50. Powles, J. G.; Evans, W. A. B.; Quirke, N. *Mol. Phys.* **1982,** *46*, 1347.
51. Rowlinson, J. S. "Liquids and Liquid Mixtures"; Butterworths: London, 1969; Ch. 6.

52. Schneider, G. M. In "Chemical Thermodynamics," McGlashan, M. L., Ed., Chemical Society: London, 1978.
53. Schneider, G. M. *Angew. Chem. Int. Ed. Eng.* **1978,** *17*, 716.
54. Streett, W. B. *Can. J. Chem. Eng.* **1974,** *52*, 92.
55. Streett, W. B.; Gubbins, K. E. "Proc. of the Eighth Symposium on Thermophysical Properties"; American Society of Mechanical Engineers: New York, 1982, p. 303.
56. Gibbons, R. M. *Mol. Phys.* **1969,** *17*, 81; *ibid.* **1970,** *18*, 809.
57. Lebowitz, J. L.; Rowlinson, J. S. *J. Chem. Phys.* **1964,** *41*, 133.
58. Pavlíček, J.; Nezbeda, I.; Boublik, T. *Czech. J. Phys.* **1970,** *27B*, 1061.
59. Boublik, T.; Nezbeda, I. *Czech. J. Phys.* **1980,** *30B*, 121.
60. *Ibid.* **1980,** *30*, 953.
61. Boublik, T. *Mol. Phys.* **1981,** *42*, 209.
62. Smith, E. B.; Lea, K. R. *Nature,* **1960,** *186*, 714.
63. Smith, E. B.; Lea, K. R. *Trans. Faraday Soc.* **1963,** *59*, 1535.
64. Alder, B. J. *J. Chem. Phys.* **1964,** *40*, 2724.
65. Rotenberg, A. *J. Chem. Phys.* **1965,** *43*, 4377.
66. Monson, P. A.; Rigby, M. *Mol. Phys.* **1980,** *39*, 977.
67. Henderson, D.; Leonard, P. J. In "Physical Chemistry, An Advanced Treatise, Liquid State"; Henderson, D., Ed., Academic Press: New York, 1971; Vol. 8B, p. 414.
68. McDonald, I. R. In "Statistical Mechanics"; Singer, K., Ed., Chemical Society: London, 1973; Vol. 1.
69. Gubbins, K. E. *AIChEJ* **1973,** *19*, 684.
70. Barker, J. A.; Henderson, D. *Rev. Mod. Phys.* **1976,** *48*, 587.
71. Henderson, D. In "Equations of State in Engineering and Research"; Chao, K. C.; Robinson, R. L., Eds.; ACS ADVANCES IN CHEMISTRY SERIES No. 182; ACS: Washington, D.C., 1979; p. 1.
72. Smith, W. R. *Can. J. Chem. Eng.* **1972,** *50*, 271.
73. Leland, T. W.; Rowlinson, J. S.; Sather, G. A. *Trans. Faraday Soc.* **1968,** *64*, 1447.
74. Mo, K. C.; Gubbins, K. E.; Jacucci, G.; McDonald, I. R. *Mol. Phys.* **1974,** *27*, 1173.
75. Smith, W. R. *Mol. Phys.* **1971,** *22*, 105.
76. Leland, T. W.; Rowlinson, J. S.; Sather, G. A.; Watson, I. D. *Trans. Faraday Soc.* **1969,** *65*, 2034.
77. McDonald, I. R. *Mol. Phys.* **1972,** *23*, 41.
78. Henderson, D.; Barker, J. A. *J. Chem. Phys.* **1968,** *49*, 3377.
79. Leonard, P. J.; Henderson, D.; Barker, J. A. *Trans. Faraday Soc.* **1970,** *66*, 2439.
80. Grundke, E. W.; Henderson, D.; Barker, J. A.; Leonard, P. J. *Mol. Phys.* **1973,** *25*, 883.
81. Lee, L. L. and Levesque, D. *Mol. Phys.* **1973,** *26*, 1351.
82. Mansoori, G. A.; Leland, T. W., Jr. *J. Chem. Phys.* **1970,** *53*, 1931.
83. Mansoori, G. A. *J. Chem. Phys.* **1972,** *56*, 5335.
84. Jonah, D. A.; Shing, K. S.; Venkatasubramanian, V.; Gubbins, K. E. Paper presented at the 74th Annual Meeting of the American Institute of Chemical Engineers, New Orleans, Nov. 1981.
85. Mansoori, G. A.; Leland, T. W. *J. Chem. Soc., Faraday Trans. II* **1972,** *68*, 320.
86. Pople, J. A. *Proc. Roy. Soc.* **1954,** *221A*, 998.
87. Stell, G.; Rasaiah, J. C.; Narang, H. *Mol. Phys.* **1974,** *23*, 393.

88. Nicolas, J. J.; Gubbins, K. E.; Streett, W. B.; Tildesley, D. J. *Mol. Phys.* **1979,** *37*, 1429.
89. Twu, C. H.; Gubbins, K. E. *Chem. Eng. Sci.* **1978,** *33*, 879.
90. Gray, C. G.; Gubbins, K. E.; and Twu, C. H. *J. Chem. Phys.* **1978,** *69*, 182.
91. Gubbins, K. E.; Twu, C. H. *Chem. Eng. Sci.* **1978,** *33*, 863.
92. Twu, C. H.; Gubbins, K. E.; Gray, C. G. *J. Chem. Phys.* **1976,** *64*, 5186.
93. Gibbs, G. M., Thesis, University of Oxford, 1979.
94. Shukla, K. P.; Singh, S.; Singh, Y. *J. Chem. Phys.* **1979,** *70*, 3086.
95. *Ibid.* **1980,** *72*, 2719.
96. Clancy, P.; Gubbins, K. E.; Gray, C. G. *J. Chem. Soc., Faraday Disc.* **1978,** *66*, 116.
97. Lobo, L. Q.; McClure, D.W.; Staveley, L. A. K.; Clancy, P.; Gubbins, K. E.; Gray, C. G. *J. Chem. Soc., Faraday Trans. II*, **1981,** *77*, 425.
98. Clancy, P.; Gubbins, K. E. *Mol. Phys.* **1981,** *44*, 581.
99. Tsang, C. Y.; Clancy, P.; Calado, J. C. G.; Streett, W. B. *Chem. Eng. Commun.* **1980,** *6*, 365.
100. Machado, J. R. S., M.S. Thesis, Cornell University 1979.
101. Machado, J. R. S.; Gubbins, K. E.; Lobo, L. Q.; Staveley, L. A. K. *J. Chem. Soc., Faraday Trans. I*, **1980,** *76*, 2496.
102. Moser, B.; Lucas, K.; Gubbins, K. E. *Fluid Phase Equil.* **1981,** *7*, 153.
103. Gubbins, K. E.; Gray, C. G.; Machado, J. R. S. *Mol. Phys.* **1981,** *42*, 817.
104. Wertheim, M. S. *Mol. Phys.* **1979,** *37*, 83.
105. Patey, G. N.; Valleau, J. P. *Chem. Phys. Lett.* **1976,** *42*, 407.
106. *Ibid.* **1978,** *58*, 157.
107. Patey, G. N.; Torrie, G. M.; Valleau, J. P. *J. Chem. Phys.* **1979,** *71*, 96.
108. Boublik, T. *Fluid Phase Equil.* **1976,** *1*, 37.
109. Boublik, T. *Coll. Czech. Chem. Commun.* **1976,** *41*, 1265.
110. Boublik, T. *Mol. Phys.* **1976,** *32*, 1737.
111. Boublik, T. *Fluid Phase Equil.* **1979,** *3*, 85.
112. Pavlíček, J.; Boublik, T. *Fluid Phase Equil.* **1981,** *7*, 1, 15.
113. Enciso, E.; Lombardero, M. *Mol. Phys.* **1981,** *44*, 725.
114. Lee, T. J.; Lee, L. L.; Starling, K. E. In "Equations of State in Engineering and Research" Chao, K. C.; Robinson, R. L., Eds., ACS ADVANCES IN CHEMISTRY SERIES No. 182; ACS: Washington, D.C., 1979; p. 125.
115. Gibbs, J. W. "Elementary Principles in Statistical Mechanics"; Yale University Press: New Haven, 1902; Eq. 540.
116. Kirkwood, J. G.; Buff, F. P. *J. Chem. Phys.* **1951,** *19*, 774.
117. Ben-Naim, A. "Water and Aqueous Solutions"; Plenum: New York, 1974; Chap. 4.
118. O'Connell, J. P. *Mol. Phys.* **1971,** *20*, 27.
119. Brelvi, S. W.; O'Connell, J. P. *AIChEJ* **1972,** *18*, 1239.
120. Gubbins, K. E.; O'Connell, J. P. *J. Chem. Phys.* **1974,** *60*, 3449.
121. Mathias, P. M.; O'Connell, J. P., in "Equations of State in Engineering and Research" Chao, K. C.; Robinson, R. L., Eds., ACS ADVANCES IN CHEMISTRY SERIES, No. 182; ACS: Washington, D.C., 1979; p. 97.
122. Mathias, P. M.; O'Connell, J. P. *Chem. Eng. Sci.* **1981,** *36*, 1123.
123. O'Connell, J. P. *Fluid Phase Equil.* **1981,** *6*, 21.

RECEIVED for review January 27, 1982. ACCEPTED for publication October 6, 1982.

5

A Model for the Calculation of Thermodynamic Properties of a Fluid

Using Hard-Sphere Perturbation Theory and the Zero-Kelvin Isotherm of the Solid

GERALD I. KERLEY

Los Alamos National Laboratory, Applied Theoretical Physics Division, Los Alamos, NM 87545

The CRIS model of fluids is reviewed and calculations using the theory are compared with experimental data. The equation of state is computed from an expansion about a hard-sphere reference system, in which the optimum hard-sphere diameter is chosen by a variational principle. All information about the intermolecular forces is obtained from the zero-Kelvin isotherm of the solid. Calculations for the rare gases, for the hydrogen isotopes and other polyatomic molecules, and for liquid iron are shown to agree well with experiment. Liberman's model for the electronic structure of a compressed atom is used to calculate contributions from thermal electronic excitation to the equation of state. These terms are shown to be important in explaining shock-wave data for xenon.

SEVERAL EXCELLENT THEORIES recently have been developed for calculating the thermodynamic properties of fluids from specified pair potentials (*1–10*). Barker and Henderson (*1*) showed that hard-sphere perturbation methods are very accurate, even at low temperatures, when the hard-sphere diameter is defined in an optimum fashion. Subsequently, Mansoori and Canfield (*2*) and Rasaiah and Stell (*3*) developed the variational principle for choosing the hard-sphere diameter. Anderson et al. (*4*) showed that perturbation theories succeed because repulsive forces, or effects of excluded volume, play the principal role in determining the equilibrium structure of dense fluids (for spherically symmetric molecules). Approaching the problem from a different point of view, Rosenfeld and Ashcroft (*5*) developed an accurate integral equation method that relies on the universality of the short-range structure in dense fluids. Other important developments include applications of fluid theory to nonspherical molecules (*1, 6, 7*) and to liquid metals (*8–10*).

0065-2393/83/0204-0107$09.00/0

Unfortunately, applications of these accurate theories to problems of practical interest are often hampered by lack of knowledge about intermolecular forces. For this reason, we have developed the CRIS model (*11, 12*), a perturbation theory of fluids in which explicit knowledge of the interaction potentials is not required. Our model retains the key concepts of fluid structure that are essential to the success of the other perturbation theories. However, the energy of a fluid molecule in the cage formed by its neighbors is estimated from the zero-Kelvin isotherm of the solid. This *cold curve* is usually easier to compute or measure than is an effective pair potential.

In this chapter we discuss the theoretical model and review the results of several calculations. First, the theory for the case of spherical molecules in the ground state is considered. The model is shown to agree with computer simulation studies on systems where the pair potentials are known (*13*). We then show how other degrees of freedom can be included in calculating equations of state. In particular, an electronic structure model due to Liberman (*14*) is useful for computing contributions from thermal electronic excitation. Rotational ordering and other perturbations of intramolecular motions are not considered in this paper. Additional theoretical problems, including treatment of vaporization, melting, and shock waves, are then discussed. The rest of the chapter compares calculations using the model with experimental data for rare gases, molecular fluids, and liquid metals.

Because of space limitations, only an outline of the main theoretical ideas is presented here. Detailed and rigorous discussions are given in the literature cited. We also note that Rosenfeld (*15*) has derived the first-order CRIS model by a method different from ours.

Outline of the CRIS Model

Consider a system of N spherical molecules, having no internal degrees of freedom, in a volume V at temperature T. The thermodynamic properties of the system are determined by the potential energy function Φ (*1*). Although Φ is a function of the positions of all N molecules, only the short range structure is important for perturbation theories. To see this fact, define coordinates $\{\boldsymbol{q}_k\}$ that specify the positions of all molecules relative to an origin fixed at the center of mass of molecule k. We write

$$\Phi = \sum_{k=1}^{N} \phi(\boldsymbol{q}_k) \tag{1}$$

where $\phi(\boldsymbol{q}_k)$, the potential energy of molecule k in the field of its neighbors, includes all pair, triplet, and higher-order interactions (*11*). This function depends only on the local structure of the fluid, i.e., the coordinates $\boldsymbol{q}$ of nearby molecules relative to the one under consideration.

For spherical molecules, the structure of dense fluids is determined primarily by the effects of excluded volume, and it is useful to express the Helmholtz free energy as a perturbation expansion about a model system, the hard-sphere fluid (*11*).

$$A_\phi(V,T,N) = A_0(V,T,N;\sigma) + \langle\Phi\rangle_0 + \Delta A_\phi \tag{2}$$

Here A_0 is the free energy for hard spheres having diameter σ; $\langle\Phi\rangle_0$, the first order correction, is an average of Φ taken in the hard-sphere system. By definition, ΔA_ϕ contains all remaining contributions to A_ϕ; these corrections are caused by differences between the structure of the real fluid and that of the hard-sphere system. The term ΔA_ϕ can be made quite small by making an optimum choice for σ.

It can be shown that the first-order approximation gives an upper bound to the true free energy of the system (*2, 3*).

$$\tilde{A}_\phi = A_0 + \langle\Phi\rangle_0 \geqslant A_\phi \tag{3}$$

Our procedure is to minimize $\tilde{A}_\phi$ with respect to σ; in that way, we find the hard-sphere system whose structure is closest to that of the real fluid. When σ is defined in this optimum fashion, first-order perturbation theory gives realistic predictions for the properties of fluids (*2, 3*). However, the correction term ΔA_ϕ must be included if quantitative results are desired (*1*). In the CRIS model, for ΔA_ϕ we use an approximate expression derived from macroscopic fluctuation arguments. We believe it to be accurate if σ is chosen by the variational principle. Detailed discussion of this term is given elsewhere (*11, 12*).

The first-order correction to A_ϕ can be written as an average of $\phi(\boldsymbol{q})$ over all configurations of neighboring molecules.

$$\langle\Phi\rangle_0 = N\langle\phi\rangle_0 = \int \phi(\boldsymbol{q})\, \overline{n}_0(\boldsymbol{q})\, d\boldsymbol{q}_1\, d\boldsymbol{q}_2 \ldots \tag{4}$$

where $\overline{n}_0(\boldsymbol{q})$ is a hard-sphere distribution function. It specifies the probability density that a molecule in the fluid will have neighbors located within differential elements $d\boldsymbol{q}_1, d\boldsymbol{q}_2, \ldots$, at positions $\boldsymbol{q}_1, \boldsymbol{q}_2, \ldots$ (*11*). Because $\phi(\boldsymbol{q})$ depends upon the short range structure of the fluid, the position of the first shell of neighbors is the most important quantity specifying the local configuration. In the CRIS model, the nearest neighbors are assumed to lie on a spherical shell, of radius R, that varies from molecule to molecule (*12*). We further assume that the coordination number ν varies with R so that the volume per molecule is fixed at the macroscopic value V/N. If there are 12 nearest neighbors in a close-packed configuration, it can be shown that

$$\nu = 6\sqrt{2}\, NR^3/V \tag{5}$$

In this approximation, only one variable, R, is required to specify the local arrangement of neighbors about a particular molecule. Equation 4 becomes

$$\langle \phi \rangle_0 = \frac{1}{N} \int \phi(R, \nu)\, \overline{n}_0(R)\, 4\pi R^2\, dR \tag{6}$$

Furthermore, the distribution function $\overline{n}_0(R)$ is given by

$$(\nu/N)\, \overline{n}_0(R) = (N/V)\, g_0(R) \tag{7}$$

where $g_0(R)$ is the contribution from the nearest neighbor shell to the radial distribution function for the hard-sphere fluid. A satisfactory working definition of this quantity can be obtained from the first peak in the radial distribution function (*12*).

The potential energy function $\phi(R, \nu)$ can be estimated from the zero-Kelvin isotherm of the solid in the following way. Let $E_c(V_s)$ be the electronic contribution to the energy per molecule for the close-packed solid at volume V_s and zero temperature. (Note that this definition does *not* include any contribution from the zero-point lattice vibrations, which are not part of the intermolecular forces.) In the solid, there are 12 nearest neighbors on a sphere of radius R, given by

$$V_s = NR^3/\sqrt{2} \tag{8}$$

In the fluid, a molecule has the same potential energy as it would have in the solid phase at the same nearest neighbor distance, except that the coordination number is reduced from 12 to ν. Hence

$$\phi(R, \nu) \cong (\nu/12)\, E_c(V_s) = (V_s/V)\, E_c(V_s) \tag{9}$$

This result is approximate because it assumes the forces between molecules to be pairwise additive. The assumption is not correct for liquid metals; however, the theory is found to give good results in practice. Further discussion of this point is given later in this chapter.

Once the free energy has been defined by the above equations, the internal energy E_ϕ and pressure P_ϕ can be computed from standard thermodynamic formulas,

$$E_{\phi} = A_{\phi} - T(\partial A_{\phi}/\partial T)_{V,N} \quad (10)$$

$$P_{\phi} = -(\partial A_{\phi}/\partial V)_{T,N} \quad (11)$$

In order to test the CRIS model, we have compared our calculations to Monte Carlo and molecular dynamics data for fluids having inverse-power and 6-12 potentials (*13*). The cold curve is obtained by summing the pair potential $U_2(R)$ over all molecules in the lattice (*16*).

$$E_c(V_s) = (1/2) \sum_i n_i \, U_2(R_i) \quad (12)$$

where n_i and R_i are the number of molecules and the radius for the i-th shell of neighbors, respectively. For the 6-12 potential, we have

$$U_2(R) = 4\varepsilon[(d/R)^{12} - (d/R)^6] \quad (13)$$

$$E_c(V_s) = (\varepsilon/2)[C_{12}\rho^{12} - 2C_6\rho^6] \quad (14)$$

where $\rho = Nd^3/V_s$, $C_{12} = 12.12188$, and $C_6 = 14.45392$.

In Figures 1 and 2, we compare our equation of state for the 6-12 fluid with Monte Carlo calculations (*17*) on three isotherms that range from the triple point to above the critical point. Agreement for both the pressure and the internal energy is very good. In fact, the CRIS model was shown to agree well with all of the available computer simulation data for thermodynamic properties, radial distribution functions, and the vapor-liquid coexistence curve (*13*).

The calculations for these model fluids involve no parameters that can be adjusted to give agreement with experiment. The success of the CRIS model in these tests shows that it retains the essential features of a good fluid theory while eliminating the need to know the intermolecular potentials explicitly.

Internal Degrees of Freedom

The model discussed above describes the translational degrees of freedom of the molecules, which interact through forces determined by the ground electronic state of the system. Internal degrees of freedom can also contribute to the equation of state. For example, the free energy is

$$A = A_{\phi} + A_{VR} + A_e \quad (15)$$

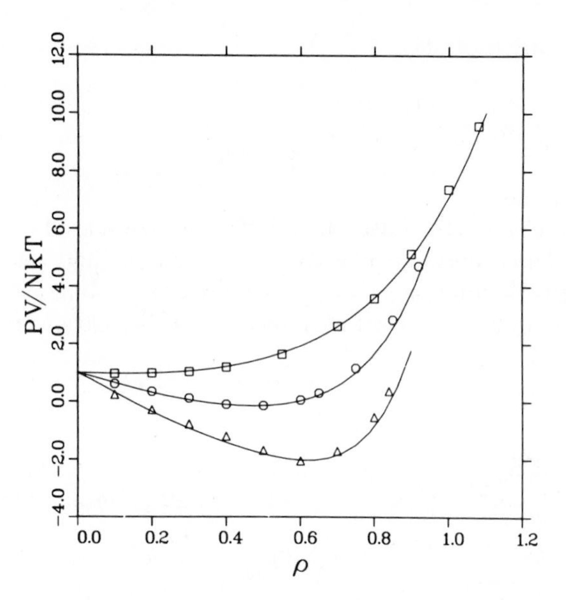

Figure 1. Equation of state for the 6-12 fluid. The solid line was calculated using the CRIS model; the discrete points are Monte Carlo results (17). *Key:* □, $T = 2.74$; ○, $T = 1.15$; *and* △, $T = 0.75$. *(Reproduced with permission from Ref. 13. Copyright 1980, American Institute of Physics.)*

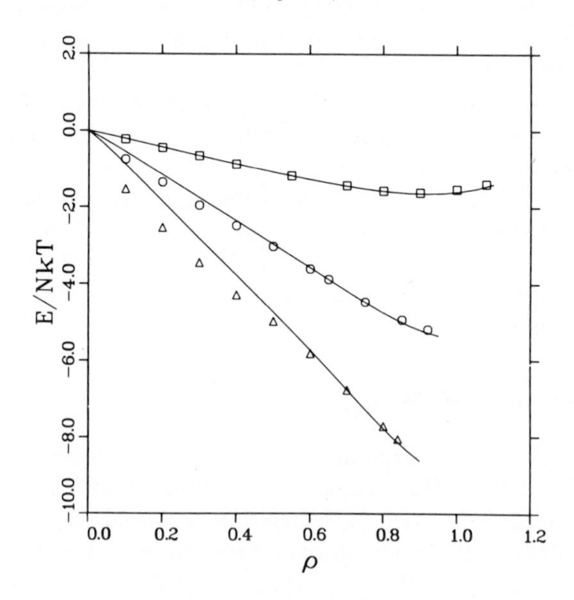

Figure 2. Excess internal energy for the 6-12 fluid. Symbols have the same meaning as in Figure 1. (Reproduced with permission from Ref. 13. Copyright 1980, American Institute of Physics.)

where A_ϕ is given by Equation 2, A_{VR} includes contributions from intramolecular vibration and rotation, and A_e includes contributions from thermal electronic excitation and ionization. In our model, A_{VR} and A_e are calculated in a "static" approximation; that is, the translational motion of the molecules is not coupled to the internal degrees of freedom.

In most applications, we have computed A_{VR} from the rigid-rotator, harmonic-oscillator approximation (*18*). To some extent, effects due to hindered rotation and vibration can be included in the definition of the cold curve for the CRIS model (*19*). This approach is most reasonable at temperatures high enough to allow free molecular rotation. Perturbations to the vibrational motion were included in our calculations of the equation of state of deuterium (*20*). These effects were found to be fairly small when compared with uncertainties in the zero-Kelvin isotherm.

Several models exist for calculating the temperature dependence of electronic structure and the corresponding contributions to the equation of state. Statistical atom theories, such as the Thomas–Fermi–Dirac (TFD) model (*21*), are often used for this purpose. A well-known problem with these theories is that they do not reproduce the electronic shell structure that is characteristic both of free atoms and of condensed matter. However, the TFD model is a good approximation at high densities, where the shell structure has been crushed by pressure ionization.

At low densities, electronic contributions to the equation of state can be computed from the theory of ionization equilibrium (*22*), using energy levels and ionization potentials of the isolated atoms and ions. However, the standard approximations made in such calculations break down at high densities where the energy levels are strongly perturbed and pressure-ionized by the forces of surrounding ions (*23*).

For intermediate densities, ranging from about 0.05 to 200 times normal solid density, we have found the INFERNO model of Liberman (*14*) to be very useful. Liberman considers an average atom, with a point nucleus at the center of a spherical cell, surrounded by an electron gas and a uniform positive charge. The charge distribution outside the sphere simulates the environment of neighboring atoms in the real system, and the sphere radius r_s is defined by the average volume per atom,

$$(4\pi/3)Nr_s^3 = V \tag{16}$$

Liberman solves the Dirac equation to obtain wave functions and energies for both the discrete bound states and the continuum free levels. The average electron charge density is computed from the wave functions by populating the energy levels according to Fermi–Dirac statistics. The screened potential and the charge distribution, which depend upon density and temperature, are required to be self-consistent. The electronic entropy is also calculated by Fermi statistics. We calculate the pressure, energy, and free energy numerically, using standard thermodynamic formulas (*24*).

Although the INFERNO model contains many approximations, it treats most of the electronic structure problems well enough to calculate the equation of state. At low densities, the atomic structure agrees well with that of the isolated atom; all of the electrons are in bound levels, and there is an insulating gap between the highest occupied state and the continuum. As the density increases, this gap narrows, and the bound levels cross into the continuum. A notable feature of the model is that the way a bound level changes into a free "resonance" is handled in a continuous fashion. The theory predicts a transition from an insulating to a metallic state and is in reasonable agreement with band theoretical calculations (*25*) for solids.

In Figure 3 we compare the INFERNO and TFD results for the electronic entropy versus temperature for aluminum at several densities. At low densities, the INFERNO calculations exhibit steps that correspond to different stages of ionization. At higher densities this structure disappears because the discrete atomic levels pass into the continuum and become broad resonances. The INFERNO and TFD models are in good agreement at high densities and at high temperatures.

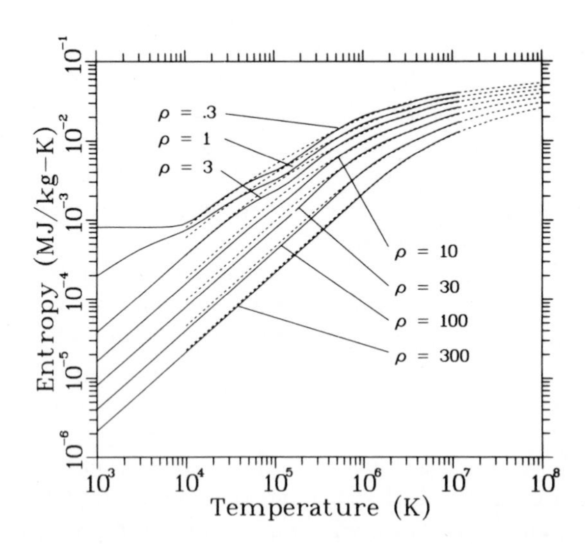

Figure 3. Electronic entropy for aluminum as a function of temperature at several densities. Solid lines were calculated using the INFERNO model; dashed lines, using the TFD model. (Reproduced from Ref. 65.)

Other Theoretical Considerations

Zero-Kelvin Isotherms. In the CRIS model, all information about the intermolecular interactions is obtained from the cold curve of the solid. Therefore, construction of this curve is the principal problem in applying the theory to a specific material. In most cases we use both experimental and theoretical results to accomplish this task.

Experimental information about the zero-Kelvin isotherm can be obtained from both static and shock wave measurements. Accurate static measurements of pressure versus volume up to 2–4 GPa have been available for some time, and recent diamond cell techniques have extended the range of static data to much higher pressures. We used such measurements on iron (*26*), extending to 95 GPa, in our calculations for liquid iron, discussed below. Methods for reducing shock-wave data to the cold curve can be very useful (*24, 27*), but they involve some approximations; we do not report such methods here.

Recently, accurate band theoretical calculations of the zero-Kelvin isotherm have been made for several materials. These calculations usually agree well with experimental measurements, although the normal solid density may be in error by a small amount. As a practical matter, we correct this discrepancy by adding a small constant term to the theoretical pressure curve. The band theoretical results of Ross and McMahan (*25*) were used in our computations for liquid xenon.

For the CRIS model, the cold curve must be specified both in compression and in tension. For the tension region we normally use an analytic expression of the following type (*24, 28*)

$$E_c(V_s) = a_1 \exp(-a_2 V_s^{1/3}) - a_3/V_s^{a_4} \tag{17}$$

where the four constants are determined from the solid binding energy, normal density, and compressibility data. Fortunately, most results using the CRIS model are not very sensitive to details of the cold curve in tension. However, the liquid density on the coexistence boundary is one exception.

At very high densities a reasonable estimate of the zero-Kelvin isotherm can be obtained from TFD theory. In many problems we have used an interpolation formula, based on TFD results, to represent the cold curve in regions where no better data are available (*24, 28*).

Vaporization. At low temperatures, isotherms calculated using the CRIS model display van der Waals loops, indicating the existence of a vapor–liquid coexistence curve and a critical point. For example, our theoretical equation of state surface for methane (*28*) is depicted in Figure 4. For temperatures above the critical point, 200 K, the pressure is a

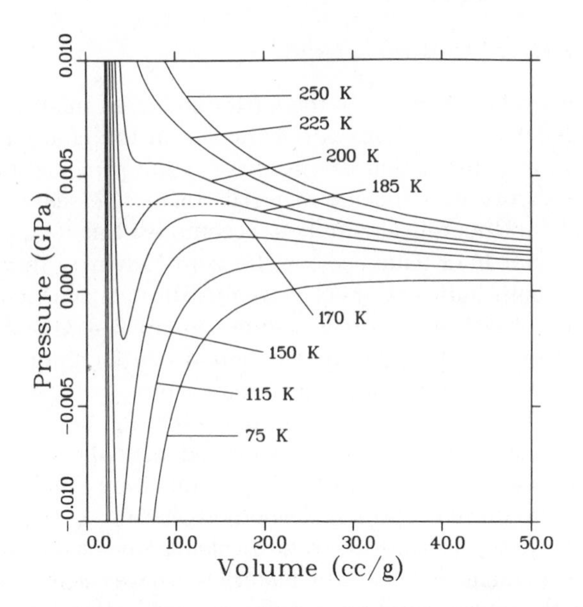

Figure 4. Equation of state for fluid methane at low densities. The dashed line is the equilibrium vapor pressure at $T = 185$ K.

monotonically decreasing function of volume, and a single fluid phase exists. Below the critical point, the fluid at equilibrium separates into a vapor–liquid mixture (29). The vapor pressure and the other properties of the liquid and vapor are determined by requiring that the two phases have equal pressures and Gibbs free energies. The equilibrium vapor pressure for the 185-K isotherm is shown by a dashed line in Figure 4.

The CRIS model was shown to give good agreement with computer simulation results for the coexistence curve of the 6-12 fluid (*13*). However, *neither* calculation gives the correct result near the critical point, where long-range density fluctuations are important (*30*). The CRIS model, like other mean field theories, overestimates the critical temperature by 5–10%.

Hugoniot Measurements. Shock wave experiments can test an equation of state model at high densities and at high temperatures that are not accessible by other methods. The Hugoniot is calculated from the standard relation (27)

$$E_H - E_0 = (1/2)(P_H + P_0)(\rho_0^{-1} - \rho_H^{-1}) \tag{18}$$

where E_H, P_H, and ρ_H are the energy, pressure and density of the shocked state, and E_0, P_0, and ρ_0 are the initial state conditions. Experiments usually measure the particle velocity U_P and shock velocity U_s

$$U_P = \sqrt{(P_H - P_0)(\rho_0^{-1} - \rho_H^{-1})} \tag{19}$$

$$U_s = U_P/(1 - \rho_0/\rho_H) \tag{20}$$

These relations can be derived from the conservation laws of mass, momentum, and energy, applied to a single steady, plane shock wave.

A few experiments have been performed in which a shock wave is reflected back into the material, compressing it for a second time (*31*). These double shock experiments do not heat the material as much as a single shock does, and they can reach higher compressions. Conditions for the reflected shock state are determined by applying Equations 18–20 a second time.

Melting. From thermodynamics, the melting line is simply defined as the pressure–temperature locus at which the solid and liquid phases have equal Gibbs free energies. In calculations for methane and for metals, we found that the Debye model (*28*) gave a good representation of the thermodynamic properties of the solid. For hydrogen and deuterium, in which anharmonic effects on the lattice vibrations are not described by such a model, we used the free volume theory for the solid (*20*).

Because melting depends upon the difference in free energy between the two phases, small errors in the equation of state can cause large errors in the predicted melting curve. All our calculations for insulating fluids have given reasonable melting predictions. However, our calculations for metals show that the CRIS model underestimates the free energy of the liquid by about 5% of the cohesive energy. We attribute most of this error to the use of Equation 9, which assumes additive forces between molecules. A reasonable melting curve can be obtained by subtracting a constant from the liquid energy in order to match the observed melting point at zero pressure. This expedient was used in our calculations for iron. However, further work, leading to an improvement of Equation 9, is needed.

Fluid Structure and Transport Properties. As noted above, the hard-sphere diameter σ used in the CRIS model is obtained by minimizing the first-order free energy expression, Equation 3. The parameter σ, which depends on both temperature and density, provides a crude description of the short range structure of the fluid.

To first-order, the radial distribution function of the real fluid is equal to that of a hard-sphere fluid (with an optimum σ). In this approximation, the structure factor can also be computed from the hard-sphere formula (*17*). In the CRIS model, corrections to the radial distribution function and structure factor due to the soft core of the interaction in the real fluid can be derived from the same arguments used to obtain the higher order correction ΔA_{ϕ} to the free energy. The CRIS model was shown to give good agreement with computer simulation results for the radial distribution functions of inverse-power and 6-12 potentials (*13*).

Dymond and Alder (*32*) have shown that reasonable predictions of transport properties of fluids can be made using hard-sphere formulas, if σ is determined from the equation of state of the real fluid. Hence the σ obtained from the CRIS model can be used to make rough estimates of these quantities. For the shear viscosity v_s, in poise, we use Dymond's formula (*33*).

$$v_s = 7.61 \times 10^5 \frac{\sqrt{WT}}{(V\eta)^{2/3}} \frac{\eta}{1 - 1.869\eta} \tag{21}$$

where W is the molecular weight, and $\eta = N\pi\sigma^3/6V$ is the packing fraction.

Corrections to the Basic Model. Quantum corrections to the CRIS model can be computed by several methods. Our approach (*24*) has been to add a quantum correction to the hard-sphere free energy A_0, in Equation 2, using the formula of Singh and Sinha (*34*). Admittedly, this procedure is not rigorous, although it is found to give reasonable results for the hydrogen isotopes, as shown later in this chapter. Alternate methods have been used by Rosenfeld (*35*) and by Fiorese (*36*). Further study of this problem is desirable.

As noted above, use of Equation 9 is not a fully satisfactory way to correct the potential energy of a molecule for the change in coordination number when going from the solid to the fluid. As a practical matter, Equation 9 must be modified at very high densities where the electrons are free and $E_c(V_s) \propto V_s^{-2/3}$. In calculations for real materials we use an interpolation formula (*24*)

$$\phi = [(1 - f)(V_s/V) + f(V_s/V)^{2/3}]E_c(V_s) \tag{22}$$

Here f, the fraction of electrons that are free, is estimated from

$$f = \exp[-(0.23 + 0.6544\,Z^{2/3})(V_s/Z)^{1/3}] \tag{23}$$

where Z is the atomic number. This expression was obtained by making a rough estimate of the number of free electrons from TFD theory. We stress that Equation 22 differs very little from Equation 9 in any of the calculations discussed in this chapter. For example, use of the modified formula does not substantially improve the free energy of the fluid near melting. This feature of our model also requires further study.

Calculations for the Rare Gases

Extensive comparisons have been made between calculations using the CRIS model and experimental data for the rare gases neon, argon, krypton, and xenon (*37*). For these substances, it is known (*1*) that an effective pair potential of the Buckingham form (*38*) gives a good description of the intermolecular forces up to moderate densities. Hence, Equation 17, with $a_4 = 2$, is a good representation of the cold curve both in tension and in moderate compression.

At high densities, the zero-Kelvin isotherm for xenon is known both from static measurements up to 11 GPa (*39*) and from band theoretical calculations up to 100 GPa (*25*). Using these data, together with INFERNO calculations at higher densities, we constructed the cold curve shown in Figure 5.

For the other rare gases, data extend to only 2 GPa (*40*), and we used a TFD interpolation formula for higher pressures (*37*). Therefore, we do not know the cold curve as well for these gases as we do for xenon. However, this uncertainty is only important in the comparisons with

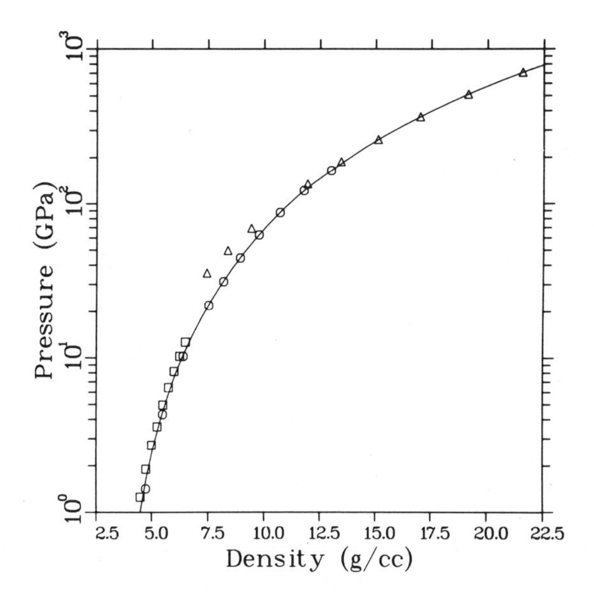

Figure 5. Zero-Kelvin isotherm for xenon. Key: ○, from Ref. 25; □, from Ref. 39; △, calculated using the INFERNO model; and —, the cold curve used in this work. (Reproduced from Ref. 65.)

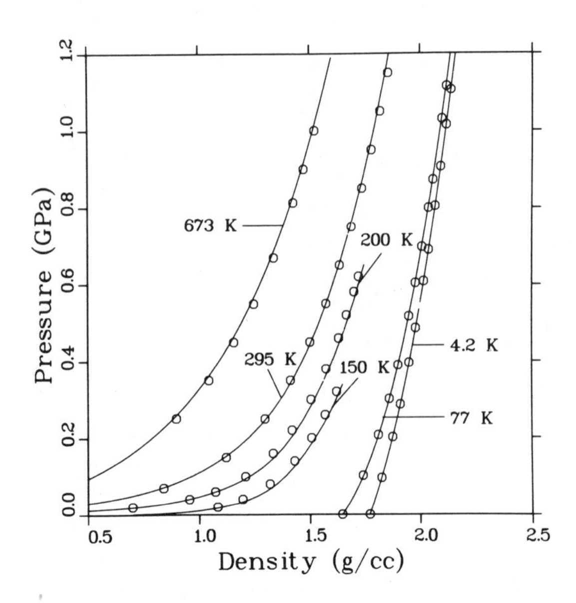

Figure 6. Equation of state for solid and liquid argon. Data for the solid at 4.2 and 77 K are from Ref. 40; data for the liquid are from Refs. 41–44; solid lines are our calculations.

shock wave measurements because the other experimental data were taken at fairly low pressures.

Equation of state data (*40–44*) for solid and liquid argon are shown in Figure 6. The same zero-Kelvin curve was used in both the solid and liquid models. The CRIS model accurately predicts the liquid compressibility and the expansion that occurs at melting and upon heating of the liquid. The other rare gases give similar results (*37*). For example, Figure 7 shows the sound speeds as a function of pressure at 298.15 K (*45*). The pressure dependence and the trends among the four rare gases are in good agreement with the measurements.

Vapor pressure curves for the rare gas liquids (*46*) are shown in Figure 8. The model accurately predicts the temperature dependence of the pressure and the trends among the four elements. The coexistence curve for krypton (*46*) is given in Figure 9. Agreement between theory and experiment is good. As noted above, the theory overestimates the critical temperature and pressure because it does not include long-range density fluctuations.

The calculated and measured radial distribution functions for argon at 85 K (*47*) are compared in Figure 10. Only the first peak is calculated by the theory (*11, 12*). The agreement is good, showing that the model correctly predicts the short-range liquid structure using no information except the solid cold curve.

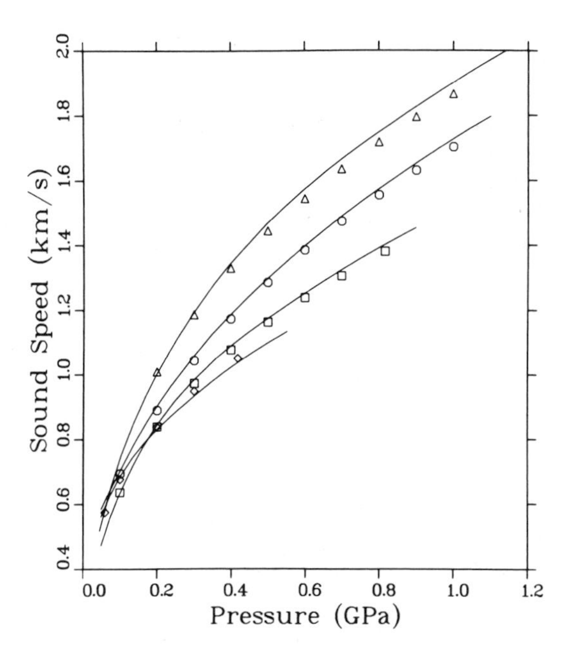

Figure 7. Sound speeds for the rare gases at 298.15 K. The solid lines are our calculations; points are experimental data (45). Key: ○, *neon;* △, *argon;* □, *krypton; and* ◇, *xenon.*

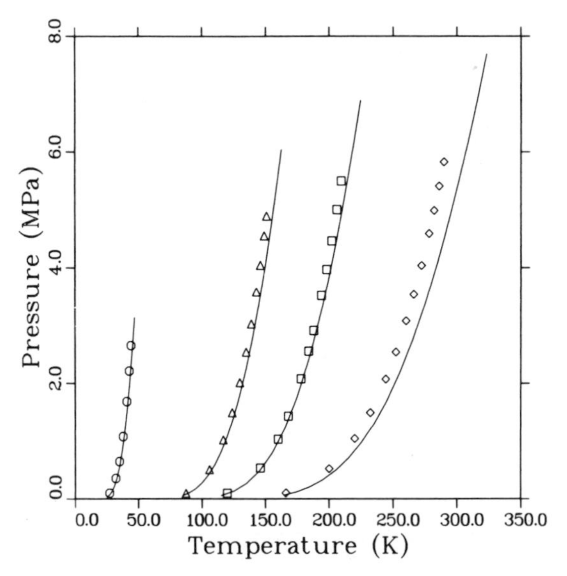

Figure 8. Vapor pressures for the rare gas liquids. The solid lines are our calculations; points are experimental data (46). Key: ○, *neon;* △, *argon;* □, *krypton; and* ◇, *xenon.*

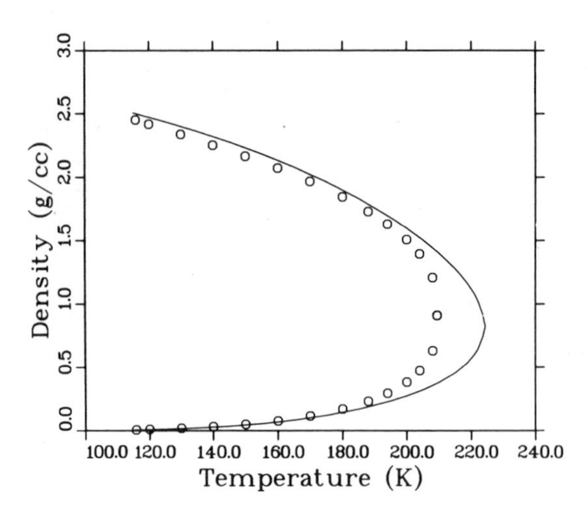

Figure 9. Temperature and density on the coexistence curve for krypton. The solid lines are our calculations; points are experimental data (46).

In Figure 11 we compare measured shear viscosities for liquid argon, krypton, and xenon (*48*) with our calculations, using only the hard-sphere formula, Equation 21. The good agreement with experiments is encouraging; if a perturbation expansion can be developed to calculate corrections to this simple model, better results might be obtained.

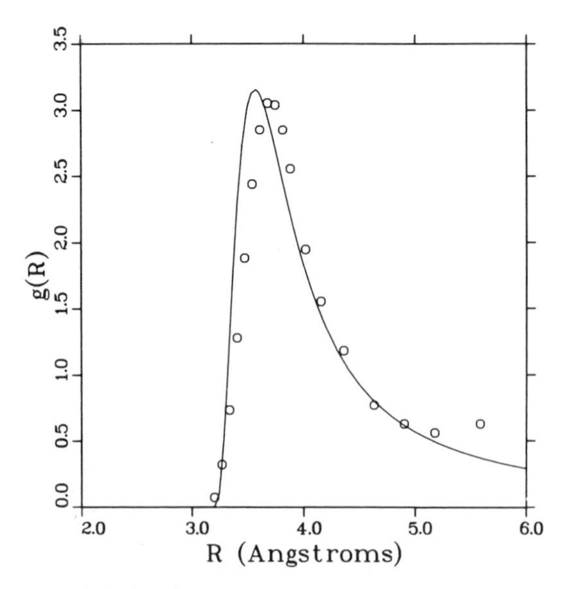

Figure 10. Radial distribution function for liquid argon at a density of 1.409 g/cm³ and a temperature of 85 K. Points are experimental data (47). *The solid line is our calculation of the first peak.*

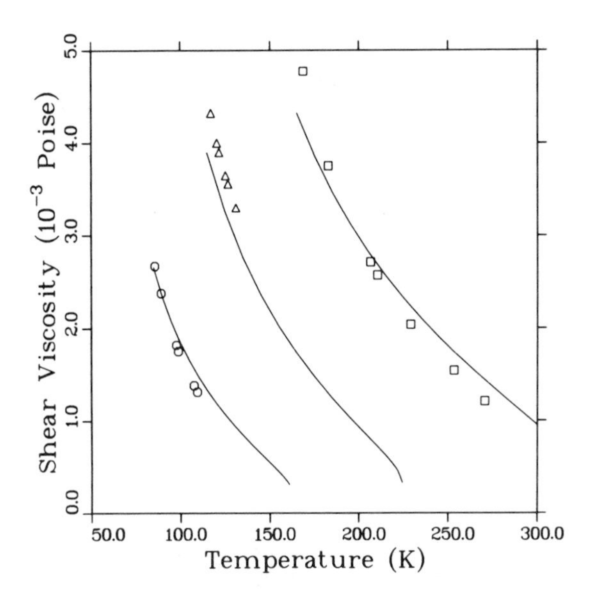

Figure 11. Shear viscosities for liquid argon, krypton, and xenon. Points are experimental data (48); *solid lines are our calculations. Key:* ○, *argon;* △, *krypton; and* □, *xenon.*

Hugoniot measurements for the rare gases are of interest because they generate both high compressions and high temperatures. Very high pressure data for argon and xenon show interesting behavior that is associated with thermal excitation of the electrons. Because the cold curve for xenon is well known, it is possible to make an a priori calculation and to compare the results with experiment as a test of the theory.

At low densities the rare gases are insulators, with a large energy gap between the closed-shell ground state configuration and the empty conduction band. Hence there is very little electronic excitation at low temperatures. Good agreement with the low-pressure shock data for argon and xenon was obtained using the CRIS model without including the electronic term (*37*).

At high densities, the rare gases are expected to become metallic (*49*). For xenon, band calculations (*25*) predict the energy gap between the $5p$ band and the conduction band to close at a density of about 12 g/cm^3. At the high temperatures reached in some of the shock wave experiments, effects of the insulator–metal transition can be observed at lower densities. As shown by Ross (*49*), narrowing of the band gap increases the energy absorbed by electronic excitation and also makes a negative contribution to the pressure; both effects soften the Hugoniot.

The INFERNO calculations give results that are similar to those predicted by Ross's model. As shown in Figure 12, the thermal electronic pressure is negative in the density range 2–10 g/cm^3, for temperatures

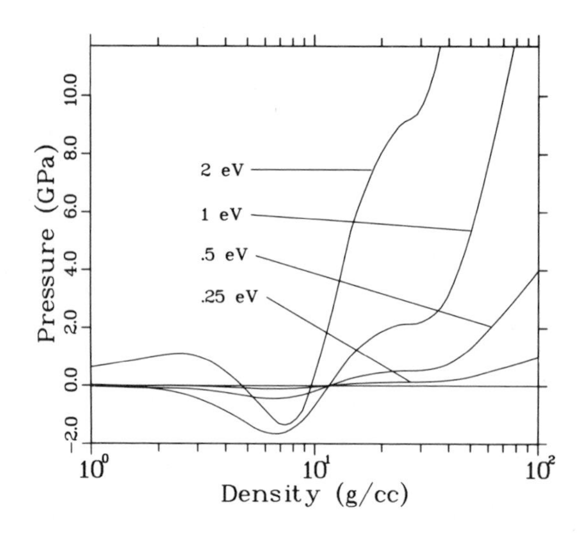

Figure 12. Thermal electronic pressure as a function of density for xenon at several temperatures. (Reproduced from Ref. 65.)

less than 3 eV. INFERNO predicts closing of the band gap to occur at about 10 g/cc, in fair agreement with the band calculations.

The Hugoniot for xenon (*50, 51*) is shown in Figure 13. Calculations in which no electronic excitation is allowed are in good agreement with experiment at pressures below 40 GPa but give poor results at the high pressures. When the TFD model is used to describe the electronic excitations, the results are better but still not satisfactory. Calculations using the INFERNO model are in excellent agreement with the experimental data. The theory gives similar results when applied to shock wave data for argon.

Calculations for the Hydrogen Isotopes

Calculations for the equations of state of hydrogen and deuterium are complicated by the existence of several phases, by the transition from the molecular form to a metallic form at high pressures, and by the effects of dissociation and ionization in the fluid phase at high temperatures. We have developed a detailed theoretical model that accounts for all of these phenomena (*20, 52*). Our calculated phase diagram for deuterium is shown in Figure 14. Separate equations of state were computed for the molecular solid, the metallic solid, and the fluid phases; the phase boundaries were determined by matching the pressures and Gibbs free energies. The fluid phase was treated as a mixture of molecular and metallic (atomic) species, and the fraction of dissociation was computed using a chemical equilibrium model. The EOS for both the molecular

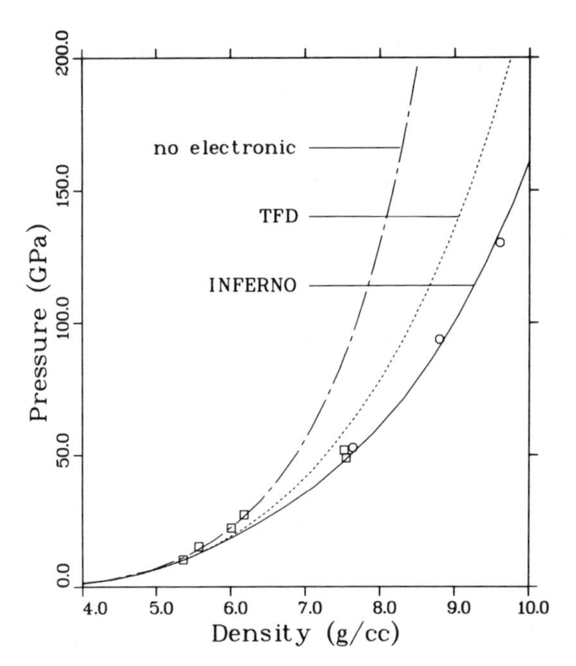

Figure 13. Hugoniot data and theoretical calculations for xenon. Key: □, from Ref. 50; and ○, from Ref. 51. The three theoretical curves are calculated using different models for thermal electronic excitation. (Reproduced from Ref. 65.)

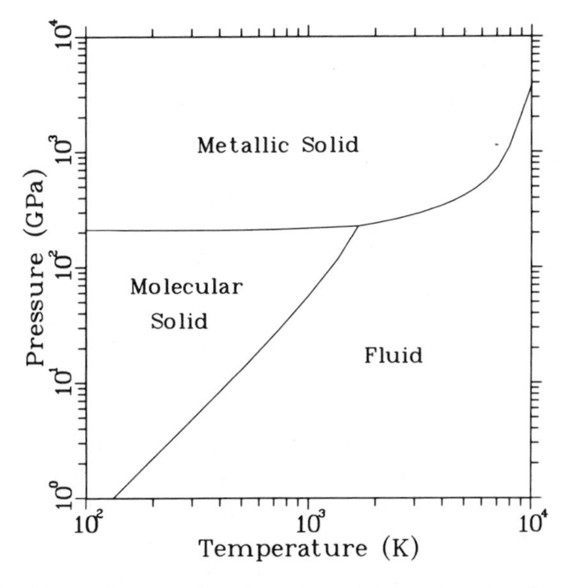

Figure 14. Phase diagram for deuterium. (Reproduced with permission from Ref. 52. Copyright 1972, Elsevier Scientific Publishing Co.)

and metallic fluids was calculated with an early version of the CRIS model (*53*) that does not include either quantum terms or corrections to first-order theory.

At present very few experimental data exist for the hydrogen isotopes in regions where either dissociation or metallization can occur. Our equation of state is consistent with experiments that report observation of the metallic transition (*54*), but those results are preliminary because the measurements did not provide complete diagnostics. Therefore, we limit the discussion in this article to calculations for the molecular fluid. For completeness we have redone the computations using our improved version of the CRIS model and recent experimental data for the cold curve. These new results differ only slightly from our earlier work (*13*, *52*).

CRIS model calculations for hydrogen have also been reported by Rosenfeld (*35*) and by Fiorese (*36*). Rosenfeld included quantum corrections but made only a first-order calculation of the equation of state. Fiorese included the higher-order perturbation corrections as well as quantum terms. Fiorese also showed that the CRIS model gave good agreement with his Monte Carlo calculations that treated the rotational degrees of freedom explicitly, using a dumbbell model for the molecules.

The zero-Kelvin isotherm for hydrogen is shown in Figure 15. For pressures up to 2 GPa we have used the measurements of Anderson and Swenson at 4.2 K (*55*). To obtain that part of the cold curve that is due to the intermolecular forces, the contribution from zero-point nuclear motion must be subtracted. We estimated this correction from the difference between the hydrogen and deuterium data. As in our previous work (*52*), the high density portion of the cold curve was taken from the band theoretical calculations of Liberman (*56*). Our fit to these two sets of data is shown by a solid line.

The results of recent high pressure diamond cell experiments on hydrogen are also shown in Figure 15. Shimizu et al. (*57*) measured the pressure dependence of the acoustic velocities for the fluid and solid phases at 300 K; the volume versus pressure was calculated by integrating the sound velocity data. Van Straaten et al. (*58*) measured the volume at 5 K from observations of the actual dimensions of the cell under pressure. These two sets of data differ by about 40% in pressure at the highest density. Our theoretical curve is in better agreement with the results of Ref. 58.

The equation of state for fluid hydrogen (*59*) and deuterium (*60*), at temperatures up to 300 K and pressures up to 2 GPa, are shown in Figures 16 and 17, respectively. For comparison, we also show the 4.2-K solid isotherms (*55*), with the zero-point pressure term included. The calculations are in good agreement with experiment, but the results for deuterium are significantly better than those for hydrogen. This suggests

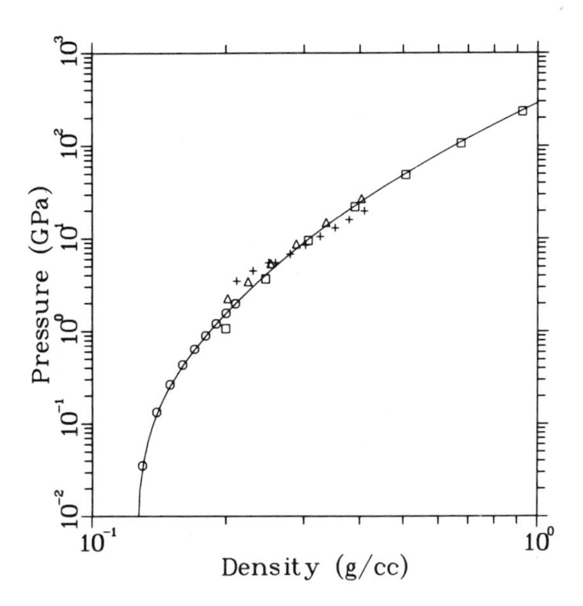

Figure 15. Zero-Kelvin isotherm for hydrogen. Key: ○, data from Ref. 55, corrected for the zero-point nuclear term; □, calculations from Ref. 56; +, from Ref. 57; △, from Ref. 58; and —, cold curve used in our calculations.

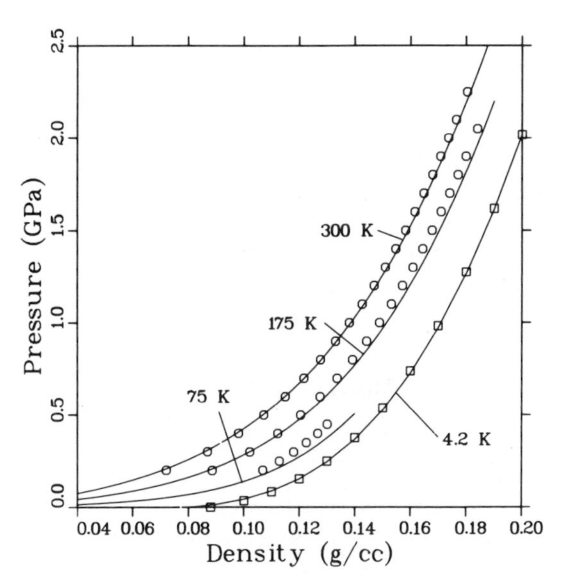

Figure 16. Equation of state for hydrogen. Key: ○, fluid hydrogen data (59); and □, solid hydrogen data (55).

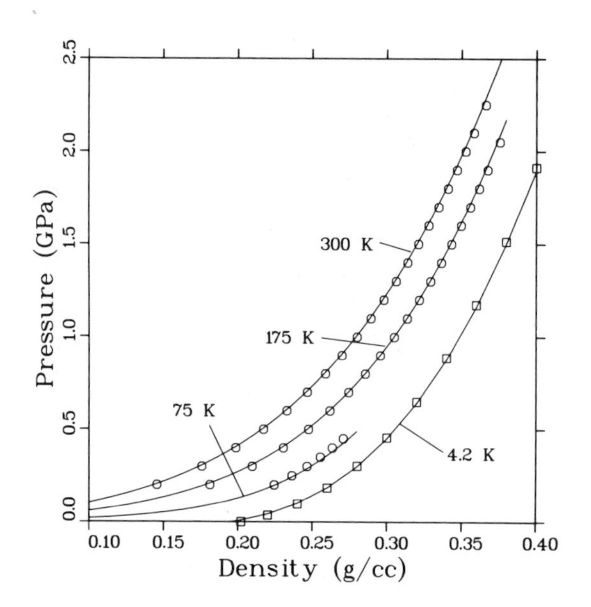

Figure 17. Equation of state for deuterium. Key: ○, *fluid deuterium data* (60); *and* □, *solid deuterium data* (55).

that our treatment of the quantum corrections to the fluid EOS is not adequate, because these terms are more important in the case of hydrogen.

The Hugoniot for liquid deuterium is shown in Figure 18. Our calculations are in very good agreement with the measurements (*31, 61,* 62). The highest shock velocity reached in these experiments corresponds to a pressure of about 20 GPa, a temperature of about 5000 K, and a compression of about 3.5 times that of the liquid at the triple point. Reflected shock Hugoniots for liquid deuterium are shown in Figure 19. The highest pressure points, at about 90 GPa, correspond to a temperature of about 7500 K and a compression of about six times normal density. Agreement between theory and experiment (*31, 61*) is very good.

Calculations for Other Molecular Fluids

The CRIS model has also been used to calculate the thermodynamic properties of methane (*28*) and nitrogen, oxygen, and carbon monoxide (*19*). The molecules were assumed to rotate freely, so that the "effective cold curve" used in the CRIS model corresponds to an average over all orientations of the molecules. The effects of dissociation and ionization were not included in the theory.

For methane, the zero-Kelvin isotherm up to about 2 GPa was

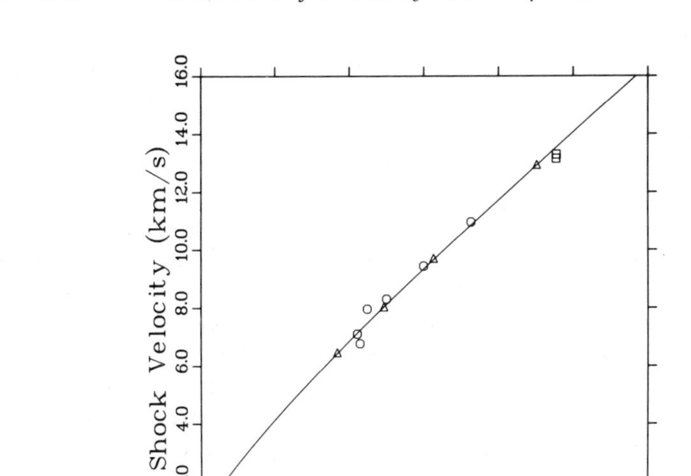

Figure 18. Single shock Hugoniot for liquid deuterium. Key: ○, from Ref. 31; □, from Ref. 61; and △, from Ref. 62.

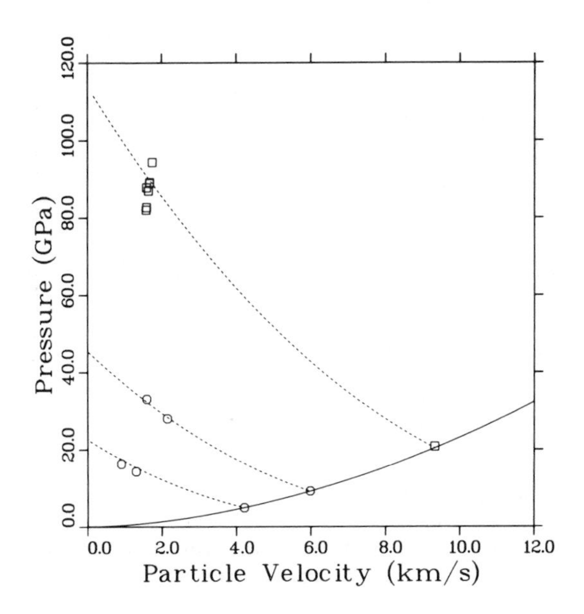

Figure 19. Reflected shock data for liquid deuterium. The principal Hugoniot is shown by a solid line, and reflected Hugoniots are shown by dashed lines. Key: ○, from Ref. 31; and □, from Ref. 61.

constructed from an effective pair potential that was obtained from second virial coefficient data (*38*). This potential is of the Buckingham form and corresponds to an average over all rotational degrees of freedom. As in the rare gases, the zero-Kelvin isotherm for this potential is given by Equation 7, with $a_4 = 2$. Using this cold curve, our calculations gave good agreement with experimental data for both the solid and fluid (*28*). For pressures above 2 GPa, the TFD interpolation formula was used for the cold curve. This formula has one adjustable parameter, which was chosen to fit shock-wave data. Good agreement was obtained for both the single shock and reflected shock Hugoniots (*28*).

A portion of the methane equation of state table is pictured in Figure 4. The coexistence curve was determined by matching the pressures and Gibbs free energies of the liquid and vapor phases, as described earlier. The calculated vapor pressure curve is compared with experimental data (*63*) in Figure 20. Also shown are isochores (*64*) in the liquid, vapor, and supercritical regions. Agreement with the measurements is very good in all cases.

Nitrogen, oxygen, and carbon monoxide molecules do not rotate freely in the solid phase. Therefore, only a rough estimate of the cold curve for use in the CRIS model can be obtained from experimental data

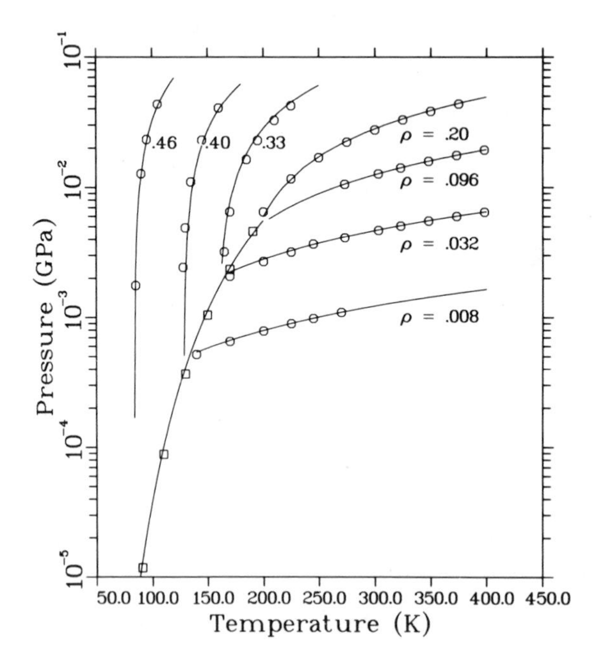

Figure 20. Vapor pressures (□) *and isochores* (○) *for methane* (63, 64). *(Reproduced with permission from Ref. 28. Copyright 1980, American Institute of Physics.)*

for the solid. In these cases, we used Equation 7, with $a_4 = 2$, to represent the zero-Kelvin isotherm, just as we had done for methane and the rare gases. The remaining three constants were determined by fitting the model to experimental data for the liquid (*19*). These calculations were not rigorous, because perturbations to the rotational and vibrational motions were included in the definition of the empirically determined cold curve. However, it is encouraging that the equation of state constructed in this fashion is accurate over the entire fluid range. Furthermore, the parameters obtained in our calculations are in reasonable agreement with data for the solid. For example, the effective "binding energies" are slightly less than the experimental ones. This result is to be expected; the difference corresponds to the energy needed to go from preferential molecular orientations in the crystal to the disordered rotational state in the liquid. A more fundamental treatment could make use of techniques discussed by other workers (*1*, *6*, *7*).

Calculations for Liquid Iron

In all of the examples that we have discussed above, it is likely that nonadditive contributions to the intermolecular forces are fairly small and that good results can be obtained using theories that employ an effective pair potential (*1*). In liquid metals, however, the valence electrons are delocalized, and the intermolecular forces are not even approximately additive. Successful results for liquid metals can be obtained using the pseudopotential method (*8–10*). The cohesive forces can be approximately separated into a free-electron part, which is independent of the positions of the ions, and a structure-dependent part, which is expressed in terms of a density-dependent pair potential. This pair potential is used to calculate the liquid structure.

The CRIS model provides an alternate theory for calculating the properties of liquid metals. Because it uses only the zero-Kelvin curve of the solid to represent the intermolecular forces, it eliminates the need to determine the effective pair potential from a theoretical calculation, and it is easy to apply in practice. We have chosen recent work on liquid iron (*65*) to illustrate the method and the results.

Equation of state calculations for iron are complicated by the existence of several solid phases (*66*). The ferromagnetic alpha phase is stable at room temperature and pressure; it transforms to the hexagonal close-packed (hcp) epsilon phase when compressed and to the face-centered cubic (fcc) gamma phase when heated. To simplify the problem, we treated iron as if it had only one solid phase, taken to be the close-packed phase. Differences in the equations of state of hcp and fcc structures are usually small and were ignored. The cold curve for the close-packed solid was also used in our CRIS model calculations for the fluid. To some extent, this decision is arbitrary, and different results would have been obtained if the cold curve for the alpha phase had been used instead. However, computer simulations show that the radial distribution function of the liquid at the melting point is quite similar to that for the close-packed solid (*12*). Furthermore, the liquid phase is not ferromagnetic. The electronic structure in liquid iron may be different from that in any of the solid phases, but a close-packed crystal should provide the best approximation.

The zero-Kelvin isotherm for close-packed iron is pictured in Figure 21. The solid line, which was used in our calculations, is a fit to the diamond cell measurements of Mao and Bell (*26*). Band-theoretical calculations (*67, 68*) are also shown.

Two assumptions in our application of the CRIS model to iron are that the fluid structure is dominated by short-range forces and that these forces are similar in the fluid and the solid. In Figure 22 we compare our calculated structure factor for molten iron with the measurements of Waseda and Suzuki (*69*). The agreement is very good, demonstrating that these key ideas lead to a good description of the short-range liquid structure. The shear viscosity of molten iron, computed from the hard-sphere formula, is shown in Figure 23. Agreement with the data of Cavalier (*70*) is fairly good.

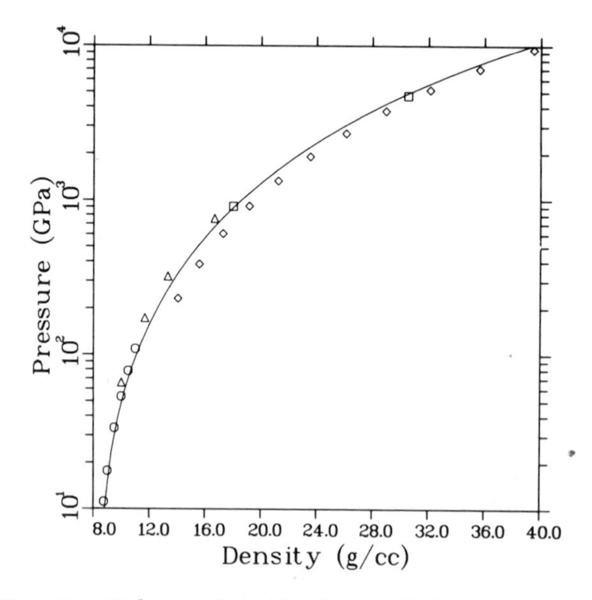

Figure 21. Zero-Kelvin isotherm for close-packed iron. Key: ○, from Ref. 26; △, from Ref. 67; □, from Ref. 68; ◇, calculated using the INFERNO model; and —, cold curve used in this work. (Reproduced from Ref. 65.)

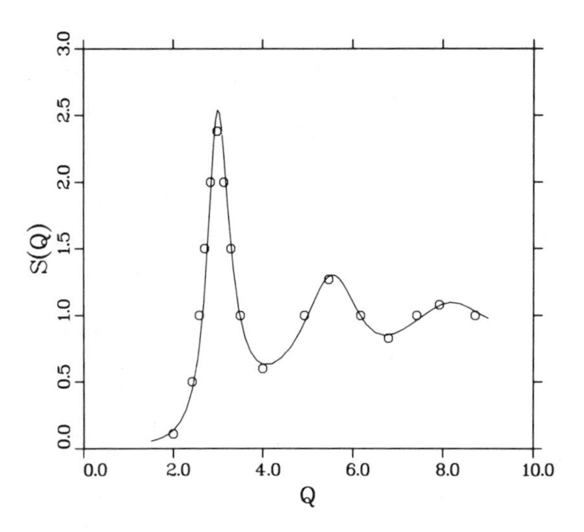

Figure 22. Structure factor for liquid iron at a density of 6.862 g/cm³ and temperature of 1893 K. Key: ○, from Ref. 69; and —, theory. (Reproduced from Ref. 65.)

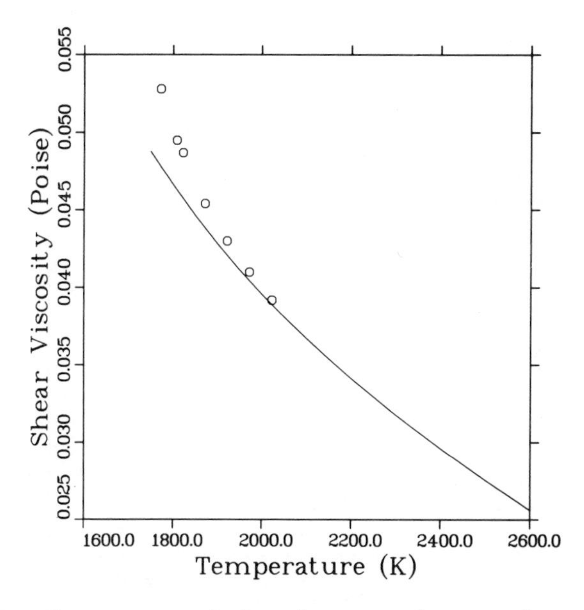

Figure 23. Shear viscosity for liquid iron as a function of temperature. Key: ○, from Ref. 70; and —, theory. (Reproduced from Ref. 65.)

Our theoretical melting curve for iron is shown in Figure 24. In this calculation, we forced agreement with the experimental melting point at zero pressure (*71*) by subtracting an empirically determined constant from the free energy of the fluid. The correction was 4.25 kcal/mol, about 4% of the solid binding energy.

Calculated Hugoniots for iron of two initial densities are also shown in Figure 24. Alpha-phase iron, having a density of 7.85 g/cm^3, transforms to the ε-phase at about 13 GPa under shock loading (*66*). According to our calculations, melting should begin at about 320 GPa. This result is in fair agreement with the value of 250 GPa obtained by Brown and McQueen (*72*). Porous α-phase iron, with an initial density of 4.8 g/cm^3, is predicted to melt at 45 GPa.

Curves of shock velocity versus particle velocity for iron of various initial densities (*73–75*) are shown in Figure 25. Agreement between the theory and the measurements for normal density iron is very good over the entire range of the close-packed solid and fluid phases, extending up to 1000 GPa. The theory also predicts the correct behavior as a function of porosity. More detail can be seen in Figure 26, which shows the shock data for an initial density of 4.8 g/cm^3. Agreement with experiment is excellent except at the lowest pressures, for which the shocked state is the α-phase.

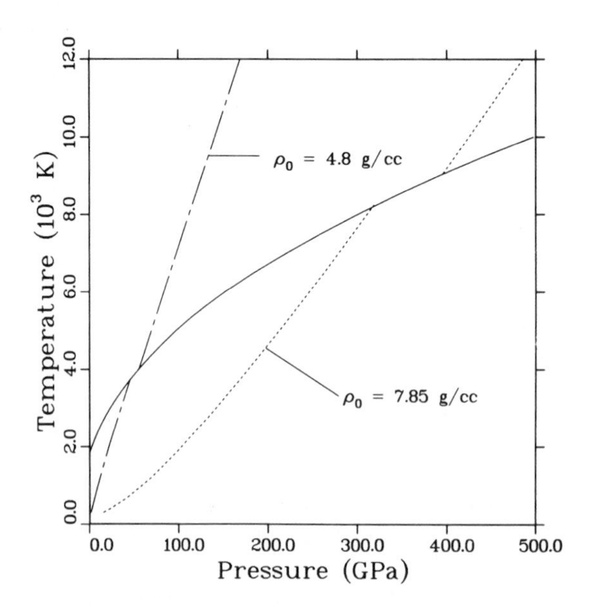

Figure 24. Theoretical melting curve and shock Hugoniots for iron. (Reproduced from Ref. 65.)

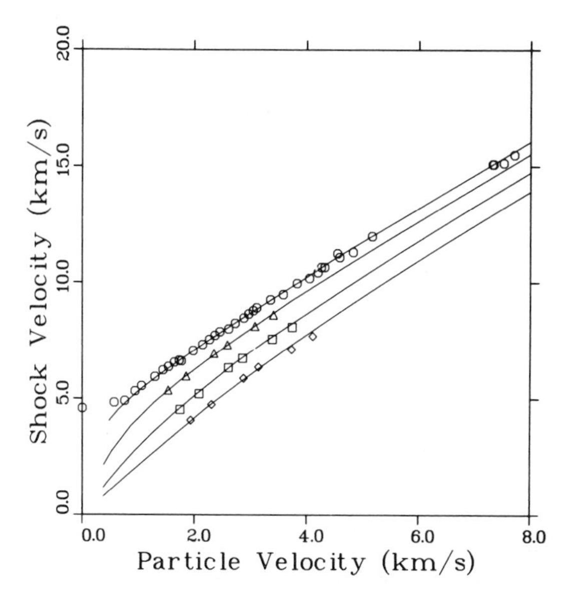

Figure 25. Shock velocity vs. particle velocity for iron at four initial densities. Points are experimental data (73–75); solid lines, theory. Key: ○, ρ_0 = *7.85 g/cm*^{3}; △, ρ_0 = *7.0 g/cm*^{3}; □, ρ_0 = *5.7 g/cm*^{3}; *and* ◇, ρ_0 = *4.4 g/cm*^{3}. *(Reproduced from Ref. 65.)*

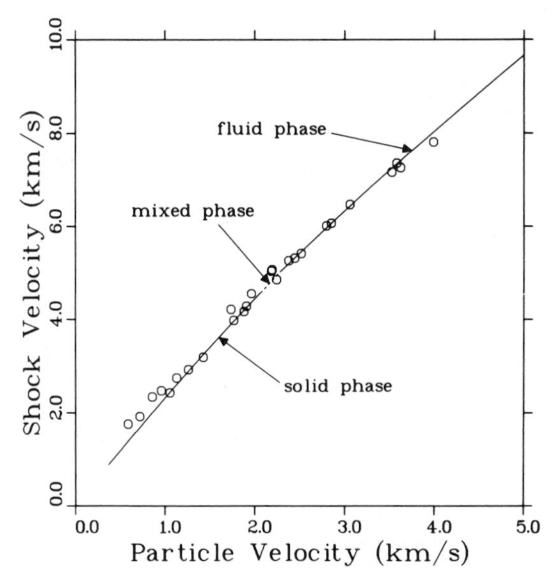

Figure 26. Hugoniot for porous iron having an initial density of 4.8 g/cm^3. *Points are experimental with permission data (73, 74); the curve is theory. (Reproduced from Ref. 65.)*

Summary

In this chapter, we have shown how the thermodynamic properties of a fluid can be computed without explicit knowledge of the interaction potentials, using only the zero-Kelvin isotherm of the solid phase. The theory was shown to agree with experimental data for simple monoatomic fluids such as the rare gases and liquid metals. It also gives good results for polyatomic molecules such as hydrogen and methane, if the molecules are freely rotating. The CRIS model can be used together with the INFERNO model of Liberman to calculate the equation of state at high temperatures where thermal excitation of the electrons becomes an important effect.

Acknowledgments

Parts of this work were done in collaboration with D. A. Liberman, J. Abdallah, Jr., R. D. Dick, and P. M. Henry. I am grateful for technical discussions with many colleagues, especially D. Chandler, H. DeWitt, W. Hoover, J. D. Johnson, J. Kincaid, R. L. Mills, and Y. Rosenfeld. Special thanks are due to J. F. Barnes and W. F. Huebner, for general direction and encouragement.

This work was supported by the U.S. Department of Energy, Offices of Basic Energy Sciences, Military Applications, and Laser Fusion.

Literature Cited

1. Barker, J. A.; Henderson, D. *Rev. Mod. Phys.* **1976,** *48*, 587.
2. Mansoori, G. A.; Canfield, F. B. *J. Chem. Phys.* **1969,** *51*, 4958.
3. Rasaiah, J. C.; Stell, G. *Mol. Phys.* **1970,** *18*, 249.
4. Anderson, H. C.; Chandler, D.; Weeks, J. D. *Adv. Chem. Phys.* **1976,** *34*, 105.
5. Rosenfeld, Y.; Ashcroft, N. W. *Phys. Rev. A* **1979,** *20*, 1208.
6. Chandler, D. *Annu. Rev. Phys. Chem.* **1978,** *29*, 44.
7. Gubbins, K. E.; Gray, C. G.; Machado, J. R. S. *Mol. Phys.* **1981,** *42*, 817.
8. Jones, H. D. *Phys. Rev. A* **1973,** *8*, 3215.
9. Ashcroft, N. W.; Stroud, D. *Solid State Phys.* **1978,** *33*, 1–81.
10. Ross, M.; DeWitt, H. E.; Hubbard, W. B. *Phys. Rev. A* **1981,** *24*, 1016.
11. Kerley, G. I. *J. Chem. Phys.* **1980,** *73*, 469.
12. Kerley, G. I. *J. Chem. Phys.* **1980,** *73*, 478.
13. Kerley, G. I. *J. Chem. Phys.* **1980,** *73*, 487.
14. Liberman, D. A. *Phys. Rev. B* **1979,** *20*, 4891.
15. Rosenfeld, Y. *J. Chem. Phys.* **1980,** *73*, 5753.
16. Hirschfelder, J. O.; Curtiss, C. F.; Bird, R. B. "Molecular Theory of Gases and Liquids"; Wiley: New York, 1954; pp. 1035–1041.
17. Verlet, L.; Weis, J. J. *Phys. Rev. A* **1972,** *5*, 939.

18. Herzberg, G. "Molecular Spectra and Molecular Structure II. Infrared and Raman Spectra of Polyatomic Molecules"; Van Nostrand Reinhold: New York, 1945, Chap. 5, pp. 501–537.
19. Kerley, G. I.; Abdallah, J., Jr. *J. Chem. Phys.* **1980,** *73*, 5337.
20. Kerley, G. I. "A Theoretical Equation of State for Deuterium", Los Alamos Scientific Laboratory, LA-4776, 1972.
21. Cowan, R. D.; Ashkin, J. *Phys. Rev.* **1957,** *105*, 144.
22. Rouse, C. A. *Astrophys. J.* **1961,** *134*, 435.
23. Stewart, J. C.; Pyatt, K. D., Jr. *Astrophys. J.* **1966,** *144*, 1203.
24. Kerley, G. I. "Users Manual for PANDA: A Computer Code for Calculating Equations of State", Los Alamos National Laboratory, LA-8833-M, 1981.
25. Ross, M.; McMahan, A. K. *Phys. Rev. B* **1980,** *21*, 1658.
26. Mao, H. K.; Bell, P. M. *J. Geophys. Res.* **1977,** *84*, 4533.
27. Rice, M. H.; McQueen, R. G.; Walsh, J. M. *Solid State Phys.* **1958,** *6*, 1–63.
28. Kerley, G. I. *J. Appl. Phys.* **1980,** *51*, 5368.
29. Hirschfelder, J. O.; Curtiss, C. F.; Bird, R. B. "Molecular Theory of Gases and Liquids"; Wiley: New York, 1954; pp. 364–367.
30. Verlet, L. *Phys. Rev. A* **1970,** *165*, 201.
31. Dick, R. D.; Kerley, G. I. *J. Chem. Phys.* **1980,** *73*, 5264.
32. Dymond, J. H.; Alder, B. J. *J. Chem. Phys.* **1966,** *45*, 4061.
33. Dymond, J. H. *J. Chem. Phys.* **1974,** *60*, 969.
34. Singh, B. P.; Sinha, S. K. *J. Chem. Phys.* **1977,** *67*, 3645.
35. Rosenfeld, Y. *J. Chem. Phys.* **1980,** *73*, 5760.
36. Fiorese, G. *J. Chem. Phys.* **1981,** *75*, 1427.
37. Kerley, G. I.; Henry, P. M. "Theoretical Equations of State for the Rare Gases", Los Alamos Scientific Laboratory, LA–8062, 1980.
38. Hirschfelder, J. O.; Curtiss, C. F.; Bird, R. B. "Molecular Theory of Gases and Liquids", Wiley: New York, 1954, pp. 180–181.
39. Syassen, K.; Holzapfel, W. B. *Phys. Rev. B* **1978,** *18*, 5826.
40. Anderson, M. S.; Swenson, C. A. *J. Phys. Chem. Solids* **1975,** *36*, 145.
41. Street, W. B.; Staveley, L. A. K. *J. Chem. Phys.* **1969,** *50*, 2302.
42. Crawford, R. K.; Daniels, W. B. *J. Chem. Phys.* **1969,** *50*, 3171.
43. Robertson, S. L.; Babb, S. E., Jr.; Scott, G. J. *J. Chem. Phys.* **1969,** *50*, 2160.
44. Liebenberg, D. H.; Mills, R. L.; Bronson, J. C. *J. Appl. Phys.* **1974,** *45*, 741.
45. Vidal, D.; Guengant, L.; Lallemand, M. *Physica, Utrecht* **1979,** *96A*, 545.
46. Vargaftik, N. B. "Tables on the Thermophysical Properties of Liquids and Gases", 2nd ed., Wiley: New York, 1975, pp. 536–584.
47. Yarnell, J. L.; Katz, M. J.; Wenzel, R. G.; Koenig, S. H. *Phys. Rev. A* **1973,** *7*, 2130.
48. Baharudin, B. Y.; Jackson, D. A.; Schoen, P. E. *Phys. Lett.* **1975,** *51A*, 409.
49. Ross, M. *Phys. Rev.* **1978,** *171*, 777.
50. Van Thiel, M. "Compendium of Shock Wave Data", Lawrence Livermore Laboratory, UCRL–50108, 1977.
51. Nellis, W. J.; Van Thiel, M.; Mitchell, A. C. *Phys. Rev. Lett.* **1982,** *48*, 816.
52. Kerley, G. I. *Phys. Earth Planet. Inter.* **1972,** *6*, 78.
53. Kerley, G. I. "A New Model of Fluids", Los Alamos Scientific Laboratory, LA–4760, 1971.
54. Hawke, R. S.; Burgess, T. J.; Duerre, D. E.; Huebel, J. G.; Keeler, R. N.; Klapper, H.; Wallace, W. C. *Phys. Rev. Lett.* **1978,** *41*, 998.

55. Anderson, M. S.; Swenson, C. A. *Phys. Rev. B* **1974,** *10*, 5184.
56. Liberman, D. A. *Int. J. Quantum Chem.* **1976,** *10*, 297.
57. Shimizu, H.; Brody, E. M.; Mao, H. K.; Bell, P. M. *Phys. Rev. Lett.* **1981,** *47*, 128.
58. Van Straaten, J.; Wijngaarden, R. J.; Silvera, I. F. *Phys. Rev. Lett.* **1982,** *48*, 97.
59. Mills, R. L.; Liebenberg, D. H.; Bronson, J. C.; Schmidt, L. C. *J. Chem. Phys.* **1977,** *66*, 3076.
60. Mills, R. L.; Liebenberg, D. H.; Bronson, J. C. *J. Chem. Phys.* **1978,** *68*, 2663.
61. Van Thiel, M.; Hord, L. B.; Gust, W. H.; Mitchell, A. C.; D'Addario, M.; Boutwell, K.; Wilbarger, E.; Barrett, B. *Phys. Earth Planet. Inter.* **1974,** *9*, 57.
62. Nellis, W. J.; Van Thiel, M.; Mitchell, A.C.; Ross, M "The Equation of State of D_2 and Xe in the Megabar Pressure Range", Eighth Symposium on Thermophysical Properties, 1981 National Bureau of Standards, Gaithersburg, MD.
63. Prydz, R.; Goodwin, R. D. *J. Chem. Thermodyn.* **1972,** *4*, 127.
64. Goodwin, R. D.; Prydz, R. *J. Res. Nat. Bur. Stand.* **1972,** Sec. A, *76*, 81.
65. Kerley, G. I. "Theoretical Model of Liquid Metals"; Eighth Symposium on Thermophysical Properties, 1981, National Bureau of Standards, Gaithersburg, MD.
66. Andrews, D. J. *J. Phys. Chem. Solids* **1973,** *34*, 825.
67. Bukowinski, M. S. T. *Phys. Earth Planet. Inter.* **1977,** *14*, 333.
68. Kmetko, E. A., unpublished data.
69. Waseda, Y.; Suzuki, K. *Phys. Stat. Sol.* **1970,** *39*, 669.
70. Cavalier, G. *Compt. Rend.* **1963,** *256*, 1308.
71. Strong, H. M.; Tuft, R. E.; Hanneman, E. E. *Metall. Trans.* **1973,** *4*, 2657.
72. Brown, J. M.; McQueen, R. G. *Geophys. Res. Lett.* **1980,** *7*, 533.
73. McQueen, R. G., private communication.
74. Marsh, S. P. "LASL Shock Hugoniot Data", 1980, University of California, Berkeley.
75. Altshuler, L. Y.; Kalitkin, N. N.; Kuzmina, L. V.; Chekin, B. S. *Sov. Phys. JETP* **1977,** *45*, 167.

RECEIVED for review January 27, 1982. ACCEPTED for publication October 10, 1982.

6

Fluids at Interfaces

JOHANN FISCHER

Ruhr-Universität, Institut für Thermo- und Fluiddynamik, D-4630 Bochum, Federal Republic of Germany

Three topics are considered in this chapter: gas adsorption on solid surfaces, the free liquid surface, and a liquid in contact with a wall. They are treated theoretically, and results are compared with experiments and simulations. A review of virial expansions and of the application of the first equation of the Born–Green–Yvon hierarchy for inhomogeneous fluids is given. For the case of gas adsorption, the structure of a fluid adsorbed on a plane surface at supercritical and subcritical temperatures is shown, together with adsorption isotherms. Adsorption in pores is dealt with in a simple model. From the study of the liquid–gas interface the coexisting densities, the surface tension, and the surface thickness are obtained. Finally, the structure of a liquid close to a wall is discussed.

THE EQUILIBRIUM PROPERTIES OF FLUIDS AT INTERFACES are of practical importance in many engineering processes. While molecular theory will hardly be able to make quantitative predictions for all real situations, it can help us at least in understanding many of the interface phenomena. In order to achieve the latter goal we will keep the model of the fluid molecules and the solid surfaces as simple as possible—we consider only Lennard–Jones or hard-sphere molecules and unstructured walls with forces perpendicular to the surface—and concentrate on some physically interesting situations.

Statistical mechanics of inhomogeneous fluids is now in development and different theoretical methods are in competition. Here, we will briefly describe the virial expansion and the use of the first equation of the Born–Green–Yvon hierarchy, as these will subsequently be used in treating the model systems. A comparison with the density functional method closes the first section.

Theory

Consider a fluid of N identical spherical particles which neither form a free boundary nor are in contact with a smooth solid surface at tem-

0065-2393/83/0204-0139$06.00/0

perature T. The potential between two fluid atoms is $u(r_{ik}) = u_{ik}$ and the potential that a fluid atom experiences from the solid surface is $u^s(\mathbf{r}_i) = u_i^s$. We are mainly interested in the local density $n(r)$, which is the same as the one-particle distribution function and is given by

$$n(\mathbf{r}_1) = (N/Z_N) \int \exp\left\{-\beta\left[\sum_i u_i^s + \sum_{i<k} u_{ik}\right]\right\} d\mathbf{r}_2 \ldots d\mathbf{r}_N \quad (1)$$

where Z_N denotes the configurational partition function and $\beta = 1/kT$.

For evaluating the local density in the case of a gas in contact with a wall we can think of a virial expansion. Technically we start from Equation 1 and use van Kampen's method (*1*), which results in (*2*)

$$n(\mathbf{r}_1) = n_b \exp\{-\beta u_1^s\}[1 + \nu_1(\mathbf{r}_1)n_b + \nu_2(\mathbf{r}_1)n_b^2 + \ldots] \quad (2)$$

where n_b is the gas density far away from the solid surface. In that expansion the coefficient ν_i describes the interaction of $(i+1)$ particles among themselves and with the wall. Expressions for ν_1 and ν_2 are given elsewhere (*2*). If the kinetic energy of the gas molecules is small compared with the adsorptive potential of the wall, then the molecules tend to sit in a layer close to the wall. Thus, even for low bulk gas densities, the local density in the adsorbed layer may become so high that one has to consider the simultaneous interaction of many particles. In that case, the expansion shown in Equation 2 can no longer be used. Concluding, we can say that a virial expansion is expected to be useful at high temperatures and low densities.

Another route for evaluating the local density $n(r)$ is the first equation of the Born–Green–Yvon hierarchy. In that connection we consider the probability of finding simultaneously two particles in the volume elements $d\mathbf{r}_1$ and $d\mathbf{r}_2$. We denote that probability as $n(\mathbf{r}_1)\, n(\mathbf{r}_2)\, g(\mathbf{r}_1, \mathbf{r}_2)\, d\mathbf{r}_1\, d\mathbf{r}_2$. The function $g(\mathbf{r}_1, \mathbf{r}_2)$ is called the pair correlation function. Now, by differentiating Equation 1 with respect to the local coordinate $\mathbf{r}_1$ and denoting that differentiation by ∇_1, we obtain, after rearranging

$$\nabla_1 \ln n(\mathbf{r}_1) = -\nabla_1 \beta u_1^s + \int (-\nabla_{-1} \beta u_{12})\, g(\mathbf{r}_1, \mathbf{r}_2)\, n(\mathbf{r}_2)\, d\mathbf{r}_2 \quad (3)$$

This is the first equation of the Born–Green–Yvon hierarchy, (the BGY equation). The first term on its right-hand side is the force exerted by the wall on a particle at $\mathbf{r}_1$. The integral is the mean force that a particle feels from the other fluid particles. These forces are balanced by the density gradient on the left-hand side. In order to calculate the local density from that equation one has to use an approximation for the pair correlation function.

Different approximations for the pair correlation function in connection with the BGY equation are possible. In a previous article (*3*) the author and a co-worker had the goal of making a physically reasonable approximation which, on the other hand, should be simple enough that numerical solutions were readily attainable. We split the potential u_{ik} between two fluid particles into a short-range repulsive part and an attractive part. Hence the mean force in Equation 3 splits into a mean repulsive and a mean attractive force. In the mean attractive force we neglect any correlations. The mean repulsive force is treated in the hard-sphere approximation, the pair correlation function being taken as that of a homogeneous hard-sphere system at a mean density, which is obtained by averaging the local density over the volume of a molecule. Contrary to the virial expansion, this is not a systematic but an ad hoc approximation scheme. The BGY method, however, has a much larger range of applications. Moreover, a recent investigation (*4*) using a more sophisticated approximation for the pair correlation function has confirmed that the above described method yields at least qualitatively correct results.

The first approach to fluids at interfaces was originated by van der Waals and is called, in its modern version, density functional theory. The basic idea is to write the Helmholtz energy A of the system as a function of the local density $n(\boldsymbol{r})$ and the direct correlation function $c(\boldsymbol{r}_1, \boldsymbol{r}_2)$

$$A = \mathcal{F}[n(\boldsymbol{r}), \ c(\boldsymbol{r}_1, \boldsymbol{r}_2)] \tag{4}$$

Instead of the direct correlation function the pair correlation function may also be used. It must be stated that all the functions used are only approximate expressions. After making suitable approximations for the correlation function, one gets an equation for the local density by minimizing the Helmholtz energy. A review of such approaches can be found elsewhere (*5*).

In comparing the BGY with the density functional approach we learn that in both methods an approximation for the correlation function has to be made. The starting equation in the BGY approach, however, is exact, while the expression for the density functional is always an approximation. On the other hand, our BGY method requires some numerical calculations. This is not necessarily the case in the density functional theory. For slowly varying density profiles, for example, gradient expansions can be made, which greatly facilitates the evaluation (*5*).

Gas Adsorption on Plane Solid Surfaces

A simple model for physical adsorption is that of spherical fluid particles in contact with a plane, structureless wall. In nature corresponding systems are that of argon or krypton adsorbed on the basal plane of graphite, for which accurate measurements have been made (*6, 8*).

We assume the fluid particles to interact through a 6-12 Lennard–Jones potential and to be in contact with a plane 3-9 wall

$$u^s(z) = \frac{1}{2} 3^{3/2} \varepsilon_{gs}[(\sigma_{gs}/z)^9 - (\sigma_{gs}/z)^3] \tag{5}$$

where the z-axis is perpendicular to the surface.

One quantity that can be calculated from theory and measured in experiments is the surface excess density Γ. It tells us how much gas per unit surface area the system contains in excess of an idealized system, where the bulk gas density would be maintained up to the wall. Usually the Gibbs dividing surface between solid and gas is defined as that surface where the wall gas potential goes through zero, $u^s(z) = 0$. Hence we have

$$\Gamma = \int_{-\infty}^{0} n(z)\, dz + \int_{0}^{\infty} [n(z) - n_b]\, dz \tag{6}$$

which, as a function of the bulk gas density n_b, is called an adsorption isotherm.

At low bulk gas densities, the local density is obtained by the first term in the virial expansion

$$n(z) = n_b \exp\{-\beta u^s(z)\} \tag{7}$$

and insertion into Equation 6 yields the low density value for Γ. With increasing density, higher order correction terms have to be calculated from Equation 2.

As it has been argued, the virial expansion is most useful at higher temperatures. There, the repulsive forces between the fluid particles are the dominating ones. It was for this case that the coefficients describing the interaction of two or three hard spheres with a 3-9 wall have been calculated explicitly (*2*). The main results of that calculation can be summarized as follows:

- At low densities the adsorption isotherm is essentially determined by the interaction of a single particle with the wall, expressed by Equation 7. The simultaneous interaction of two and three particles with the wall leads only to minor corrections in the adsorption isotherm.
- The experimental results for the system argon–graphite (*6*) could

be reproduced with high accuracy if the parameters for the wall–particle potential were properly chosen (ε_{gs}/k = 1108 K, σ_{gs} = 0.191 nm).

- At high temperatures the coefficient describing the interaction of two particles with the wall can be determined from an experimental adsorption isotherm only with great inaccuracy.
- For that special model the virial expansion for Γ has the form of a geometric series. This suggested casting the adsorption isotherm into the mathematical form

$$\Gamma = Bn_b/(1 + qn_b) \tag{8}$$

even at rather high bulk densities. Generally, if we look for a representation of the surface excess density as a function of the pressure p instead of the bulk density, a useful expression is (9)

$$\Gamma = P(p)/(1 + qp) \tag{9}$$

where $P(p)$ denotes a simple polynomial of the pressure.

At higher bulk densities the virial expansion breaks down, at least in the sense that the higher order virial coefficients become too complicated to be calculated. That breakdown strongly depends on the temperature. At temperatures much higher than the critical temperature of the gas, virial expansions can still be valid at pressures of several bars, while at subcritical temperatures, virial expansions may break down at near-vacuum conditions. For such cases we have solved the BGY equation. The calculations were done again for the system argon–graphite at both supercritical and subcritical temperatures. We took the same wall–particle potential as in the virial expansion and the usual Lennard–Jones potential for argon (ε/k = 119.8 K, σ = 0.3405 nm).

Solutions of the BGY equation for the local density were obtained in the supercritical region at several states in a temperature–density grid and are shown in Figure 1. These results have also been evaluated to yield layer coverages, which are compiled in Table I. We learn that at lower temperatures the first adsorbed layer is quickly filled, while at higher temperatures this filling occurs much more slowly due to the higher kinetic energy of the molecules. Moreover, we observe that a second and third layer are already formed before the first layer is completely filled. Adsorption isotherms have also been calculated by using Equation 6. Comparison with experimental results (6) in Figure 2 shows qualitative agreement. One reason for the discrepancies seen at higher densities could be the approximation scheme for the pair correlation function. More probably, an unfortunate choice for the wall-particle po-

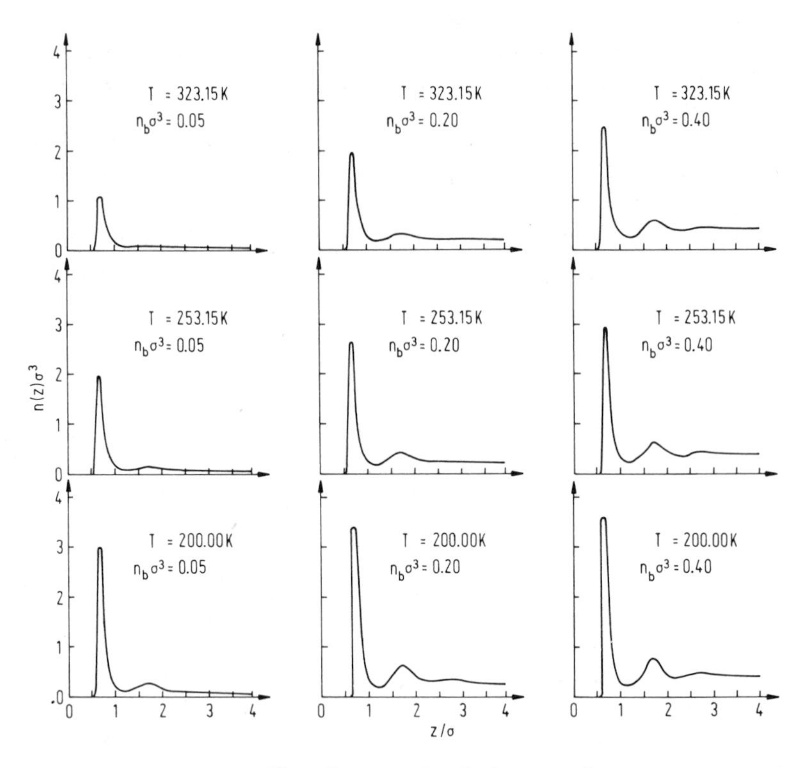

Figure 1. Density profiles of argon adsorbed on graphite at supercritical temperatures.

tential or changes in the intermolecular potential between the fluid particles caused by the solid surface can be the sources of the discrepancies.

At subcritical temperatures, one interesting problem is the structure of the adsorbed film and its behavior in the case where the bulk gas density approaches the dew line. It is important to mention that for the argon–graphite system, computer simulations (*10*) have been made at two subcritical temperatures with the same model potentials described above. In solving the BGY equation (*3*) we first wanted to compare our results with the simulation results. Another aim was to study the approach to the dew line. For that purpose we had to know exactly the coexisting gas density within our model. As will be shown in the next section, we can obtain that value from an eigensolution of the BGY equation for the liquid–gas interface. At a temperature of 120 K the dew density was found to be $n_b\sigma^3 = 0.0207$, which is somewhat lower than the experimental value. At that temperature, density profiles for the adsorbed gas were calculated for the bulk gas density, $n_b\sigma^3$, equal to 0.0200 and 0.01919. The results are reproduced in Figure 3. The lower density

Table I. Layer Coverages for Argon Adsorbed on Graphite

Temperature (K)	*Degree of Coverage*		
	$n_b\sigma^3 = 0.05$	$n_b\sigma^3 = 0.20$	$n_b\sigma^3 = 0.40$
323.15	$\theta^1 = 0.27$	0.48	0.59
	$\theta^2 = 0.10$	0.31	0.51
	$\theta^3 = 0.07$	0.27	0.50
253.15	$\theta^1 = 0.43$	0.58	0.65
	$\theta^2 = 0.14$	0.37	0.53
	$\theta^3 = 0.08$	0.30	0.50
200.00	$\theta^1 = 0.61$	0.68	0.71
	$\theta^2 = 0.23$	0.49	0.58
	$\theta^3 = 0.11$	0.39	0.53

Note: The layers are defined by the distances 1.1 σ, 2.2 σ, and 3.3 σ. Values given are the relative degrees of coverage θ^i referring to a triangular packing of Lennard–Jones atoms with nearest neighbor distance $2^{1/6}$ σ. In the absence of any wall-induced structure, the degree of coverage would be $\theta = 1.2\ n_b\sigma^3$.

corresponds to the simulation run at $kT/\varepsilon = 1.002$ (*10*) and shows good qualitative agreement. It is also interesting to learn from Figure 3 that in approaching the dew point, the first two layers adjacent to the wall remain unchanged while the transition zone between the third layer and the bulk gas tends to form a plateau. The onset of bulk condensation may be explained by considering that the film represented by that transition zone rapidly increases in density and extends into the gas volume. It should be mentioned that we were not able to find BGY solutions for bulk densities higher than 0.0207.

For the system krypton–graphite, experimental values (*7*) are also available at 253.15 K, which is a lower reduced temperature (with respect to the critical temperature) than in the case of argon. Therefore, we also performed calculations for that system using the 3-12 wall potential suggested elsewhere (*11*) and a usual 6-12 potential for krypton (ε/k = 165.2 K, σ = 0.366 nm). It turned out that at high bulk densities the calculated values for the adsorption isotherms were as much as 40% higher than the experimental ones, which continues the trend already observed for argon. Obviously, there is a relation to the previous finding (*11*) that with the same wall potential, the experimental results could only be explained by a weakening of the dispersion forces of the krypton atoms close to the wall. A more detailed discussion of that question will be given elsewhere (*8*).

Gas Adsorption in Pores

If we apply adsorption in technical processes, we are interested in high surface areas and hence use porous adsorbing materials. The effect

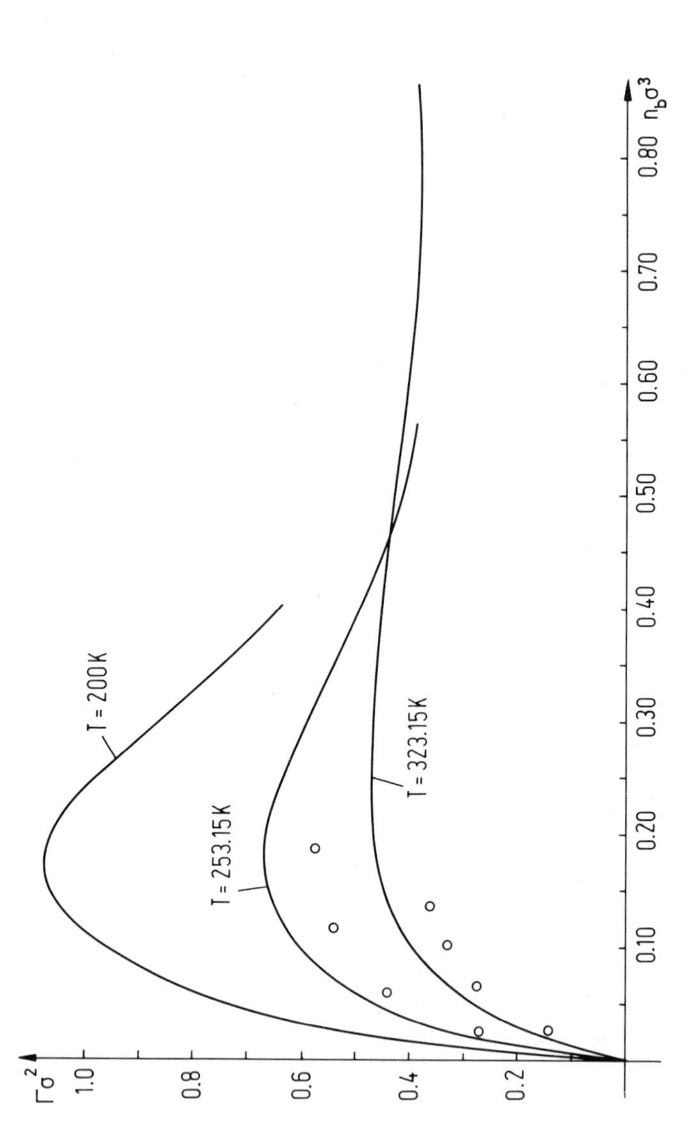

Figure 2. *Comparison of theoretical and experimental adsorption isotherms of argon on graphite. Key:* ○, *experiment; and* —, *theory.*

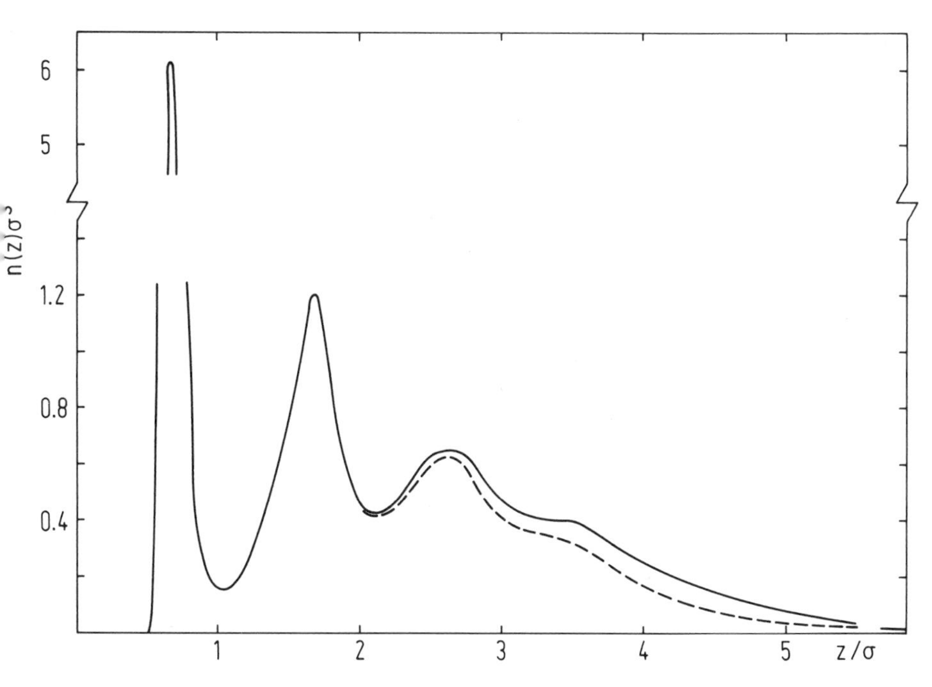

Figure 3. Density profiles of argon adsorbed on graphite at densities close to the dew point ($n_b\sigma^3$ = *0.0207). Temperature, 120 K. Key:* —, $n_b\sigma^3$ = *0.02000; and* ---, $n_b\sigma^3$ = *0.01919. (Reproduced with permission from Ref. 3. Copyright 1980, American Institute of Physics.)*

of pores, however, is not only that of enlarging the surface area. In pores, the gas–wall potential may be changed considerably by purely geometrical reasons. In order to study that effect in more detail we investigated a very simple model (*12*) corresponding to krypton adsorbed in porous carbon. We assume the pores to be circular cylinders and the carbon atoms to be smeared out uniformly over the cylinder surface with a given surface density. For the interaction of one carbon and one krypton atom we take a 6-12 potential with the parameters suggested by Steele (*13*). Now, we are able to calculate a wall potential. The most significant result is that the attractive well becomes strongly deeper with decreasing pore size as a consequence of the pore curvature. An example is shown in Figure 4. For the limit of low bulk densities, the amount of adsorbed gas corresponding to Henry's constant can be calculated from the first term in the virial expansion, Equation 7. For high bulk densities, it seems reasonable to assume that the surface of the pore has the same coverage as a plane surface, which can be calculated by analogy to Table I. From this limiting value an interpolation for mean bulk densities can be made by using Equation 8. In order to allow comparison with experiment, we assume that in a real adsorbent all the pores are cylinders of equal radii. From given experimental values for the surface area and

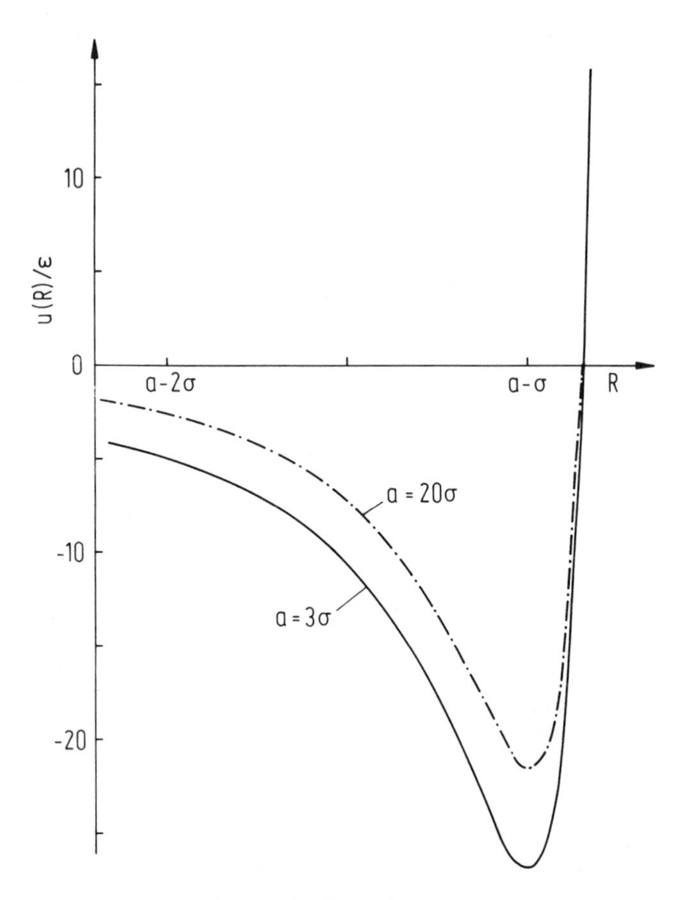

Figure 4. The potential in a cylindrical pore for two pores of different pore size a. *Generally the potential has its minimum at a distance approximately σ away from the sites of the solid atoms. Due to the curvature at small pore sizes, this minimum can become very deep. The model corresponds to krypton adsorbed by porous carbon.*

the pore volume we calculate the pore size for the model. A comparison (*12*) of a calculated with an experimental adsorption isotherm is shown in Figure 5. In spite of all the simplifications in the model, the predictions are surprisingly good.

The Free Liquid Surface

Investigations of the liquid–gas interface aim at understanding the structure and predicting the surface tension. Beyond that, however, one may speculate whether vapor–liquid phase equilibria should not also be determined by the situation in the interface. Of course it should not

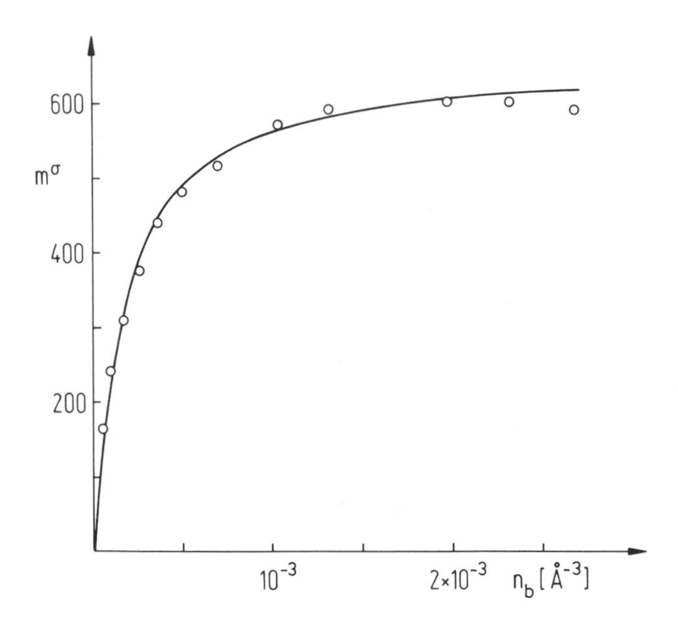

Figure 5. Excess amount of adsorbed krypton on porous carbon. The comparison shows experimental and theoretical values at 298.15 K, given in milligrams krypton per gram of carbon, for an adsorbent with a surface of 1100 m²/g and a micropore volume of 0.42 cm³/g. Key: ○, experiment; and —, theory.

be doubted that equal chemical potential is a condition for phase equilibrium. On the molecular level, however, the molecules in the gas phase "know" about the situation in the liquid phase only by the mediation of the molecules in the interface. Subsequent results will give support to the idea that phase equilibria can also be obtained by considering the interface without using the chemical potential explicitly.

The model fluid consists again of Lennard–Jones atoms and comparison is always made with argon. We are looking now for nontrivial solutions of the homogeneous BGY equation, as the liquid–gas interface will be considered in the absence of any external forces. The most interesting point is that the homogeneous BGY equation turns out to be an eigenvalue equation, which means that at a given temperature a solution could be found only for one definite value of the bulk liquid density. The corresponding eigensolution for $n(\mathbf{r})$ gives the density profile in the interface and the coexisting gas density.

Bubble- and dew-point densities obtained from the eigensolutions of the BGY equation (*3, 14*) are compared in Figure 6 with simulation results, using the condition of equal chemical potential (*15*), and with experimental results for argon (*16*). In spite of the approximations in the

BGY equation, the coexisting densities are qualitatively correct. Note that the chemical potential has not been used explicitly.

The density profiles always show a monotonic decrease from the bulk liquid to the bulk gas density. From those profiles we have calculated the surface thickness d (3) and obtained $d = 1.41\ \sigma$ for $T = 91$ K and $d = 3.20\ \sigma$ for $T = 135$ K. These values are compared in Reference 14 with simulation results (*17*) and experimental values (*18*). Taking into account the inherent difficulties in both the latter methods, the agreement is reasonable.

Surface tensions γ can be obtained (*3*, *14*) from the density profiles using the same approximation for $g(\boldsymbol{r}_1, \boldsymbol{r}_2)$ as in the BGY equation. The results can be correlated (*14*) by

$$\gamma\sigma^2/\varepsilon = 2.33[1 - T/167.4]^{1.30} \tag{10}$$

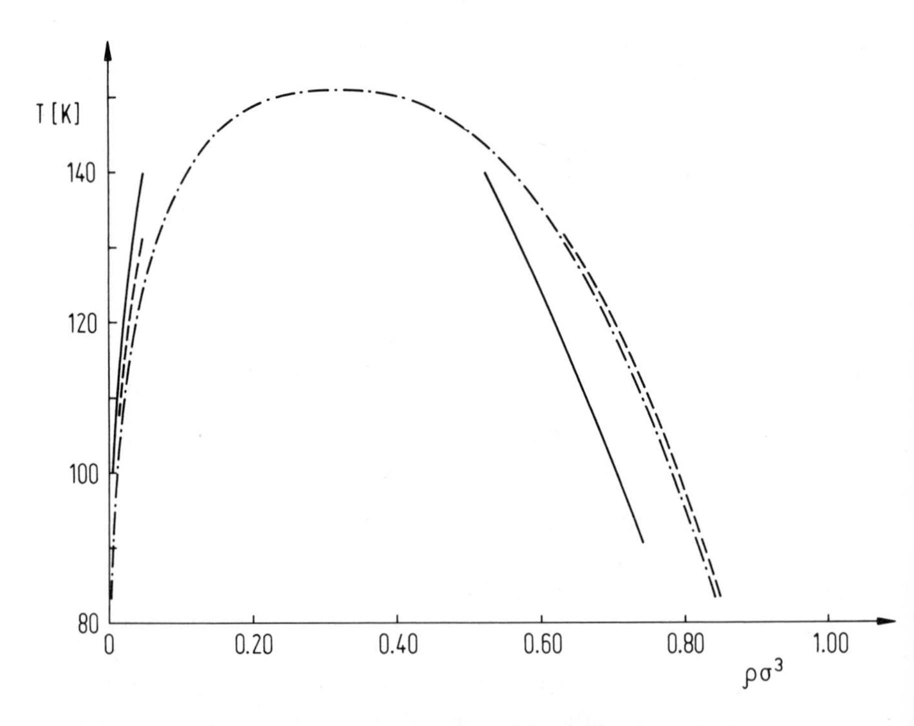

Figure 6. Coexistence curves determined by different methods. Key: experimental data for argon (16) *(— · — · —); Lennard–Jones system with equal chemical potential from simulations* (15) *(— — — —); Lennard–Jones system with interface investigations using the BGY equation (————). (Reproduced with permission from Ref. 14. Copyright 1981, Berichte der Bunsengesellschaft für Physikalische Chemie.)*

Comparison with experimental results (*19*) shows good agreement at low temperatures, but with increasing temperature the calculated values are much too high, which is reflected in a calculated critical temperature of 167.4 K. On the other hand, at all temperatures our results are in reasonable agreement with simulation results (*17*). The reason for these discrepancies is not yet clear.

Liquids in Contact with a Wall

The BGY equation can also be solved for liquids in contact with a wall (*3, 14*). In a previous paper (*14*) a systematic study was made for a Lennard–Jones liquid in contact with a 3-9 wall. The parameters of the wall potential were $\varepsilon_{gs}/k = 265$ K and $\sigma_{gs} = 0.234$ nm, so that it is much more shallow than in the case of gas adsorption. Density profiles were calculated again in a temperature and density grid. As can be seen in Reference 14, the bulk density has a considerable influence on the structure of the liquid close to the wall, while the influence of the temperature is negligible.

It is interesting to note (*3*) that with the same wall-particle potential a hard-sphere fluid is more strongly adsorbed than a Lennard–Jones fluid. This result agrees with findings of Snook and van Megen (*20*) and can be explained by the fact that for Lennard–Jones particles, the attractive "background" force of the bulk liquid balances the attractive wall force. This may help us in understanding adsorption from liquid solutions.

Acknowledgment

Part of this work was supported by the Deutsche Forschungsgemeinschaft under grant Fi 287/1.

Literature Cited

1. Van Kampen, N. G. *Physica* **1961,** *27*, 783.
2. Fischer, J. *Mol. Phys.* **1977,** *34*, 1237.
3. Fischer, J.; Methfessel, M. *Phys. Rev. A:* **1980,** 22, 2836.
4. Nieminen, R. M.; Ashcroft, N. W. *Phys. Rev. A:* **1981,** *24*, 560.
5. Davis, H. T.; Scriven, L. E. *Adv. Chem. Phys.* **1982,** *49*, 357.
6. Specovius, J.; Findenegg, G. H. *Ber. Bunsenges. Phys. Chem.* **1978,** *82*, 174.
7. Blümel, S.; Köster, F.; Findenegg, G. H. *Faraday Trans. II* **1982,** *78*, 1753.
8. Fischer, J.; Findenegg, G. H., to be published.
9. Fischer, J.; Findenegg, G. H.; Specovius, J. *Chem. Ing. Tech.* **1978,** *50*, 41.
10. Rowley, L. A.; Nicholson, D.; Parsonage, N. G. *Mol. Phys.* **1976,** *31*, 365.
11. Putnam, F. A.; Fort, T., Jr. *J. Phys. Chem.* **1977,** *81*, 2164.

12. Fischer, J.; Bohn, M.; Findenegg, G. H.; Körner, B. *Chem. Ing. Tech.* **1982,** *54*, 763; Engl. Trans., *Ger. Chem. Eng.*, in press.
13. Steele, W. A. "The Interaction of Gases with Solid Surfaces"; Pergamon Press: Oxford, 1974; p. 13.
14. Fischer, J. *Ber. Bunsenges. Phys. Chem.* **1981,** *85*, 1041.
15. Adams, D. J. *Mol. Phys.* **1976,** *32*, 647.
16. Angus, S.; Armstrong, B.; Eds. "International Thermodynamic Tables of the Fluid State, Argon"; Butterworths: London, 1971.
17. Chapela, G. A.; Saville, G.; Thompson, S. M.; Rowlinson, J. S. *J. Chem. Soc., Faraday Trans. II.* **1977,** *73*, 1133.
18. Beaglehole, D. *Phys. Rev. Lett.* **1979,** *43*, 2016.
19. Stansfield, D. *Proc. Phys. Soc., London* **1958,** *72*, 854.
20. Snook, I. K.; van Megen, W. *J. Chem. Phys.* **1979,** *70*, 3099.

RECEIVED for review January 27, 1982. ACCEPTED for publication September 9, 1982.

Fluctuations of Local Fluxes in Fluids: Simulation Versus Langevin Theory

J. KIEFER[1] and P. B. VISSCHER
University of Alabama, Department of Physics and Astronomy, University, AL 35486

Numerical simulations of flux fluctuations in a fluid have been performed. The results for momentum fluxes are in qualitative disagreement with the standard (Langevin–Landau–Lifshitz) theory of fluctuating hydrodynamics, which assumes stochastic flux correlations are local. Since the theoretical predictions for number-density fluctuations have been verified by light-scattering experiments, the difficulty appears to be related to the part of the momentum flux that does not couple to the density. We have calculated the stochastic fluxes explicitly from molecular dynamics data for soft spheres by using a recently developed discrete formulation of hydrodynamics. A possible explanation of the nonlocal correlations is described; this involves renormalization–group techniques related to those used in the theory of critical phenomena.

FLUCTUATIONS IN FLUIDS are calculated in this chapter by using a recently introduced, exactly renormalizable discrete formulation of hydrodynamics. Several descriptions of the discrete formulation of hydrodynamics have been published previously. Originally introduced as a discrete analogue of continuum hydrodynamics (*1*), it can also be thought of as an adaptation to fluid mechanics of the renormalization–group methods used in the theory of critical phenomena (*2*, *3*). However, an important advantage of the discrete formulation is that fluctuations can be treated more accurately than they can in continuum theories. Therefore, in the present paper we will approach discrete hydrodynamics as a discrete analogue of the stochastic Langevin equations used in fluctuating hydrodynamics (*4*, *5*). We will begin by reviewing the usual (Langevin–Landau–Lifshitz) form of these equations. We then show how the discrete formulation of hydrodynamics may be regarded as a concretization (for

[1] Current address: St. Bonaventure University, Department of Physics, St. Bonaventure, NY 14778

0065-2393/83/0204-0153$06.00/0

application to a real fluid) of the continuum Langevin equation. We show how this discrete Langevin equation can be directly determined by simulation. The results for the stochastic fluxes are then used to study the validity of the assumptions usually made in the Langevin equation.

In particular we find the assumption (due to Landau and Lifshitz) that there are no correlations between fluxes at different points to be qualitatively incorrect for the momentum fluxes.

Langevin–Landau–Lifshitz Equations of Fluctuating Hydrodynamics

The Langevin equation for the stochastic force on a Brownian particle (6) was first extended to describe fluctuations in fluids by Landau and Lifshitz (7). For simplicity, we will first discuss a Langevin equation for a single degree of freedom, say a diffusing density $\rho(\boldsymbol{r}, t)$. The usual deterministic equation of motion is

$$\left(\frac{\partial \rho}{\partial t}\right)^{m} = -\nabla \cdot \boldsymbol{j}^{m} \tag{1}$$

where the flux vector is

$$\boldsymbol{j}^{m}(\boldsymbol{r}, t) = -D \nabla \rho(\boldsymbol{r}, t) \tag{2}$$

and D is the diffusivity. The superscript m indicates that Equations 1 and 2 give only the mean values of $\partial\rho/\partial t$ and $\boldsymbol{j}$ {for a given function $\rho(\boldsymbol{r}, t)$}. The Langevin approach involves adding a stochastic term $(\partial\rho/\partial t)^{s}$ describing deviations from the mean

$$\frac{\partial \rho}{\partial t} = \left(\frac{\partial \rho}{\partial t}\right)^{m} + \left(\frac{\partial \rho}{\partial t}\right)^{s} \tag{3}$$

It is usually assumed (4), following Landau and Lifshitz (7), that there is a field $\boldsymbol{j}^{s}$ such that

$$\left(\frac{\partial \rho}{\partial t}\right)^{s} = -\nabla \cdot \boldsymbol{j}^{s} \tag{4}$$

with the additional properties

$$< \boldsymbol{j}^{s} > = 0 \tag{5}$$

$$< j^{s}_{i'}(\boldsymbol{r}', t')\, j^{s}_{i}(\boldsymbol{r}, t) > = A\delta(\boldsymbol{r}' - \boldsymbol{r})\delta(t' - t)\delta_{i'i} \tag{6}$$

for some constant A. It is important to understand why the stochastic flux $\boldsymbol{j}^s$ need be introduced; i.e., why $(\partial\rho/\partial t)^s$ cannot itself be assumed to have simple δ-function correlations as in Equation 6. This is because the requirement that the total number of particles be constant implies a constraint on the fluctuations

$$\int d^3\boldsymbol{r} < \frac{\partial\rho}{\partial t}(\boldsymbol{r}', t')^s \frac{\partial\rho}{\partial t}(\boldsymbol{r}, t)^s > = 0 \tag{7}$$

This constraint is automatically satisfied by any $(\partial\rho/\partial t)^s$ that is obtained as a divergence; Equation 4 in particular gives

$$< \frac{\partial\rho}{\partial t}(\boldsymbol{r}', t')^s \frac{\partial\rho}{\partial t}(\boldsymbol{r}, t)^s > = A\delta(t' - t)\nabla^2\delta(\boldsymbol{r}' - \boldsymbol{r}) \tag{8}$$

However, without the Laplacian ∇^2, Equation 8 would violate the conservation law constraint. We will reconsider this question, with benefit of hindsight, in the last section of this chapter.

The simplest statement of the Langevin equation for a conserved density is therefore

$$\frac{\partial\rho}{\partial t} = \nabla \cdot \boldsymbol{j} \tag{9}$$

$$\boldsymbol{j} = \boldsymbol{j}^m + \boldsymbol{j}^s \tag{10}$$

together with Equation 2 for the mean flux $\boldsymbol{j}^m$ and Equations 5 and 6 for the stochastic flux $\boldsymbol{j}^s$.

We may now generalize the above Langevin equations to describe a fluid. Essentially, there are five conserved densities instead of one; these are number, energy, and momentum (with three components), which we will denote by ρ_N, ρ_E, and ρ_P. There are therefore three versions of Equation 9, involving three fluxes $\boldsymbol{j}_N$, $\boldsymbol{j}_E$, and $\boldsymbol{j}_P$, the momentum flux being a 3×3 tensor. These equations were first proposed by Landau and Lifshitz (7), and were analyzed and extended to allow a bulk viscosity by Fox and Uhlenbeck (5); we will refer to them as the Langevin–Landau–Lifshitz equations.

Fox and Uhlenbeck (5) have given explicit expressions for the coefficient A in Equation 6, for each of the hydrodynamic fluxes. Their expressions for the fluctuations follow from a fluctuation–dissipation theorem if one makes the Landau–Lifshitz assumption (Equation 6) of complete independence of the fluxes at different space and time points. For the shear momentum flux, for example, they obtain

$$A = 2k_B T\eta \tag{11}$$

where η is the shear viscosity, k_B is the Boltzman constant, and T is the absolute temperature. Their result for longitudinal momentum flux is

$$A = 2k_B T\{(4/3)\eta + \zeta\} \tag{12}$$

where ζ is the bulk viscosity.

Discrete Hydrodynamics

How could one test the applicability of the Langevin–Landau–Lifshitz theory of continuum fluctuating hydrodynamics to a real fluid? It is first necessary to establish what the theory means in a real fluid of particles having finite size, in terms of quantities one could actually measure in a simulation (or, for that matter, in a laboratory experiment). It is by no means obvious how to do this; we present below what appears to us to be the most straightforward such concretization of the formal Langevin–Landau–Lifshitz theory. In a molecular dynamics simulation one can measure discrete analogues of the densities; that is, the number, energy, and momentum contents of finite cells, at a finite number of discrete times, say multiples of some interval τ. Let us divide our system into cubical cells of width W, each labeled by its center l. Denote the content of cell l at time $m\tau$ by $c(l,m)$. In a fluid we need number, energy, and momentum contents c_N, c_E, c_P. As in the previous section, however, we shall consider a simple diffusive system for simplicity. There is then only one content, c_N, and we may omit the subscript. The discrete analogue of the flux is the transfer, the number of particles (or amount of momentum, and so forth) crossing a square face separating two cells during the interval between two of our discrete times, say $m\tau$ and $m\tau + \tau$. Labeling each face by its midpoint f, and the time interval by its midpoint $(m + 1/2)\tau$, we may denote the corresponding transfer by $x(f,m + 1/2)$.

The Langevin–Landau–Lifshitz theory, like any hydrodynamic theory, is supposed (*4*) to describe a fluid on a scale that is large compared to molecular sizes. Thus the current $\boldsymbol{j}$ and the density ρ are limits of the transfer x and the content c (with suitable factors of W and τ) as $W, \tau \rightarrow \infty$. The meaning of Equations 9 and 10, as applied to a real fluid, is that if we replace $\boldsymbol{j}$ and $\partial\rho/\partial t$ by x and the content change $\Delta c(l) = c(l,m+1) - c(l,m)$, the resulting dynamic equation is correct in the limit $W, \tau \rightarrow \infty$. Although we can only test it for finite W and τ, we can hope to make inferences about the probable limit.

The above considerations do not tell us, however, what to do with $\boldsymbol{j}^m$. In the formal continuum theory one simply postulates a form for it (in terms of ρ). It is supposed to represent the mean flux for a given instantaneous density profile. This can be made precise in a discrete formulation by defining an ensemble of fixed contents $c(l,0)$ at $t = 0$, and denoting a mean transfer in this ensemble by $[x(f,1/2)]$. Then $[x(f,1/2)]$ is the natural analogue of $\boldsymbol{j}^m$. The analogue of Equation 2 for $\boldsymbol{j}^m$ should express the conditional mean transfer $[x(f,1/2)]$ as a linear function of the contents $c(l,0)$

$$[x(f,1/2)] = \sum_l [x(f,1/2)]_{c(l,0)}\, c(l,0) \qquad (13)$$

where the $[x(f,1/2)]_{c(l,0)}$ factors are simply constant coefficients about which we have not yet made any assumptions. Evidently they contain information about the diffusivity D. They are uniquely defined if we regard Equation 13 as the first term in a power-series expansion of the mean flux.

One can see from Equation 10 that the discrete analogue of the stochastic flux $\boldsymbol{j}^s$ is the difference $x(f,1/2) - [x(f,1/2)]$. Our analogue of Equation 6 should describe the moments of this quantity in the equilibrium ensemble. However, since the Langevin theory assumes that the probability distribution of the stochastic flux $\boldsymbol{j}^s(\boldsymbol{r},0)$ is independent of the density $\rho(\boldsymbol{r}',0)$, the moments in the equilibrium ensemble are the same as those in each ensemble of fixed $\rho(\boldsymbol{r}',0)$. Therefore, our fluctuation equation could equally well describe the moments in the ensemble of fixed contents $c(l,0)$,

$$[\{x(f',1/2) - [x(f',1/2)]\}\{x(f,1/2) - [x(f,1/2)]\}]$$

which we prefer because these are exactly the cumulant moments (8) and our fluctuation equation takes the form

$$[x(f',1/2)\,x(f,1/2)]^c = \text{function of } f',f \qquad (14)$$

This choice also results in an attractive unification of the two equations, Equations 13 and 14, which comprise our discrete analogue of the Langevin equation. They describe the first two cumulant moments of the probability distribution of the transfers $x(f,1/2)$ in the ensemble of fixed $c(l,0)$. From this point we will omit the c superscript; all moments will be cumulants.

A number of possible generalizations of the basic discrete Langevin equations (Equations 13 and 14) now present themselves. First, it is possible that a more accurate prediction of the mean transfer (Equation 13) could be made by taking into account some or all of the previous transfers $x(f,-1/2)$, $x(f,-3/2)$, . . . as well as the present contents $c(l,0)$. Let us denote by $[\ ]^1$ a mean in an ensemble in which the terms $x(f,-1/2)$ (for all f), as well as $c(l,0)$, are fixed. It is therefore a function of the values chosen for $c(l,0)$ and $x(f,-1/2)$. Similarly $[\ ]^2$ is a mean in an ensemble in which $c(l,0)$, $x(f,-1/2)$, $x(f,-3/2)$ are fixed; note that specifying $c(l,-1)$ would be redundant since it is determined by $c(l,0)$ and $x(f,-1/2)$. In general $[\ ]^M$ involves fixing transfers at M different times. The conditional means discussed previously can now be denoted $[\ ]^0$. The necessary generalization of the equation of motion for $[x]^0$ (Equation 13) is

$$[x(f,1/2)]^M = \sum_l [x(f,1/2)]^M_{c(l,0)}\, c(l,0) + \sum_{0>m>-M} [x(f,1/2)]^M_{x(f',m)}\, x(f',m) \qquad (15)$$

and Equation 14 becomes

$$[x(f',1/2)\,x(f,1/2)]^M = \text{function of } f',\, f,\, M \qquad (16)$$

Our principal interest is in the case $M = \infty$, for which all previous transfers are constrained and the equation of motion, Equation 15, involves all previous times. This is because the theory for that case is exactly renormalizable (*see* the section on Numerical Results).

The continuum analogue of the non-Markovian theory represented by Equation 15 with $M = \infty$ is referred to as "generalized hydrodynamics" and has been heavily studied by many authors (*9*, *10*).

The linear theory described by Equations 15 and 16 can easily be generalized to the nonlinear case. Each equation can be thought of as the lowest-order nonvanishing term in a power-series expansion of a conditional cumulant moment in terms of the "history variables" $c(l,0)$, $x(f,-1/2)$, ... In Equation 16 only the constant term is present, and in Equation 15 the constant term vanishes and the linear term is given. In the most general formulation of discrete hydrodynamics (*1*, *8*) all orders are allowed in these power series. The restriction to the first and second cumulant moment corresponds to assuming Gaussian fluctuations (*4*); this restriction also is removed in the general theory. In dealing with a fluid on a fairly macroscopic scale, the linear Gaussian case is sufficient; this is acceptable because the important fixed points under a scale-coarsening (renormalization) transformation (*3*, *11*) are linear and Gaussian. Even on the rather microscopic scale (four-particle cells) used in our simulation work, both non-Gaussian and nonlinear terms have been calculated (*12*) and found not to be numerically important. Therefore, in this chapter we consider only the linear, Gaussian parts of the equation of motion; i.e., Equations 15 and 16.

Before calculating, in the next section, the actual values of the parameters in the discrete equations of motion (Equations 15 and 16), we discuss what the continuum Langevin–Landau–Lifshitz theory of fluctuating hydrodynamics (Equations 2 and 6) predicts for them. The predictions about the mean transfers (Equation 15) are the same as those of the ordinary deterministic equations of hydrodynamics, and have been discussed in detail previously (*12–14*). Our interest here is in the Langevin fluctuations (Equation 6) of the stochastic flux component j_i^s (i = X, Y, or Z). Regarding j_i^s as being the large-cell limit of the transfer per unit time per unit area, across a face normal to the i direction, $x^s/\tau W^2$, we see that the assumption of independence of the j_i^s terms at different positions translates into an assumption of independence of the x terms.

$$[x(f',1/2)x(f,1/2)]^M = AW^2\tau\delta_{ff'} \tag{17}$$

(The factor $W^2\tau$ provides a normalization equivalent to that of Equation 6.) In a Markovian theory this cumulant is independent of M, since constraining M previous transfers makes no difference if nothing depends on them. This predicted independence of M is borne out fairly well by our numerical results covered in the next section.

Of course we cannot expect to find exact independence of transfers across different faces for finite W, even if the Langevin theory were correct. However, the Langevin theory or various generalizations (*9, 10*) certainly imply that the correlations should be clearly dominated by a local part for reasonable W and τ. As we see in the simulation this appears to be definitely false for the momentum fluxes.

Calculating Discrete Equations of Motion

We would now like to calculate explicitly the parameters defining the discrete equations of motion; i.e., the coefficients $[x(f,1/2)]^M_{c(l,0)}$ and $[x(f,1/2)]^M_{x(f',m)}$ of Equation 15 and the constants $[x(f',1/2)x(f,1/2)]^M$. The former describe the dependence of the expected transfer on the history of the system, as represented by the "history variables" $c(l,0)$ and $x(f',m)$. We will schematically denote any of these history variables by h, and suppress all space and some time indices. In previous discrete hydrodynamics work (*12, 14*) we have considered only the most general equation of motion having $M=\infty$ (i.e., allowing $x(1/2)$ to depend on all previous transfers). We obtained $[x]_h$ from equilibrium averages by multiplying Equation 15 (with $M=\infty$) by a history variable h' and averaging over the equilibrium ensemble

$$\langle xh'\rangle = \sum_h [x]_h^\infty \langle hh'\rangle \tag{18}$$

This is a set of linear equations from which $[x]_h^\infty$ can be calculated if the equilibrium averages $\langle xh\rangle$, $\langle hh'\rangle$ are known. However, the number of coupled equations is the number of $[x]_h$ terms we wish to calculate. This was 35 in Reference 14, but would be much larger if high accuracy were desired. For the present calculations we have used a more efficient method in which $[x]_h^M$ is calculated recursively from $[x]_h^{M-1}$, the equation of motion in which history variables are constrained at one fewer past times.

The necessary recursion equations can be obtained by multiplying Equation 15 by $x'(-M + 1/2)$ and averaging it over the "$M-1$" ensemble, which is larger than the "M" ensemble because $x(-M + 1/2)$ is not constrained

$$[x(1/2)x'(-M + 1/2)]^{M-1} = \sum_{x(-M+1/2)} [x(1/2)]^M_{x(-M+1/2)} \times [x(-M + 1/2)x'(-M + 1/2)]^{M-1} \quad (19)$$

This is again a coupled system of linear equations, one for each $[x]_h^M$, but now the h terms are all at a single time, so there are many fewer. The other $[x]_h^M$ terms (for h terms at times m, $-M + 1/2 < m \leqslant 0$) are determined by

$$[x(1/2)]_h^{M-1} = [x(1/2)]_h^M + \sum_{x(-M+1/2)} [x(1/2)]^M_{x(-M+1/2)}[x(-M + 1/2)]_h^{M-1} \quad (20)$$

which are not coupled at all; Equation 20 is obtained by averaging Equation 15 over the larger $M-1$ ensemble and picking out terms proportional to h. For the next stage of the recursion we must know fluctuations in the M ensemble, which are obtained by multiplying Equation 15 by $x(m)$ for $m = 1/2$ or $m < -M$, and averaging

$$[x(1/2)\,x(m)]^{M-1} = [x(1/2)\,x(m)]^M + \sum_{x(-M+1/2)} [x(1/2)]^M_{x(-m+1/2)}[x(-M + 1/2)\,x(m)]^{M-1} \quad (21)$$

This determines the fluctuation moment on the right side of Equation 19, provided one realizes the latter is equivalent, by time reflection in $t = (1-M)/2$, to $[x(1/2)x(1/2)]^{M-1}$.

The recursion equations (Equations 19–21) give the lowest-order power series coefficients for cumulant moments in a smaller ensemble from those in a larger ensemble. Generalizations of such equations to all orders are given in Reference 3, but are not needed here.

The lowest M for which the recursion relations in the form of Equations 19–21 make sense is $M = 1$. We must initially know the cumulants for $M = 0$; i.e., for the ensemble in which only the $c(0)$ terms are fixed. However, these may be calculated from the equilibrium cumulants by equations essentially identical to Equations 19 and 21 in which $M = 0$ and $x(-M + 1/2)$, the variables we are newly constraining, are replaced by $c(0)$. The cumulants for $M = -1$, of the form $[x(1/2)c(0)]^{-1}$ and $[c(0)c'(0)]^{-1}$, are averages in an ensemble in which even $c(0)$ is not fixed; i.e., the full equilibrium ensemble. Thus the recursion may begin with molecular dynamics data on fluctuations in the equilibrium ensemble ($M = -1$) and all other coefficients (for $M = 0, 1, 2, \ldots$) can be calculated.

We have obtained equation-of-motion coefficients from the equilibrium correlations obtained from simulation, by solving Equations 19–21. In principle, the number of equations is infinite, because the number of times (m's) is infinite. Elaborate schemes for choosing an appropriate truncation of the system of equations have been used previously (*12, 14*). In the present work we have used a simplification of the scheme of Reference 14. Our program starts with a list of tolerances within which we would like to know the equation-of-motion coefficients $[x]_h^M$ and $[x'x]^M$. These determine (assuming nearly diagonal matrices in Equation 19) the tolerances for each equation. Imposing this tolerance on each term in the sums in Equations 19–21 implies a tolerance for each factor. Each cumulant that exceeds its tolerance as obtained in any of these ways is regarded as important, and its "neighbors" (obtained by shifting h or x by one unit in space or time) are added to the truncation. We begin with a small "start-up" set of 183 cumulants (*14*), and repeat this truncation expansion procedure (with decreasing tolerances) until adequate convergence is obtained. The results given here required about 1000 cumulants.

Molecular Dynamics Simulation

The soft sphere fluid has been studied extensively using Monte Carlo (*15, 16*) and nonequilibrium molecular dynamics techniques (*17*), and therefore we have chosen this model as an example of the application of discrete hydrodynamics.

The soft sphere interaction potential is central and has the form

$$\phi(r) = \varepsilon \left(\frac{\sigma}{r} \right)^{12} \tag{22}$$

The separation distance between two particles is r, ε defines the energy scale, and σ is the effective hard core radius. It is convenient to define energy, length, and mass units so that $\varepsilon = \sigma = m = 1$ where m is the particle mass.

The state of the fluid system is specified by the mass density, ρ, and the absolute temperature T. However, since the potential function has a power law form, we can select $k_B T = \varepsilon = 1$ and the properties of the system at any other T can then be obtained via a scale transformation (*16*). The mass density (particle density, since $m = 1$) is specified by defining a dimensionless reduced density $\rho_r \equiv \rho\sigma^3/\sqrt{2}$. Then $\rho_r \leqslant 0.2$ represents the dilute fluid region where kinetic theory is applicable. Previous discrete hydrodynamics work has addressed the case of $\rho_r = 0.6$ (*12*). Here the viscosity, for instance, is about 50% higher than that predicted by the Enskog theory (*17*). At $\rho_r = 0.6$ discrete hydrodynamics yields a viscosity in apparent agreement with nonequilibrium molecular dynamics calculation (*12, 17*). The results discussed in this chapter refer to a reduced density of 0.8, just less than the freezing density of 0.813.

For the sake of economy we have chosen a system of 32 particles in a cube with periodic boundary conditions. The cube is divided into eight small cubical cells of side W, each containing on average four particles. Since we set $\rho_r = 0.8$, we must have $W = 1.523\sigma$. The periodic system is then a cube of width $2W$.

The molecular dynamics simulation is performed by integrating Newton's law numerically for each of the 32 particles in the periodic cube.

The integration algorithm used is that due to Verlet (*18*), which involves the particle positions $\boldsymbol{r}_i$ and velocities $\boldsymbol{v}_i$ as follows:

$$\boldsymbol{v}_i(t + (1/2)\Delta t) = \boldsymbol{v}_i(t - (1/2)\Delta t) + \Delta t\, \boldsymbol{F}_i(t)/m \tag{23}$$

and

$$\boldsymbol{r}_i(t + \Delta t) = \boldsymbol{r}_i(t) + \Delta t\, \boldsymbol{v}_i(t + (1/2)\Delta t) \tag{24}$$

where $F_i(t)$ is the force acting on particle i at time t, and Δt is the integration interval.

The data recorded on magnetic tape consist of small cell (width W) particle number, total energy, and momentum contents recorded at times $m\tau$, and the net amount of these quantities transported or transferred from one cell to a neighboring cell during the time interval $[(m-1)\tau, m\tau]$. Here m is an integer. The contents and transfers are calculated as described in detail elsewhere (*1*).

We have computed discrete equations of motion from our molecular dynamics averages by using two different discrete intervals. We previously reported (*14*) results for a time interval = 0.1981. This was chosen so that a sound wave just crosses the cell in time τ such that $\tau = W/v_s$, where the sound velocity $v_s = 7.69$ is obtained from the equation of state (*15, 16*) for $\rho_r = 0.8$. We report here results for $\tau = 0.1981$ and also for half that, $\tau = 0.0990$.

Computational economy motivates the use of as large an integration interval Δt as possible. The integration interval must be small enough, however, that numerical errors in computing the particle positions and velocities are not serious. We have used an interval $\Delta t_0 = 0.0099$ (1/10 of the smaller τ). This produces adequate total energy conservation (one part in 1000). The most relevant criterion for our purposes, however, is the convergence of the microcanonical averages of products of cell variables that are used in Equations 20 and 21 with $M = 0$. Theoretically, the errors in these averages should be of order Δt^2 in the Verlet algorithm. We have looked at several of them for increasing Δt (multiples of $\Delta t_0/2$), averaged over 2250 Δt_0. Within the statistical uncertainty, we detect no significant changes in the averages up to $\Delta t = 2.5\ \Delta t_0$. For $\Delta t = 3\Delta t_0$, our molecular dynamics code broke down (particles went further than W during Δt). This is strong evidence that our Δt was adequately small. The integration is started with the particles in a face-centered cubic lattice. Initial velocities are randomly specified such that the initial kinetic energy is about 3.0 $[k_BT]$. After 50 integration steps at this high temperature, the system is quenched slowly (over 100 Δt) until the average kinetic energy is about 1.5; the total energy is then 5.943 as it should be according to the equation of state (*15, 16*). From this point, data are taken from a simulation of 18,000 Δt_0. The average kinetic energy over that period is 1.486 ± 0.008.

Although the reduced density $\rho_r = 0.8$ is less than the infinite fluid freezing density, the simulated fluid froze three times in the course of the 18,000 integration steps because of the small size of the system and periodic boundary conditions. When this occurred, the simulation was stopped, and at a point preceding the phase change the system was perturbed by integrating for 400 steps of length $(1/2)\Delta t_0$. Then the simulation was resumed with steps Δt_0.

A cell variable (a content or a transfer) or a product of cell variables was averaged over rotations and translations consistent with the symmetry of the product (*19*), and over four separate time segments of length 4500 Δt. The statistical uncertainty of an overall average (over 18,000 Δt) was obtained by computing the standard deviation of the mean of the four time segments.

Numerical Results

We have calculated the discrete equation-of-motion coefficients $[x]^M$ and $[x'x]^M$ by using the recursion equations (Equations 19–21). Equation 19 is a matrix equation in which the matrix is the covariance matrix of the transfers at time $-M + 1/2$, denoted $x(-M + 1/2)$. Because of time-reversal symmetry, this is the same as the covariance matrix of $x(1/2)$ in the same ensemble, in which the contents $c(0)$ and (if $M>1$) $M-1$ sets of transfers $x(-1/2), \ldots x(-M + 3/2)$ are constrained. It was our original hope that the Landau–Lifshitz assumption of uncorrelated stochastic transfers (Equation 17) would be approximately correct, because then the covariance matrix would be nearly diagonal and the system of equations well conditioned. Unfortunately the opposite is the case; the transfers are highly correlated, so much so that the matrix is very nearly singular. Such singularities are common in the literature of linear regression (*20*) (Equation 19 is essentially a linear regression equation) and are referred to as "multicollinearity." This occurs when the transfers are nearly linearly dependent; i.e., when there is a linear combination of the transfers that fluctuates much less than the individual transfers do. One symptom of multicollinearity is that the smallest eigenvalue of the matrix is much smaller than the diagonal elements; in our case, it is as low as one tenth of the diagonal elements. We have encountered this problem previously (*14*), but it occurred in the covariance matrix of all history variables, in which its physical significance was not obvious. In our present formulation we deal with much smaller matrices involving only transfers at time $-M + 1/2$. We find that the eigenvectors having small eigenvalues correspond almost exactly to the sum of the six transfers into each cell. That is, the nonfluctuating linear combination is the content change

$$\Delta c(l) = c(l,1) - c(l,0) \tag{25}$$

Some of our numerical results demonstrating these strong correlations are presented in Tables I and II. Table I gives the cumulant moments of the momentum transfers and content changes in various ensembles. It can be seen that the Langevin–Landau–Lifshitz predictions (Equation 17, labeled LLL in the table) are fairly good for the mean square fluctuations of the transverse and longitudinal momentum transfers x_T and x_L. However, the predictions (zero) for their correlations are quite wrong. We give the largest correlation $[x_L x_T]$ in the table; the correlations between two different x_L terms or two different x_T terms are smaller. The seriousness of the discrepancy is seen most clearly by comparing the first and second columns in Table I. The second column gives the value that $[\Delta c^2]$ would have if the transfers fluctuated independently (since Δc is a sum of four x_T terms and two x_L terms). It is very far from the actual value of $[\Delta c^2]$ (first column) even in the equilibrium ensemble, and is wrong by a factor of 10 in the most constrained ensembles. The latter are precisely the ensembles whose fluctuations should correspond most closely to the Langevin notion of "stochastic flux", since non-Markovian as well as Markovian deterministic effects have been subtracted out; as described in the section on Discrete Hydrodynamics. (*See also* Figure 1.)

One might note that the discrepancy in $[\Delta c^2]$ is less for the smaller

Table I. Fluctuations of Z-Momentum Transfers and Content Changes

Ensemble	$[\Delta c^2]^M$	$4[x_T^2]^M + 2[x_L^2]^M$	$[x_T^2]^M$	$[x_L^2]^M$	$[x^T x^L]^M$
τ = 0.198					
Equilibrium					
(M = −1)	8.00	35.40	4.16	9.38	−1.59
M = 0	3.40	32.73	4.00	8.36	−1.60
M = 1	3.05	27.30	3.43	6.78	−1.21
M = 2	2.66	26.87	3.37	6.69	−1.21
Uncertainties[a]	±0.02	±0.42	±0.07	±0.14	±0.05
LLL Predictions[b]	33.00	33.00	4.96	6.61	0.00
τ = 0.099					
Equilibrium					
(M = −1)	6.23	14.08	1.63	3.78	−0.46
M = 0	3.52	12.55	1.57	3.14	−0.50
M = 1	3.04	10.43	1.34	2.55	−0.33
M = 2	2.00	10.24	1.32	2.47	−0.33
Uncertainties[a]	±0.02	±0.14	±0.02	±0.05	±0.02
LLL Predictions[b]	16.50	16.50	2.48	3.30	0.00

Note: Content change is $\Delta c = c(1) - c(0)$, and $x_L(1/2)$ and $x_T(1/2)$ are the longitudinal and transverse transfers across the faces shown in Figure 1. M = 0, 1, and 2 are successively more constrained ensembles. The top and bottom halves of the table give large-τ and small-τ results respectively. For comparison, the equilibrium mean square content is $\langle c^2 \rangle = 3.54 \pm 0.02$.

[a] All quoted uncertainties are standard deviations of the mean for several runs.

[b] The Langevin–Landau–Lifshitz (LLL) prediction is from Equations 11, 12, and 17; we use $\eta = 5.4$ (*17*) and $\zeta \ll \eta$ (*21*).

Table II. Fluctuations of Energy and Number Transfers and Content Changes

Ensemble	$[\Delta c_E^2]^M$	$6[x_E^2]^M$	$[\Delta c_N^2]^M$	$6[x_N^2]^M$
τ = 0.198				
Equilibrium	24.7	81.9	1.37	1.62
M = 0	11.6	70.8	0.87	1.43
M = 1	11.2	69.6	0.86	1.32
M = 2	11.1	69.2	0.83	1.28
Uncertainties	±0.2	±1.0	±0.02	±0.05
τ = 0.099				
Equilibrium	19.4	35.3	0.98	1.04
M = 0	11.2	28.6	0.72	0.94
M = 1	10.9	27.1	0.72	0.89
M = 2	10.3	26.4	0.71	0.86
Uncertainties	±0.2	±0.6	±0.01	±0.03

Note: Content change Δc_E is the sum of six transfers x_E across the faces of the cell. In equilibrium ensemble, $\langle c_E^2 \rangle = 11.5 \pm 0.2$ and $\langle c_N^2 \rangle = 0.98 \pm 0.01$.

τ used and suggest that our τ's are simply too large to allow us to identify the flux with the transfer divided by τ and the area. Indeed, it is clear the Langevin–Landau–Lifshitz prediction (Equation 17) for $[\Delta c^2]$ cannot possibly be right for very large τ, since it is proportional to τ while the actual $[\Delta c^2]$ (which is equal to $[c(1)^2]$ if $M \gg 0$; i.e., if $c(0)$ is constrained) is bounded by $\langle c(1)^2 \rangle$ independently of τ (since constraining a variable decreases the fluctuations of the others in any Gaussian distribution). However, we do not believe decreasing τ is physically sensible. For the smaller $\tau = 0.099$ we have used, the transport of shear momentum out of a cell ($[x]_c$ in Figure 2) is already very small, and any smaller τ would make it difficult to extract information on the viscosity (*12*). Furthermore, $\tau = 0.099$ is very small by any physical criterion; the distance a particle moves at the thermal velocity $(kT/m)^{1/2}$ during this time is only 0.065 times the cell width W, or 0.092 times the close-packed interparticle spacing. Even a sound wave moves only half the cell width.

Further evidence that the strong correlations we have found are unavoidable is provided by work that has been done (*3*) on the properties of discrete equations of motion under scale-coarsening transformations. These are similar to the renormalization–group transformations used in

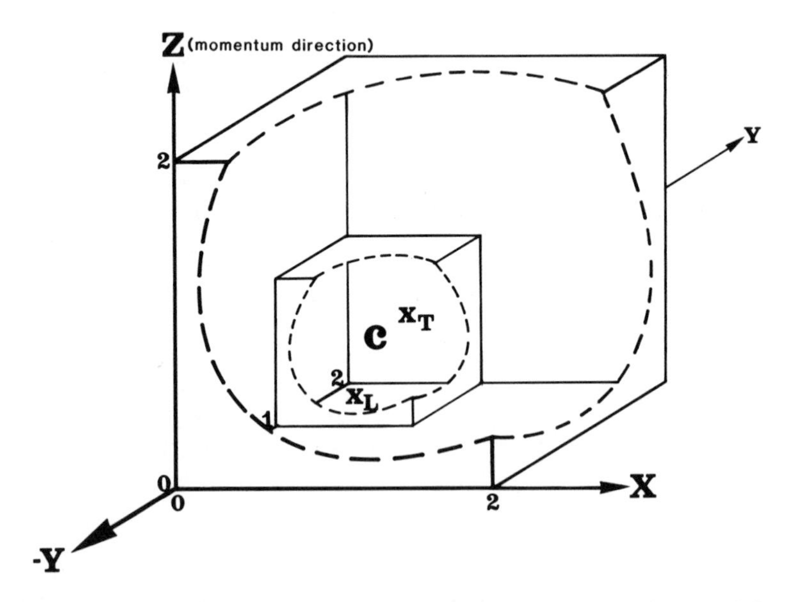

Figure 1. The molecular dynamics system (large cube) with one of the small cells used in analyzing the fluid motion. The cell is labeled c *since Table I refers to its Z-momentum content* c_{PZ}. *The Z-momentum transfer across the bottom face is* x_L *(longitudinal), and that across the back face is* x_T *(transverse). The length unit along the axes is* W, *the cell width.*

$\tau = 0.198$

$[x_T]_c$

Y = 3/2

0.090 ± 0.003	0.032 ± 0.001
0.103 ± 0.003	0.017 ± 0.001

$[x_T]_x$ (transverse)

Y = 2

−0.181 ± 0.010 −0.053 ± 0.010	−0.064 ± 0.007 −0.027 ± 0.010
0.194 ± 0.010 0.043 ± 0.010	−0.018 ± 0.009 −0.036 ± 0.009

Y = 1

−0.046 ± 0.007 −0.004 ± 0.007	0.036 ± 0.007 −0.041 ± 0.008
−0.059 ± 0.007 −0.004 ± 0.006	−0.032 ± 0.006 0.009 ± 0.010

$[x_T]_x$ (longitudinal)

Y = 3/2, Z = 0

0.013 ± 0.001 0.001 ± 0.003	−0.003 ± 0.001 0.011 ± 0.002

Figure 2. Discrete equation-of-motion coefficients $[x_T(1/2)]_{c(0,l)}$ *and* $[x_T(1/2)]_{x(m,l)}$.

Here $x_T(1/2)$ *is the transfer of Z-momentum across the face entered at* $f = (1/2, 2, 1/2)$ *and labeled* x_T *in Figure 1; it is* $x_{PZ}(f,1/2)$ *in the notation of the section on calculations. The four-compartment squares are cross sections of the system (the large cube in Figure 1) in the plane of the figure. The first square gives the dependences on the Z-momentum contents,* $[x_T]_{c(l,0)}$, *for the four cells whose centers* l *lie on the plane* Y = 3/2. *The next two squares give the dependences on previous transverse transfers,* $[x_T]_{x(f,-1/2)}$ *(upper number in each compartment) and* $[x_T]_{x(f,-3/2)}$ *(lower number) for faces whose centers* f *lie on the planes* Y = 2 *and* Y = 1. *There are only two inequivalent longitudinal transfers, that labeled* x_L *in Figure 1 (centered at* f = {1/2, 3/2, 0}) *and the one to its right* {f = (3/2, 3/2, 0)}. *The effects* $[x_T]_{x(f,m)}$ *of these transfers are given in the final two-compartment rectangle, which may be thought of as lying horizontally on the bottom of Figure 1.*

Continued on next page

$\tau = 0.099$

$[x_T]_c$

$Y = 3/2$

+0.037 ± 0.001	+0.006 ± 0.002
+0.077 ± 0.002	+0.009 ± 0.001

$[x_T]_x$ (transverse)

$Y = 2$

−0.124 ± 0.020 −0.050 ± 0.010	−0.035 ± 0.005 −0.030 ± 0.006
0.218 ± 0.010 0.027 ± 0.009	−0.017 ± 0.005 −0.001 ± 0.004

$Y = 1$

−0.098 ± 0.009 −0.037 ± 0.004	−0.047 ± 0.006 −0.026 ± 0.005
−0.051 ± 0.006 −0.044 ± 0.004	−0.022 ± 0.044 −0.038 ± 0.003

$[x_T]_x$ (longitudinal)

$Y = 3/2,\ Z = 0$

0.009 ± 0.001 −0.011 ± 0.002	0.046 ± 0.003 −0.004 ± 0.003

Figure 2. Continued. *Discrete equation-of-motion coefficients* $[x_T(1/2)]_{c(0,l)}$ *and* $[x_T(1/2)]_{x(m,l)}$.

the theory of critical phenomena. The ultimate objective of that work is to see if the equations of continuum hydrodynamics may be understood as a fixed point of a coarsening transformation; i.e., an equation of motion that stays the same when the time and space scales are coarsened, and that various different small-scale discrete equations of motion of real fluids approach when the scale becomes large. The coarsening behavior has not yet been investigated for fluids, but work (unpublished) on a two-dimensional diffusive system indicates that as the time and space scales increase, the transfer correlations increase in such a way that the ratio of the mean-square content change to the mean-square transfer becomes arbitrarily small. This suggests that the small values we have found for this ratio in our fluid represent generic behavior, and that the violation of the Langevin–Landau–Lifshitz prediction that it implies becomes more rather than less severe on larger scales.

Table II gives data on the fluctuations of energy and number. It can be seen that the Langevin–Landau–Lifshitz prediction of uncorrelated transfers again fails for the energy transfers; the content change it predicts (second column of Table II) is too large by as much as a factor of six. The transfer correlations are less severe for number transfers, which is consistent with the success of the theory in predicting density fluctuations measured in inelastic light scattering experiments (*21*). We do not give a numerical fluctuating-hydrodynamics prediction for the energy transfer, since Reference 4 does not (a sort of "temperature-flux" is given instead).

The order-of-magnitude discrepancy in the momentum–content change fluctuation is particularly distressing because the content (i.e. momentum density) is more directly physically measurable than the transfer (i.e. momentum flux). The discrepancy calls into question our decision to regard the transfer as the primary fluctuating variable (i.e., to use Equation 14), and then derive the content change fluctuations from it. This was done (*1, 2, 12, 14*) for the same reason that the Langevin–Landau–Lifshitz continuum theory regarded the flux rather than the density as having the simplest fluctuations; that is, some nonlocal correlations in the content changes are forced by the conservation law $\sum_{l} \Delta c(l) = 0$. This was particularly awkward in older formulations of discrete hydrodynamics (*1*) which parameterized probability distribution functions directly, but it is not a serious complication when they are parameterized through moments such as Equation 14. As we mentioned above, the strongly nonlocal flux correlations are of interest not only in relation to the Langevin–Landau–Lifshitz theory, but also for the practical reason that they make the discrete hydrodynamic equation of motion very difficult to calculate. They have severely limited the success of transport coefficient calculations based on this method (*2, 12, 14*). The fact that the smallest eigenvalues of the covariance matrix are 10 times smaller than the diagonal elements has the effect of magnifying small errors in the matrices by a factor of 10. This increases the number of terms which must be included in the truncated Equation 19; to reduce truncation errors below the statistical errors quoted here we had to use essentially all possible cumulants for $M = 0, 1, 2$. Fortunately, in a $2 \times 2 \times 2$ cell system there are only about 1000. Clearly, however, it would be better to work with variables that are more nearly independent. The content changes are essentially the longitudinal part of the momentum flux; one can define other linear combinations of transfers that are transverse and do not couple to the density. Hydrodynamic equations based on such variables would probably be simpler and easier to calculate (the matrices would not be so singular) and should be considered in future work in this area.

Figure 2 gives a few of the coefficients in the deterministic part of the equation of motion (Equation 15). To save space we give only $[x_T]_h$ where x_T is a transverse momentum transfer (of Z-momentum in the Y-direction) and h is a content or transfer of Z-momentum. These are exactly the coefficients one would need to calculate the viscosity (*12–14*). However, the present results do not give a good estimate of the viscosity—it is lower by a factor of two than that of Ashurst and Hoover (*17*). This may be because the numerical problems described above made it impossible to go beyond $M = 2$, or because a 32-particle system is too small, or (less likely) because 4-particle cells are too small and the viscosity will renormalize as the cell size is coarsened (*3, 8*).

In conclusion, it appears from the present work that the Langevin assumption of uncorrelated flux fluctuations is not valid in a fluid, at least not for momentum fluxes. Proper calculation of the equation of motion, and therefore the transport properties, would be greatly facilitated by a better understanding of the fluctuations.

Acknowledgments

For support, the authors would like to thank the National Science Foundation, under grant CHE-70906649, and the University of Alabama Research Grants Committee, under project number 1082.

Literature Cited

1. Visscher, P. B. *J. Stat. Phys.* **1978,** *18*, 59.
2. Visscher, P. B.; Kiefer, J. In "Proceedings of the 8th Symposium on Thermophysical Properties"; Sengers, J. V., Ed., American Society of Mechanical Engineers: New York; 1982, to be published.
3. Visscher, P. B. *J. Stat. Phys.* **1981,** *25*, 211.
4. Fox, R. F. *Physics Reports* **1978,** *48*, 179 (review).
5. Fox, R. F.; Uhlenbeck, G. E. *Phys. Fluids* **1970,** *13*, 1893.
6. Chandrasekhar, S. *Rev. Mod. Phys.* **1943,** *15*, 1.
7. Landau, L. D.; Lifshitz, E. M. "Fluid Mechanics"; Addison-Wesley: New York; 1959,
8. Visscher, P. B. *Physica* **1979,** *97A*, 410.
9. Boon, J. P.; Yip, S. "Molecular Hydrodynamics"; McGraw: New York, 1980.
10. Mori, H. *Progr. Theor. Phys.* **1965,** *33*, 423.
11. Visscher, P. B. *Physica* **1981,** *108A*, 152.
12. Begum, S.; Visscher, P. B. *J. Stat. Phys.* **1979,** *20*, 641.
13. Visscher, P. B. *J. Chem. Phys.* **1978,** *68*, 750.
14. Kiefer, J.; Visscher, P. B. *J. Stat. Phys.* **1982,** *27*, 389.
15. Hoover, W. G.; Ross, M.; Johnson, K. W.; Henderson, D.; Barker, J. A.; Brown, B. C. *J. Chem. Phys.* **1970,** *52*, 4931.
16. Hoover, W. G.; Gray, S. G.; Johnson, K. W. *J. Chem. Phys.* **1971,** *55*, 1128.
17. Ashurst, W. T.; Hoover, W. G. *Phys. Rev.* **1975,** *11A*, 658.
18. Verlet, L. *Phys. Rev.* **1967,** *159*, 98.

19. Visscher, P. B. *J. Stat. Phys.* **1979,** *20*, 629.
20. Draper, N. R.; Smith, H. "Applied Regression Analysis"; Wiley: New York, 1966.
21. Hoover, W. G.; Ladd, A. J. C.; Hickman, R. B. *Phys. Rev.* **1980,** *21A*, 1756.
22. Berne, B. J.; Pecora, R. "Dynamic Light Scattering"; Wiley: New York, 1976.

RECEIVED for review January 27, 1982. ACCEPTED for publication September 24, 1982.

8

Equations of State of Nonspherical Hard-Body Systems

TOMÁŠ BOUBLÍK

Czechoslovak Academy of Sciences, Institute of Chemical Process Fundamentals, 16502 Prague, Czechoslovakia

Derivation of the hard-body equation of state within the scaled particle theory for fluids of fused hard-sphere molecules and hard convex bodies is briefly summarized, and the modified equations of state are discussed. Computer data for virial coefficients and the compressibility factor of fluids assuming hard convex bodies and hard interaction site models of different types are used to verify the applicability of these equations.

THE EQUILIBRIUM BEHAVIOR OF MOLECULAR FLUIDS, that is, systems with spherically unsymmetrical intermolecular forces, has received increasing interest over the past few years. Because the structure of liquids is affected largely by the short-range repulsive forces, knowledge of the behavior of hard nonspherical bodies (with shape and size corresponding to the structure of the molecules considered) forms the basis of our understanding of the equilibrium behavior of real fluids just as knowledge of hard spheres does in case of simple fluids.

Repulsive forces of polyatomic molecules have been described essentially in two ways:

1. as a sum of the site–site interactions (dependent on the respective site–site distances) in the interaction site model (ISM).
2. by the generalized Kihara potential, in which intermolecular forces are assumed to depend only on the shortest surface-to-surface distance of hard convex cores, ascribed to given molecules.

Accordingly, hard interaction site models (HISM) (in effect fused hard spheres) and hard convex bodies (convex bodies parallel to the cores) have been considered in theoretical and simulation studies.

In this work we give a brief outline of the derivation of the hard-body equations of state and of the expression for the contact value of distribution functions. We then apply two of the discussed equations to

0065-2393/83/0204-0173$06.00/0

determine virial coefficients and the compressibility factor of the hard-convex-body and hard-interaction-site-model systems (pure fluids and mixtures) for which computer data are at our disposal.

Equation of State from the Extended Scaled Particle Theory

The scaled particle theory (SPT) formulated for simple fluids by Reiss et al. (*1*), yielded the first reliable equation of state of hard spheres, which is identical with the Percus–Yevick (c) [PY(c)] expression obtained several years later. In this theory a variable diameter is taken as a coupling parameter (similar to the relationship for the chemical potential) that makes it possible to express the reversible work connected with the introduction of a particle into the system under study. Several exact relations were found for this reversible work (for the particle diameter equal to zero or infinity) and an expansion in powers of the reciprocal diameter of the test particle was used for the interpolation. A weak point of the theory, which was extended also to mixtures, is the fact that it yields distribution functions only at the closest approach distance.

An extension of the scaled particle theory to a class of molecular fluids—hard convex bodies—was given first by Gibbons (*2, 3*) and re-derived by the present author (*4*); recently it was shown (*5*) that the same formalism can be used also for the HISM systems. It appears that both the HISM and the hard convex body pair-potentials can be written in terms of a single variable—the shortest surface-to-surface distance, s, where

$$u(s) = \infty \qquad \text{for } s \leq 0 \qquad (1)$$
$$= 0 \qquad s > 0$$

In the case of HISM, s is equal to a minimum of all the site–site distances, $s = \min(r_{12}^{\alpha\gamma} - \sigma^{\alpha\gamma})$, where $\sigma^{\alpha\gamma}$ is the characteristic distance. Then, in general relationships for pressure, the chemical potential, and further thermodynamic functions of molecular fluids

$$PV/NkT = 1 - (\rho/6kT) \int [R_{12}(\partial u/\partial R_{12})] g_m(\boldsymbol{R}_{12}, \boldsymbol{\omega}_1, \boldsymbol{\omega}_2) d\boldsymbol{R}_{12} d\boldsymbol{\omega}_1 d\boldsymbol{\omega}_2 \qquad (2)$$

$$(\mu - \mu^*)/kT = (\rho/kT) \int_a^b \int (\partial u/\partial \xi) g_m(\boldsymbol{R}_{12}, \boldsymbol{\omega}_1, \boldsymbol{\omega}_2, \xi)\, d\boldsymbol{R}_{12} d\boldsymbol{\omega}_1 d\boldsymbol{\omega}_2 d\xi \qquad (3)$$

and so forth, where R_{12} is the center-to-center distance, $\boldsymbol{\omega} = \theta\varphi\phi$ stands for the orientation coordinates, $\int d\boldsymbol{\omega} = 1$, and ξ is the coupling parameter that scales the test particle size while its shape remains unchanged. It is possible to express the derivatives with respect to R_{12} and ξ in terms of derivatives with respect to s, and the volume element $d\boldsymbol{R}_{12}$ in terms of s and the angle coordinates characterizing the geometry of the given pair of molecules, 1 and 2. When the coordinate system is fixed in particle 1 it can be written

$$PV/NkT = 1 - (\rho/6kT) \int \langle \boldsymbol{R}_{12}\boldsymbol{\nu} \rangle (\partial u/\partial s) g^{\mathrm{av}}(s) S_{1+s+2} ds \tag{4}$$

and

$$(\mu - \mu^*)/kT = -\ln(1 - \rho V_2) - (\rho/kT) \int_0^1 \int \langle \boldsymbol{R}_1 \boldsymbol{\nu} \rangle (\partial u/\partial s) g^{\mathrm{av}}(s, \lambda) S_{\lambda 1+s+2} ds d\lambda \tag{5}$$

Here λ is the dilatation coefficient employed as the coupling parameter, $\boldsymbol{\nu}$ is the unit vector in the direction of $\boldsymbol{s}$ and g^{av} is the weighted average (over surface area) correlation function, S_{1+s+2} is the mean surface area, given as a locus of the center of molecule 2 when it moves around molecule 1 with the given distance s. This surface area is (*4–6*)

$$S_{1+s+2} = S_{1+2} + 8\pi \mathcal{R}_{1+2} s + 4\pi s^2 \tag{6}$$

$$S_{1+2} = S_1 + 8\pi \mathcal{R}_1 \mathcal{R}_2 + S_2 \tag{7}$$

and

$$\mathcal{R}_{1+2} = \mathcal{R}_1 + \mathcal{R}_2 \tag{8}$$

where V_i, S_i and $\mathcal{R}_1$ stand for volume, surface area, and the $(1/4\ \pi)$ multiple of the mean surface integral, respectively.

Because $(\partial u/\partial s)$ possesses properties of the Dirac δ-function for hard body systems, it holds that

$$PV/NkT = 1 + \frac{1}{6} \rho \sigma^{\mathrm{av}} g^{\mathrm{av}}(0)\ S_{1+2} \tag{9}$$

and

$$(\mu - \mu^*)/kT = -\ln(1 - \rho V_i) + \rho \int_0^1 \langle \boldsymbol{R}_1 \boldsymbol{\nu} \rangle g^{\mathrm{av}}(0, \lambda) S_{\lambda 1+2} d\lambda \tag{10}$$

where $\sigma^{\mathrm{av}} = \langle \boldsymbol{R}_{12}^{c} \boldsymbol{\nu} \rangle = \langle \boldsymbol{R}_1 \boldsymbol{\nu} \rangle + \langle \boldsymbol{R}_2 \boldsymbol{\nu} \rangle$.

To derive the equation of state it is useful to introduce the average correlation function $G(\lambda)$ weighted over volume elements; then

$$PV/NkT = 1 + \frac{1}{2}G(1)V_{1+2} \tag{11}$$

The mean volume V_{1+2} of two hard bodies at contact is (3–5)

$$V_{1+2} = V_1 + S_1\mathcal{R}_2 + \mathcal{R}_1S_2 + V_2 \tag{12}$$

Equation 12 holds exactly for hard convex bodies, whereas for HISM systems a small volume Δv_{12} is neglected; Δv_{12} can be calculated exactly (in special cases) or approximately (5), but it can not be factored into contributions of HISM bodies 1 and 2. Because of this, the second virial coefficient of HISM from the resulting equation of state differs from the exact one (unless Δv_{12} is added) and the difference indicates the accuracy of this approximation.

For the chemical potential we have similarly

$$(\mu - \mu^*)/kT = -\ln(1 - \rho V_i) + \int_0^1 G(\lambda)(3\lambda^2V_1 + 2\lambda S_1\mathcal{R}_2 + \mathcal{R}_1S_2)d\lambda \tag{13}$$

The knowledge of $G(\lambda)$ in a relatively narrow interval $\lambda\varepsilon\langle 0, 1\rangle$ suffices for the determination of the equation of state and the chemical potential; instead, the values of $G(\lambda)$ for $\lambda = 0$ and $\lambda = \infty$ and $(\partial G/\partial\lambda)$ for $\lambda = 0$ are at our disposal. To make use of these relations a suitable three-constant interpolation formula (an analogue of the expansion by Reiss et al.) was considered. After some rearrangement it follows that

$$G(1) = \frac{1}{1 - y} + \frac{3\alpha(1 + \alpha)y}{(1 + 3\alpha)(1 - y)^2} + \frac{3\alpha^2y^2}{(1 + 3\alpha)(1 - y)^3} \tag{14}$$

where

$$y = \rho V_i \quad \text{and} \quad \alpha = \mathcal{R}_iS_i/3V_i \tag{15}$$

By substituting Equation 14 into Equation 11

$$\frac{PV}{NkT} = \frac{1}{1 - y} + \frac{3\alpha y}{(1 - y)^2} + \frac{3\alpha^2y^2}{(1 - y)^3} \tag{16}$$

It is obvious that the equation of state depends only on the packing fraction, y, and the nonsphericity parameter, α. Because of the close interrelation of G and g^{av} the theory yields also the expression for the contact value of the latter function

$$g^{av}(0) = \frac{1}{1-y} + \frac{3\alpha y}{2(1-y)^2} + \frac{3\alpha^2 y^2}{2(1 + 4\pi\mathcal{R}_i/S_i)(1-y)^3} \tag{17}$$

The equation of state, Equation 16, is an analogue of the scaled particle theory or PY(c) expression for hard spheres; it is known that the most often used Carnahan–Starling equation of state of a hard-sphere system can be obtained as a sum of 2/3 of the PY(c) and 1/3 of the PY(v) expressions, or—as the present author pointed out (*7*)—by multiplying the last term of Equation 14 (for $\alpha = 1$) by a factor of 2/3. By reducing this term in the same way in the general relationship we obtained (*8*)

$$\frac{PV}{NkT} = \frac{1}{1-y} + \frac{3\alpha y}{(1-y)^2} + \frac{3\alpha^2 y^2 - \alpha^2 y^3}{(1-y)^3} \tag{18}$$

This equation of state, Equation 18, can be considered as an extension of the Carnahan–Starling relationship to a general case of the nonspherical hard-body system. The corresponding values of G and g^{av} follow from Equations 14 and 17 by the above-mentioned reduction; for the reduced virial coefficients, $B_k^* = B_k/V_i^{k-1}$, it holds that

$$\begin{aligned} B_2^* &= 1 + 3\alpha \\ B_3^* &= 1 + 6\alpha + 3\alpha^2 \\ B_4^* &= 1 + 9\alpha + 8\alpha^2 \end{aligned} \tag{19}$$

and so forth. Equation 18 was extended also to mixtures of hard nonspherical bodies:

$$\frac{PV}{NkT} = \frac{1}{1-v} + \frac{rs}{\rho(1-v)^2} + \frac{qs^2(3-v)}{9\rho(1-v)^3} \tag{20}$$

where $v = \rho\Sigma x_i V_i$ is the packing fraction, $r = \rho\Sigma x_i \mathcal{R}_i$, $s = \rho\Sigma x_i S_i$, and $q = \rho\Sigma x_i \mathcal{R}_i^2$. [The two-dimensional equation was also derived (*9*)].

Equations 18 and 20 are generally valid for all the hard-body fluids and fluid mixtures. As shown below, Equation 18 gives a good prediction of the *P–V–T* behavior of the HISM systems; the accuracy of values of higher virial coefficients and the compressibility factor of hard convex bodies for $\alpha > 1.2$ is less satisfactory. Therefore, modified equations of state of convex bodies were developed.

Equations of State of Hard Convex Bodies

All the modified equations of state proposed for hard convex body systems start with Equation 18, which can be written

$$\frac{PV}{NkT} = \frac{1}{1-y} + \frac{C_2 y}{(1-y)^2} + \frac{C_3 y^2 + C_4 y^3}{(1-y)^3} \tag{21}$$

Nezbeda (*10*) determined C_2–C_4 by fitting the pseudoexperimental data of virial coefficients and the compressibility factors to simple relationships in terms of α. Then

$$\frac{PV}{NkT} = \frac{1}{1-y} + \frac{3\alpha y}{(1-y)^2} + \frac{(\alpha^2 + 4\alpha - 2)y^2 - \alpha(5\alpha - 4)y^3}{(1-y)^3} \tag{22}$$

Equation 22 and its extension to mixtures by Pavlíček et al. (*11*) yield very good results even for extreme values of α (and y); however, the functional dependence of the higher virial coefficients on α is rather strange. For example

$$\begin{aligned} B_3^* &= -1 + 10\alpha + \alpha^2 \\ B_4^* &= -5 + 25\alpha - 2\alpha^2 \end{aligned} \tag{23}$$

Similarly, the expressions for the contact values of the correlation function g^{av} and g_{ij}^{av} and the equation of state of mixtures possess complicated forms.

In another version Nezbeda et al. (*12*) proposed to determine the coefficients C_2–C_4 from the pseudoexperimental data of virial coefficients by employing the relationships

$$\begin{aligned} C_2 &= B_2^* - 1 \\ C_3 &= B_3^* - 2B_2^* + 1 \\ C_4 &= B_4^* - 3(B_3^* - B_2^*) - 1 \end{aligned} \tag{24}$$

which follow from the low-density expansion of Equation 21. This method, as well as a similar variant of Barboy and Gelbart (*13*), (who considered an expansion in $y/(1-y)$, so that the term $C_4y^3/(1-y)^4$ appeared in their equation of state) predict the P–V–T behavior of different hard body systems, which are not necessarily convex, with good accuracy provided the values of the second to fourth virial coefficients are available. This fact limits the applicability of these equations considerably, especially in the case of mixtures.

Recently (*14*), on the basis of an inequality proposed by Kihara and Miyoshi (*15*) for the third virial coefficient (giving the upper and lower limits), the present author formulated the following approximations for B_3^* and B_4^*

$$B_3^* = 1 + 6\alpha + 3\alpha^2 \qquad B_4^* = 1 + 14\alpha + 3\alpha^2 \tag{25}$$

From Equations 21 and 24 we can obtain

$$\frac{PV}{NkT} = \frac{1}{1-y} + \frac{3\alpha y}{(1-y)^2} + \frac{3\alpha^2 y^2(1-2y) + 5\alpha y^3}{(1-y)^3} \tag{26}$$

For mixtures, it holds that

$$\frac{PV}{NkT} = \frac{1}{1-v} + \frac{rs}{\rho(1-v)^2} + \frac{qs^2(1-2v) + 5rsv^2}{3\rho(1-v)^3} \tag{27}$$

where the variables r, s, q, and v have the same meaning as in Equation 20. Relatively simple expressions for g^{av} and g_{ij}^{av} are available, also (*14*).

Naumann et al. (*16*) began with the lower limit of the Kihara–Miyoshi inequality and introduced a further geometric parameter, τ

$$\tau = 4\pi R_i^2/S_i \tag{28}$$

Their equation of state, obtained in a semiempirical way (*16, 17*) is

$$\frac{PV}{NkT} = \frac{1}{1-y} + \frac{3\alpha y}{(1-y)^2} + \frac{1.5\alpha^2(1/\tau + 1)y^2 - 0.5\alpha^2(5 - 3/\tau)y^3 + 7\alpha^2(1/\tau - 1)y^4}{(1-y)^3} \tag{29}$$

For $\tau = 1$ the expression reduces to Equation 18. The corresponding third and fourth virial coefficients from Equation 29 are

$$B_3^* = 1 + 6\alpha + 1.5\alpha^2(1/\tau + 1) \qquad B_4^* = 1 + 9\alpha + \alpha^2(6/\tau + 2) \tag{30}$$

Equation 29 gives theoretical values of the compressibility factor of hard spherocylinders in good agreement with pseudoexperimental results (*16*) and comparable with predictions from Equations 22 or 26. The introduction of the parameter τ is, however, theoretically unjustified (within the scaled particle theory), and the known data for higher virial coefficients and the compressibility factors for different hard-body systems do not indicate the necessity of introducing a further nonsphericity parameter in addition to α.

Equations 18 and 26 fulfill well our claims for the sound theoretical basis and sufficient generality of the hard-body equations of state. In the following section their applicability is shown.

Virial Coefficients and Compressibility Factor for Hard Convex Bodies

In order to test the equation of state of hard convex bodies, Equation 26, we shall first consider the values of the higher virial coefficients. Computer data of virial coefficients for hard convex body systems are relatively abundant; in addition to data for prolate spherocylinders, known in a broad range of length-to-breadth ratios, values for several kinds of oblate spherocylinders and ellipsoids of revolution are at our disposal. Thus, the most important types of shape (from the point of view of structures of real molecules) are included and the comparison of theoretical results with pseudoexperimental data gives a stringent test of the approximations used in Equation 25. Moreover, this test can reveal any dependence on another parameter by comparing the virial coefficient data for convex bodies that have different shapes but the same value of the nonsphericity parameter α.

Hard prolate spherocylinders—convex bodies parallel to rods—were studied most thoroughly (*18–21*). Their geometry can be suitably characterized by the length-to-breadth ratio, γ. If $\sigma/2$ is the thickness, it holds that

$$\begin{aligned} \mathcal{R}_i &= (\gamma + 1)\sigma/4 \\ S_i &= \gamma\pi\sigma^2 \\ V_i &= (3\gamma - 1)\pi\sigma^3/12 \end{aligned} \tag{31}$$

and

$$\alpha = \gamma(\gamma + 1)/(3\gamma - 1) \tag{32}$$

Oblate spherocylinders—convex bodies parallel to circles—can be characterized by the ratio of the basic circle diameter d and the breadth

σ, $\Phi = d/\sigma$ (22). Then

$$\mathcal{R}_i = (\pi\Phi/4 + 1)\sigma/2$$
$$S_i = (\Phi^2 + \pi\Phi + 2)\pi\sigma^2/2 \qquad (33)$$
$$V_i = (6\Phi^2 + 3\pi\Phi + 4)\pi\sigma^3/24$$

and

$$\alpha = \frac{(\pi\Phi + 4)(\Phi^2 + \pi\Phi + 2)}{(12\Phi^2 + 6\pi\Phi + 8)} \qquad (34)$$

Ellipsoids of revolution were studied by Freasier and Bearman (23); the characteristic parameters are length of axis of revolution, a, and length of the other axis, b; then $m = b/a$. The geometric functionals can be determined from the relationships

$$\mathcal{R}_i = [m + \arccos m/\sqrt{1 - m^2}]\, a/4 \qquad \text{for} \quad m < 1$$
$$\mathcal{R}_i = [m + \ln (m + \sqrt{m^2 - 1})/\sqrt{m^2 - 1}]\, a/4 \qquad \text{for} \quad m > 1$$
$$S_i = \pi\left[1 + \frac{m^2}{\sqrt{1 - m^2}} \ln\left(\frac{1 + \sqrt{1 - m^2}}{m}\right)\right]\frac{a^2}{2} \qquad \text{for} \quad m < 1 \qquad (35)$$
$$S_i = \pi\left[1 + \frac{m^2}{\sqrt{m^2 - 1}} \arccos\left(\frac{1}{m}\right)\right]\frac{a^2}{2} \qquad \text{for} \quad m > 1$$
$$V_i = \pi m a^3/6$$

In Table I a comparison of theoretical and pseudoexperimental values of the third, fourth, and fifth virial coefficients is given for the above three types of hard convex bodies (the second virial is known exactly). The standard errors of the pseudoexperimental data are estimated to be 0.2–0.4% for the third, 1–2% for the fourth, and 4% for the fifth virial coefficient. It is obvious that agreement is very good in all three cases. For given values of α, no dependence on any further parameter can be traced.

Simulation studies of hard convex bodies have been performed to date only in systems of hard prolate spherocylinders (24–29). In Table II the values of the compressibility factor calculated from Equation 26 for the spherocylinders of $\gamma = 2$ and $\gamma = 3$ are compared with the simulation data. Full agreement within the estimated errors is found in all cases except for the highest value of the packing fraction and $\gamma = 3$, where the difference exceeds by 0.5% the estimated error of this experimental point.

Table I. Viral Coefficients of Hard Convex Bodies

		B_3^*		B_4^*		B_5^*		
		Prolate Spherocylinders						
γ	α	*Theory*	*Monte Carlo*	*Theory*	*Monte Carlo*	*Theory*	*Monte Carlo*	*Refs.*
1.0	1.00	10.00	10.00	18.00	18.36	28.00	28.31	20
1.2	1.02	10.19	10.19	18.31	19.47	28.42	—	21
1.4	1.05	10.61	10.64	19.01	19.26	29.35	—	18
1.6	1.09	11.16	11.30	19.92	21.35	30.56	—	21
1.8	1.15	11.81	11.84	20.97	21.50	31.93	—	18
2.0	1.20	12.52	12.34	22.12	22.34	33.40	31.9	10, 19
			12.54		22.50			20
2.5	1.35	14.51	14.30	25.28	26.06	37.35	—	10, 19
3.0	1.50	16.75	16.20	28.75	28.00	41.50	36.80	10, 19, 20
			16.27		29.15			
4.0	1.82	21.83	20.43	36.37	31.90	50.09	39.70	19, 20
			20.48					
		Oblate Spherocylinders						
Φ	α	*Theory*	*Monte Carlo*	*Theory*	*Monte Carlo*	*Theory*	*Monte Carlo*	
1.0	1.13	11.60	11.66	20.63	21.28	31.48	—	22
1.5	1.23	12.97	13.02	22.85	23.49	34.32	—	22
2.0	1.35	14.54	14.62	25.32	26.15	37.40	—	22
3.0	1.59	18.11	18.03	30.82	30.14	43.91	—	22
		Ellipsoids of Revolution						
a/b	α	*Theory*	*Monte Carlo*	*Theory*	*Monte Carlo*	*Theory*	*Monte Carlo*	
1.5	1.06	10.72	10.69	19.20	19.73	29.61	29.88	23
2.0	1.18	12.25	12.09	21.69	21.57	32.85	31.87	23
0.67	1.06	10.72	10.73	19.20	19.62	29.61	29.51	23
0.50	1.18	12.25	12.30	21.69	22.81	32.85	33.18	23

Using the Monte Carlo method, studies have been performed of mixtures of hard convex bodies, i.e., the system of mixed hard spheres and prolate spherocylinders ($\gamma = 2$). Monson and Rigby (*30*) considered two equimolar mixtures. In the first case (mixture A) the thickness of the spherocylinder was equal to the radius of the sphere. In the second case (mixture B) both the hard bodies possessed the same volume. In a simulation study (*31*) the system corresponding to mixture A was followed at three concentrations (mixture C).

It is obvious from Table III that Equation 27 yields the compressibility factor of mixtures in full accord with Monte Carlo data. Good agreement of the theoretical and pseudoexperimental values of the average correlation functions at contact was also found (*14*).

Table II. Compressibility Factor of Hard Prolate Spherocylinders

	$\gamma = 2$		$\gamma = 3$		
y	*Theory*	*Monte Carlo*	*Theory*	*Monte Carlo*	*Refs.*
0.20	2.67	2.65 ± 0.02(2.69)	3.09	3.07 ± 0.03	24, 27
0.30	4.56	4.48 ± 0.07	5.48	5.40 ± 0.10	24
0.40	8.08	8.20 ± 0.10(8.10)	9.89	9.60 ± 0.10	24, 25, 27
0.45[a]	10.71	10.74 ± 0.24	13.44	13.00 ± 0.16	25, 28
0.50	15.20	15.20 ± 0.20	18.50	18.00 ± 0.40	24, 25

[a] Rounded value for $\gamma = 2$.

Virial Coefficients and Compressibility Factor of Fused Hard Spheres

A considerable number of computer studies has been devoted to ISM systems with soft-sphere or Lennard–Jones interactions. From HISM the linear models have been considered. Virial coefficients are available for hard homonuclear and heteronuclear dumbbells (*21, 32–36*). The hard

Table III. Compressibility Factor of Mixtures of Hard Spheres (1) and Hard Prolate Spherocylinders (2) of $\gamma = 2$

	Mixture A		Mixture B	
v	*Theory*	*Monte Carlo*[a]	*Theory*	*Monte Carlo*[a]
0.20	2.51	2.50 ± 0.06	2.53	2.52 ± 0.04
0.30	4.19	4.10 ± 0.05	4.26	4.20 ± 0.05
0.40	7.36	7.31 ± 0.07	7.49	7.39 ± 0.06
0.45	9.97	9.87 ± 0.10	10.16	10.22 ± 0.10

v	*Theory*	*Monte Carlo*[b]
	Mixture C, $x_1 = 0.20$	
0.33	5.13	5.17 ± 0.10
0.44	9.68	9.89 ± 0.20
0.50	14.43	14.34 ± 0.40
	Mixture C, $x_1 = 0.50$	
0.31	4.49	4.52 ± 0.08
0.42	8.11	8.07 ± 0.15
0.48	11.74	11.59 ± 0.23
	Mixture C, $x_1 = 0.71$	
0.30	4.02	4.03 ± 0.07
0.40	6.96	7.02 ± 0.12
0.45	9.81	9.70 ± 0.21

[a] From Ref. 30.
[b] From Refs. 11, 31.

homonuclear dumbbells are characterized by a reduced length $l^* = L/\sigma$; a volume and a surface area of hard dumbbells can be calculated from the relationships

$$V_i = \pi\left(1 + \frac{3}{2}l^* - \frac{1}{2}l^{*3}\right)\sigma^3/6$$
$$S_i = \pi(1 + l^*)\sigma^2 \qquad (36a)$$

while in a method consistent with Equation 18

$$\mathcal{R}_i = (2 + l^*)\sigma/4 \qquad (36b)$$

then

$$\alpha = (1 + l^*)(2 + l^*)/(2 + 3l^* - l^{*3}) \qquad (37)$$

Values of the second to fourth virial coefficients of hard homonuclear dumbbells are compared with pseudoexperimental data in the first part of Table IV. In the lower part of the table, virial coefficients of heteronuclear dumbbells are listed. In addition to $l^* = L/\sigma_1$, these bodies are characterized by the ratio $\sigma_2/\sigma_1 = \gamma$. For the first three models of Table IV, γ equals 1.5; for the fourth, γ equals 1.8; and in the last case, γ equals 1.2.

Table IV. Virial Coefficients of Fused Hard-Sphere Bodies

		B_2^*		B_3^*		B_4^*		
l*	α	*Theory*	*Exact[a] (Monte Carlo)*	*Theory*	*Monte Carlo*	*Theory*	*Monte Carlo*	*Refs.*
			Homonuclear Dumbbells					
0.2	1.02	4.06	4.06	10.22	10.22	18.47	19.43	21
0.4	1.07	4.21	4.21	10.87	10.94	19.83	20.35	18
0.6	1.16	4.48	4.48	12.01	12.11	22.22	22.98	18
0.8	1.30	4.89	4.87	13.82	14.04	26.11	27.61	21
1.0	1.50	5.50	5.44	16.75	16.93	32.50	34.88	33
			Heteronuclear Dumbbells					
0.75	1.11	4.33	4.40 (4.32)	11.37	11.48	20.88	21.61	38
0.5	1.03	4.10	4.13 (4.10)	10.40	10.40	18.84	19.31	38
1.0	1.23	4.69	4.80 (4.65)	12.93	12.92	24.21	24.80	38
0.9	1.09	4.28	4.34 (4.25)	11.14	11.12	20.40	20.91	38
0.6	1.12	4.37	4.43 (4.35)	11.50	11.51	21.16	21.70	38

[a] From Ref. 37.

Defining (39) the quantity a

$$a_{\pm} = \frac{1}{2} l^* \pm (\gamma^2 - 1)/8l^* \tag{38}$$

we can write

$$V_i = \pi[1 + \gamma^3 + 3(\gamma^2 a_+ + a_-) - 4(a_+^3 + a_-^3)]\sigma^3/12$$

$$S_i = \pi\left[\gamma\left(\frac{1}{2}\gamma + a_+\right) + \left(\frac{1}{2} + a_-\right)\right]\sigma^2 \tag{39}$$

$$\mathcal{R}_i = \frac{1}{4}[\gamma - 1 + l^* + (\gamma - 1)^2/4l^* + 2]\sigma$$

and the parameter α follows from Equation 15. The virial coefficients calculated from the relationships in Equation 15 are again in very good accord with the pseudoexperimental data.

In Table V theoretical values of the compressibility factor are compared with data (40) for hard homonuclear dumbbells of l^* equal to 0.6

Table V. Compressibility Factor of Hard Dumbbells

	Homonuclear Dumbbells $l^* = 0.6$		$l^* = 1.0$	
y	*Theory*	*Monte Carlo*[a]	*Theory*	*Monte Carlo*[a]
0.105	1.63	1.63	1.80	1.79
0.157	2.11	2.13	2.44	2.46
0.209	2.77	2.78	3.33	3.36
0.262	3.66	3.67	4.57	4.62
0.314	4.89	4.95	6.31	6.40
0.366	6.63	6.69	8.82	8.95
0.419	9.15	9.23	12.49	12.64
0.445	10.82	10.89	14.97	15.12
0.471	12.88	12.87	18.02	18.06

Heteronuclear Dumbbells at y = *0.4084*

Type	γ	α	*Theory*	*Monte Carlo*[b]	*Type*	γ	α	*Theory*	*Monte Carlo*[b]
VI	0.5	1.30	9.74	10.0	IX	0.5	1.02	7.47	7.8
VII	0.5	1.17	8.65	8.9	X	0.67	1.31	9.88	10.1
VIII	0.5	1.08	7.92	8.3	XI	0.84	1.29	9.73	9.9

[a] From Ref. 40.
[b] From Ref. 41.

and l^* equal to 1.0. The agreement in this case is perfect. Table V brings also a comparison of theoretical and Monte Carlo values for heteronuclear dumbells at a constant packing fraction y equal to 0.4084, and γ equal to 0.5, 0.67, and 0.84. The final Monte Carlo data (*41*) are estimated to be accurate only to within about 7%. It is obvious that the theoretical compressibility factors agree well within this uncertainty with the pseudoexperimental compressibility factors. In addition to dumbells, hard triatomics were studied (*42*). Three models (related to carbon disulfide) were considered, all at y equal to 0.4697 and with the reduced distance of the outside sites l^* equal to 0.897. For the three models, the ratios of diameters of the central and the outside spheres were 0.857, 1.0, and 1.2. The compressibility factors from Equation 18 are 14.36, 13.54, and 12.42. In comparison, the Monte Carlo data are 14.84, 12.84 and 12.88.

The present author has also performed simulations in systems of linear and nonlinear triatomics (*43*). Both of the models considered are formed by equal spheres with the central-to-outside site distance $l^*/2 = 0.5$. In the case of linear triatomics the theoretical compressibility factor 12.69 at $y = 0.4533$ compares well with the pseudoexperimental one, 12.88. In the case of nonlinear ($\sphericalangle = \pi/2$) triatomics at $y = 0.3981$, the theoretical value is 8.19 and the Monte Carlo result is 8.34. Taking into account the lower accuracy of simulation results for triatomics in comparison with dumbells, the agreement can again be considered to be very good.

Conclusions

In summary, it can be said that the equations of state, Equations 18 and 26, and the corresponding expressions for mixtures, represent an optimum description of the P–V–T behavior of fused hard-sphere and hard convex body systems. Both these equations have a sound theoretical background and reduce in the special case to the Carnahan–Starling equation (*44*). They yield reliable prediction of virial coefficients, and they are sufficiently general and accurate. Their extension to mixtures possesses a simple form, and comparison with the available pseudoexperimental data reveals the reliability of these expressions in the description of the equilibrium behavior of mixtures. It is believed that these equations will be useful in the characterization of the molecular fluid behavior, just as the Carnahan–Starling equation proved to be for simple fluids.

Literature Cited

1. Reiss, H.; Frisch, H. L.; Lebowitz, J. L. *J. Chem. Phys.* **1959,** *31*, 369.
2. Gibbons, R. M. *Mol. Phys.* **1969,** *17*, 81.
3. Ibid. **1970,** *18*, 809.

4. Boublík, T. *Mol. Phys.* **1974,** *27,* 1415.
5. Boublík, T. *Mol. Phys.* **1981,** *44,* 1369.
6. Kihara, T. *Adv. Chem. Phys.* **1963,** *5,* 147.
7. Boublík, T. *J. Chem. Phys.* **1970,** *53,* 471.
8. Boublík, T. *J. Chem. Phys.* **1975,** *63,* 4084.
9. Boublík, T. *Mol. Phys.* **1975,** *29,* 421.
10. Nezbeda, I. *Chem. Phys. Lett.* **1976,** *41,* 55.
11. Pavlíček, J.; Nezbeda, I.; Boublík, T. *Czech. J. Phys.* **1979,** *29B,* 1061.
12. Nezbeda, I.; Pavlíček, J.; Labík, S. *Colln. Czech. Chem. Commun.* **1979,** *44,* 3555.
13. Barboy, B.; Gelbart, W. M. *J. Chem. Phys.* **1979,** *71,* 3053.
14. Boublík, T. *Mol. Phys.* **1981,** *42,* 209.
15. Kihara, T.; Miyoshi, K. *J. Statist. Phys.* **1975,** *13,* 337.
16. Naumann, K. -H.; Chen, Y. P.; Leland, T. W. *Ber. Bunsenges. Phys. Chem.* **1981,** *85,* 1029.
17. Naumann, K. -H., unpublished data.
18. Rigby, M. *J. Chem. Phys.* **1970,** *53,* 1021.
19. Nezbeda, I. *Czech. J. Phys.* **1976,** *26B,* 355.
20. Monson, P. A.; Rigby, M. *Mol. Phys.* **1978,** *35,* 1337.
21. Nezbeda, I.; Smith, W. R.; Boublík, T. *Mol. Phys.* **1979,** *37,* 985.
22. Nezbeda, I.; Boublík, T. *Czech. J. Phys.* **1977,** *27B,* 953.
23. Freasier, B. C.; Bearman, R. J. *Mol. Phys.* **1976,** *32,* 551.
24. Monson, P. A.; Rigby, M. *Chem. Phys. Lett.* **1978,** *58,* 122.
25. Vieillard-Baron, J. *Mol. Phys.* **1974,** *28,* 809.
26. Boublík, T.; Nezbeda, I.; Trnka, O. *Czech. J. Phys.* **1976,** *26B,* 1081.
27. Rebertus, D. W.; Sando, K. M. *J. Chem. Phys.* **1977,** *67,* 2585.
28. Nezbeda, I.; Boublík, T. *Czech. J. Phys.* **1978,** *28B,* 353.
29. Nezbeda, I. *Czech. J. Phys.* **1980,** *30B,* 601.
30. Monson, P. A.; Rigby, M. *Mol. Phys.* **1980,** *39,* 977.
31. Boublík, T.; Nezbeda, I. *Czech. J. Phys.* **1980,** *30B,* 121.
32. Rigby, M. *J. Chem. Phys.* **1970,** *53,* 1021.
33. Freasier, B. C.; Jolly, D.; Bearman, R. *Mol. Phys.* **1976,** *31,* 255.
34. Freasier, B. C. *Mol. Phys.* **1980,** *39,* 1273.
35. Rigby, M. *Mol. Phys.* **1978,** *35,* 1337.
36. Chen, Y. D.; Steele, W. A. *J. Chem. Phys.* **1969,** *50,* 1428.
37. Isihara, A. J. *J. Chem. Phys.* **1951,** *19,* 397.
38. Jolly, D.; Freasier, B. C.; Bearman, R. *Chem. Phys. Lett.* **1977,** *46,* 75.
39. Nezbeda, I.; Boublík, T. *Czech. J. Phys.* **1977,** *27B,* 1071.
40. Tildesley, D. J.; Streett, W. B. *Mol. Phys.* **1980,** *41,* 85.
41. Streett, W. B.; Tildesley, D. J. *J. Chem. Phys.* **1978,** *68,* 1275.
42. Streett, W. B.; Tildesley, D. J. *J. Chem. Soc., Faraday Disc.* **1978,** *66,* 27.
43. Boublík, T., *Czech. J. Phys.* (in press).
44. Carnahan, N. F.; Starling, K. E. *J. Chem. Phys.* **1969,** *51,* 635.

RECEIVED for review January 27, 1982. ACCEPTED for publication September 27, 1982.

9

Modeling of Simple Nonpolar Molecules for Condensed Phase Simulations

C. S. MURTHY and K. SINGER

Royal Holloway College, Department of Chemistry, Egham Hill, Egham, Surrey TW20 0EX, England

I. R. McDONALD

University of Cambridge, Department of Physical Chemistry, Lensfield Road, Cambridge CB2 1EP, England

A number of simple potential models have been tested for their usefulness in the simulation of the condensed phases of nitrogen, carbon dioxide, fluorine, and chlorine. The properties studied systematically include the lattice energy and lattice spacing of the crystal, the zone center lattice vibrational frequencies, thermodynamic properties of the liquid, and the temperature dependence of the second virial coefficient. In particular cases, more stringent tests of an intermolecular potential are also considered, involving the orientational order of the α-phase and the cubic–tetragonal phase transition in nitrogen, the librational Grüneisen parameters, and the molecular tilts in the halogen crystals. For nitrogen and carbon dioxide, models consisting of site–site Lennard–Jones potentials plus point quadrupolar interactions account moderately well for the properties listed above. For fluorine and chlorine, these simple models are adequate for the liquid and gas but fail to describe satisfactorily the properties of the solid phase.

THE STUDY OF FLUIDS AND SOLIDS BY COMPUTER SIMULATION requires as basic input simple and realistic interaction potentials. Ideally, the interaction potential should account for all observable properties. In practice, sophisticated potential models that account explicitly for many-body forces and include large numbers of interaction sites would be difficult to construct and computationally very expensive to incorporate into a simulation. The motivation is therefore strong for searching for pairwise-additive effective potentials that can account for a wide range of properties.

0065-2393/83/0204-0189$06.00/0

For small, nonpolar molecules the most widely used models consist of short-ranged, Lennard–Jones (usually 6-12) site–site potentials plus electrostatic interactions based either on a point quadrupole moment or a set of fractional point charges. We shall refer to such models as $n\mathrm{LJ}+Q$ or $n\mathrm{LJ}+q$, respectively, where n denotes the number of sites. In some cases, Buckingham (6-exponential) rather than Lennard–Jones potentials have been used. Models of this general type have been used extensively in lattice dynamics calculations (*1–3*) and in computer simulations of the condensed phases (*4–7*). Until recently, however, little effort had been made to obtain effective pair potentials that could account for a wide range of experimental data of both solid and liquid. Reasonably successful attempts to remedy this situation have been described for nitrogen and carbon dioxide (*8, 9*). In this chapter we summarize and extend this work and also discuss some recent results for fluorine and chlorine. We assess the adequacy of models proposed by ourselves and others to account for the structure and lattice energy of the solid, lattice vibrational frequencies, thermodynamic properties of the liquid, and the second virial coefficient of the gas (*10*).

Our attention is focused mainly on $n\mathrm{LJ}+Q$ and $n\mathrm{LJ}+q$ models, with n equal to 2 or 3, though for nitrogen other semiempirical (*11*) and ab initio (*12, 13*) potentials are also discussed. The $n\mathrm{LJ}+Q$ (or $n\mathrm{LJ}+q$) models are characterized by four parameters: the Lennard–Jones constants ε and σ, the separation l of the Lennard–Jones sites, and Q, the quadrupole moment of the molecule.

Nitrogen and carbon dioxide are conveniently treated together, partly because the low temperature solids have the same structure (cubic, space group Pa3). This work is described in the first section below. The results for fluorine and chlorine are then described, and a summary of our conclusions follows.

Nitrogen and Carbon Dioxide

Potential Models. Numerous potential models have been proposed both for nitrogen and for carbon dioxide. The characteristic parameters for a number of these models are given in Table I. The potentials listed in the table fall into two groups. The first consists, with the exception of one ab initio model, of empirical intermolecular potentials fitted by workers other than ourselves to properties of the phases listed in the "Source" column in Table I. The second group, MSKM for nitrogen and the three MSM models for carbon dioxide, were constructed by us to fit as closely as possible the lattice parameter a of the Pa3 crystal, the lattice energy W_L at 0 K, and the second virial coefficient at a low temperature (75 K for nitrogen, 260 K for carbon dioxide). The experimental values of W_L were estimated from the measured enthalpies of sublima-

Table I. Intermolecular Potentials for Nitrogen and Carbon Dioxide

Model[a]	*Source*	ε/k (*K*)	σ (Å)	l (Å)	$-Q$ (*DÅ*)	$-W_L$ (*kJ/mol*)[b]
			Nitrogen			
RGA	Gas	38.0	3.308	1.098	0	7.8
KD	Solid	33.4	3.383	0.923	1.112	8.2
CP	Liquid	35.3	3.314	1.101	1.48	8.6
TC	Solid	35.3	3.380	0.900	0.90	8.3
RG	Solid, gas	*See text*		1.100	1.40	8.3
MSKM	Solid, gas	36.4	3.318	1.098	1.17	8.3
BV[c]	ab initio	*See text*		1.094	1.33	7.5
Experiment				1.098[d]	1.4[e]	8.3[f]
			Carbon Dioxide			
MSB	Lattice, gas	129.7	2.980	2.38	3.89	28.4
STS	Liquid	163.6	2.989	2.37	0	22.8
MSM-A1	Solid, gas	136.7	2.979	2.32	3.66	28.6
MSM-A2	Solid, gas	130.0	3.033	2.10	3.69	28.8
MSM-C	Solid, gas	29.0(C–C)	2.785	2.32	3.85	28.6
		83.1(O–O)	3.014			
Experiment				2.32[g]	4.3[h]	28.6[i]

[a] MSM-C is a $3LJ+Q$ model; all other models are $2LJ+Q$ or $2LJ+q$ except RG and BV. Models KD, RG, and BV use fractional charge representation of the electrostatic interaction.

[b] W_L is calculated at the experimental lattice parameter.

[c] Two-centered Buckingham model.

[d] From Ref. 16.

[e] From Ref. 43.

[f] From Ref. 19.

[g] From Ref. 44.

[h] From Ref. 45.

[i] From Ref. 9.

tion. In the fit, account was taken of the zero-point energy and its variation with volume. In cases where this was not done, the potential is described in the tables as fitted to the lattice rather than to the solid.

Table I includes six models proposed by others for nitrogen. Of these, the only one that ignores the quadrupole moment of the molecule, and that hence may be termed a 2LJ model, is the potential RGA (*14*). In this case, the parameters were obtained by requiring the spherically averaged potential to reduce to the isotropic Lennard–Jones potential derived from gas phase properties. Our own attempts to construct a 2LJ model led to essentially the same parameter values as those in model RGA, which are also close to those deduced by Huler and Zunger (*15*) from a detailed analysis of static and dynamic properties of the ordered solid phases (α and γ). Many other 2LJ models have been put forward as descriptive of nitrogen (*16*), and in spite of their very different origins, the parameter values proposed are remarkably similar. We find, however, that 2LJ models do not successfully reproduce the range of properties mentioned above. In addition, it is known from lattice dynamics calculations that 2LJ models give rise to lattice mode instabilities in the γ-phase (*17*). It is also known from molecular dynamics simulations (*8*) that they lead to much too great a degree of orientational disorder in the cubic α-phase. For these reasons, we make no comparison here of their merits; model RGA is included only as an example of this class of potentials.

Model CP (the authors' quadrupolar model) (*18*) was derived by fitting to the experimental internal energy and equation of state of the liquid. Models KD (*19*) and TC (the authors' model A) (*20*) were fitted to the structure and lattice energy of the Pa3 crystal (the α-phase) and to the measured lattice vibrational frequencies. In the case of model KD, the fitted experimental quantities included the intensities of peaks in the inelastic neutron scattering spectrum. The special feature of these two models, in contrast to model CP, for example, is that the site–site separation is about 20% shorter than the internuclear distance. However, the larger values obtained for σ indicate that the overall length of the molecule is comparable with that deduced from the 0.002 electron density contour (*21*). The results for model RG, and also for models MSM-A2 and MSM-C for carbon dioxide, show that the introduction of a third site at the bond center is similar in effect to a shortening of the site–site separation. The parameters of functions of the exponential type characterizing the short range repulsions in the several versions of model RG (*11*) were obtained by fitting to the experimental PV isotherm from 0 to 3.6 kbar at 4.2 K and to molecular beam data, in addition to the quantities a and W_L.

Model BV (the authors' model I′, for which the repulsive and attractive sites are coincident with the nuclei) (*12*), was obtained by fitting the results of ab initio calculations to a two-site Buckingham potential supplemented by point charge interactions.

In the case of carbon dioxide, model STS (*6*) was parameterized to liquid-state thermodynamic properties, and model MSB (*10*) to the second virial coefficient and static lattice data.

Thermodynamic Properties of the Liquid. The calculated values of W_L are given in Table I. Table II contains some results for the configurational internal energy U and pressure P obtained by molecular dynamics simulations of the two liquids. Overall, the agreement between calculation and experiment is consistent with the way in which the potentials were parameterized. For example, models RGA and BV were not based on any experimental solid-state data and the results for W_L are accordingly poor. Models CP (for nitrogen) and STS (for carbon dioxide) were both derived by fitting to liquid state properties; the fact that model STS is the less successful is not surprising, given the absence of any quadrupolar interaction. For nitrogen, liquid phase simulations have also been reported for one of the RG models (RG-5) (*see* the discussion in Reference 8) and for the ab initio potential of Jönsson et al. (*13*), with results that are about as good as those obtained for model CP.

The main conclusion to be drawn is the unsurprising one that calculation of thermodynamic properties is not a sensitive test of an intermolecular potential. However, it is notable that all the models of Table I that were fitted to solid-state properties work well for the liquid. However, the converse is not true, as can be seen in the results for model CP and, in particular, model STS. These are not isolated examples, and a body of evidence now exists that suggests that careful parameterization to the energy and structure of molecular crystals can be expected to yield a satisfactory potential for the liquid, but that the opposite route is less likely to be successful.

Lattice Vibrational Frequencies. In previous articles (*8, 9*) we have discussed the results of lattice dynamics calculations in the Pa3 structure for several of the models listed in Table I. Here we focus attention on the seven zone center modes, of which four are translational and three are librational in character; these are the modes for which most experimental information is available. Our earlier work has shown that all potentials considered give generally satisfactory results for the translational modes. This implies that it is relatively easy to model correctly the isotropic part of the intermolecular potential. However, the librational modes, all of which are Raman active, are sensitive to the anisotropy of the potential and are correspondingly more difficult to fit. Results for the librational modes are given in Table III. For carbon dioxide, the best agreement with experiment is provided by model MSM-C, but it is reasonably good even for those potentials (MSB and STS) that were parameterized without appeal to any solid-state data. For ni-

Table II. Molecular Dynamics Results for Liquid Nitrogen and Liquid Carbon Dioxide

Nitrogen[a]

ρ (*mol/dm*^{3})	*Property (unit)*	*RGA*	*KD*	*CP*	*TC*	*MSKM*	*Experiment*[b]
30.10	T (K)	67.5	68.5	68.2	66.5	69.8	67.6
	−U (kJ/mol)	5.54	5.39	5.37	5.67	5.42	5.40
	P (kbar)	−0.04	−0.03	0.04	−0.14	0.05	0.0
25.09	T (K)		99.2		94.8	97.9	97.1
	−U (kJ/mol)		4.28		4.52	4.31	4.26
	P (kbar)		0.01		−0.06	0.01	0.006

Carbon Dioxide[c]

ρ (*mol/dm*^{3})	*Property (unit)*	*MSB*	*STS*	*MSM-A1*	*MSM-A2*	*MSM-C*	*Experiment*[d]
26.63	T (K)	213.8	218	219.2	224.8	220.4	218.0
	−U (kJ/mol)	14.3	14.4	14.4	14.2	14.0	14.3
	P (kbar)	−0.18	0.24	−0.03	−0.00	−0.09	0
21.12	T (K)	276.3	273	272.8	276.9	269.0	273.0
	−U (kJ/mol)	10.5	11.0	10.7	10.6	10.6	10.9
	P (kbar)	0.04	−0.01	−0.05	−0.02	−0.04	0

[a] Reproduced with permission from Ref. 8. Copyright 1980, Taylor and Francis, Ltd.
[b] From Ref. 46.
[c] Reproduced with permission from Ref. 9. Copyright 1981, Taylor and Francis, Ltd.
[d] From Ref. 47.

trogen, on the other hand, the situation is much less satisfactory. The one 2LJ model considered (RGA) yields much too low a frequency for the upper T_g mode, but the quadrupolar models, apart from KD and TC, all seriously overestimate the librational frequencies. In models KD and TC, the fit to the frequencies was achieved by reducing both the site–site separation and the quadrupole moment of the molecule. This has the consequence, as computer simulations have shown (*8*), that the molecules become orientationally disordered at temperatures in the region of 30 K. Experimentally, there is a transition at 35.6 K to a hexagonal close-packed structure, the β-phase, in which the molecules undergo rotational diffusion. In the α-phase, however, although there is a large amplitude librational motion, orientational order persists up to the α–β transition temperature. The reduction of the effective bond length in models KD and TC therefore succeeds in resolving the immediate problem, at the expense of introducing others.

The reason for the difficulty in fitting the librational frequencies in nitrogen is at present unclear, but it may stem at least in part from the harmonic approximation used in the lattice dynamics calculations. Some evidence that this is so is given by the fact that there is apparently no similar problem for carbon dioxide, in which the librational amplitude is much smaller than in nitrogen. Estimates (*11, 25–27*) of the anharmonic corrections for certain models are included in Table III. The general effect is to improve the agreement with experiment. However, there are discrepancies even in sign between the corrections calculated for different models by different groups, and more work on this interesting question is clearly needed.

Prediction of the volume dependence of the lattice frequencies, characterized by the Grüneisen parameters $(d \ln \nu_i / d \ln V)_T$, would provide a good test of a potential model if extensive and accurate experimental data were available. Unfortunately, this is not the case at present. There have been suggestions in the past that a change from a Lennard–Jones 6-12 to a 6-9 potential would lead to improved agreement with the available experimental data on the Grüneisen parameters. While this may be true for a 2LJ model (*22*), there appears to be no case for such a change when allowance is made for quadrupolar interactions. For carbon dioxide, model MSM-C gives results in good agreement with the recently measured (*23*) Grüneisen parameters for the zone center librational modes. In the case of nitrogen (*24*), the Grüneisen parameters are well reproduced by the quadrupolar models of Table I. As we have already discussed, an acceptable model of nitrogen should also account for the persistence of orientational order in the α-phase up to the temperature of the plastic crystal transition. In this respect, model MSKM is much superior to either the KD or TC potentials (*8*). Overall, we consider MSKM to be the best of the models we have constructed.

Table III. Zone Center Librational Frequencies for Nitrogen and Carbon Dioxide

				α-Nitrogen				
Mode	*RGA*	*KD*	*CP*	*TC*	*RG*	*MSKM*	*BV*	*Experiment*
E_g	33.6(40.2)[a]	35.4	45.2	31.8	37.5(32.9)[b]	41.6	40.8(39.5)[c]	32.2[d]
T_g	37.8(42.5)[a]	42.9	55.5	38.1	47.7(37.5)[b]	50.0	50.7(48.5)[c]	36.3[d]
T_g	45.9(51.8)[a]	63.3	82.6	54.4	75.2	71.5	64.3(70.3)[c]	59.6[d]

			Carbon Dioxide			
Mode	*MSB*	*STS*	*MSM-A1*	*MSM-A2*	*MSM-C*	*Experiment*
E_g	69	67	69	67	68	72[e]
T_g	96	88	93	87	89	88[e]
T_g	144	139	137	125	129	128[e]

Note: Values in parentheses include anharmonic effects. Data are in cm^{-1}.
[a] From Ref. 25.
[b] From Ref. 11.
[c] From Ref. 27.
[d] From Ref. 19.
[e] From Ref. 23.

Second Virial Coefficients. Work reported earlier for nitrogen and carbon dioxide (*8*, *9*) showed that models that are reasonably satisfactory for the liquid and solid phases also give good results for the second virial coefficient, at least at high temperatures where the behavior is determined by the size of the molecular hard core. At low temperatures, there are some deviations; the calculated values fall typically about 5% above the experimental ones. Since the second virial coefficient is a pair property, the discrepancies at low temperatures provide some measure of the importance of many-body forces in the condensed phases, to which the potentials have been tailored. Monson and Rigby (*28*, *29*) have calculated that the dominant three-body interaction contributes about 6% to the static lattice energy of nitrogen and carbon dioxide, a percentage that is comparable with that found for argon (*30*).

Although the calculation of second virial coefficients provides a useful check, in practice it may be of limited value in parameterizing a model. The problem lies in the fact that the anisotropy of the potential, particularly of the electrostatic interactions, has its effect mainly at very low temperatures. This is the region where experimental data on molecular systems are scarcest and, in general, least reliable.

The Gamma Phase of Nitrogen. At least four distinct crystalline forms of nitrogen are known (*16*). This poses a considerable challenge in potential modeling, since a satisfactory model should be capable of explaining the variety of behavior observed in different phases. For this reason, there is interest in seeing how far models fitted to properties of the α-phase are able to describe the structure and dynamics of the ordered γ-phase. This has a tetragonal structure and is stable at pressures greater than about 3.5 kbar; the α–β–γ triple point is at $P = 4.6$ kbar, $T = 46$ K.

In Table IV we list the γ-phase zone center frequencies for models that, apart from TC-B, have already been considered in the discussion of the α-phase. To reproduce the librational frequencies in the γ-phase, Thiéry and Chandrasekharan (*20*) found it necessary to increase either the site–site separation or the quadrupole moment adopted in their model A (model TC in Table I, model TC-A in Table IV); for model TC-B, they chose $\varepsilon/k = 38.7$ K, $\sigma = 3.300$ Å, $l = 1.098$ Å and $Q = -0.90$ DÅ. Thiéry and Chandrasekharan (*20*) conclude that it is not possible to fit the librational frequencies in both phases with a single $2LJ + Q$ model. They do show that if allowance is made for anharmonic corrections along the lines of Raich and Gillis (*11*), model TC-A gives results in satisfactory agreement with experiment. However, model TC-A was parameterized in part by fitting to the frequencies of the α-phase without regard for anharmonicity, so this result may be fortuitous. Until the reliability of different anharmonic corrections is established, it will be difficult to assess the merits of models either for the α- or the γ-phase.

Table IV. Zone Center Lattice Mode Frequencies for γ-Nitrogen

Mode	*RGA*	*KD*	*CP*	*TC-A*	*TC-B*	*MSKM*	*RG*	*BV*	*Experiment*
E_g	53.1(52.8)[a]	48.7	62.6	46.4	56.7	59.8	50.5(58.8)[b]	57.9(56.5)[c]	54.1[d]
B_{1g}	87.6(90.0)[a]	74.1	101.9	66.7	93.1	97.6	74.8(99.8)[b]	86.5(85.2)[c]	97.1[d]
A_{2g}	100.2(106.5)[a]	96.1	131.2	86	112.5	121.4	105.1	109.7(107.1)[c]	
E_u	63.4(67.0)[a]	66.6	74.7	66.9	64.0	65.2	58.3	72.0(69.3)[c]	65[e]
B_{1u}	105.9(119.6)[a]	102.8	100.5	104.1	104.1	104.2	103.1	110.3(107.4)[c]	

Note: Values in parentheses include anharmonic effects. Data are in cm^{-1}.

[a] From Ref. 26.
[b] From Ref. 11.
[c] From Ref. 27.
[d] From Ref. 49.
[e] From Ref. 31.

Table IV shows that most of the models considered give good agreement with a recent infrared measurement of the translational E_u frequency (*31*). This again suggests that the models describe the isotropic part of the potential reasonably well.

Filippini et al. (*17*) have pointed out that although it is possible to find a 2LJ model that predicts correctly the zone center frequencies, imaginary frequencies are found elsewhere in the zone, showing the γ-phase to be unstable in such a model. They were able to eliminate this difficulty empirically by making the repulsive term of the 2LJ potential anisotropic. Even with this refinement, however, it did not prove possible to explain the occurrence of the α–γ transition. Though we have not found any such instabilities for model MSKM (*32*), modification of the potential is again necessary in order to describe the α–γ transition. The same appears to be true of the three-center models of Raich and Gillis (*11*) and of the Kihara type of potential (*3*).

Fluorine and Chlorine

Molecular fluorine and chlorine differ in many respects from nitrogen and carbon dioxide, but from the point of view of modeling them by simple potentials, two facts in particular should be noted: First, the quadrupole moments of fluorine and chlorine are positive, whereas those of nitrogen and carbon dioxide are negative. This reflects the fact that the electron distribution is very different in the two cases. Second, the stable, low temperature phases of the halogens are not cubic; chlorine crystallizes in the orthorhombic Cmca (*33, 34*) structure and α-fluorine is monoclinic (*35*), C2/c or C2/m. Molecules are tilted in opposite directions in alternate layers, towards nearest neighbors in Cmca and C2/c, or next-nearest neighbors in C2/m. The differences in electron distribution and in crystal structure must, of course, be related.

English and Venables (EV) (*36*) have made a thorough study of the most stable crystal structures for a series of diatomic molecular solids described by models of both the 2LJ and 2LJ + Q type. We have included their potentials in the present work, partly as a check on our own calculations. Singer et al. (*6*) have modeled the interactions in fluorine and chlorine by 2LJ potentials (STS) and used these in molecular dynamics simulations. In addition, they have reported results for liquid bromine. Because chlorine has a large quadrupolar moment (*37, 38*), we have also developed (*39*) a 2LJ + Q model by fitting to thermodynamic properties of the liquid model MS. Kobashi and Klein (*40*) have used a 2LJ + q model in a study of the lattice frequencies of α-fluorine (model KK).

We have used the models listed above as starting points in the search for effective pair potentials. It turns out that none of the models is able to account for all the main features of the low-temperature solid, namely the lattice symmetry, unit cell dimensions, molecular tilts and lattice frequencies. Our conclusions are similar to those of English and Venables (*36*). Details of the potentials are given in Table V; in the case of model KK, we have made the calculations for a 2LJ + Q potential rather than the published (*40*) 2LJ + q version.

The difficulties encountered in modeling noncubic systems are highlighted by the results given in Table VI. The table contains results obtained by unconstrained minimization of the energy, and by constraining either the cell parameters a, b, c and the angle between the a and c axes (the monoclinic angle β) or the molecular tilts to have the experimental values. The constraints lead to results that are of interest in their own right; they are also necessary if the lattice dynamics calculations are to be performed with the correct lattice structure.

The results for fluorine are shown in Table VI. The 2LJ model STS, which gives good results for liquid state properties and for the second virial coefficient, correctly stabilizes the monoclinic structure and gives lattice constants which differ by 5–10% from the experimental values. The calculated lattice energy is also in good agreement with experiment, but the molecular tilts are too large and the lattice frequencies (*40*) (not shown) are also poor. Model KK gives better results for the lattice frequencies, but the molecular tilts are even larger than for model STS. With constraints on the molecular tilts, model STS gives a good value for the lattice energy and lattice constants with errors of 5–10% for C2/m, and gives an equally accurate value for W_L and about 5% errors in the lattice constants for C2/c. The tilt-constrained model KK is marginally better for C2/c; it gives even better lattice constants but a poor lattice energy for C2/m. If constraints are placed on the cell dimensions a, b, and c, neither the lattice energy nor the molecular tilts are satisfactorily predicted by either model. Both potentials give good results for the monoclinic angle, except in the case of unconstrained minimization for the C2/m structure with model KK. The results for the EV potentials are poor, but they show the same general trend: a 2LJ interaction stabilizes the monoclinic structures, whereas a 2LJ + Q model favors the Pa3 structure.

The results reported for chlorine in Table VI are even worse. Unconstrained minimization does not lead to the correct orthorhombic Cmca structure for any model, as would be expected on the basis of the results for fluorine. Nor do constrained minimizations lead to satisfactory results for the lattice energy, cell dimensions, or molecular tilts. In addition, one of the translational lattice mode frequencies is imaginary for both the STS and MS models. The only feature which is mildly encouraging is the fact that models STS and MS both yield satisfactory results for liquid-state thermodynamic properties and for the temperature dependence of the second virial coefficient.

Table V. Intermolecular Potentials for Fluorine and Chlorine

Model[a]	*Source*	ε/k (*K*)	σ (Å)	*l* (Å)	Q (*D*Å)	$-W_L$ (*kJ/mol*)[b]
			Fluorine			
EV	Lattice	119.5	2.960	1.480	0	17.1, 15.6[c]
STS	Liquid	52.8	2.825	1.426	0	8.8, 8.5
KK	Lattice	60.5	2.809	1.416	1.2	9.1, 8.4
(Expt)				1.42[d]	0.88[e]	9.6[f]
			Chlorine			
EV	Lattice	282.5	3.365	2.120	0	39.0
STS	Liquid	178.3	3.332	2.199	0	25.1
MS	Liquid	164.3	3.374	2.010	3.60	26.2
(Expt)				1.99[b]	5.0 ± 0.5[g], 3.6[h]	31.3[f]

[a] All models either 2LJ or 2LJ + Q.
[b] W_L is calculated at the experimental lattice cell parameters.
[c] Values correspond to C2/m structure.
[d] From Ref. 44.
[e] From Ref. 48.
[f] Our estimates.
[g] From Ref. 37.
[h] From Ref. 38.

Table VI. Optimum Structures for Fluorine and Chlorine

Model	Structure	Cell Dimensions (Å) a	b	c	Molecular Tilts[a] (degrees) θ	φ	$-W_L$ (kJ/mol)	Monoclinic Angle, β (degrees)
				Fluorine				
EV	Pa3	5.23					20.7	
	Cmca	5.34	3.22	8.23	0	(90)	20.8	
	C2/c	5.48	3.17	8.86	0	(90)	21.6	114.4
EV-*Q*	Pa3	5.21					21.6	
(*Q* = 1.0 DÅ)	Cmca	5.10	4.73	6.00	45.1	(90)	20.6	
	C2/c	5.47	3.39	8.61	23.1	(90)	20.6	116.5
STS	Pa3	4.99					9.1	
	Cmca	5.14	3.08	7.89	0	(90)	9.2	
	C2/c	5.23	3.02	10.42	0	84	9.5	132.1
		(5.50)	(3.28)	(10.01)	25.3	80.9	8.9	(134.7)
		5.22	3.13	10.19	(18)	(90)	9.4	132.3
	C2/m	5.23	3.02	10.42	0	0	9.5	132.0
		(5.50)	(3.28)	(10.01)	22.0	−25.5	8.8	134.7
		5.29	3.02	10.43	(11)	(−11)	9.5	132.7
KK	Pa3	4.92					12.2	
	Cmca	4.92	4.38	5.66	48.5	(90)	11.5	
	C2/c	4.93	4.35	7.52	48.7	90	11.5	131.0
		(5.50)	(3.28)	(10.01)	31.0	81.8	10.2	(134.7)
		5.31	3.14	10.21	(18)	(90)	9.5	133.4
	C2/m	7.49	3.05	11.90	57.0	−65.5	10.8	153.0
		(5.50)	(3.28)	(10.01)	29.8	−30.9	10.0	(134.7)
		5.37	3.10	10.40	(11)	(−11)	8.9	133.6

Experiment[b]	C2/c	5.50	3.28	10.01	18	90	9.6	134.7
	C2/m	5.50	3.28	10.01	11	−11		134.7
Chlorine								
EV	Pa3	6.14					44.3	
	C2/c	6.22	3.66	10.81	15.5	(90)	46.0	112.9
	Cmca	6.11	3.75	10.06	17.5	(90)	44.7	
EV-Q	Pa3	6.14					44.6	
(Q = 1.0 DÅ)	C2/c	6.22	3.69	10.76	17.9	(90)	45.6	113.2
	Cmca	6.09	3.81	9.95	20.5	(90)	44.6	
STS	Pa3	6.08					28.0	
	C2/c	6.16	3.62	10.71	15.5	(90)	29.0	112.9
	Cmca	6.04	3.71	9.97	17.6	(90)	28.2	
		(6.24)	(4.48)	(8.26)				
		5.82	4.41	8.79	(35.2)	(90)	27.4	
MS	Pa3	6.04					32.1	
	C2/c	6.04	5.20	9.34	51.1	(90)	30.4	130.3
	Cmca	6.04	5.20	7.11	51.1	(90)	30.3	
		(6.24)	(4.48)	(8.26)				
		5.51	5.23	7.94	(35.2)	90	28.1	
Experiment[c]	Cmca	6.24	4.48	8.26	35.2	90	31.3	

Note: Values in parentheses fixed during minimization.

[a] If (θ_i, ϕ_i) represent the polar angles of the molecular axis on the i-th sublattice (i = 1, 2), then by symmetry, $\phi_1 = \phi_2 = 0°$ and θ_1 and θ_2 are independent of each other for C2/m; $(\theta_1, \phi_1) = (\theta_2, -\phi_2)$ for C2/c.

[b] From Ref. 35.

[c] From Ref. 33.

Discussion

Our work indicates that it is advantageous to base the search for effective condensed-phase potentials on the properties of the low temperature solid. This is economical, because much can be done through minimization of the lattice energy and by lattice dynamics calculations before resorting to computationally expensive simulations. More important is the fact that potentials that reproduce the low temperature crystal properties, at least of the simpler molecular systems, work well for the liquid, while the converse is not, in general, true. This obviously implies that certain properties of the solid state are much more sensitive to details of the potential than is the case in the liquid, where substantial averaging occurs, and that the overall effect of many-body forces is similar in solid and liquid phases. Calculations of the second virial coefficient are less useful, but these also are cheap and easy to execute.

Models of the 2LJ + Q (or 2LJ + q) type give a fair description of a wide range of properties of both nitrogen and carbon dioxide, and are certainly superior to the 2LJ type. A main reason for their success is that they correctly stabilize the Pa3 structure found experimentally in these systems. The difficulties inherent in the use of such simple models become apparent when attention is turned to the noncubic structures. They are already to be seen in the case of γ-nitrogen, but are much more evident for the halogens; solid chlorine poses a particularly severe problem. Suggestions have been made in the past that some form of "chemical" bonding characterizes the halogen crystals (*36*, *41*), but this remains to be convincingly demonstrated. A possible refinement of the simple models is the introduction of anisotropic dispersion (AD) forces between molecular centers. The basis for this suggestion lies in lattice dynamics calculations for solid oxygen and fluorine (*2*, *40*). In the case of chlorine, the AD + Q type of model has been shown (*32*) to destabilize the Pa3 lattice relative to the observed Cmca structure, but the problem of unstable lattice modes has not yet been overcome. Inclusion of AD forces has also not been able to explain the α–γ transition in nitrogen (*42*, *32*). Solid acetylene has a low-temperature crystal structure similar to that of chlorine; it is significant that attempts to develop a simple potential model for acetylene (*52*) have so far also been unsuccessful.

We have concentrated here on the empirical route to potential modeling. This is inevitable, since the construction of accurate pair potentials from ab initio potential energy surfaces is for the present an unrealistic goal for the type of system we have discussed. However, such calculations may be helpful in suggesting realistic functional forms (*12*, *13*, *50*). Both points are well illustrated by the recent work of Berns and van der Avoird on nitrogen (*12*). They fitted two models to the calculated potential energy surface; model I contained four and model I′ only two short-range interaction sites, with the electrostatic interactions represented in each case by four point charges per molecule. Model I was found to give a better fit both to the potential energy surface (*12*) and to the experimental properties (*27, 51*) of nitrogen, but neither model is as satisfactory as the empirical potentials KD and MSKM.

The results described here for nitrogen and carbon dioxide and elsewhere (*7*) for carbon disulfide probably bring us close to the limits of what can be achieved with nLJ+Q or similar models. Rather than stressing their deficiencies, it is worth pointing out that these crude representations of the intermolecular potential are in many respects surprisingly successful. However, even for a molecule as apparently simple as carbon tetrachloride, recent calculations (*53*) based on 5LJ models have shown how difficult it is to account for details of the structure and dynamics of either liquid or solid phase.

Further advances could well rest on the use of more realistic descriptions of the electrostatic interactions. An interesting scheme whereby this could be achieved has recently been described by Stone (*54*). It should be noted that in several of the models detailed in Table I, either Q or l, or both, are treated as adjustable parameters. Though there is some theoretical justification for this, the physical meaning of an effective quadrupole moment or an effective bondlength is obscure. The use of such concepts has an empirical value, but it may well not be necessary if more details of the molecular charge distribution were to be incorporated into the potential model.

Key to Models Discussed in Text

BV = Ref. 12
CP = Ref. 5
EV = Ref. 36
KD = Ref. 19
KK = Ref. 40
MS = Ref. 39
MSB = Ref. 10
MSKM = Ref. 8
MSM = Ref. 9
RG = Ref. 11
RGA = Ref. 14
STS = Ref. 6
TC = Ref. 20
nLJ = n-site Lennard–Jones potential
nLJ + Q = As nLJ, with point quadrupole
nLJ + q = As nLJ, with point fractional charges

Literature Cited

1. Schnepp, O.; Jacobi, N. "Dynamical Properties of Solids", Horton, G. K.; Maraduddin, A. A., Eds.; North Holland: Amsterdam, 1975; Vol. 2, p. 151.
2. Kobashi, K.; Klein, M. L.; Chandrasekharan, V. *J. Chem. Phys.* **1979,** *71*, 843.
3. Kobashi, K.; Kihara, T. *J. Chem. Phys.* **1980,** *72*, 378.
4. Barojas, J.; Levesque, D.; Quentrec, B. *Phys. Rev.* **1973,** *7A*, 1092.
5. Cheung, P. S. Y.; Powles, J. G. *Mol. Phys.* **1975,** *30*, 921.
6. Singer, K.; Taylor, A.; Singer, J. V. L. *Mol. Phys.* **1977,** *33*, 1757.
7. Tildesley, D. J.; Madden, P. A. *Mol. Phys.* **1981,** *42*, 1137.
8. Murthy, C.S.; Singer, K.; Klein, M. L.; McDonald, I. R. *Mol. Phys.* **1980,** *41*, 1387.
9. Murthy, C. S.; Singer, K.; McDonald, I. R. *Mol. Phys.* **1981,** *44*, 135.
10. MacRury, T. B.; Steele, W. A.; Berne, B. *J. Chem. Phys.* **1976,** *64*, 1288.
11. Raich, J. C.; Gillis, N. S. *J. Chem. Phys.* **1977,** *66*, 846.
12. Berns, R. M.; van der Avoird, A. *J. Chem. Phys.* **1980,** *72*, 6107.
13. Jönsson, B.; Karlstron, G.; Romano, S. *J. Chem. Phys.* **1981,** *74*, 2856.
14. Raich, J. C.; Gillis, N. S.; Anderson, A. B. *J. Chem. Phys.* **1974,** *61*, 1399.
15. Huler, E.; Zunger, A. *Phys. Rev.* **1975,** *12B*, 5878.
16. Scott, T. A. *Phys. Rep.* **1976,** *27C*, 89.
17. Filippini, G.; Gramaccioli, C. M.; Simonetta, M.; Suffritti, B. *Mol. Phys.* **1978,** *35*, 1659.
18. Cheung, P. S. Y.; Powles, J. G. *Mol. Phys.* **1976,** *32*, 1383.
19. Kjems, J. K.; Dolling, G. *Phys. Rev.* **1975,** *11B*, 1639.
20. Thiéry, M. M.; Chandrasekharan, V. *J. Chem. Phys.* **1977,** *67*, 3659.
21. Bader, R. F. W.; Henneker, W. H.; Cade, P. E. *J. Chem. Phys.* **1967,** *46*, 3341.
22. Gibbons, T. G.; Klein, M. L. *J. Chem. Phys.* **1974,** *60*, 112.
23. Schmidt, J. W.; Daniels, W. B. *J. Chem. Phys.* **1980,** *73*, 4848.
24. Powell, B. M.; Dolling, G.; Piseri, L.; Martel, P. "Neutron Inelastic Scattering"; IAEA. 1972; p. 207.
25. Kobashi, K. *Mol. Phys.* **1978,** *36*, 225.
26. Kobashi, K.; Chandrasekharan, V. *Mol. Phys.* **1978,** *36*, 1645.
27. Luty, T.; van der Avoird, A.; Berns, R. M. *J. Chem. Phys.* **1980,** *73*, 5305.
28. Monson, P. A.; Rigby, M. *Mol. Phys.* **1980,** *39*, 1163.
29. Ibid. **1981,** *42*, 249.
30. Barker, J. A. "Rare Gas Solids"; Klein, M. L.; Venables, J. A., Eds.; Academic Press: New York, 1975; Vol. 1.
31. Fondere, F.; Obriot, J.; Varteau, P. L.; Allavena, M.; Chakround, H. *J. Chem. Phys.* **1981,** *74*, 2675.
32. Murthy, C. S.; Righini, R., unpublished data.
33. Collin, R. L. *Acta Crystallogr.* **1952,** *5*, 431; Ibid. **1956,** *9*, 537.
34. Donohue, J.; Goodman, S. H. *Acta Crystallogr.* **1965,** *18*, 568.
35. Meyer, L.; Barrett, C. S.; Greer, S. C. *J. Chem. Phys.* **1968,** *49*, 1902.
36. English, C. A.; Venables, J. A. *Proc. R. Soc. London* **1974,** *340A*, 57.
37. Emrich, R. J.; Steele, W. *Mol. Phys.* **1980,** *40*, 469.
38. Williams, J. H.; Amos, R. D. *Chem. Phys. Lett.* **1980,** *70*, 162.
39. Murthy, C. S.; Singer, K., unpublished data.
40. Kobashi, K.; Klein, M. L. *Mol. Phys.* **1980,** *41*, 679.

41. Pasternak, A.; Anderson, A.; Leech, J. W. *J. Phys. C. Solid State Phys.* **1978,** *11*, 1563.
42. Raich, J. C.; Mills, R. L. *J. Chem. Phys.* **1971,** *55*, 1811.
43. Buckingham, A. D.; Disch, R. L.; Dunmar, D. A. *J. Am. Chem. Soc.* **1968,** *90*, 3104.
44. Herzberg, G., "Molecular Spectra and Molecular Structure"; Van Nostrand: Princeton, NJ; 1950; p. 541.
45. Buckingham, A. D.; Disch, R. L. *Proc. R. Soc. London* **1963,** *273A*, 275.
46. Rowlinson, "Liquids and Liquid Mixtures"; Butterworths: London; 1969; pp. 48 and 60.
47. Din, F., "Thermodynamic Functions of Gases"; Butterworths: London; 1956; p. 102.
48. Stogryn, D. E.; Stogryn, A. D. *Mol. Phys.* **1966,** *11*, 371.
49. Thiéry, M. M.; Fabre, D. *Mol. Phys.* **1976,** *32*, 257.
50. Klein, M. L.; McDonald, I. R.; O'Shea, S. F. *J. Chem. Phys.* **1978,** *69*, 63.
51. Murthy, C. S., unpublished data.
52. Klein, M. L.; McDonald, I. R. *Chem. Phys. Lett.* **1981,** *80*, 76.
53. McDonald, I. R.; Bounds, D. G.; Klein, M. L. *Mol. Phys.* **1982,** *45*, 521.
54. Stone, A. J. *Chem. Phys. Lett.* **1981,** *83*, 233.

RECEIVED for review January 27, 1982. ACCEPTED for publication September 13, 1982.

10

Effects of Molecular Anisotropy

FRIEDRICH KOHLER

Ruhr-Universität, Institut für Thermo- und Fluiddynamik, D-4630 Bochum, Federal Republic of Germany

NICHOLAS QUIRKE[1]

Royal Holloway College, Chemistry Department, Egham, Surrey, England

The effects of molecular anisotropy considered in this chapter are the molecular shape (the anisotropy parameter being the elongation of two-center Lennard–Jones fluids), the dipole moment and the quadrupole moment. An attempt is made to scale density and temperature of two-center Lennard–Jones fluids in such a way that a comparison with the law of corresponding states is possible. With respect to electric moments, it is observed that their effect on thermodynamic and structural properties is less on two-center Lennard–Jones fluids than on spherical fluids. This is investigated in some detail.

Although considerable progress has been achieved in understanding the behavior of molecular liquids (*1–4*), we are still far from having a complete picture of the way in which the molecular shape and electric moments contribute to thermodynamic and structural properties. For example, our understanding of the deviations from the law of corresponding states has not improved since Rowlinson's work in 1954 (*5*).

One important difficulty is that of scaling temperature and density when comparing experimental and theoretical results. While theoreticians use the characteristic parameters of the pair potential for scaling (e.g., the depth of the potential ε and the zero potential separation σ), experimentalists use critical data. At present a sound correlation between these different approaches exists only for one-center Lennard–Jones liquids. In the first section of this chapter we suggest ways of extending this correlation to two-center Lennard–Jones liquids.

The next section of this chapter gives a critical review of a computationally fast thermodynamic perturbation theory treatment of two-center Lennard–Jones liquids. While the Helmholtz energies are predicted accurately, some details of the structural properties are still missing. The subsequent two sections are devoted to the problem of treating molecules with electric moments within the framework of perturbation theory. The treatment given is in some respects preliminary

[1] Current address: University of Maine, Department of Chemistry, Orono, ME 04469

0065-2393/83/0204-0209$07.50/0

and the results, for the moment, qualitative. One section deals with the thermodynamic properties, while the final section deals with those structural properties that are related to the static dielectric constant. We hope that the approach outlined in these sections will form the basis for future work on this topic.

Scaling Parameters for Two-Center Lennard–Jones Fluids

In this section we suggest methods of scaling densities in three regions—regions of low densities, critical densities and liquid densities. Low density scaling is considered first. Table I contains second virial coefficients for various elongations $L = l/\sigma$. This extends the table given by Wojcik et al. (*6*) for a limited temperature range[2]. In Figure 1, these results are plotted against the reduced temperature T/T_B, where T_B is the Boyle temperature. It can be seen that the curves for higher elongations become progressively steeper. A more detailed comparison is provided by Figure 2, where the second virial coefficients are reduced by an effective σ^3_{2CLJ} in such a way that $B/N_A\sigma^3_{2CLJ}$ is the same for all elongations at $T/T_B = 0.3$.

This scaling produces a single curve for all elongations in the temperature range $0.3 < T/T_B < 1.05$; at lower temperatures the reduced curves begin to spread (Figure 3), with the higher elongations having the more negative second virial coefficients. This might at first suggest that the parameters L and σ could be determined separately from such a plot of experimental second virial coefficients in the low temperature region ($T/T_B < 0.3$). However, this is questionable for two reasons: (1) low-temperature second virial coefficients are in most cases subject to large errors and (2) the two-center Lennard–Jones model potential cannot accurately reproduce the low-temperature second virial coefficients of real substances. Returning to the problem of low density scaling, Table II shows $T_{B,2CLJ}/T_{B,1CLJ}$ and Figure 4 shows $\sigma_{2CLJ}/\sigma_{1CLJ}$ using the values for effective σ obtained by equalizing the reduced second virial coefficients at $T/T_B = 0.3$. By chance, the plot in Figure 4 is almost a straight line. Figure 4 shows a similar plot for hard dumbbells obtained using the Boublík–Nezbeda equation of state (*7*), which gives

$$\frac{2}{3}\pi N_A\sigma^3_{1CLJ} = \frac{\pi}{6}N_A\,\sigma^3_{2CLJ}\left(1 + \frac{3}{2}L - \frac{1}{2}L^3\right) \cdot\left(1 + \frac{3\,(1 + L)(2 + L)}{2 + 3L - L^3}\right) \tag{1}$$

$$\sigma^3_{1CLJ}/\sigma^3_{2CLJ} = 1 + \frac{3}{2}L + \frac{3}{8}L^2 - \frac{1}{8}L^3$$

[2] Note added in proof: A table listing the second virial coefficients for the elongations 0.1, 0.2, 0.3, 0.4, 0.5, and 0.6 is contained in Maitland, G. C.; Rigby, M.; Smith, E. B.; Wakeham, W.A. *Intermolecular Forces*, Clarendon Press, Oxford 1981.

Table I. Reduced Second Virial Coefficients of Two-Center Lennard–Jones Fluids

T^*	$\frac{T^*}{T^*_B}$	$\frac{B}{N_A \sigma^3}$
	L = 0.3292	
0.8	0.0908	−102.631
0.9	0.1022	−71.376
1.0	0.1135	−53.365
1.25	0.1419	−31.189
1.55	0.1759	−19.905
2.0	0.2270	−12.276
2.6430	0.3000	−7.404
3.0	0.3405	−5.871
5.2860	0.6000	−1.745
8.0	0.9081	−0.246
8.75	0.9932	−0.017
8.8100	1.0000	0.000
9.00	1.0216	0.050
9.2505	1.0500	0.113
	L = 0.505	
0.8	0.1202	−71.037
0.9	0.1352	−51.371
1.0	0.1502	−39.503
1.1	0.1652	−31.692
1.2	0.1803	−26.213
1.3	0.1953	−22.181
1.4	0.2103	−19.101
1.5	0.2253	−16.677
1.6	0.2403	−14.724
1.7	0.2554	−13.118
1.8	0.2704	−11.775
1.9	0.2854	−10.637
1.9971	0.3000	−9.686
2.1	0.3155	−8.812
2.2	0.3305	−8.071
2.3	0.3455	−7.417
2.4	0.3605	−6.835
2.5	0.3755	−6.316
2.6	0.3906	−5.848
2.7	0.4056	−5.426
3.28	0.4927	−3.602
3.9992	0.6000	−2.224
4.1	0.6159	−2.069
6.5	0.9764	−0.074
	L = 0.505 (continued)	
6.65	0.9989	−0.003
6.6570	1.0000	0.000
6.9813	1.0500	0.140
7.0	1.0515	0.148
	L = 0.63	
0.75	0.1330	−70.251
1.0	0.1774	−33.674
1.25	0.2217	−20.856
1.5	0.2661	−14.522
1.6911	0.3000	−11.518
2.0	0.3548	−8.326
2.5	0.4435	−5.305
2.81	0.4985	−4.108
3.3823	0.6000	−2.595
3.51	0.6227	−2.339
5.25	0.9313	−0.263
5.62	0.9970	−0.011
5.6371	1.0000	0.000
5.9190	1.0500	0.166
6.0	1.0644	0.211
	L = 0.793	
0.6	0.1271	−113.160
0.7	0.1483	−69.746
0.8	0.1695	−48.612
0.9	0.1907	−36.501
1.0	0.2119	−28.773
1.1	0.2330	−23.457
1.35	0.2860	−15.458
1.4160	0.3000	−14.059
1.5	0.3178	−12.543
1.75	0.3708	−9.229
2.0	0.4237	−7.019
2.8321	0.6000	−3.097
4.0	0.8474	−0.775
4.7201	1.0000	0.000
4.755	1.0074	0.031
4.9561	1.0500	0.197
5.0	1.0593	0.232

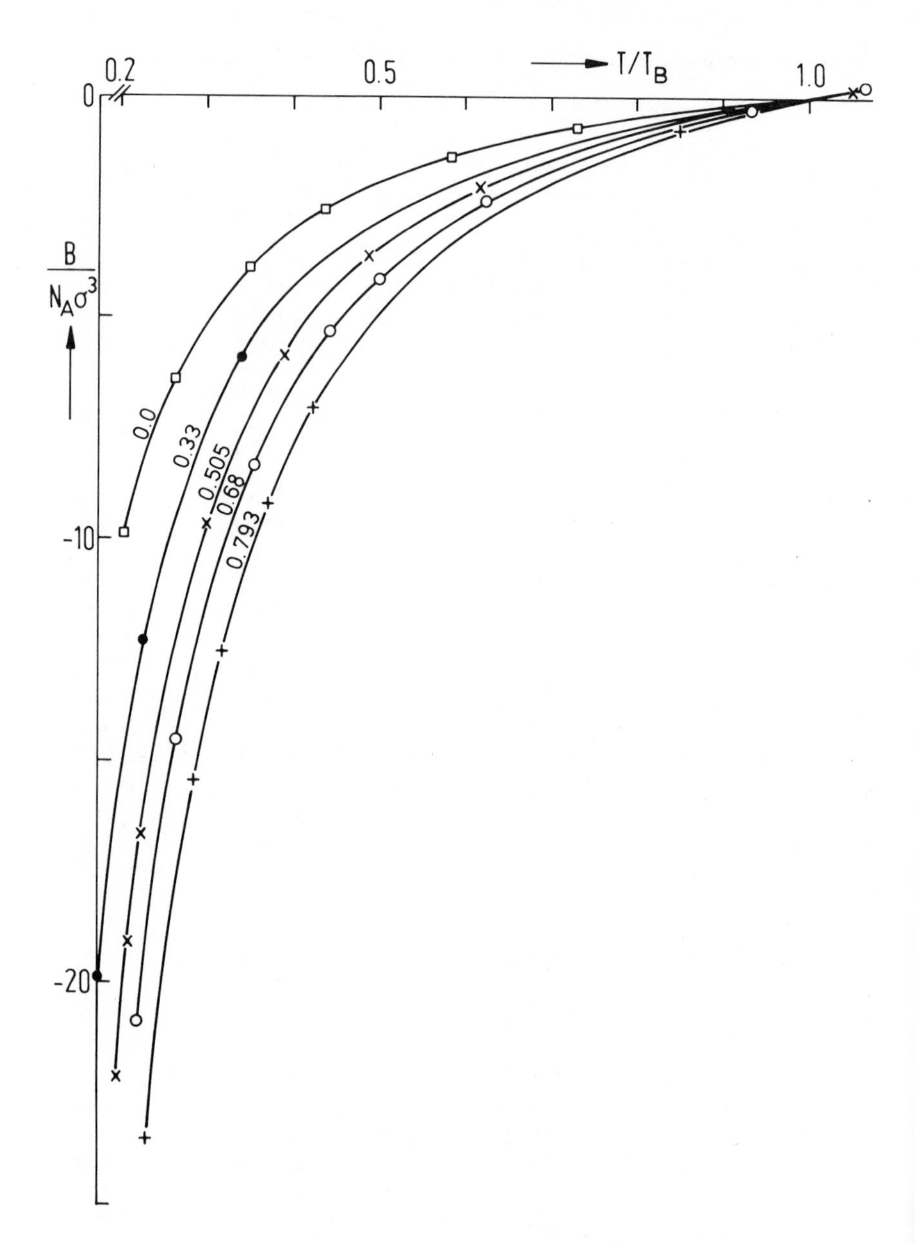

Figure 1. The reduced second virial coefficients, $B/N_A\sigma^3$*, of one-center and two-center Lennard–Jones fluids, plotted against the temperature reduced by the Boyle temperature, with the elongation* L *as parameter.*

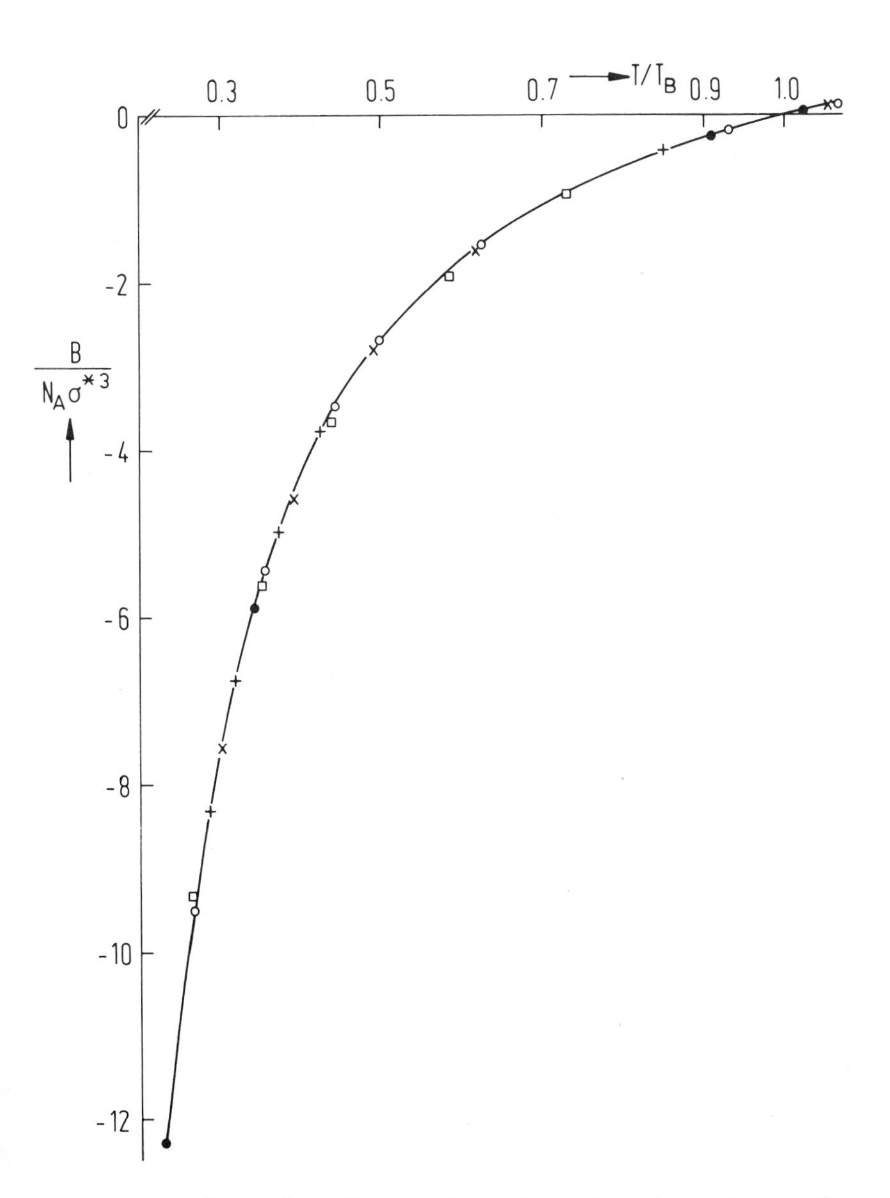

Figure 2. Plot similar to Figure 1, but with an adjusted σ_{2CLJ} *that equalizes all* $B/N_A\sigma^3_{2CLJ}$ *at* $T/T_B = 0.3$*. The notation of the points corresponds to Figure 1.*

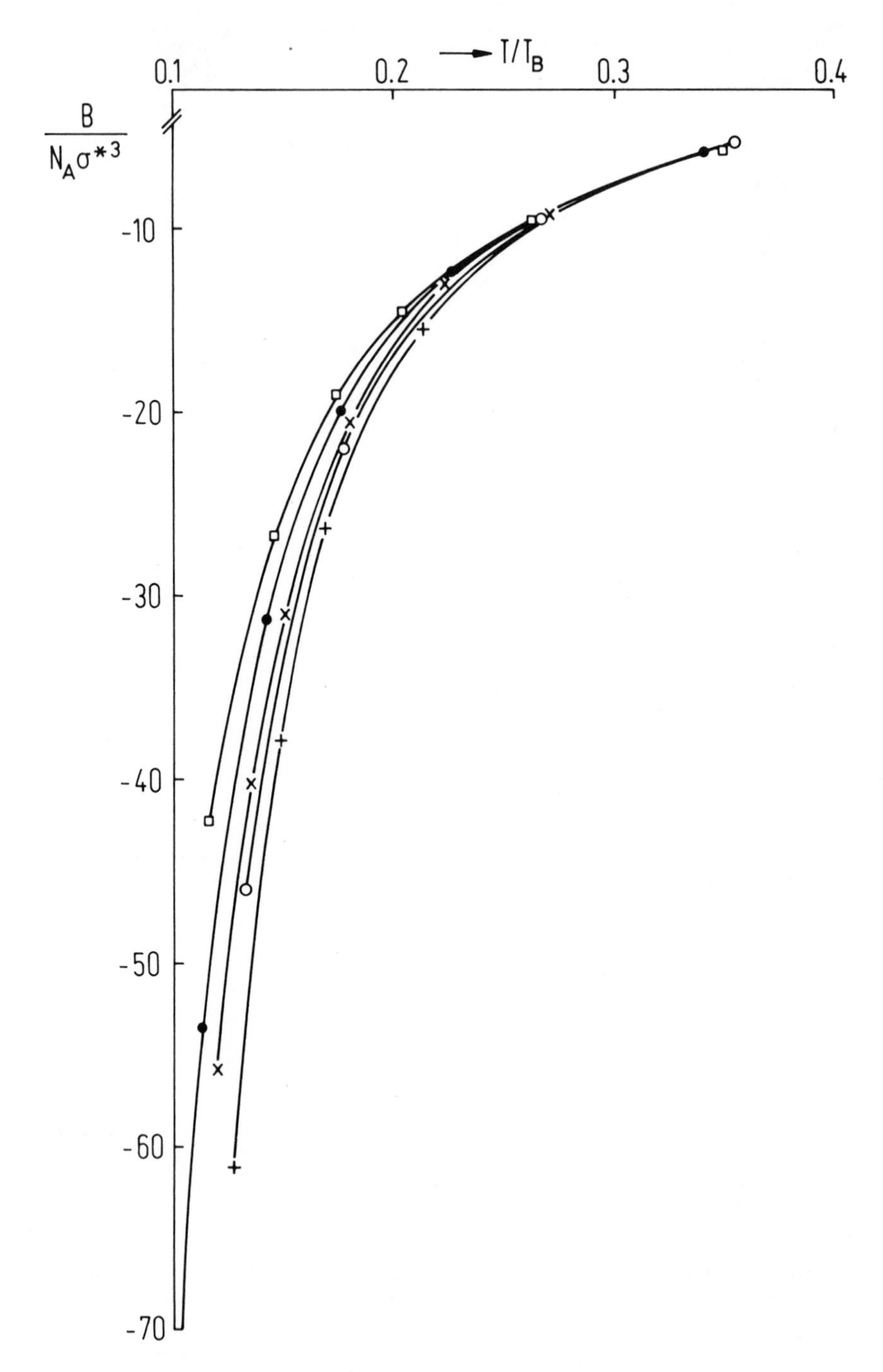

Figure 3. Plot similar to Figure 2, for low values of T/T_B.

Table II. Boyle Temperatures and Calculated Critical Temperatures of Two-Center Lennard–Jones Fluids Compared with the "United Atom"

L	$T^*_{B,2CLJ}/T^*_{B,1CLJ}$	$T^*_{c,2CLJ}/T^*_{c,1CLJ}$	$(T^*_{c,2CLJ}/T^*_{c,1CLJ})_{exp}$
0	1	1	1
0.329	0.644	0.676	0.633
0.505	0.487	0.524	0.525
0.630	0.412	0.451	0.450
			0.437
0.793	0.345	0.386	0.358

Note: Critical temperatures are calculated by Equation 3 with $a = 0.15$. Experimental quantities are calculated from experimental critical temperatures of liquids for which certain parameters L and ε have been used successfully.

The agreement between the 2CLJ curves and the hard dumbbell curves can be improved for large L if the reduced second virial coefficients are set equal at higher temperatures, $T/T_B = 0.6$ or 1.05 rather than 0.3 (*see* Figure 4), but no such improvement can be achieved at small L. This small discrepancy between two-center Lennard–Jones fluids and hard dumbbell fluids is probably related to the different temperature dependence of the effective sphere radius for hard spheres and hard dumbbells. Figure 5 shows results obtained from the perturbation theory reviewed in the next section. We have not been able to scale these results.

Turning now to higher densities, we consider an approximate method of scaling the critical densities based on the generalized van der Waals model (*8–10*)

$$\frac{p}{\rho NkT} = \left(\frac{p}{\rho NkT}\right)_{\text{Hard Fluid}} - \frac{a\rho}{NkT} \tag{2}$$

Applying the critical conditions $\frac{\partial p}{\partial \rho} = 0$ and $\frac{\partial^2 p}{\partial \rho^2} = 0$, two equations for the two unknown ρ_c and $A_c = a/T_c$ are obtained. Using the Carnahan–Starling equation for the hard sphere fluid and the Boublík–Nezbeda equation (*7*) for the hard dumbbell fluid, effective values for σ_{2CLJ} can be found, which when used to reduce the critical densities of Table III make them all equal. Figure 4 shows the resulting values $\sigma_{2CLJ}/\sigma_{1CLJ}$.

For densities in the liquid range, we have attempted to scale the orthobaric density curve (effectively zero pressure densities) given by Wojcik et al. (*6*) as a function of elongation. In order to bring them into a form comparable to the law of corresponding states, we had to assume

$$\left(\frac{T_c}{T_B}\right)_{2CLJ} = \left(\frac{T_c}{T_B}\right)_{1CLJ} \times (1 + aL) \tag{3}$$

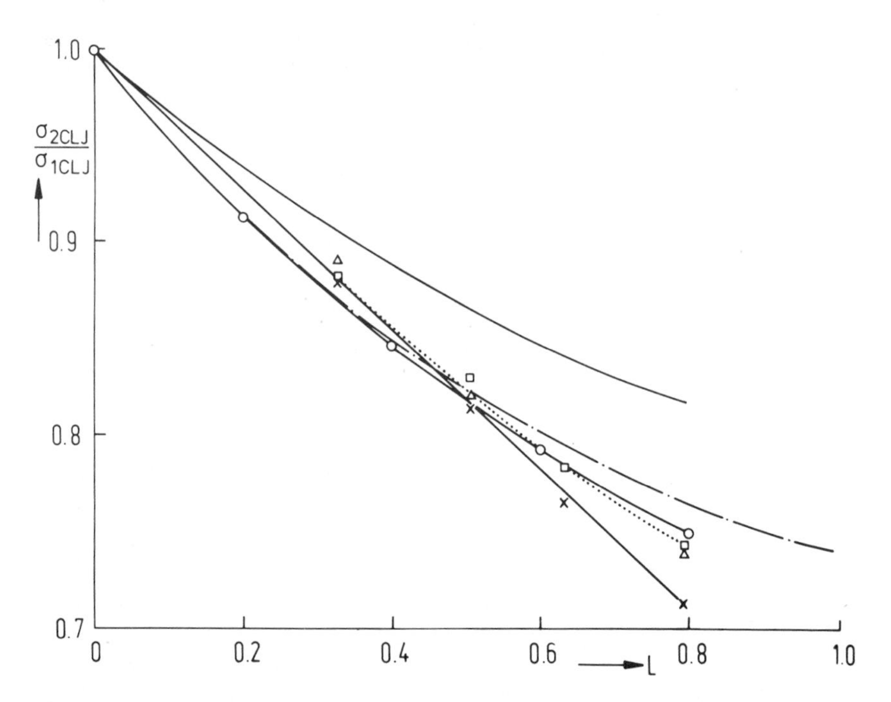

Figure 4. Various scalings for $\sigma_{2CLJ}/\sigma_{1CLJ}$. Key: ×——×, based on equalization of reduced second virial coefficients at T/T$_B$ *= 0.3; . . . ., based on equalization of second virial coefficients at* T/T$_B$ *= 0.6 (□) and* T/T$_B$ *= 1.05 (△); ○——○, based on the relation between hard dumbbells and hard spheres (Equation 1); —·—·—, based on critical densities derived from a generalized van der Waals model (Equation 2); ———, based on scaling orthobaric densities.*

with $a = 0.15$ and the $\sigma_{2CLJ}/\sigma_{1CLJ}$ scaling curve given in Figure 4. However, the results of this scaling, given in Figure 6, are not perfect. Though we consider only the temperature range given by Wojcik et al. (*6*), some of the curves show inconsistent behavior near the ends of the temperature range. This behavior is such that it is difficult to explain on the basis of errors in scaling approximations. Further, from the known deviations from the law of corresponding states, the reduced density should be highest for the largest elongation at low temperatures, which is not the case so far. It could be achieved by making the parameter a in Equation 3 larger, which would in turn lead to a $\sigma_{2CLJ}/\sigma_{1CLJ}$ curve a little lower than that in Figure 4. It would be impossible, however, to make a so large that the $\sigma_{2CLJ}/\sigma_{1CLJ}$ curves would be coincident with the results from lower densities. It is probable that the scaling procedures described here require data of a higher accuracy than are presently available.

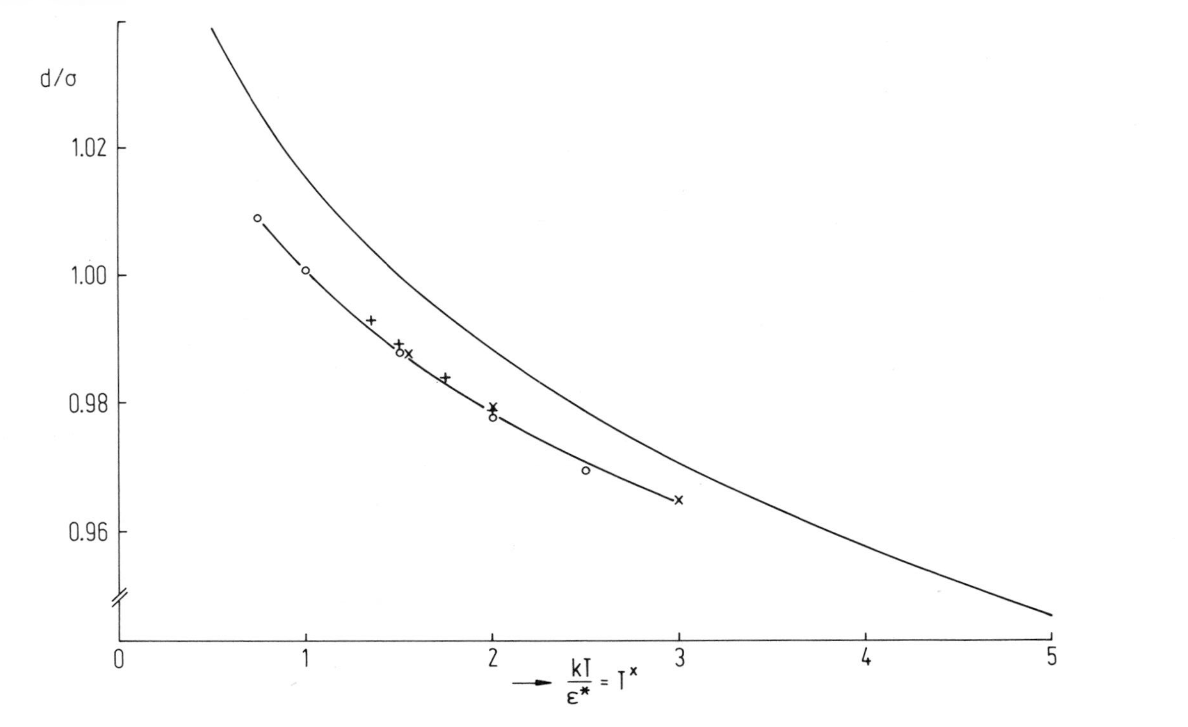

Figure 5. The reduced diameter d/σ *for equivalent hard spheres and hard dumbbells plotted against* kT/ε, *with* ε *and* σ *being depth and zero potential separation, respectively, of either the one-center or the two-center Lennard–Jones potential. Upper curve, one center (BH1 treatment); lower curve, two centers. Key:* ×, L = *0.329;* ○, L = *0.63; and* +, L = *0.793. For details of the BH1 treatment see Smith, W. R. In* Statistical Mechanics *Vol. 4, Singer, K, Ed; Specialist Periodical Report, Chemical Society, London 1973, p. 95. (Reproduced from Ref. 11. Copyright 1981, American Chemical Society.)*

Table III. Critical Densities of the Generalized van der Waals Model

L	y_c
0.2	0.12929
0.4	0.12612
0.6	0.12116
0.8	0.11443
1.0	0.10577

Note: Critical densities are reduced by the volume of the hard body. Generalized van der Waals model is given by Equation 2.

With the same scaling factors that we have applied for the orthobaric densities; i.e., the values of $T^*_{c,2CLJ}/T^*_{c,1CLJ}$ from Equation 3, shown in Table II, and the curve of $\sigma_{2CLJ}/\sigma_{1CLJ}$ shown as the upper curve in Figure 4, we plot in Figure 7 the results of residual Helmholtz energies derived by Fischer (*11*) on the basis of perturbation theory discussed below. As Figure 7 shows, the scaling is quite good, but not perfect. Curves for higher elongations are steeper at high densities, and the two lowest curves are a bit too far apart. However, the results indicate that our scaling factors are not far from the best values. We hope that this discussion, enabling better comparisons, will induce more accurate calculations on two-center Lennard–Jones fluids.

A short concluding remark should be made on the physical sense of Equation 3. The work of Rowlinson (*5*) has clearly shown that T_c/T_B of anisotropic molecules is increased in comparison with the same ratio for spherical molecules. Furthermore, for carbon dioxide, $(T_c/T_B)_{CO_2}$ = 0.427 (*12*), which is 1.16 times that of argon, $(T_c/T_B)_{Ar}$ = 0.367 (*13*). The comparison with real substances is, of course, hindered partly by experimental uncertainties and partly by the unknown effects of the quadrupole moments of real substances. Therefore, it would be of much interest to extend the computation of critical points from one-center (*14*) to two-center Lennard–Jones liquids. Some work in this direction, although primarily concerned with the coexistence line, has been reported recently (*15*).

Perturbation Expansions for Two-Center Lennard–Jones Liquids

The perturbation expansion is a generalization to molecular liquids of the Weeks–Chandler–Andersen (WCA) (*16*) expansion for atomic liquids. This procedure assumes that the structure of the dense liquid is primarily determined by the repulsive forces. Those parts of the pair potential responsible for repulsive forces are separated out and used as a reference potential. The properties of the reference fluid interacting

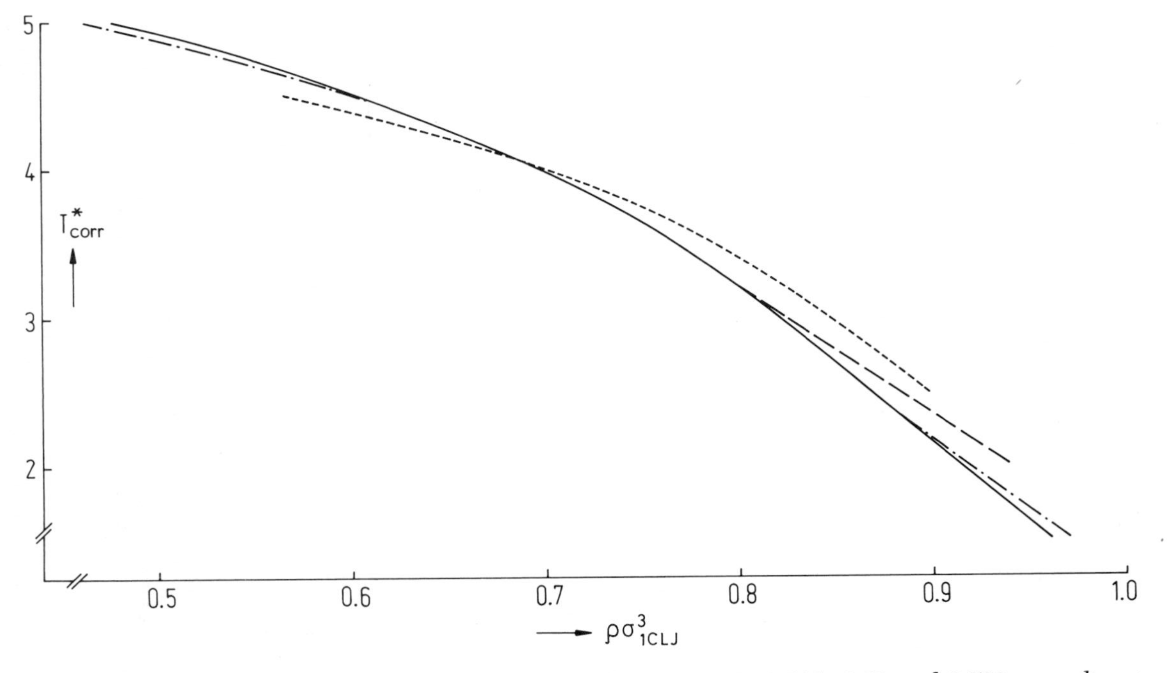

Figure 6. Orthobaric densities calculated for L *equal to 0, 0.329, 0.505, 0.63 and 0.793 according to Ref. 6, plotted against reduced temperatures* T^*_{corr}. *The densities are scaled with* $\sigma_{2CLJ}/\sigma_{1CLJ}$ *ratio corresponding to the upper curve of Figure 4, the temperatures are* $T^*_{corr} = (kT/\varepsilon_{2CLJ})(T_{B,1CLJ}/T_{B,2CLJ})/(1 + 0.15\ L)$. *There are only slight deviations from the curve for* L = *0.793 (———). The largest deviation is for* L = *0.329 (----). Other lines are for* L = *0 (– – –);* L = *0.505 for high* T^*_{corr} *(—·—·—) and* L = *0.505 and* L = *0.63 for low* T^*_{corr} *(—·—·—).*

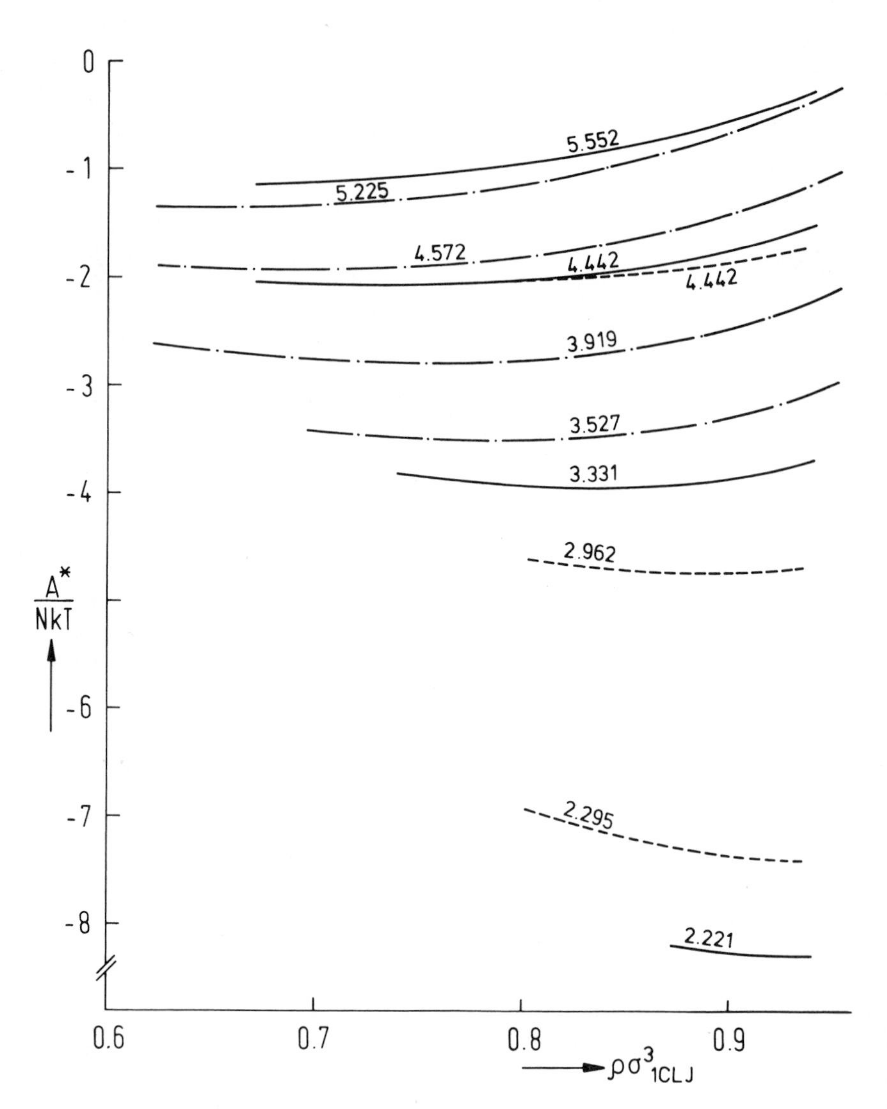

Figure 7. The residual Helmholtz energies of two-center Lennard–Jones liquids plotted against reduced densities. Reduced temperatures are given as parameter. Scaling of densities and temperatures as in Figure 6. Key: – – –, L = *0.329:* ———, L = *0.63;* —·—·—, L = *0.793.*

by the reference potential are obtained as those of an equivalent hard-body fluid. Earlier attempts (*17–19*) at a generalization of the WCA approach were hindered by the lack of a suitable hard reference fluid. Current work has been made possible by the advent of an analytic expression for the thermodynamic properties of hard convex bodies and hard dumbbells (*7*). Two versions of the generalized WCA expansion for two-center Lennard–Jones liquids have been developed, one based on the molecular potential (*11*, *20*) and another on the site–site potential (*21*). In the molecular approach two developments combine to increase the speed of the calculations. The simplified solution of the Percus–Yevick equation for potentials of finite range introduced by Baxter (22), and an analytical expression for part of the Boltzmann factor of the hard dumbbell potential (23).

In the treatment based upon the molecular potential, the first stage is to perform the WCA division of the potential. This is carried out for each orientation $\Omega_1\ \Omega_2$ of the molecular pair, as illustrated in Figure 8.

Denoting the minimum coordinates at each orientation $r_{\min}(\Omega_1\ \Omega_2)$ and $u_{\min}(\Omega_1\ \Omega_2)$, we have then for the reference potential

$$\begin{aligned} u_{\mathrm{ref}}(r, \Omega_1, \Omega_2) &= u(r, \Omega_1, \Omega_2) - u_{\min} & r &\leq r_{\min}(\Omega_1\ \Omega_2) \\ u_{\mathrm{ref}}(r, \Omega_1, \Omega_2) &= 0 & r &> r_{\min}(\Omega_1\ \Omega_2) \end{aligned} \tag{4}$$

and for the perturbation

$$\begin{aligned} u_{\mathrm{pert}}(r, \Omega_1, \Omega_2) &= u_{\min}(\Omega_1\ \Omega_2) & r &\leq r_{\min}(\Omega_1\ \Omega_2) \\ u_{\mathrm{pert}}(r, \Omega_1, \Omega_2) &= u(r, \Omega_1, \Omega_2) & r &> r_{\min}(\Omega_1\ \Omega_2) \end{aligned} \tag{5}$$

The perturbation expansion leads to the following expression for the Helmholtz energy of the molecular liquid

$$A = A_{\mathrm{ref}} + 2\pi\rho \int r^2\, dr\, \langle y_{\mathrm{ref}} e^{-\beta u_{\mathrm{ref}}}\, u_{\mathrm{pert}} \rangle_{\Omega_1, \Omega_2} \tag{6}$$

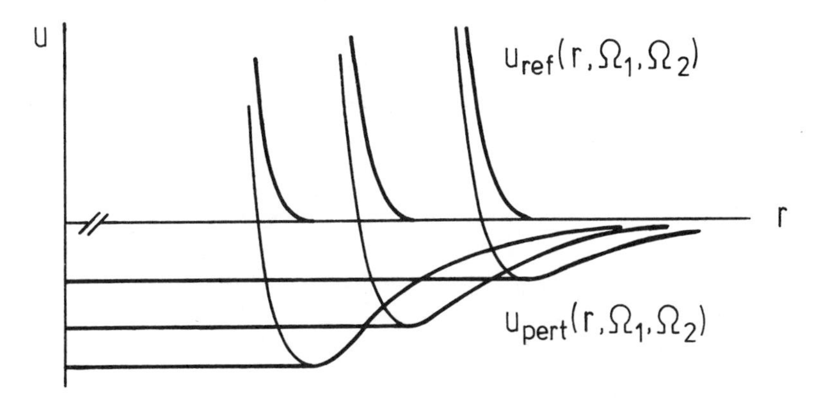

Figure 8. Decomposition of the molecular potential for different orientations in the molecular perturbation theory (after Kohler, Ref. 4).

where the brackets denote angle-averaging and y is the background correlation function $y = g/e^{\beta u}$. The next problem is to obtain the properties of the reference fluid in terms of the adjusted hard-body fluid. In a fashion completely analogous to the WCA treatment of atomic liquids to first order we obtain,

$$A_{\text{ref}} = A_d + 2\pi\rho \int r^2\, dr\, \langle y_d(e^{-\beta u_{\text{ref}}} - e^{-\beta u_d})\rangle \tag{7}$$

where the subscript d refers to a hard dumbbell fluid with diameter d. This implies that $A_{\text{ref}} = A_d$ when

$$\int r^2\, dr\, \langle y_d(e^{-\beta u_{\text{ref}}} - e^{-\beta u_d})\rangle = 0 \tag{8}$$

and is the prescription for adjusting the hard dumbbell diameter according to Kohler, Quirke, and Perram (*20*). The second hard dumbbell parameter, the elongation, has been kept equal to the elongation l_m of the molecular liquid in all calculations to date. Now we come to a crucial assumption, which makes Equations 6 and 8 tractable and which has proved to be a reasonable approximation for some previous cases (*24*, *25*), that the background correlation function is effectively angle independent and equal to its angle average

$$y(r, \Omega_1, \Omega_2) = \langle y(r, \Omega_1, \Omega_2)\rangle = y_{000}(r) \tag{9}$$

(For an early recognition of some cases where this approximation is poor, see Ref. 26. A more thorough investigation of this approximation, giving methods of improvement, is in progress by W. A. Steele.)

This assumption takes y_d out of the angle-averaging bracket in Equation 8, $y_d = y_{\text{ref}}$ being the background correlation function for an assembly with interaction potential $\bar{u}(r) = -kT \ln \langle e^{-\beta u_d}\rangle$. The function y_d is obtained by solving the Percus–Yevick equation for $\bar{u}(r)$ (*20*), giving the final result (cf Equation 6)

$$A = A_d + 2\pi\rho \int r^2\, dr\, y_d(r)\, \langle u_{\text{pert}}\, e^{-\beta u_{\text{ref}}}\rangle \tag{10}$$

An alternative formulation is to expand the hard dumbbell Helmholtz energy about that of the reference fluid (*11*)

$$A_d = A_{\text{ref}} + 2\pi\rho \int r^2\, dr\, y_{\text{ref}}(r)\, (\langle e^{-\beta u_{\text{ref}}} - e^{-\beta u_d}\rangle) \tag{11}$$

with y_{ref} determined from $\langle e^{-\beta u_{ref}} \rangle$, which leads to

$$A = A_d + 2\pi\rho \int r^2 \, dr \, y_{ref}(r) \langle u_{pert} \, e^{-\beta u_{ref}} \rangle \tag{12}$$

These two formulations give essentially the same results. Pressures calculated by differentiating the theoretical Helmholtz energy have been found to be in very good agreement with computer simulation results for elongations from $L = 0.33$ (simulating nitrogen) to $L = 0.793$ (simulating carbon dioxide) (*11*). A more direct test is the comparison of the Helmholtz energy with results from computer simulations. This has become possible because fast and economical Monte Carlo methods (working with a small number of particles) have been developed for obtaining differences in Helmholtz energy between a reference system and the system in question (*27, 28*). In Table IV, we have collected all the results for the Helmholtz energy of one state of the nitrogen simulation (*10, 29, 30*). This extends a similar table given previously (*10*). Different computer methods, starting from different reference states, agree remarkably well with each other and with results from different perturbation theories. This good agreement extends also to other temperatures and densities typical for simulated liquid nitrogen (*30*).

We have included in Table IV the results of the perturbation method for two-center Lennard–Jones liquids developed by Tildesley (*21*) in the site–site coordinate frame, which is again a generalization of the WCA approach. Instead of dividing the full molecular potential at each ori-

Table IV. Configurational Helmholtz Energy of Simulated Liquid Nitrogen at $T^* = 3.0$, $\rho^* = 0.70$

Method	A*/NkT	*Reference Fluid*	*Reference*
Theory			
Equation 10	−3.00	Hard dumbbell	20
Equation 12	−3.06	Hard dumbbell	11
Tildesley	−3.11	Hard dumbbell	32
Simulation			
Bennett (32)[a]	−3.07	Argon	10
Virtual Overlap (32)	−3.03	Hard dumbbell	29
Marquee (32)	−3.06	Hard dumbbell	29
Bennett (32)	−3.09	Hard dumbbell	30
Marquee (32)	−3.09	Hard dumbbell	30
Bennett (108)	−3.04	Hard dumbbell	30

[a] Numbers in parentheses are the numbers of particles used in the computer simulations.

entation into repulsive and attractive components, the site–site potential is divided as in the case of the spherical Lennard–Jones potential along the site–site distances (R_{ss}). Then the four site–site repulsive Lennard–Jones potentials are recombined to produce the two-center Lennard–Jones reference system.

Equation 6 now separates into four equivalent terms, each containing one of the truncated site–site potentials and a $y_{ref}(R_{ss})$ term. This is the site–site background correlation function defined by

$$y_{ref}(R_{ss}) = e^{\beta u_{ref}(R_{ss})} g_{ref}(R_{ss}) \tag{13}$$

The properties of the repulsive two-center Lennard–Jones reference fluid are expressed, as before, in terms of a hard dumbbell fluid, the structure of which is given by a solution of a RISM (reference interaction site model) equation (2). This theory predicts the Helmholtz energy and site–site distribution function for a two-center Lennard–Jones liquid. The pressures obtained using this expansion are also in a good agreement with simulation (*21*).

Considering now the structural predictions of the perturbation methods, the centers correlation functions predicted by the molecular approach (Equation 9)

$$g_{000}(r) = y_{000} \langle e^{-\beta u_{ref}} \rangle \tag{14}$$

are in good agreement (*11*, *20*) with the corresponding computer simulation results. However, higher spherical harmonic radial coefficients, i.e., averages of g(12) over spherical harmonics ($l \neq 0$), which are calculated using the assumption of an angle independent y

$$g_{11'm}(r) = \frac{\langle e^{-\beta u_{ref}} Y_l^m(\Omega_1) Y_{l'}^{-m}(\Omega_2) \rangle}{\langle e^{-\beta u_{ref}} \rangle} \tag{15}$$

are found to be erroneous at large separations (r). This is because the reference potential is zero for all orientations $r > l + 2^{1/6}\sigma$ and therefore $g(r, \Omega_1, \Omega_2)$ is angle independent outside this range. The true $g(r, \Omega_1, \Omega_2)$ is angle dependent well beyond this separation. The short range angle dependence of $g(r, \Omega_1, \Omega_2)$ from Equation 15 means that it cannot be used to calculate structural properties of the molecular liquid, such as the site–site correlation function, which even for small R_{ss} depend on the value of $g(r, \Omega_1, \Omega_2)$ over a range of center–center separations r, including those where the angle dependence is erroneous. The angle dependence of g (12) can be somewhat improved by using the exponential of the full molecular potential (25). The alternative formulation of the perturbation theory in terms of site–site functions just discussed yields $g_{ss}(R_{ss})$ as its only structural prediction. These are in very good agreement with simulation (*21*). Part of the angle dependence of $g(12)$ can be obtained from $g_{ss}(R_{ss})$ using simple site superposition approximations of a type proposed recently (*31*). This enables properties such as the pressure and mean squared torque, which depend upon integrals of the angular variation of $g(r, \Omega_1, \Omega_2)$, to be calculated from a knowledge of $g_{ss}(R_{ss})$ alone. Not all the angular information can be reconstructed in this manner, for example, the light scattering factor G_2 is not predicted accurately (*31*).

In conclusion, the thermodynamic properties of two-center Lennard–Jones liquids are predicted accurately by generalizations of the WCA perturbation expansion using a hard dumbbell reference fluid. The structural predictions are limited to the unweighted angle average of the angular correlation function $g(r, \Omega_1, \Omega_2)$ in the coordinate frame employed. Using the molecular approach, we obtain the centers correlation function, and using the site–site approach, the site–site correlation function. These structural predictions can be extended to include part, but not all, of the angle dependence of $g(12)$ by the use of further approximations.

Thermodynamic Effects of Adding Electric Moments to Two-Center Lennard–Jones Liquids

In this section we consider a rapid method of calculating the thermodynamic effects of electric moments, within the framework of the perturbation theory outlined in the section above. The aim is to obtain qualitative trends rather than accurate numerical results.

The problems involved in treating the long range multipole forces have usually been considered for the case where the molecular shape is spherical. Two perturbation expansions, relating the properties of the fluid with electric moments to a reference fluid interacting with a spherical potential, have been tried. The first (*33, 34*) obtains a reference potential from the angle averaged potential

$$u_{\text{ref}}(r) = \langle u(r, \Omega_1, \Omega_2)\rangle \tag{16}$$

The second (*35–37*) angle-averages the Boltzmann factor

$$e^{-\beta u_{\text{ref}}(r)} = \langle e^{-\beta u(r,\, \Omega_1, \Omega_2)}\rangle \tag{17}$$

In both cases the resulting reference fluid has then to be treated by the usual WCA (*16*) perturbation method. The reference potential of Equation 17, unlike that of Equation 16, contains contributions from the electric moments and is therefore somewhat closer to the full potential. However, it still gives Helmholtz energies that are far too negative, whereas Equation 16 gives Helmholtz energies that are far too positive. The calculation of higher terms in the expansions requires a knowledge of three-body and higher correlation functions of the reference fluid, which have to be obtained from simulation or by superposition approximation. For moderate to large electric moments, the expansion series have very poor convergence. However, the expansion based on Equation 16, when combined with the use of a Padé approximant, works well (38, 39), except for short-range force cases noted in Ref. 34. The Padé approximant procedure is impractical for two-center Lennard–Jones liquids and in this section we seek another method in the reference fluid. From the methods discussed above we obtain an effective dipole potential of $\frac{1}{3}\beta\mu^4/r^6$ for small values of μ^2/r^3. The same result can be obtained, within the framework of the perturbation theory of the last section, by splitting the dipole–dipole interaction into u_{DD} for attractive pair configurations and $u_{DD}e^{-\beta u_{DD}}$ for repulsive pair configurations. However, for small distances r, the Boltzmann factor $e^{-\beta u_{DD}}$ cuts out too much of the repulsive potential, making the resulting reference potential too negative, in a similar fashion to u_{ref} of Equation 17 or the first perturbation term of the expansion based on Equation 16. It is then necessary to replace the Boltzmann factor by a function that does not tend to zero for large values of βu_{DD}. A convenient choice is the Langevin function $L(\beta u_{DD})$, which is linear for small βu_{DD} and approaches a constant for large values of βu_{DD}. For spherical molecules with electric moments we have tried the following reference potential,

$$u_{ref}(r, \Omega_1, \Omega_2) = u_{1CLJ}(r) + u_{DD}\, e^{-3/2L(2\beta u_{DD})\,\theta\,(\beta u_{DD})} \quad (18)$$

The step function $\theta\,(\beta u_{DD})$ indicates that the exponential is applied only for positive values of βu_{DD}. We have postponed a detailed analysis of the perturbation expansion about the reference potential of Equation 18, and have calculated the Helmholtz energies for the reference fluid only. We compared these results with those for two-center Lennard–Jones liquids with electric moments using the analogous reference potential

$$u_{ref} = u_{2CLJ}(r, \Omega_1, \Omega_2) + u_{DD}(r, \Omega_1, \Omega_2)\, e^{-3/2L(2\beta u_{DD})\,\theta\,(\beta u_{DD})} \quad (19)$$

A reference potential for quadrupole two-center Lennard–Jones liquids can be obtained by replacing u_{DD} by u_{QQ}. The properties of these nonspherical reference fluids can now be calculated using the perturbation theory outlined in the section above. Because our results apply only to the reference fluid for liquids containing electric moments, and no detailed analysis has justified the neglect of higher order terms, the results must be considered to be preliminary.

Table V presents the residual Helmholtz energies reduced by N/β. Table VI gives the differences between $\beta A/N$ values for liquids with and without moments. The results for $L = 0$ and $\mu^{*2} = 0.8696$ and 3.478 are slightly more negative than those of Verlet and Weis (*40*), who have $\beta A/N = -0.535$ and -4.40. This indicates that not all of the effects of the electric moments are incorporated into the reference fluids defined by Equations 18 and 19. From Table VI we see that the contribution of the dipole moments is reduced in going from $L = 0$ to $L = 0.5$, as explained later, after which it remains approximately fixed. The effects of elongation on the contribution of quadrupole moments are more complicated. The quadrupole moments have been chosen so that the multipole potential has approximately the same effective influence in the dipolar and quadrupolar liquids. This is achieved by requiring that the potentials obtained from the angle-averaged Boltzmann factor for the dipole–dipole and quadrupole–quadrupole potentials have the same value at contact ($r = \sigma$)

$$-\frac{1}{3}\,\mu^{*4}\,kT = -\frac{7}{5}Q^{*4}\,kT \tag{20}$$

Table VI shows that the effect of elongation is to reduce considerably the contribution of the quadrupolar energy to $\beta A/N$ at $L = 0.5$, but that there is a much smaller reduction for $L = 0.8$. We also note that the interference between elongation and electric moment is much more dependent upon the size of the quadrupole moment than of the dipole moment. The origin of this difference between dipole and quadrupole

Table V. Values of the Residual Reduced Helmholtz Energies A^*/NkT for One-Center and Two-Center Lennard–Jones Liquids with Added Dipole or Quadrupole Moments

L	T*	$\rho\sigma^3$	*Without Moment*	*With* $\mu^{*2} = 0.8696$	*With* $\mu^{*2} = 3.478$	*With* $Q^* = 0.4243$	*With* $Q^{*2} = 1.697$
0	1.15	0.85	−1.70	−2.30	−6.83	−2.39	−5.21
0.5	2.3	0.4517	−2.08	−2.60	−6.56	−2.44	−4.90
0.8	1.587	0.3068	−2.08	−2.60	−6.45	−2.60	−5.56

Note: T^* equals kT/ε_{1CLJ} or kT/ε_{2CLJ}. T^* and $\rho\sigma^3$ are varied according to the low-density scaling factors described earlier in order to make the values of A^*/NkT without moments comparable. The reduced moments are defined by $\mu^{*2} = \dfrac{\mu^2}{\sigma^3 kT}$ and $Q^{*2} = \dfrac{Q^2}{\sigma^5 kT}$ with $\sigma/\mathring{A} = 3.4$

Table VI. Contribution of the Electric Moments to A^*/NkT

L	$\mu^{*2} = 0.8696$	$\mu^{*2} = 3.478$	$Q^{*2} = 0.4243$	$Q^{*2} = 1.697$
0.0	−0.60	−5.13	−0.69	−3.51
0.5	−0.52 (−29%)[a]	−4.48 (−29%)	−0.36 (−51%)	−2.82 (−34%)
0.8	−0.52 (−29%)	−4.37 (−30%)	−0.52 (−38%)	−3.48 (−19%)

Note: Calculated for the states given in Table V.

[a] Values in parentheses indicate the percentage reduction compared with the contribution in the $L = 0$ case, referred to A^*/NkT without moment.

contributions to $\beta A/N$ probably lies in the different symmetry of the two potentials. The quadrupolar potential, like the two-center Lennard–Jones potential, is invariant to a reflection of a molecule about the normal to the molecular axis, whereas the dipole–dipole interaction changes sign. For each allowed pair configuration of the molecules, the dipole–dipole potential can be attractive or repulsive while the quadrupole–quadrupole potential must be one or the other. The total contribution of the quadrupolar potential is therefore very dependent on the range of separations for which certain strongly attractive (T-shaped) or repulsive (parallel or end-to-end) orientations are allowed by the shape of the two-center Lennard–Jones molecules. This range will be different for different elongations, making the quadrupole contribution more elongation dependent than that of the dipole, as we see in Table VI. The above discussion can also be used to explain the variation of the quadrupolar contributions to $\beta A/N$ shown in Table VI. The two elongations are $L = 0.5$ and $L = 0.8$, where for scaling reasons (discussed earlier) we have set $\sigma_{L=0.5} = 2.754$ Å and $\sigma_{L=0.8} = 2.421$ Å ($\sigma_{L=0} = 3.4$ Å). For $L = 0.5$, the attractive T-shaped configuration and the repulsive end-to-end configuration become important for $r_T > 3.355$ Å and $r_E > 4.131$ Å, respectively. For $L = 0.8$ we have $r_T > 3.187$ Å and $r_E > 4.357$ Å. The range of r for which the T-shaped configuration will predominate is larger for $L = 0.8$. In changing from $L = 0$ to $L = 0.5$, in the presence of quadrupole moments, the quadrupole–quadrupole potential is forced to assume the repulsive parallel orientation near contact, causing the total quadrupolar contribution to the Helmholtz energy to fall. Increasing the elongation to $L = 0.8$ for the scaled two-center Lennard–Jones potential considered here has the effect of increasing the range for which attractive T-shaped orientations are allowed. The quadrupolar contribution to $\beta A/N$ is therefore increased again, as shown in Table VI.

Effects of Elongation and Quadrupole Moments on the Static Dielectric Constant and Related Structural Properties

In this section we present qualitative results for the effect of elongation and the quadrupole moment on the dielectric constant (ε), Kirkwood g-factor (g_K) and the structural correlation functions h_Δ and $f(r, \Delta)$ of dipolar two-center Lennard–Jones liquids. Our results are based on the assumption that the background correlation function $y(r, \Omega_1, \Omega_2)$ is angle independent, as discussed in an earlier section.

The static dielectric constant of a liquid can be obtained from the Kirkwood g_K-factor using the relationship

$$\frac{(\varepsilon - 1)(2\varepsilon + 1)}{9\varepsilon} = \frac{4\pi}{9} \rho^* \mu^{*2} g_K \tag{21}$$

where

$$g_K = 1 + 4\pi \frac{\rho^*}{3} \int h_\Delta(r)\, r^2\, dr \tag{22}$$

The function $h_\Delta(r)$ gives the average value of the cosine of the angle between the axis of molecules 1 and 2, Δ, in the liquid, at each separation r

$$h_\Delta(r) = \langle \Delta(\Omega_1, \Omega_2)\, g(r, \Omega_1, \Omega_2) \rangle \tag{23}$$

In order to elucidate the various effects of elongation and electric moment we introduce a new function $f(r, \Delta)$, which gives the probability density for finding a pair of molecules whose axes are inclined at an angle arc cos Δ to each other, at the separation r

$$h_\Delta(r) = \int d\Delta\, \Delta\, f(r, \Delta) \tag{24}$$

In order to obtain $h_\Delta(r)$ we use the approximation

$$g(r, \Omega_1, \Omega_2) = y(r)\, e^{-\beta u(r,\Omega_1,\Omega_2)} \tag{25}$$

where $y(r)$ is calculated by solving the Percus–Yevick equation for the reference potential

$$\bar{u}(r) = -kT \ln \langle e^{-\beta u(r,\Omega_1,\Omega_2)} \rangle \tag{26}$$

as discussed in the section on perturbation expansions. The potential $u(r, \Omega_1, \Omega_2)$ is the full molecular potential containing the two-center Lennard–Jones term plus dipole–dipole, dipole–quadrupole, and quadrupole–quadrupole interactions. Equation 25 has been found to give ε and $h_\Delta(r)$ in good agreement with simulation for hard-sphere dipoles (*41*). For the moderate dipole moments considered here (*see* Table VII), the predicted dielectric constants for the Stockmayer fluid (spherical

Table VII. Reduced Parameters for 2CLJ Liquids with Electric Moments

State	T^*	L	ρ^*	μ^{*2}	Q^{*2}	g_K	ε
A	0.75	0.0	0.72	0.75	—	1.03	4.13
B	0.75	0.0	0.72	0.75	0.37	0.73	3.13
C	2.46	0.2	0.58	0.75	—	0.64	2.44
D	2.46	0.2	0.58	0.75	0.37	0.46	1.99
E	1.74	0.4	0.45	0.75	—	0.22	1.34

Note: Temperatures, densities, and electric moments were selected according to the same principles as in Table V.

Lennard–Jones plus dipole) are about 15% too low compared to simulation results (*42*). Although we do not expect to obtain exact numerical results from Equation 25, we believe that the qualitative trends predicted will be reliable. This is especially useful because computer simulation results are not yet available for two-center Lennard–Jones liquids containing dipoles and quadrupoles. Results are becoming available for dipolar hard dumbbells, and these will be referred to below (*43, 44*). Table VII shows the states considered, chosen so as to make the liquids roughly comparable. Figure 9 shows the behavior of ε with respect to that of an

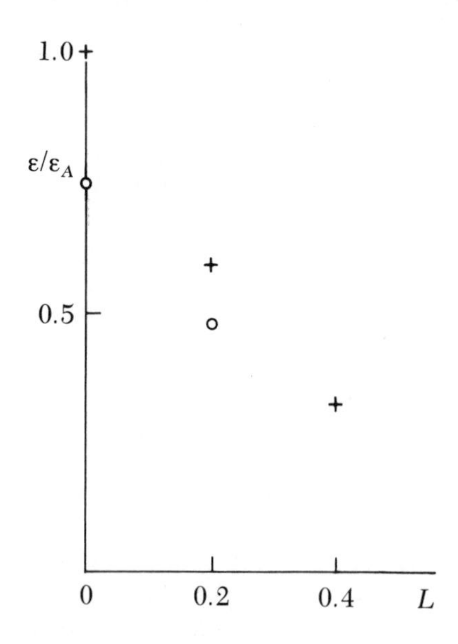

Figure 9. Dielectric constants for 2CLJ liquids with dipole moment μ^{*2} = 0.75, relative to the Stockmayer fluid (state A *of Table VII). Key: +, with zero quadrupole moment; ○, with* Q^{*2} *= 0.37.*

equivalent Stockmayer fluid as the elongation is increased. The trend is clearly to reduce ε as L is increased. The effect of the quadrupole is to reduce ε, but by a smaller amount for $L = 0.2$ than for $L = 0$. The effect on g_K is similar, as can be seen in Table VII. As far as the dielectric constant is concerned, a quadrupole moment has the same effect as increasing elongation. In order to understand these trends it is useful to plot $h_\Delta(r)$ as is done in Figure 10. We see that the Stockmayer fluid has a peak, increasing the elongation or imposing a quadrupole moment produces a trough, which becomes deeper and wider as L increases. For $L \neq 0$ and $Q^* = 0$, small positive peaks occur around $R^* > 1 + L$. In this region the orientation of molecules is no longer hampered by the shape of the molecule and the curves show the expected Stockmayer fluid behavior. In the presence of the quadrupole, these small peaks disappear. The reduction in g_K and hence ε with increasing elongation shown in Figure 9 can therefore be traced back to the increasing depth of the trough in $h_\Delta(r)$, as has been pointed out previously for dipolar hard dumbbells (*24*). Recent simulation results for $h_\Delta(r)$ of dipolar hard dumbbells also show this trend, but the positive peaks were found to be much larger than those of Figure 10 (*44*). This quantitative error is a

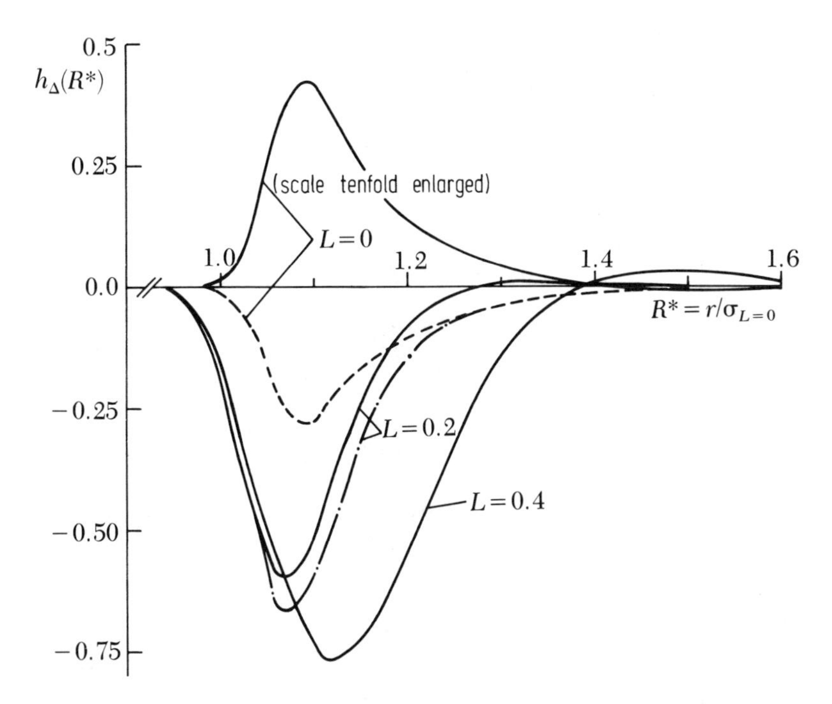

Figure 10. The correlation function $h_\Delta(r)$ *for the five states of Table VII (all with* $\mu^{*2} = 0.75$*). Key ———, without quadrupole moment; --- and –·–·–, with* $Q^{*2} = 0.37$*. The scale for the top curve is enlarged tenfold.*

feature of Equation 25. We expect therefore that computer simulation of the liquids studied in this section would produce higher dielectric constants and g_K values, but show the same trends with L as given in Figure 9.

Figure 11 shows the function $f(r, \Delta)$ normalized by the centers correlation function $g(r) = \langle g(r, \Omega_1, \Omega_2)\rangle$ for a separation of $r/\sigma_{L=0} \sim 1.1$. This corresponds to the center of the troughs in Figure 10. In the absence of a dipole, $f(r, \Delta)$ is symmetrical about $\Delta = 0$ and $h_\Delta(r) = 0$. If a dipole is placed at the center of a spherical Lennard–Jones molecule, $f(r, \Delta)$ becomes asymmetrical. Two orientations are favored, the antiparallel and the head-to-tail, giving $\Delta = -1$ and $\Delta = +1$ respectively. In Figure 11, $f(r, \Delta)$ has small peaks at these values. The head-to-tail orientation has a more negative potential energy and therefore the peak

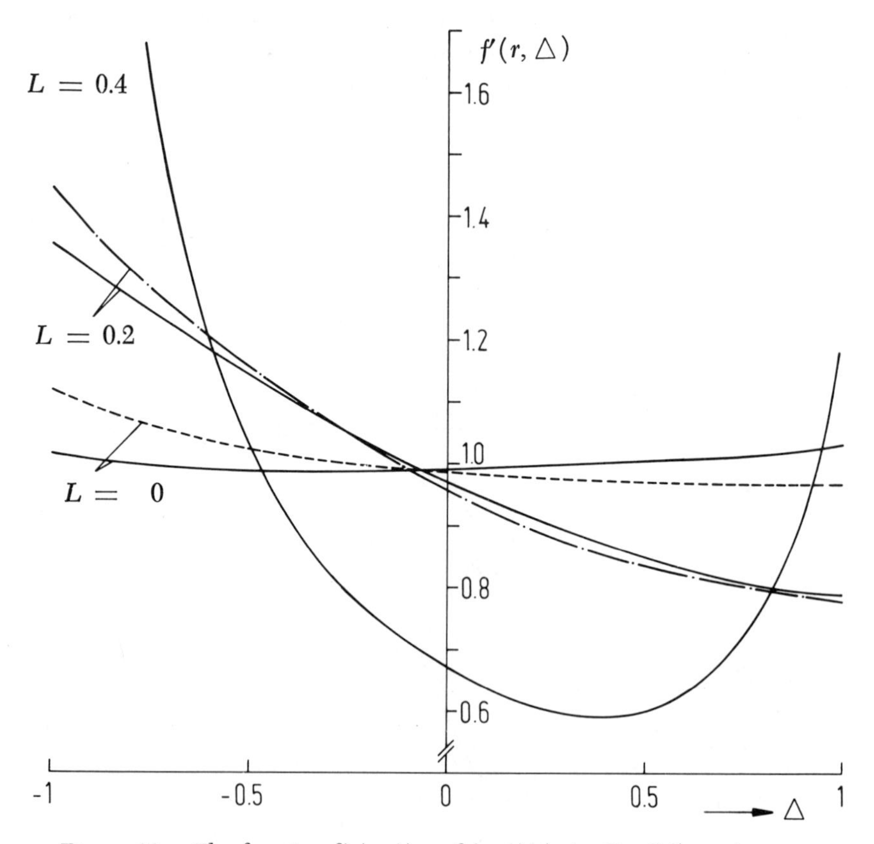

Figure 11. The function f′ (r, Δ) = f (r, Δ)/⟨g (r, Ω_1, Ω_2)⟩ *at* $r/\sigma_{1CLJ} \sim$ *1.1 for the five states of Table VII. Curves correspond to those in Figure 10.*

in $f(r, \Delta)$ is somewhat larger on the $\Delta = +1$ side, leading to a positive peak in $h_\Delta(r)$. Increasing the elongation to $L = 0.2$ leads to a much larger value of $f(r, \Delta)$ at $\Delta = -1$ and a much reduced value of $f(r, \Delta)$ at $\Delta = +1$. This is due to the shape of the two-center Lennard–Jones molecule reducing the probability of head-to-tail orientations and favoring antiparallel orientations which are much easier to pack in the dense liquid. The peak in $f(r, \Delta)$ is now on the negative side and as a consequence $h_\Delta(r)$ is negative around contact for all $L \neq 0$. Increasing L to 0.4 produces an even higher value for $f(r, -1)$ but also increases $f(r, +1)$ above the result for $L = 0.2$; this is due to the two-center Lennard–Jones potential forcing some of the molecules to adapt the parallel orientation ($\Delta = +1$) even though it is unfavorable for the dipoles. (In the absence of a dipole the two-center Lennard–Jones $f(r, \Delta)$ curve is symmetrical about $\Delta = 0$ with two equal peaks at $\Delta = n \pm 1$.) The effect of the quadrupole, like that of elongation, is to reduce the probability of head-to-tail orientations for the dipole moments. The increase near $f(r, -1)$ is because of the existence of a range of antiparallel orientations which are favorable to each of the three multipole interactions present. (Note that there are many possible orientations corresponding to a fixed value of Δ). This again leads to a negative $h_\Delta(r)$ around contact.

Acknowledgments

The authors wish to thank G. Esper and R. Lustig for their help with the calculations reported in section 2 and to J. G. Powles and D. J. Tildesley for communicating their results prior to publication.

Literature Cited

1. Streett, W. B.; Gubbins, K. E., *Annu. Rev. Phys. Chem.* **1977,** *28*, 373.
2. Chandler, D. *Annu. Rev. Phys. Chem.* **1978,** *29*, 441.
3. Wertheim, M. S., *Annu. Rev. Phys. Chem.* **1979,** *30*, 471.
4. Papers referring to the topic "Liquids" at the Bunsen-meeting 1981, *Ber. Bunsenges. Phys. Chem.* **1981,** *85*, 937–1082.
5. Rowlinson, J. S. *J. Chem. Soc., Trans. Faraday* **1954,** *50*, 647.
6. Wojcík, M.; Gubbins, K. E.; Powles, J. G. *Mol. Phys.* **1982,** *45*, 1209.
7. Boublík, T.; Nezbeda, I. *Chem. Phys. Lett.* **1977,** *46*, 315.
8. Longuet-Higgins, H. C.; Widom, B. *Mol. Phys.* **1964,** *8*, 549.
9. Kohler, F. "The Liquid State"; Verlag Chemie: Weinheim, 1972; p. 99.
10. Kohler, F.; Freasier, B. C. *Proc. 6th Int. Conf. on Thermodyn.*, Merseburg (GDR) 1980, 33.
11. Fischer, J. *J. Chem. Phys.* **1980,** *72*, 5371.
12. "International Thermodynamic Tables of the Fluid State Carbon Dioxide"; Pergamon Press: New York; Vol. 3, 1976.
13. Dymond, J. H.; Smith, E. B. "The Virial Coefficients of Pure Gases and Mixtures"; Clarendon Press: Oxford, 2nd ed., 1980.
14. Adams, D. J. *Mol. Phys.* **1979,** *37*, 211.

15. Powles, J. G. *Mol. Phys.* **1980,** *41*, 715.
16. Weeks, J. D.; Chandler, D.; Andersen, H. C. *J. Chem. Phys.* **1971,** *54*, 5237.
17. Sandler, S. I. *Mol. Phys.* **1974,** *28*, 1207.
18. Mo, K. C.; Gubbins, K. E. *J. Chem. Phys.* **1975,** *63*, 1490.
19. Boublík, T. *Fluid Phase Equilib.* **1977,** *1*, 37.
20. Kohler, F.; Quirke, N.; Perram, J. W. *J. Chem. Phys.* **1979,** *71*, 4128.
21. Tildesley, D. J. *Mol. Phys.* **1980,** *41*, 341.
22. Baxter, R. J. *Aust. J. Phys.* **1968,** *21*, 563.
23. Oelschlaeger, H., reported in Ref. 20.
24. Quirke, N. reported in Kohler, F. *Pure Appl. Chem.* **1979,** *51*, 1637.
25. Quirke, N.; Perram, J. W.; Jacucci, G. *Mol. Phys.* **1980,** *39*, 1311.
26. Wang, S. S.; Egelstaff, P. A.; Gray, C. G.; Gubbins, K. E. *Chem. Phys. Lett.* **1974,** *24*, 453.
27. Valleau, J. P.; Torrie, G. In "Modern Theoretical Chemistry"; Berne, B. J., Ed.; Plenum: New York, Vol. 5, 1977, p. 169.
28. Bennett, C. H. *J. Comput. Phys.* **1976,** *22*, 245.
29. Jacucci, G.; Quirke, N. *Mol. Phys.* **1980,** *40*, 1005.
30. Quirke, N.; Jacucci, G. *Mol. Phys.* **1982,** *45*, 823.
31. Quirke, N.; Tildesley, D. J. *Mol. Phys.* **1982,** *45*, 811.
32. Tildesley, D. J. unpublished data.
33. Rowlinson, J. S. "Liquids and Liquid Mixtures," Butterworths: London, 2nd ed.; 1969, p. 270.
34. Gray, C. G.; Gubbins, K. E.; Twu, C. H. *J. Chem. Phys.* **1978,** *69*, 183.
35. Cook, D.; Rowlinson, J. S. *Proc. Roy. Soc.* **1953,** *219A*, 405.
36. Perram, J. W.; White, L. R. *Mol. Phys.* **1972,** *24*, 1133.
37. Ibid. **1974,** *28*, 527.
38. Stell, G.; Rasaiah, J. C.; Narang, H. *Mol. Phys.* **1972,** *23*, 393.
39. Ibid. **1974,** *27*, 1393.
40. Verlet, L.; Weis, J.-J. *Mol. Phys.* **1974,** *28*, 665.
41. de Leeuw, S.; Perram, J. W.; Quirke, N.; Smith, E. *Mol. Phys.* **1981,** *42*, 1197.
42. Pollock, E. L.; Alder, B. J. *Physica* **1980,** *102A*, 1.
43. Morriss, G. P.; Cummings, P. T. *Mol. Phys.* **1982,** *45*, 1099.
44. Morriss, G. P., unpublished data.

RECEIVED for review January 27, 1982. ACCEPTED for publication September 14, 1982.

11

The Reference Average Mayer-Function (RAM) Perturbation Theory for Molecular Fluids

WILLIAM R. SMITH

University of Guelph, Department of Mathematics and Statistics, Guelph, Ontario, Canada N1G 2W1

IVO NEZBEDA

Czechoslovak Academy of Sciences, Institute of Chemical Process Fundamentals, Suchdol, 16502 Prague 6, Czechoslovakia

The RAM perturbation theory for molecular fluids is based on an expansion about a simple fluid whose Mayer function is the angular average of the Mayer function of the molecular fluid. It is one of the few theories that potentially can compute the full angular-dependent pair correlation function, g(12), of a wide range of molecular fluid models. The basis of the theory is reviewed as well as its accuracy in predicting the structural and thermodynamic properties of a number of molecular fluid models, including hard and soft diatomics, hard convex-body models, multipolar simple fluids, and hard triatomics. The results are compared with those obtained from computer simulations of these fluids. The RAM theory, especially when used to compute reduced correlation functions, $g(12)/g_{000}(r_{12})$, *produces quantitatively accurate results for a wide range of fluid models, even in cases of relatively large anisotropy. It is somewhat less satisfactory for the centers pair correlation junction,* g_{000}, *but still quite accurate.*

In simple fluids composed of molecules whose pair potentials are spherically symmetric (i.e., atoms), the two most successful classes of theoretical approaches to date are perturbation theories and theories based on the use of integral equations. (This statement views computer simulations as "experiments" on model fluids with specified pair potentials, with which theoretical results are to be compared.) In the case of molecular fluids, whose pair potentials depend on intermolecular orientation as well as on distance, there is no general consensus as to which

0065-2393/83/0204-0235$12.25/0

type of theoretical approach is to be preferred. Indeed, a widespread view seems to be that each different type of fluid model dictates the use of an associated special type of theoretical approach, and the organization of this volume reflects that viewpoint. However, in this review, we will discuss a general theoretical approach that can be applied in principle to any molecular fluid model. Thus, the present chapter could logically have appeared as part of a number of chapters of this book.

Perturbation theories have been successfully employed for about the last 15 years to calculate the structural and thermodynamic properties of simple fluids and their mixtures. The basic theoretical developments occurred in the late 1960s (*1*) and early 1970s (*2*) and the topic has been reviewed several times (*3–5*). The application of integral equation theories to simple fluids began prior to 1970, with the derivations of the Born–Green–Yvon (BGY) (*6*), hypernetted-chain (HNC) (*7*), Percus–Yevick (PY) (*8*), and mean spherical approximation (MSA) (*9*) approaches. Developments in the last decade have concentrated on determining the general characteristics of the pair potential which favor a particular form of integral equation theory, as well as refinements of the basic approaches (*10*).

In the case of molecular fluids, direct extensions of the forms of perturbation theory successfully used for simple fluids have been only moderately successful, and only for restricted classes of fluid models (*11*). (The use of perturbation theory for molecular fluids actually predates its use for simple fluids (*12*), but it has only been in the past decade that detailed numerical calculations have been possible.) Integral equation approaches are very unwieldy numerically for general models, but for suitably restricted classes of models can even be solved analytically (*13*).

A less direct extension of the simple-fluid integral equation approaches is the RISM theory (*14*), which applies to models consisting of atomic sites. Originally proposed as a type of extension of the Percus–Yevick theory for hard spheres, RISM is at best qualitatively accurate (*15*), and furnishes only partial information concerning the fluid structure (the pair correlation function for the atomic sites).

The RAM (reference average Mayer-function) theory, discussed in this review, is a general approach that applies to any molecular fluid model. It is a perturbation theory approach, and has its origins in attempts to unify the different types of perturbation theory approaches for simple fluids (*16*). We will demonstrate in this review that quite accurate structural and thermodynamic predictions are possible using this theory, even in cases of relatively large anisotropy. The models considered to date using this theory include atoms with imbedded point multipoles (hard-sphere or Lennard–Jones atoms with imbedded point dipoles or quadrupoles), site-interaction models (hard-sphere and Lennard–Jones diatomics, and linear hard-sphere triatomics), and hard-body convex models (hard spherocylinders).

In this chapter, we first discuss the theoretical basis and relevant equations for the RAM theory in the section on general theory, wherein certain more general theoretical aspects of molecular fluids are also discussed. We then describe in detail the numerical results that have been obtained to date in the application of the theory. Finally, we discuss the accuracy of the results and the underlying reasons, as well as the possibilities for future research.

General Theory

RAM Theory. Any perturbation theory involves a choice of two main ingredients: the reference system and the form of the expansion. These choices are dictated both by convergence and by practical numerical considerations. Thus, the properties of the reference system should be "close to" those of the system of interest, and the calculation of the terms in the perturbation expansion should be more easily performed than the calculation of the properties of the system of interest itself. These considerations still leave a wide range of possibilities, and we maintain as much generality as possible at the outset, following the original derivation given some time ago by Smith (*16*). The theory is a general one, but the specific expressions given in the following sections apply only to linear molecules.

We denote by $u(12)$ the pair potential of the molecular fluid of interest, and by $v(12;\gamma)$ a smooth function of the parameter γ, with the properties

$$v(12;0) = v_0(12) \tag{1}$$

$$v(12;1) = u(12) \tag{2}$$

where v_0 is some (at this point arbitrary) reference potential, and the notation (12) denotes the dependence on the positions and orientations of molecules 1 and 2. We next expand the properties of the system of interest about $\gamma = 0$ in terms of a functional (also, at this point arbitrary) $S[v]$, which in general satisfies

$$S[v(12;\gamma)] = S[v_0(12)] + \gamma\{S[u(12)] - S[v_0(12)]\} \tag{3}$$

This is equivalent to a functional Taylor expansion about the reference system in terms of S.

For the Helmholtz free energy, A, and the pair correlation function, $g(12)$, this yields, after setting γ to unity

$$A = A_0 + [\delta A]_0 + [\delta^2 A]_0/2! + \ldots \tag{4}$$

$$g(12) = g_0(12) + [\delta g(12)]_0 + \ldots \tag{5}$$

where

$$[\delta A]_0 = \frac{\rho^2}{2} \iint g_0(12) v_\gamma(12) d\boldsymbol{R}_1 d\boldsymbol{R}_2 \tag{6}$$

$$\begin{aligned}
[\delta^2 A]_0 = & \frac{\beta}{2}\rho^2 \iint g_0(12)[\beta v_\gamma^2(12) - v_{\gamma\gamma}(12)] d\boldsymbol{R}_1 d\boldsymbol{R}_2 \\
& - \frac{\beta}{2}\rho^3 \iiint g_0(123)[v_\gamma(12)v_\gamma(13) \\
& + v_\gamma(12)v_\gamma(23)] d\boldsymbol{R}_1 d\boldsymbol{R}_2 d\boldsymbol{R}_3 \\
& - \frac{\beta}{4}\rho^4 \iiiint [g_0(1234) \\
& - g_0(12)g_0(34)] v_\gamma(12) v_\gamma(34) d\boldsymbol{R}_1 d\boldsymbol{R}_2 d\boldsymbol{R}_3 d\boldsymbol{R}_4 \\
& - \frac{\rho^2}{4N}\left(\frac{\partial \rho}{\partial P}\right)_0 \left\{\frac{\partial}{\partial \rho} \iint \rho^2 g_0(12) v_\gamma(12) d\boldsymbol{R}_1 d\boldsymbol{R}_2\right\}^2
\end{aligned} \tag{7}$$

$$\begin{aligned}
[\delta g(12)]_0 = & -\beta g_0(12) v_\gamma(12) - \beta\rho \int g_0(123)[v_\gamma(13) + v_\gamma(23)] d\boldsymbol{R}_3 \\
& - \frac{\beta}{2}\rho^2 \iint [g_0(1234) - g_0(12)g_0(34)] v_\gamma(34) d\boldsymbol{R}_3 d\boldsymbol{R}_4 \\
& + \frac{1}{2N}\left(\frac{\partial \rho}{\partial P}\right)_0 \left\{\frac{\partial}{\partial \rho}[\rho^2 g_0(12)]\right\} \left\{\frac{\partial}{\partial \rho} \iint {}_3\rho^2 g_0(34) v_\gamma(34) d\boldsymbol{R}_3 d\boldsymbol{R}_4\right\}
\end{aligned} \tag{8}$$

where

$$v_\gamma(12) = \left.\frac{\partial v(12;\gamma)}{\partial \gamma}\right|_{\gamma=0} = \{S[u(12)] - S[v_0(12)]\} \Big/ \left(\frac{\partial S}{\partial v}\right)_{v=v_0} \tag{9}$$

$$v_{\gamma\gamma}(12)_{r=v_0} = \left.\frac{\partial^2 v(12;\gamma)}{\partial \gamma^2}\right|_{\gamma=0} = -\left(\frac{\partial^2 S}{\partial v^2}\right)_{v=v_0} v_\gamma^2(12) \Big/ \left(\frac{\partial S}{\partial v}\right)_{v=v_0} \tag{10}$$

In Equations 4–10, subscript 0 denotes the reference fluid, N is the number of particles, ρ the density, P the pressure and $g(12)$, $g(123)$, and $g(1234)$ are the two-, three-, and four-body distribution functions, respectively. The integrations in the above expressions are over all angular orientations of particles and over all spatial volume elements. We denote this by writing

$$d\mathbf{R}_i = dr_i d\boldsymbol{\omega}_i \tag{11}$$

where $d\boldsymbol{\omega}_i$ denotes the orientational integration.

As an alternative to expanding g(12), we may also consider the expansion of $y(12) = \exp\,[\beta u(12)]g(12)$. This yields

$$y(12) = y_0(12) + [\delta y(12)]_0 + \ldots \tag{12}$$

where

$$[\delta y(12)]_0 = -\exp\,[\beta v_0(12)]\Big\{\beta\rho\int g_0(123)[v_\gamma(13) + v_\gamma(23)]d\mathbf{R}_3 + \frac{\beta}{2}\rho^2\iint[g_0(1234) - g_0(12)g_0(34)]v_\gamma(34)d\mathbf{R}_3 d\mathbf{R}_4 - \frac{1}{2N}\left(\frac{\partial\rho}{\partial P}\right)_0\left\{\frac{\partial}{\partial\rho}[\rho^2 g_0(12)]\right\} \cdot \left\{\frac{\partial}{\partial\rho}\iint\rho^2 g_0(34)v_\gamma(34)d\mathbf{R}_3\,d\mathbf{R}_4]\right\} \tag{13}$$

Since $y(12)$ is generally expected to be less sensitive to the potential than g(12), we will make considerable use of Equations 12 and 13 (the y-expansion) in the following. However, we will see that Equations 5 and 8 (the g-expansion) are also useful in certain situations.

To this point, Equations 3–13 are merely formalism. To perform calculations, one must make a choice of both S[·] and the reference system. To try to optimize the convergence of the expansion, two obvious methods present themselves for the latter. The first is to choose a reference system that, either on physical or mathematical grounds, one expects to have properties similar to those of the system of interest. The other is to choose a reference system on the basis of mathematical considerations, and then determine the parameters of that reference system by annulling low-order terms in the expansion. The latter usually involves setting to zero Equation 6. These two strategies need not be independent. For example, Kohler et al. (*17*) expanded the properties of a Lennard–Jones diatomic system modeling nitrogen about a "similar" system, one whose molecules have the same repulsive forces. They further expanded the properties of this system about those of a hard-sphere diatomic with sphere diameter determined by annulling Equation 6.

The structure and properties of molecular fluids are generally very different from those of simple fluids. One might therefore expect that the most appropriate choice of reference system would be one that is nonspherical. This approach has been followed by Mo and Gubbins (*18*), Tildesley (*19*) and Sandler (*20*), in addition to Kohler et al. (*17*). Since the properties of a nonspherical reference system can be as difficult to compute as those of the system of interest itself, the use of a spherically symmetric reference system is very attractive, especially in view of the fact that many very accurate methods are available for computing the structural and thermodynamic properties of such fluids (*3–5, 10*). However, some workers (*19–21, 55*) have claimed that such an approach is doomed to failure. We will subsequently see that, provided some of the "essential character" of the molecular fluid is incorporated within it, the use of a spherically symmetric reference system can furnish good results for both the structure and the thermodynamics for systems exhibiting quite markedly large anisotropies.

The choice of expansion functional is not motivated by such obvious considerations as that of the reference system, and basically two choices have been made. Pople (*12*) and Gubbins et al. (*11*) have chosen

$$S[u] = u \tag{14}$$

They annulled $[\delta A]_0$ to give a reference system characterized by

$$v_0(r_{12}) = \langle u(12) \rangle_{\boldsymbol{\omega}_1, \boldsymbol{\omega}_2} \tag{15}$$

where $r_{12} = |\boldsymbol{r}_1 - \boldsymbol{r}_2|$ and $\langle \cdot \rangle_{\boldsymbol{\omega}_1, \boldsymbol{\omega}_2}$ denotes an unweighted average over the orientations, $\boldsymbol{\omega}_1$ and $\boldsymbol{\omega}_2$, of molecules 1 and 2. This approach is suggested by potentials of the form

$$u(12) = v_0(r_{12}) + \Delta u(12) \tag{16}$$

where v_0 is the hard-sphere or Lennard–Jones potential and $\Delta u(12)$ is of multipolar type. Good results are obtained for such systems for only relatively small anisotropies, but the Padé approximant to the Helmholtz free energy using $[\delta^2 A]_0$ and $[\delta^3 A]_0$ produces remarkably accurate results. The pair correlation function, however, is not blessed with such serendipity. The usefulness of Equation 15 appears to be limited to this rather special class of potentials. It will not be useful, for example, for hard-sphere diatomics, since the reference system will be one of hard spheres of diameter equal to the maximum distance of possible overlap. Indeed, Equation 15 is not likely to be useful for any interaction site model.

Sung and Chandler (*22*) and Steele and Sandler (*23*), in their study of hard-sphere diatomics, have chosen

$$S[u] = \exp[-\beta u(12)] \equiv e(12) \tag{17}$$

They chose the reference system to be one of hard spheres,

$$\begin{aligned} v_0(r_{12}) &= \infty \qquad \text{for } r_{12} < d \\ v_0(r_{12}) &= 0 \qquad \text{for } r_{12} > d \end{aligned} \tag{18}$$

and determined the hard-sphere diameter, d, by annulling $[\delta A]_0$. Since hard spheres are not necessarily "similar" to many systems of interest, good results were obtained only for systems with relatively small anisotropies.

The RAM theory attempts to incorporate some of the molecular fluid's anisotropy into the reference system, while maintaining its spherical symmetry. Recognizing the fact that the angular integrations involving $u(12)$ may be factored out of $[\delta A]_0$ if g_0 is spherically symmetric, the reference system may be chosen to annul $[\delta A]_0$ by setting

$$\iint v_\gamma(12) d\boldsymbol{\omega}_1 d\boldsymbol{\omega}_2 = 0 \tag{19}$$

or, equivalently, by means of

$$S[v_0(r_{12})] = \iint S[u(12)] d\boldsymbol{\omega}_1 d\boldsymbol{\omega}_2 \tag{20}$$

This choice has the added virtue of always also annulling the most numerically difficult terms in Equations 7, 8, and 13, those involving the four-body distribution function, as well as the terms involving the compressibility of the reference fluid.

The RAM theory uses Equation 17 for the expansion functional, resulting in the following expressions for v_0 and Equations 6–8 and 13:

$$v_0(r_{12}) = -(1/\beta)\ln\langle\exp[-\beta u(12)]\rangle_{\boldsymbol{\omega}_1,\boldsymbol{\omega}_2} \tag{21}$$

$$[\delta A]_0 = 0 \tag{22}$$

$$\begin{aligned} [\delta^2 A]_0 = {} & \frac{\beta}{2}\rho^2 \iint g_0(r_{12})[\beta v_\gamma^2(12) - v_{\gamma\gamma}(12)] d\boldsymbol{R}_1 d\boldsymbol{R}_2 \\ & - \frac{\beta}{2}\rho^3 \iiint g_0(\boldsymbol{r}_1, \boldsymbol{r}_2, \boldsymbol{r}_3)[v_\gamma(12)v_\gamma(13) \\ & + v_\gamma(12)v_\gamma(23)] d\boldsymbol{R}_1 d\boldsymbol{R}_2 d\boldsymbol{R}_3 \end{aligned} \tag{23}$$

$$\begin{aligned} [\delta g(12)]_0 = {} & -\beta g_0(r_{12}) v_\gamma(12) \\ & - \beta\rho \int g_0(\boldsymbol{r}_1, \boldsymbol{r}_2, \boldsymbol{r}_3)[v_\gamma(13) + v_\gamma(23)] d\boldsymbol{R}_3 \end{aligned} \tag{24}$$

$$[\delta y(12)]_0 = -\beta\rho \exp[\beta v_0(r_{12})] \int g_0(\boldsymbol{r}_1, \boldsymbol{r}_2, \boldsymbol{r}_3)[v_\gamma(13) + v_\gamma(23)] d\boldsymbol{R}_3 \tag{25}$$

From Equation 21, the Mayer-function (and the Boltzmann factor) of the reference system is equal to the angle-averaged value of this quantity for the system of interest. Equation 21 was apparently first suggested some time ago by Rushbrooke (*24*) and by Cook and Rowlinson (*25*). Perram and White (*26*) have also considered this choice of reference fluid, as have also Verlet and Weiss (*27*) and Steinhauser and Bertagnolli (*28*).

To illustrate the type of reference system potential produced by Equation 21, we show in Figure 1 $v_0(r)$ for a homonuclear hard diatomic (HOHD) fluid with sphere diameter σ and sphere separation L. It is seen to be a purely repulsive potential of finite range. It has a hard core at the minimum distance of closest approach, $(\sigma^2 - L^2/2)^{1/2}$, and a decaying repulsive outer shell that vanishes at the maximum distance of possible overlap, $\sigma + L$. The term $\exp[-\beta v_0(r_{12})]$ measures the fraction of the total angular volume available for rotation to two molecules a distance r_{12} apart. Thus, at the minimum distance of closest approach, the molecules are restricted to adopt the crossed configuration, and $\exp[-\beta v_0(r_{12})] = 0$, yielding the hard core in Figure 1. As r_{12} increases, the molecules enjoy more rotational freedom, $\exp[-\beta v_0(r_{12})]$ increases, and $v_0(r_{12})$ decreases. Finally, beyond $r_{12} = \sigma + L$, the molecules can rotate completely freely, $\exp[-\beta v_0(r_{12})] = 1$, and $v_0(r_{12}) = 0$.

By using Equation 17, Equations 9 and 10 become

$$v_\gamma(12) = (1 - \exp\{-\beta[u(12) - v_0(r_{12})]\})/\beta \quad (26)$$

$$v_{\gamma\gamma}(12) = \beta v_\gamma^2(12) \quad (27)$$

These yield the final form of Equations 23–25:

$$[\delta^2 A]_0 = -\frac{\rho^2}{2\beta}\iiint \Delta e(12)[\delta y(12)]_0 d\boldsymbol{R}_1\, d\boldsymbol{R}_2\, d\boldsymbol{R}_3 \quad (28)$$

$$[\delta g(12)]_0 = \Delta e(12) y_0(r_{12}) + \rho\, \exp[-\beta v_0(r_{12})]\int y_0(\boldsymbol{r}_1, \boldsymbol{r}_2, \boldsymbol{r}_3)\{e_0(r_{13})\Delta e(23) + e_0(r_{23})\Delta e(13)\} d\boldsymbol{R}_3 \quad (29)$$

$$[\delta y(12)]_0 = \rho\int y_0(\boldsymbol{r}_1, \boldsymbol{r}_2, \boldsymbol{r}_3)\{e_0(r_{13})\Delta e(23) + e_0(r_{23})\Delta e(13)\} d\boldsymbol{R}_3 \quad (30)$$

where

$$y_0(\boldsymbol{r}_1, \boldsymbol{r}_2, \boldsymbol{r}_3) = \exp\{\beta[v_0(r_{12}) + v_0(r_{13}) + v_0(r_{23})]\} g_0(\boldsymbol{r}_1, \boldsymbol{r}_2, \boldsymbol{r}_3) \quad (31)$$

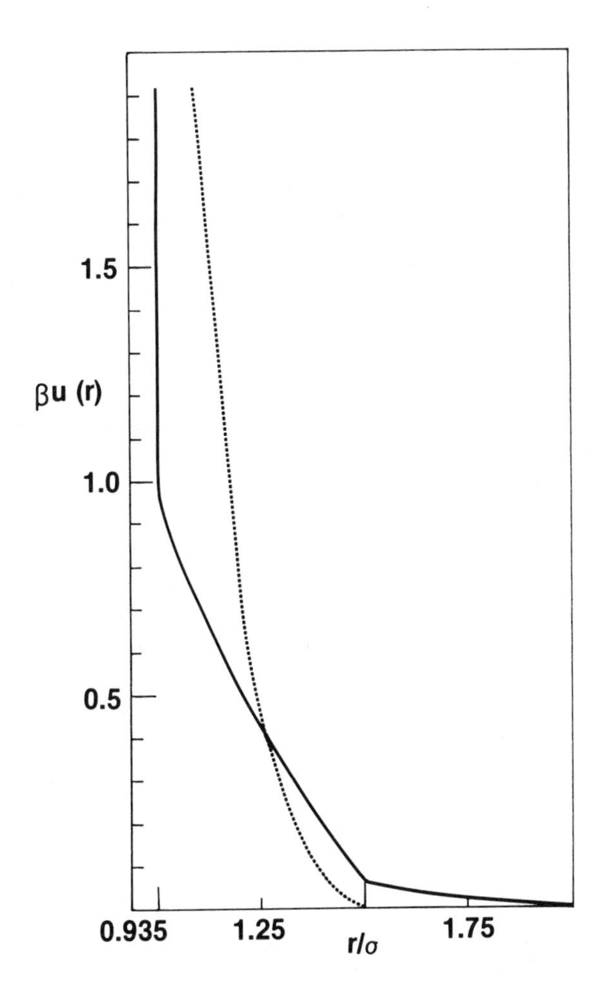

Figure 1. Reference system pair potentials for the homonuclear hard diatomic system with L/σ = 0.5. Key: . . . . ., from Equation 21 using a center-of-mass molecular coordinate system; and ———, from using a coordinate system in which the reference points in each molecule are taken at the centers of hard spheres. (Reproduced with permission from Ref. 45. Copyright 1979, North-Holland Publishing Co.)

$$\Delta e(ij) = e(ij) - e_0(r_{ij}) \tag{32}$$

We immediately note that in first-order, because of Equation 21, the angular averages of $[\delta g(12)]_0$ and $[\delta y(12)]_0$ vanish in their respective expansions. This means that, to first-order terms,

$$\langle g(12)\rangle_{\omega_1,\omega_2} \equiv g_{000}(r_{12}) = g_0(r_{12}) \tag{33}$$

using the g-expansion, and

$$\langle y(12)\rangle_{\omega_1,\omega_2} \equiv y_{000}(r_{12}) = y_0(r_{12}) \tag{34}$$

using the y-expansion, where X_{000} denotes the 000-spherical harmonic coefficient of the quantity X (*See next section*). It should be noted, however, that if the y-expansion is used to calculate g(12), via

$$g(12) = \exp[-\beta u(12)](y_0(r_{12}) + [\delta y(12)]_0 + \ldots) \tag{35}$$

then Equation 33 holds only in zeroth-order.

Finally, in order to numerically evaluate the above expressions, we must make some approximation to the triplet term, $y_0(\mathbf{r}_1, \mathbf{r}_2, \mathbf{r}_3)$. In all the calculations made to date, we have used the superposition approximation

$$y_0(\mathbf{r}_1, \mathbf{r}_2, \mathbf{r}_3) = y_0(r_{12})\, y_0(r_{13})\, y_0(r_{23}) \tag{36}$$

This type of approximation in similar integrals produces accurate results (29).

Final expressions suitable for numerical computation have been obtained to date only for linear molecules. Expanding $\Delta e(ij)$ in spherical harmonics and simplifying the results, Equations 29 and 30 become

$$[\delta g(12)]_0 = \Delta e(12) y_0(r_{12}) + e_0(r_{12})[\delta y(12)]_0 \tag{37}$$

$$[\delta y(12)]_0 = 2\pi\rho y_0(r_{12}) \sum_{l>0} (2l+1)^{1/2} J_l(r_{12})[P_l(\cos\theta_1^{12}) + P_l(\cos\theta_2^{12})] \tag{38}$$

where

$$J_l(r_{12}) = \int_0^\infty r_{13}^2\, y_0(r_{13}) e_{l00}(r_{13}) dr_{13} \cdot \int_{-1}^{1} P_l(x) h_0\{(r_{12}^2 + r_{13}^2 - 2r_{12}r_{13}x)^{1/2}\} dx \tag{39}$$

Subscript ij denotes that the polar axis is along the line joining the centers of particles i and j, $e_{l00}(r_{ij})$ is the $l00$ spherical harmonic coefficient of $e(ij)$, and P_l is a Legendre polynomial.

We see from these expressions that, to first-order, $y(12)$ is independent of the azimuthal angle ϕ_{12}, and that, using either the g or the

y expansion, the dependence of g(12) on ϕ_{12} arises only from the Boltzmann factor.

The final expressions for g(12) are given, to first-order, by

$$g_{RAMG}(12) = y_0(r_{12})\{e(12) + e_0(r_{12})y_1(12)\} \quad (40)$$

$$g_{RAMY}(12) = y_0(r_{12})\{e(12) + e(12)y_1(12)\} \quad (41)$$

where

$$y_1(12) = [\delta y(12)]_0/y_0(r_{12}) \quad (42)$$

Orientational Structure of Molecular Fluids. REDUCED PAIR CORRELATION FUNCTIONS. Since, even for linear molecules, g(12) is a function of four variables (one distance and three angles), it is impractical to compute and store it over a densely spaced set of points in any computer simulation of such a system. Two representative subsets of g(12) have thus been computed, spherical harmonic expansion coefficients, $g_{klm}(r_{12})$, and radial slices through the g(12) surface at fixed orientations, $g_\Omega(r)$.

The spherical harmonic expansion of g(12) for linear molecules, using molecule-fixed coordinates in which the z axis is taken to be the line joining two reference points, one in each molecule, is given by (*48*)

$$g(12) = 4\pi \sum_{k,l,m} g_{klm}(r_{12}) Y_{k\overline{m}}(\boldsymbol{\omega}_1) Y_{lm}(\boldsymbol{\omega}_2) \quad (43)$$

where $\overline{m} = -m$ and Y_{lm} is a spherical harmonic, defined by

$$Y_{lm}(\theta,\phi) = \left[\frac{(2l+1)(l-|m|!)}{4\pi(l+|m|!)}\right]^{1/2} P_{lm}(\cos\theta)\, e^{im\phi} \quad (44)$$

P_{lm} is the associated Legendre polynomial, defined by

$$P_{lm}(\cos\theta) = \frac{\sin^m\theta}{2^l l!} \frac{d^{l+m}}{d(\cos\theta)^{l+m}} (\cos^2\theta - 1)^l \quad (45)$$

for $l,m \geqslant 0$, and

$$P_{l,-m} = (-1)^m P_{lm} \qquad (\text{for } l > 0)$$

and (46)

$$P_{lm} = 0 \qquad (\text{for } l < 0)$$

Since P_{lm} is defined in terms of the derivative of order $l+m$ of a polynomial of degree $2l$, it vanishes unless $|m| \leq l$. The spherical harmonics $\{Y_{ij}\}$ form an orthonormal set over the sphere, satisfying

$$\int_0^{2\pi} d\phi \int_{-1}^{1} d(\cos\theta)\, Y_{i\bar{m}}(\theta,\phi) Y_{kn}(\theta,\phi) = \delta_{ik}\delta_{mn} \tag{47}$$

where δ_{ij} is the Kronecker delta.

The spherical harmonic coefficients are thus given by

$$g_{klm}(r_{12}) = \iint g(12) Y_{k\bar{m}}(\boldsymbol{\omega}_1) Y_{lm}(\boldsymbol{\omega}_2)\, d\boldsymbol{\omega}_1\, d\boldsymbol{\omega}_2 / 4\pi$$

$$g_{klm}(r_{12}) = \frac{1}{2} \int_0^{2\pi} d\phi_{12} \int_{-1}^{1} d(\cos\theta_1) \cdot \int_{-1}^{1} d(\cos\theta_2) g(12) Y_{k\bar{m}}(\theta_1, 0) Y_{lm}(\theta_2, \phi_{12}) \tag{48}$$

If the reference points are at the centers of mass, and the molecules are symmetric with respect to this point, then, since

$$g(r_{12}, \theta_1, \theta_2, \phi_{12}) = g(r_{12}, \theta_1, \pi-\theta_2, 2\pi-\phi_{12}) = g(r_{12}, \pi-\theta_1, \theta_2, 2\pi-\phi_{12}) \tag{49}$$

only even values of k and l yield nonvanishing harmonic coefficients and

$$g_{klm}(r_{12}) = g_{lkm}(r_{12}) \tag{50}$$

Finally, using further symmetry considerations, the coefficients may be evaluated numerically via

$$g_{klm}(r_{12}) = 4\int_0^{\pi} d\phi_{12} \int_0^1 d(\cos\theta_1) \int_0^1 d(\cos\theta_2) g(12) Y_{k\bar{m}}(\theta_1, 0) Y_{lm}(\theta_2, \phi_{12}) \tag{51}$$

Examination of typical computer simulation results for g_{klm}, illustrated for homonuclear hard-sphere diatomics (HOHD) in Figure 2, shows that the initial maximum (or minimum) in g_{klm} occurs at approximately the same location as the first peak in $g_{000}(r)$. In addition, in computer simulations, computations may be arranged in such a way that the primary statistics calculated in the course of determining g_{klm} are actually the ratios of g_{klm} to g_{000} (*30–37*) via

$$g_{klm}(r) = 4\pi g_{000}(r) \langle Y_{k\bar{m}} Y_{lm} \rangle_{\text{shell}} \tag{52}$$

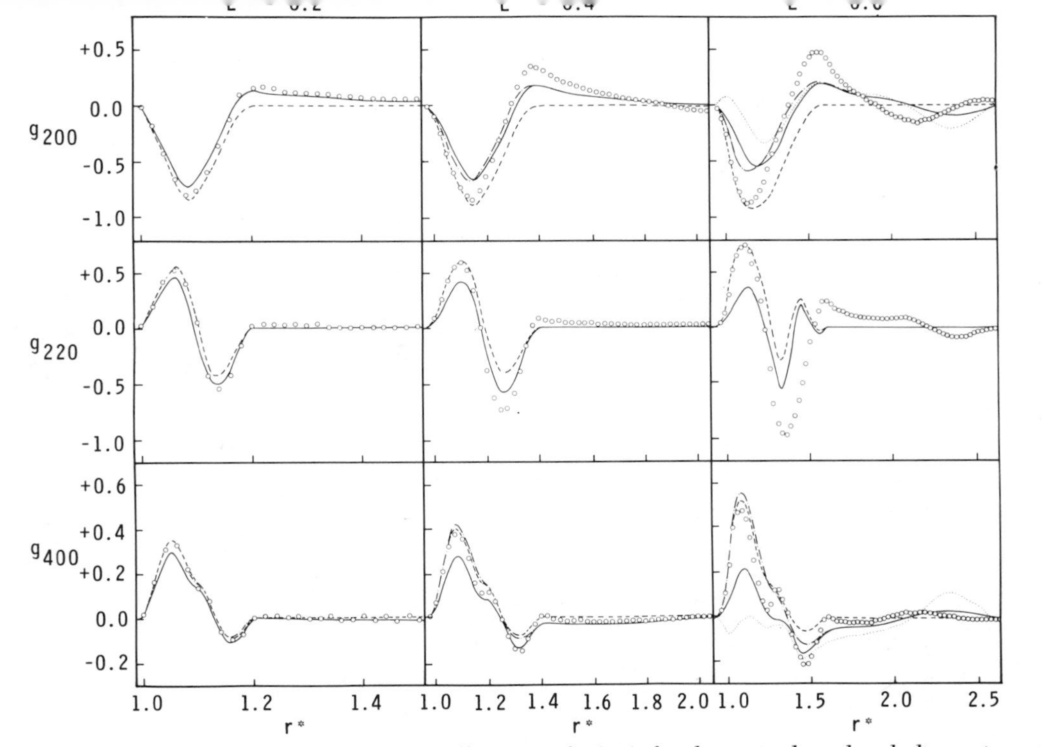

Figure 2. Spherical harmonic expansion coefficients of g(12) *for homonuclear hard diatomics at the indicated elongations,* $L^* = L/\sigma$, *and the density* $N\sigma^3/V = 0.5$ $r^* = r/\sigma$. *Key:* ○, *computer simulation results* (30); ---, *zeroth-order RAMY theory;* –·–·, *first-order RAMG theory;* ——, *first-order RAMY theory; and* . . . ., *first-order BLIPY theory* (38). *For* g_{220}, *RAMG(0) and RAMG(1) are identical. At* $L^* = 0.2$, *the RAMY(1) and RAMG(1) are essentially numerically equal. (Reproduced with permission from Ref. 38. Copyright 1980, Taylor and Francis, Ltd.)*

where $\langle X(12)\rangle_{shell}$ denotes an average over all orientations in a narrow radial shell at the distance r from a central molecule. More precisely,

$$\langle X(12)\rangle_{shell} = \frac{\int d\boldsymbol{\omega}_1 \int d\boldsymbol{\omega}_2 X(r_{12},\boldsymbol{\omega}_1,\boldsymbol{\omega}_2)g(12)}{\int d\boldsymbol{\omega}_1 \int d\boldsymbol{\omega}_2 g(12)} \tag{53}$$

Finally, Melnyk and Smith (38) have shown that the ensemble averages in Equation 52 satisfy some exact asymptotic results. For these reasons, it is appropriate to investigate the behavior of the reduced spherical harmonic coefficients, $g^*_{klm}(r)$, defined by (38)

$$g^*_{klm}(r) = g_{klm}(r)/g_{000}(r) \tag{54}$$

Some typical results for the reduced harmonic coefficients are shown in Figure 3, where it is seen that they are much more smoothly varying functions of r than the unreduced coefficients, especially at small distances.

Reduced full pair correlation functions, $g^*(12)$, may be similarly defined, via

$$g^*(12) = g(12)/g_{000}(r_{12}) \tag{55}$$

In Figure 4a and 4b, we show some results for g^*_Ω for HOHD, obtained from the computer simulation results of Cummings et al. (39).

One of the most important aspects of the reduced correlation functions is the fact that they obey some exact asymptotic results for molecules that are restricted to adopt a unique relative orientation at closest approach. For example, HOHD are restricted to adopt the crossed orientation ($\theta_1 = \theta_2 = \phi_{12} = \pi/2$) at the minimum closest approach distance, r_c, where

$$r_c = (\sigma^2 - L^2/2)^{1/2} \tag{56}$$

In this case,

$$g^*_{klm}(r) = \frac{4\pi \int\int g(12) Y_{k\bar{m}}(\boldsymbol{\omega}_1) Y_{lm}(\boldsymbol{\omega}_2) d\boldsymbol{\omega}_1 d\boldsymbol{\omega}_2}{\int\int g(12) d\boldsymbol{\omega}_1 d\boldsymbol{\omega}_2} \tag{57}$$

As r approaches r_c from above, the numerator and denominator of Equation 57 both approach zero. However, the mean-value theorem for integrals allows the cancellation of the g(12) terms in the limit of closest approach, yielding (38)

$$\lim_{r \downarrow r_c} g^*_{klm}(r) = 4\pi\, Y^X_{k\bar{m}}\, Y^X_{lm} \tag{58}$$

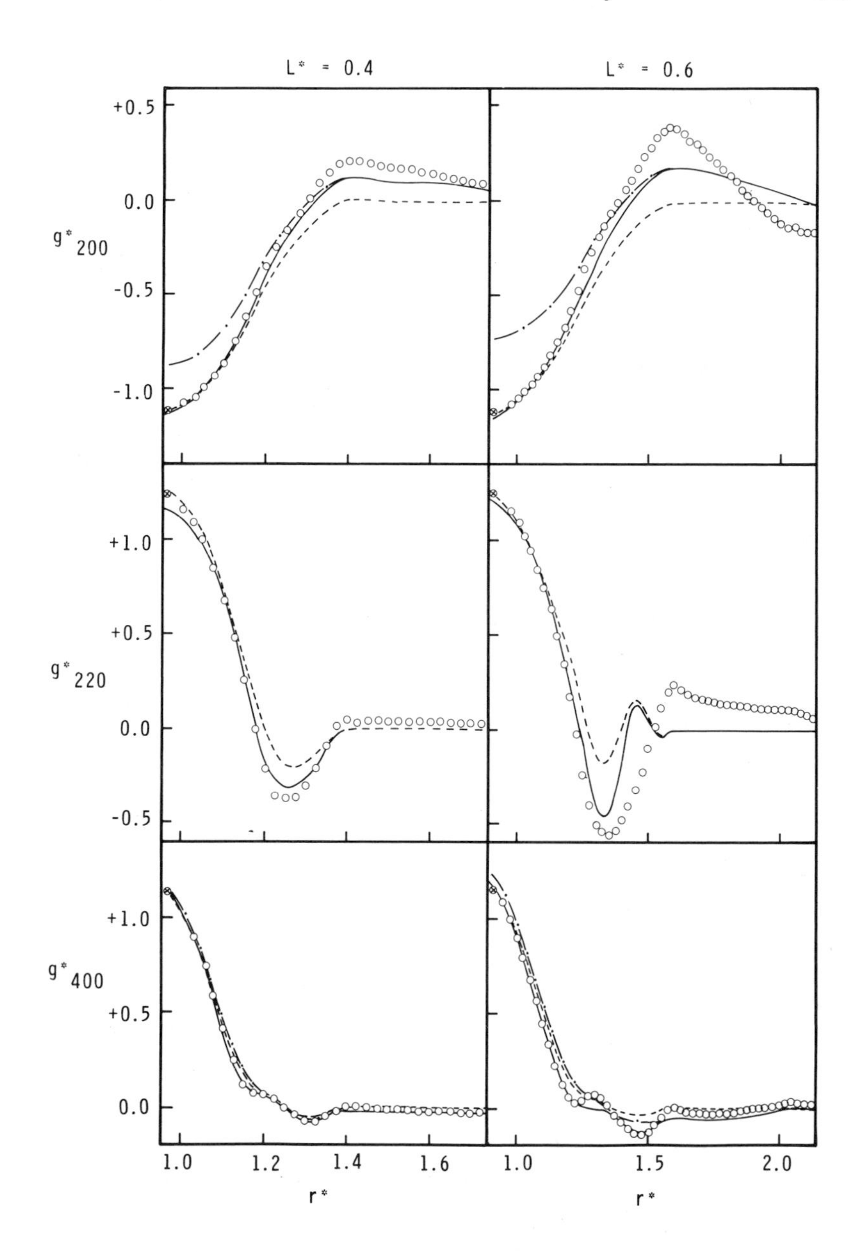

Figure 3. Reduced spherical harmonic expansion coefficients for the homonuclear hard diatomic fluid at the same density for two of the elongations in Figure 2. Key: ⊗, asymptotic limit of Equation 58; other symbols as in Figure 2. (Reproduced with permission from Ref. 38. Copyright 1980, Taylor and Francis, Ltd.)

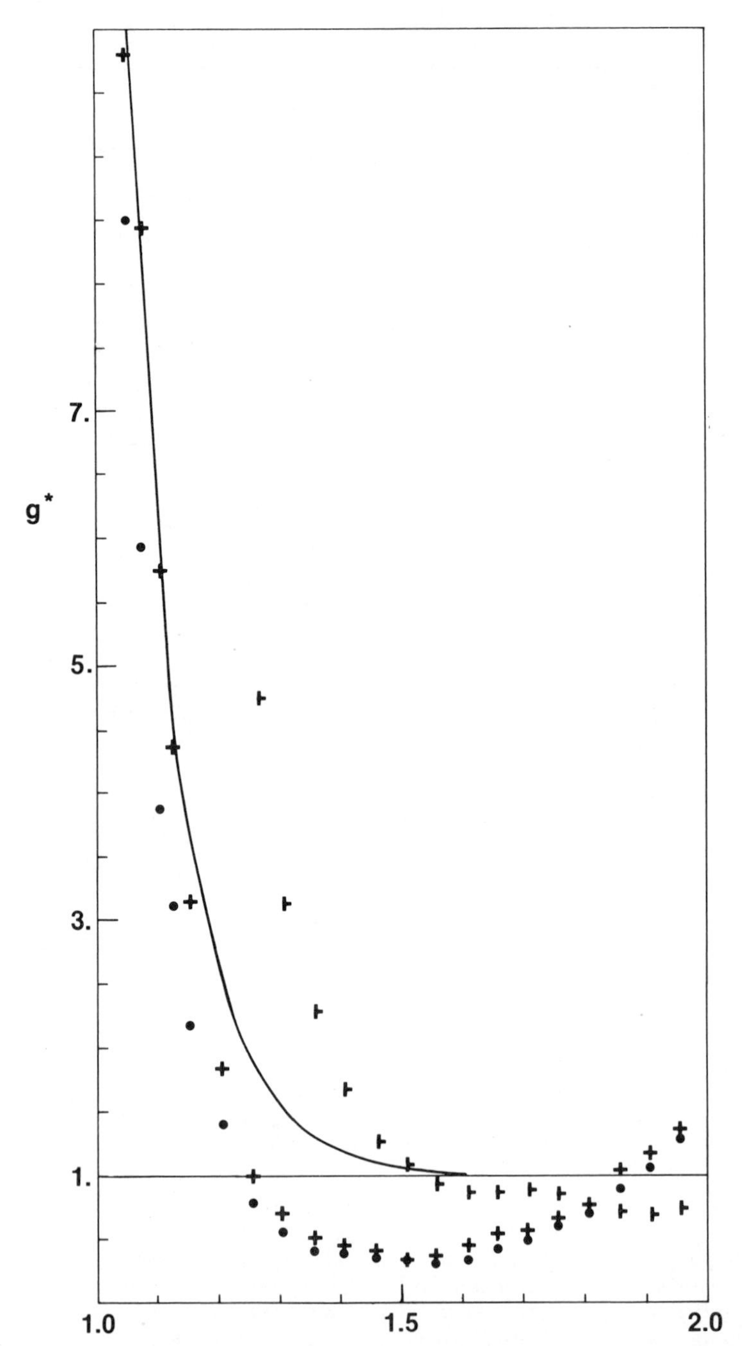

Figure 4a. Reduced g(12) as a function of distance for homonuclear hard diatomics at L* = *0.6 and* ρ* = *0.5. Key:* ●, *crossed orientation;* +, *parallel orientation;* ⊢, *T-shaped orientation; and* ——, $(e_{000})^{-1}$.

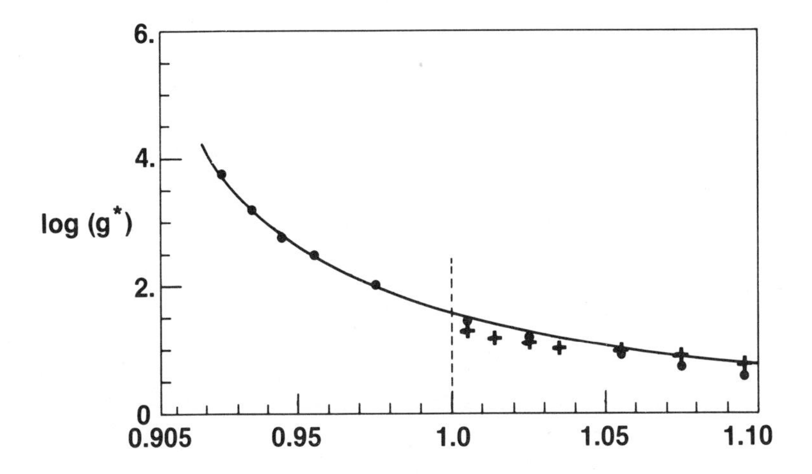

Figure 4b. Results at small distances on a logarithmic scale. Key: ----, closest approach distance in the parallel orientation; other symbols as in graph 4a.

where X denotes the crossed orientation. Equation 58 is an exact result, and hence may be used to test the accuracy of computer simulation results; Figure 3 shows the limiting values, with which the simulation results appear to agree.

A related result may be obtained for $g_{\Omega}^{*}(r)$ (*40*). For HOHD, we have

$$g_{X}^{*}(r) = \frac{16\pi^{2}y_{X}(12)}{\int\int e(12)y(12)d\boldsymbol{\omega}_{1}d\boldsymbol{\omega}_{2}} \tag{59}$$

Since, for small values of r, the only contributions to the integral in the denominator arise from values of $y(12)$ near the crossed orientation, we have

$$g_{X}^{*}(r) \approx [e_{000}(r)]^{-1} \; (r \rightarrow r_{c}), \tag{60}$$

which means that $g_{X}^{*}(r)$ approaches the quantity on the right side asymptotically. It is seen in Figure 4b that Equation 60 is remarkably accurate for values of r greater than r_c, indicating that the dependence of $y(12)$ on relative orientation is not overly strong at small distances. In addition, the close agreement of the crossed and parallel g_{Ω} results in Figure 4b indicates that $y(12)$ is relatively independent of ϕ_{12} at $\theta_1 = \theta_2 = \pi/2$.

Even when the molecules are not restricted to adopt a unique orientation at the minimum closest approach distance, approximate results analogous to Equations 58 and 60 may be obtained. For example, for hard spherocylinders (*41*), if we neglect the dependence of $y(12)$ on ϕ_{12}, then Equation 58 becomes

$$\lim_{r \downarrow 1} g^{*}_{klm}(r) \approx \delta_{m0}\{(2k+1)(2l+1)\}^{1/2} P_k(0) P_l(0) \tag{61}$$

and Equation 60 continues to hold approximately. In Figures 5 and 6 are shown some typical g^{*}_{klm} and g^{*}_{Ω} for hard spherocylinders, using the computer simulation results of Nezbeda (*37*). Equation 61 is obeyed quite well, further indicating that the dependence of $y(12)$ on ϕ_{12} is not overly strong at small distances.

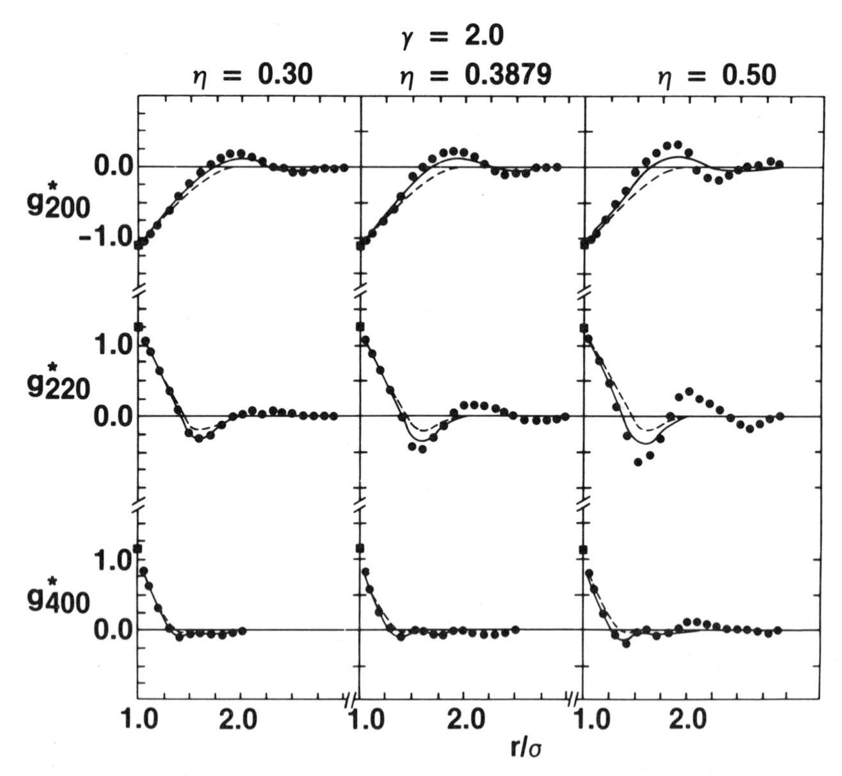

Figure 5. Reduced spherical harmonic expansion coefficients for the hard spherocylinder fluid. Key: ●, computer simulation results (37, 65); ---, RAMY(0) results; ——, RAMY(1) results; ■, the approximation of Equation 61. (Reproduced with permission from Ref. 41. Copyright 1981, Taylor and Francis, Ltd.)

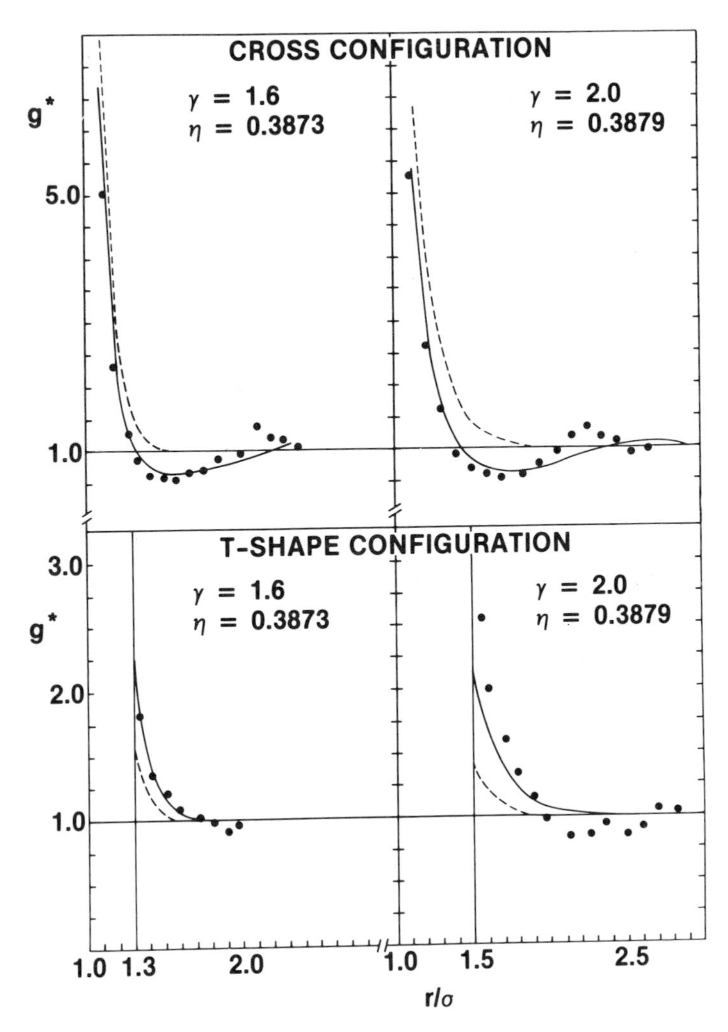

Figure 6. Radial slices through the reduced g(12) *surface for the hard spherocylinder fluid. Key:* ●, *computer simulation results of Nezbeda* (37); ---, *RAMY(0) results; and* ——, *RAMY(1) results. (Reproduced with permission from Ref. 41. Copyright 1981, Taylor and Francis, Ltd.)*

For non-hard-core molecules, similar approximate results may be obtained; their degree of accuracy depends on the particular molecular model considered. Further results of this type will be discussed in the section on comparison of RAM theory with computer simulation.

RAM THEORY RESULTS. The full g(12) for any intermolecular separation and orientation may be calculated from Equations 40 or 41 and 42 of the previous section. The spherical harmonic coefficients are given by (38)

$$g_{\mathrm{RAMG},klm}(r) = y_0(r)\{e_{klm}(r) + 2\pi\rho e_0(r)\delta_{m0}[\delta_{k0}J_l(r) + \delta_{l0}J_k(r)]\} \quad (62)$$

$$g_{\mathrm{RAMY},klm}(r) = y_0(r)[e_{klm}(r) + y_{1,klm}(r)] \quad (63)$$

where RAMG and RAMY denote, respectively, the RAM theory results using the g and y expansions, and

$$y_{1,klm}(r) = \rho \sum_{i>0} J_i(r)\left[\sum_{j=|i-k|}^{i+k} e_{jlm}(r)Q_{kijm} + \sum_{j=|i-l|}^{i+l} e_{jkm}(r)Q_{lijm}\right] \quad (64)$$

In the above, r denotes r_{12}, and the quantity Q is given by

$$Q_{nijm} = 4\pi^{3/2}\int Y_{n\bar{m}}(\boldsymbol{\omega})Y_{io}(\boldsymbol{\omega})Y_{jm}(\boldsymbol{\omega})d\boldsymbol{\omega} \quad (65)$$

Q can be expressed directly in terms of factorials (42, 43), or in terms of Clebsch–Gordon coefficients

$$Q_{nijm} = 2\pi\left|\frac{(2i+1)(2j+1)}{(2n+1)}\right|^{1/2} C(jin,m0m)C(jin,000) \quad (66)$$

using the convention of Rose (43).

As special cases of the above results, we have

$$g_{\mathrm{RAMG},000}(r) = g_0(r) \quad (67)$$

$$g_{\mathrm{RAMY},000}(r) = g_0(r)\left[1 + 4\pi\rho\sum_{i>0} e^*_{i00}(r)J_i(r)\right] \quad (68)$$

where $e^*_{i00} = e_{i00}/e_{000}$.

For completeness, we note that the spherical harmonics of $y(12)$ are given in the RAMY theory by

$$y_{\mathrm{RAMY},klm}(r) = y_0(r)\delta_{m0}\{\delta_{k0}\delta_{l0} + 2\pi\rho[\delta_{k0}J_l(r) + \delta_{l0}J_k(r)]\} \quad (69)$$

No computer simulation results are available for y_{klm}, although these would be useful, especially in view of the poor convergence of the expansion of $g(12)$ in the repulsive region of the potential (30–37).

We note that $g_{\mathrm{RAMG},klm}$ yields a nonzero first-order contribution to only g_{k00} (as is also true for $y_{\mathrm{RAMY},klm}$). However, $g_{\mathrm{RAMY},klm}$ gives a nonzero first-order contribution to all g_{klm}. If r is greater than the range of the angular-dependent part of the potential, then $g_{\mathrm{RAMG},klm}$ and $g_{\mathrm{RAMY},klm}$ become identical, since at these distances the only nonvanishing harmonic coefficient of $e(12)$ is e_{000}. In general, for all g_{klm} except those of the form g_{k00}, the zeroth-order RAMY and the first-order RAMG results are identical.

The asymptotic results discussed previously, in addition to testing computer simulation results, can also be used as a basis for testing theories. In the case of HOHD, for the zeroth-order RAMG theory, since $g(12) = g_0(r_{12})$, the asymptotic results are not obeyed. For the zeroth-order RAMY theory, we have

$$g^{(0)}_{\mathrm{RAMY}}(12) = e(12)y_0(r_{12}) \tag{70}$$

and hence that

$$\lim_{r\downarrow r_c} g^{*\,(0)}_{\mathrm{RAMY},klm}(r) = \lim_{r\downarrow r_c} \left|\frac{e_{klm}(r)}{e_{000}(r)}\right| = \lim_{r\downarrow r_c} e^*_{klm}(r) \tag{71}$$

which can be shown to be identical to the exact result of Equation 58. Similarly, Equation 60 is obeyed. In view of both these results, we expect the zeroth-order RAMY theory to yield good results at small r for $g(12)$ in the case of HOHD. This is generally true, as shown later in this chapter.

For the first-order forms of the theory, we have

$$g^{*\,(1)}_{\mathrm{RAMG},klm}(r) = e^*_{klm}(r) + 2\pi\rho\delta_{m0}g_0(r)[\delta_{k0}J_l(r) + \delta_{l0}J_k(r)] \tag{72}$$

$$g^{*\,(1)}_{\mathrm{RAMY},klm}(r) = \frac{e^*_{klm}(r) + y_{1,klm}(r)/e_{000}(r)}{1 + y_{1,000}(r)/e_{000}(r)} \tag{73}$$

where $y_{1,klm}$ is given by Equation 64. We see that the first-order RAMG theory cannot in general satisfy Equation 58 unless the second term on the right side of Equation 72 vanishes as r approaches r_c. Since this term is always zero except for the $k00$ harmonic coefficients, the asymptotic result is obeyed for coefficients other than those of the form $k00$. This is a further consequence of the fact pointed out earlier that, for all g_{klm} except those of the form g_{k00}, the zeroth-order RAMY and the first-order RAMG results are identical. The first-order RAMY theory will satisfy Equation 58 only if

$$\lim_{r\downarrow r_c} y_{1,klm}(r) = \lim_{r\downarrow r_c} [e^*_{klm}(r)\, y_{1,000}(r)] \tag{74}$$

and this is unlikely to be true in general. Even though these expressions do not satisfy the asymptotic results exactly, they may approximate them well numerically. We will examine this later.

Similarly, for HOHD, the zeroth-order RAMY theory satisfies Equation 60 exactly. The zeroth-order RAMG theory does not satisfy this result, and the first-order RAMG theory is again identical to the zeroth-order RAMY theory for g_X. For the first-order RAMY theory,

$$g_X^{*(1)}(r) = \frac{[1+4\pi\rho \sum_{i>0} (2i+1)^{1/2} J_i(r) P_i(0)]}{[e_{000}(r) + 4\pi\rho \sum_{i>0} e_{i00}(r) J_i(r)]} \tag{75}$$

If the second terms in the numerator and denominator of Equation 75 are small, then the asymptotic result will be obeyed approximately.

Numerical results for the tests of Equations 58 and 60 are given later in this chapter.

Alternative Molecular Coordinate Systems for Molecular Fluids. The relative position of two molecules is described by the distance apart of suitably chosen reference points in the molecules and the orientations of one or more intramolecular axes with respect to the line joining the two reference points. For homonuclear diatomics, for example, most computer simulation studies to date have used the coordinate system in which the reference points are chosen to be at the molecular center of mass. The relative molecular positions are described by the center-of-mass distance, the angles θ_i made by the molecular axes of each molecule with respect to the line joining the centers of mass, and the angle ϕ_{12} between the planes formed by the molecular axes and the center-of-mass line.

However, the location of the reference points in each molecule need not be restricted to the centers of mass, and it is clear that any convenient locations provide a valid coordinate system and a g(12) with respect to this coordinate system. For any of these coordinate systems, g(12) may be obtained from that in any other by means of appropriate transformations. Although the RAM theory was originally developed with the center-of-mass coordinate system in mind, g(12) discussed in the previous sections of this paper may be taken to refer to an arbitrary coordinate system with arbitrarily chosen reference points in each molecule, provided that these lie on the axis of rotational symmetry.

In general, for any physical problem, an appropriately chosen coordinate system may considerably simplify subsequent calculations. For example, if the moment of inertia of a rigid body is not calculated with respect to the principal axes, it has, in general, nine nonvanishing components rather than three. For site-interaction molecules, a convenient choice of location of the molecular reference points is at the centers of atomic sites. This gives rise to a pair correlation function denoted as $G^{\alpha\beta}(12)$, wherein the reference points are at atomic sites α and β in each molecule. Such a coordinate system was first suggested by Nezbeda (*44*). Like $g(12)$, $G^{\alpha\beta}(12)$ may be expanded in spherical harmonics, and one of its properties is

$$\langle G^{\alpha\beta}(12)\rangle_{\omega_1,\omega_2} \equiv G^{\alpha\beta}_{000}(r) = g_{\alpha\beta}(r) \tag{76}$$

where $g_{\alpha\beta}$ is the conventional site–site distribution function, whose calculation is addressed by the RISM theory (*14*), and for which computer simulation results are available (*30–36*). We note that, in general, even for homonuclear diatomic molecules, many of the symmetry properties are associated with the choice of reference points at the molecular centers of mass; for example $G^{\alpha\beta}_{klm} \neq G^{\beta\alpha}_{klm}$ in general. One of the advantages of using such a site-centered coordinate system for site-interaction molecular fluids is that the equation of state is directly related to a single spherical harmonic coefficient, in contrast to the case for the center-of-mass system, as we shall see in the next section.

All the RAM theory expressions given earlier remain formally unchanged in the site-centered coordinate system, as do the general expressions for the spherical harmonic coefficients of $G^{\alpha\beta}(12)$. Only the details of the reference system potential, $v_0(r)$, defined in Equation 21, are different. In Figure 1 are shown the different $v_0(r)$ for the HOHD system using the center-of-mass and the site-centered coordinate system (*45*, *46*). Note especially the apparent cusp in v_0 for the latter coordinate system, which produces a cusp in $g_0(r)$. Another interesting feature of v_0 is its finite value at the hard core. This property is responsible for the complicated dependence of g_0 on density (*47*).

Equation of State. The equation of state is given in general by (*48*)

$$\beta P/\rho = 1 - \frac{\beta N}{6}\iint G_m(q_1, q_2)\,\frac{\partial u(q_1, q_2)}{\partial \boldsymbol{R}_{12}}\,dq_1\,dq_2 \tag{77}$$

where G_m is a general pair correlation function referring to a coordinate system with the position and orientation of molecule i being denoted by q_i. Then $u(q_1, q_2)$ is the pair potential in that coordinate system, and R_{12} is the distance between two reference points of the molecules 1 and 2.

If the center-of-mass coordinate system is used, and the pair correlation function and the derivative of the potential are expanded in spherical harmonics, one obtains in general (35)

$$\beta P/\rho = 1 - 2\pi\beta\rho/3 \sum_{k,l,m} \int_0^\infty \frac{du_{klm}}{dr}\, g_{klm}(r) r^3\, dr \tag{78}$$

Alternatively, the equation of state can be computed by using a site-centered coordinate system, which yields the much simpler result (*49*)

$$\beta P/\rho = 1 - 2\pi\beta\rho/3 \sum_{\alpha,\beta} \int_0^\infty \frac{du_{\alpha\beta}}{dr}(r)\{r g_{\alpha\beta}(r) - 3^{1/2}[R_\alpha G_{100}^{\alpha\beta}(r) - R_\beta G_{010}^{\alpha\beta}(r)]\} r^2\, dr \tag{79}$$

where R_i is the intramolecular distance between the interaction site and the center of mass of molecule *i*. For hard-sphere interaction site models (ISM), Equation 79 becomes

$$\beta P/\rho = 1 + 2\pi\rho/3 \sum_{\alpha,\beta} \sigma_{\alpha\beta}^3 g_{\alpha\beta}(\sigma_{\alpha\beta})\{1 - 3^{1/2}[R_\alpha G_{100}^{*\alpha\beta}(\sigma_{\alpha\beta}) - R_\beta G_{010}^{*\alpha\beta}(\sigma_{\alpha\beta})]/\sigma_{\alpha\beta}\} \tag{80}$$

Equations 79 and 80 involve only a single harmonic coefficient of $G^{\alpha\beta}(12)$, in contrast to Equation 78, which requires all harmonic coefficients of g(12) as well as of the potential.

The equation of state may be calculated by means of the RAM perturbation theory by substituting the appropriate results in Equations 78 or 79. Since the main contribution in Equation 79 arises from the region of small *r*, the reduced zeroth-order RAM theory may be usefully employed, which produces the especially simple result given by

$$\beta P/\rho = 1 - 2\pi\beta\rho/3 \sum_{\alpha,\beta} \int_0^\infty \frac{du_{\alpha\beta}(r)}{dr} g_{\alpha\beta}(r)\{r - 3^{1/2}[R_\alpha e_{100}^{*\alpha\beta}(r) - R_\beta e_{010}^{*\alpha\beta}(r)]\} r^2\, dr \tag{81}$$

Equation 81 requires only the site–site distribution functions, $g_{\alpha\beta}$.

Finally, the equation of state may be calculated by differentiating the Helmholtz free energy. In the RAM theory, this is given from the results in the section on RAM theory by (*41*)

$$\beta(A - A_0)/N = -4\pi^2\rho^2 \sum_{k>0} \int_0^\infty r^2 \Delta e_{k00}(r) y_0(r) J_k(r)\, dr \tag{82}$$

which, in the first-order RAM theory, is numerically equivalent to

$$\beta(A - A_0) = -\frac{\rho^2}{4} \iint \Delta e(12) \exp[\beta u(12)] g(12)\, d\mathbf{R}_1 d\mathbf{R}_2 \tag{83}$$

Finally, Smith and Nezbeda (*41*) employed the approximation

$$g(12) = g^*_{\mathrm{RAMY}}(12)\, g_0(r_{12}) \tag{84}$$

in Equation 83, where g^*_{RAMY} is the reduced total pair correlation function obtained from the first-order RAMY theory, yielding

$$\beta(A - A_0) = -4\pi\rho^2 \sum_{k>0} \int_0^\infty r^2 e_{k00}(r)\, y_0(r) J_k(r) \frac{g_0(r)}{g^*_{\mathrm{RAMY},000(r)}}\, dr \tag{85}$$

In a later section in this chapter, we will show results obtained from these different routes for the equation of state, where they will be compared with computer simulation results for various molecular fluid models.

Comparison of RAM Theory with Computer Simulation Results

Results for $g_{000}(r)$. Smith et al. (*50*) calculated $g_{000}(r)$ using zeroth-order RAMY (which is identical to first-order RAMG) for the fluid of hard spheres with imbedded point quadrupoles. They used the Barker–Henderson perturbation theory for simple fluids (*3–5*) to compute $g_0(r)$. Good results were obtained up to very large reduced quadrupole moments. Figure 7a shows the dependence of $v_0(r)$ on the quadrupole moment, and Figure 7b shows some typical results for $\bar{g}(r) \equiv g_{000}(r)$. Smith (*51*) performed similar calculations for Lennard–Jones atoms with imbedded point dipoles, quadrupoles (LJQQ), and anisotropic overlap forces, and obtained similar good agreement up to large anisotropies.

Nezbeda and Smith (*52*) and Labik et al. (*53*) considered the calculation of g_{000} for HOHD, Smith and Nezbeda (*41*) studied hard spherocylinders (HSC), and Melnyk et al. (*54*) and Quirke et al. (*55*) considered Lennard–Jones diatomics (LJD). In all of these cases, the result $g_0 = g_{000}$ was shown to be quite accurate.

In the cases of HOHD and HSC, convenient analytical approximations may be obtained for $v_0(r)$. For HOHD, Oelschlaeger (*56*) showed that

$$\exp[-\beta v_0(r)] = 2(1 + L - r)^2(2 - L + r)/(3L^2 r)$$

for

$$\frac{1}{2}L + \sqrt{1 - L^2/4} = r_1 \leqslant r \leqslant 1 + L \tag{86}$$

where r is measured in units of σ. For small r, the following approximate result may be obtained (57) which is asymptotically valid as r approaches r_c

$$\exp[-\beta v_0(r)] = (2/3\pi)(B/LR)^2 \sin^{-1}(2B/L^2); \; r_c \leq r < 1 \qquad (87)$$

where

$$B = R^2 - (1 - L^2/2) \qquad (88)$$

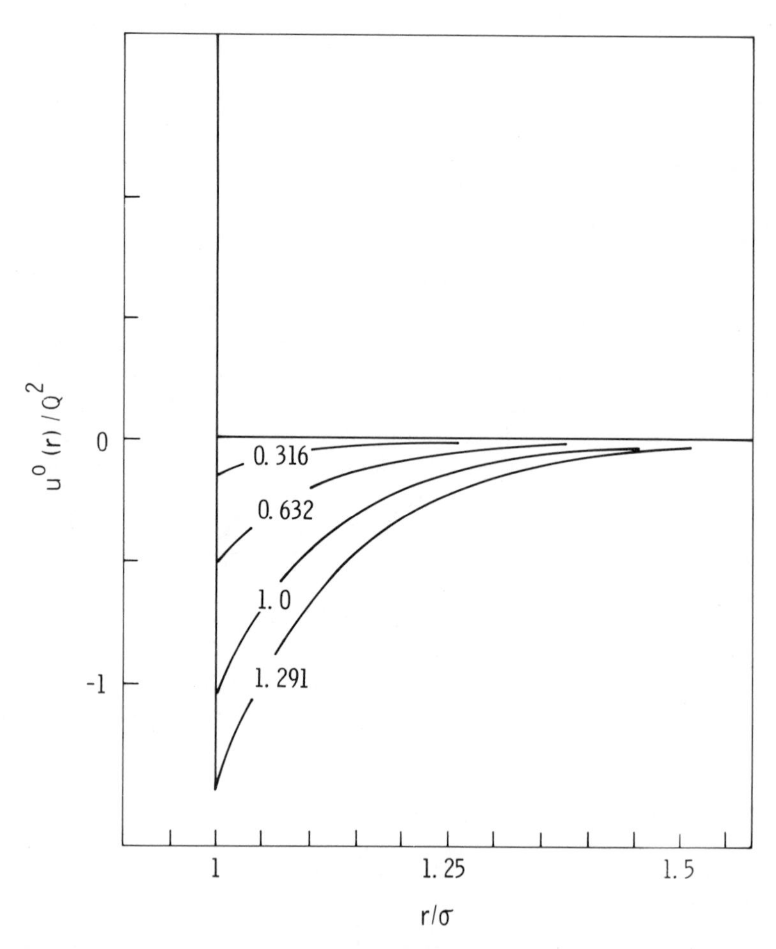

Figure 7a. Reference system potential u^0 *for the fluid of hard spheres with imbedded point quadrupoles at the indicated reduced quadrupole moments,* Q*.

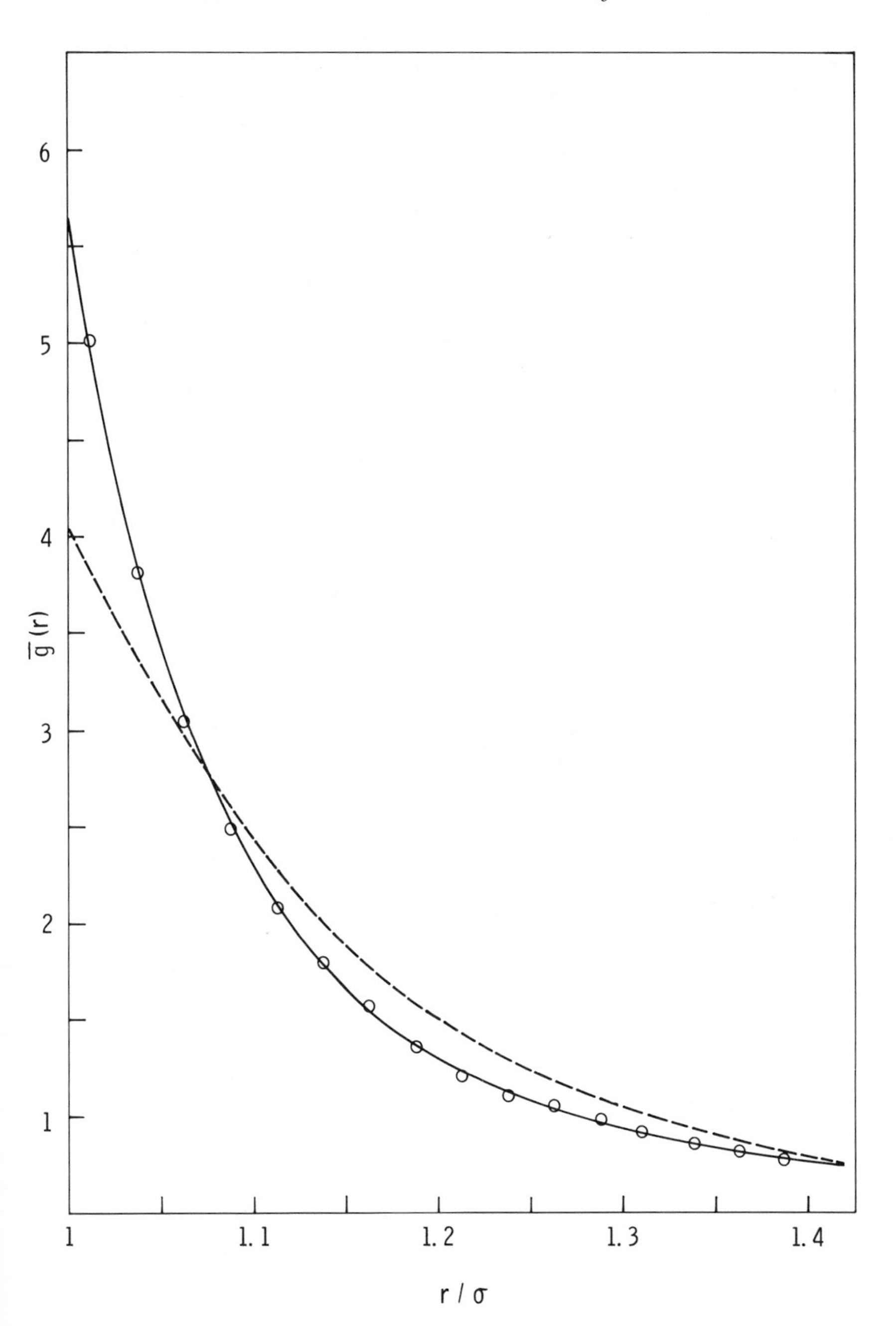

Figure 7b. Results for $\overline{g}(r) \equiv g_{000}(r)$. The reduced quadrupole moment, Q^*, is 1.000. Key: ○, experimental Monte Carlo results; ——, theory and ---, hard spheres.

Kohler et al. (58) employed Equation 86 and constructed an approximation that went smoothly to zero at r_c and that agreed with Equation 86 and its first two derivatives at r_1. This yields

$$\exp[-\beta v_0(r)] = a_2\Delta r^2 + a_3\Delta r^3/\Delta r_1 + a_4\Delta r^4/\Delta r_1^2 \tag{89}$$

where

$$\begin{aligned} \Delta r_1 &= r - r_c \\ \Delta r_1 &= r_1 - r_c \\ a_2 &= \delta/2 - 3\gamma + 6\mu \\ a_3 &= -\delta + 5\gamma - 8\mu \\ a_4 &= \delta/2 - 2\gamma + 3\mu \\ \delta &= r(2 + 3L - L^3 + r_1^3)/(3L^2r_1^3) \\ \gamma &= [4(1 - r_1^3) + 6L(1 + r_1^2) - 2L^3)/(3L^2r_1^2) \\ \mu &= (r_1)^{-2}[1 - 2(1 + L - r_1)^2(2 - L + r_1)(3L^2r_1)] \end{aligned} \tag{90}$$

Equation 89 may be used to obtain $\exp[-\beta v_0(r)]$ over the entire range without appreciable error.

For hard spherocylinders with ratio of length to breadth of γ, over the upper part of the range we have (57)

$$\begin{aligned} \exp[-\beta v_0(r)] = {} & \frac{r^2 - L - 1}{rL} \\ & + \frac{2}{rL^2}[(1 + L - r)(L^2/4 - 1) - 1/3(r - L/2)^3 \\ & + 1/3(1 + L/2)^3] \end{aligned} \tag{91}$$

for

$$L/2 + (1 + L^2/4)^{1/2} = r_2 \leqslant r \leqslant \gamma$$

where $L = \gamma - 1$ and r is measured in units of σ. For small r, we have the asymptotically correct approximation

$$\exp[-\beta v_0(r)] = (2/\pi)(r^2 - 1)\tan^{-1}(r) \tag{92}$$

for

$$1 \leqslant r < (1 + L^2/4)^{1/2}$$

Typical results for g_{000} are shown in Figure 8, where it is seen that $g_{000}(r)$ is given quite accurately by $g_0(r)$, except in the case of $\gamma = 2$ and the extremely high density $\eta = 0.50$. As noted previously, RAMY(0) and RAMG(1) both yield $g_0(r) = g_{000}(r)$. The main conclusion to be drawn from these studies is that the approximation

$$g_{000}(r) = g_0(r) \tag{93}$$

is very accurate, and that this result is yielded by the RAMY(0) and RAMG(1) expressions. These studies also found that the first-order RAMY results generally produce a deterioration in the accuracy.

Results for $g_{klm}(r)$. In this section, we discuss the results of the various forms of the theory, and draw some general conclusions.

In Figure 2, we show some selected $g_{klm}(r)$ for HOHD. It is seen that RAMY(0) is quite accurate at small r. This is a consequence of the fact that RAMY(0) obeys the asymptotic results discussed in the section on orientational structure of molecular fluids. RAMG(1) is also quite accurate, and is everywhere superior to RAMY(1), except for $g_{220}(r)$ and $g_{400}(r)$ at large r. Recall that, except for g_{k00}, RAMG(1) $\equiv$ RAMY(0).

From the discussion above where it was noted that RAMY(1) gives poor results for g_{000}, we conclude that this contributes to its poor performance at small r in Figure 2. When the reduced harmonic coefficients are considered, as shown in Figure 3, the change in accuracy is dramatic. RAMY(1) is superior to all other forms of the theory, and produces good agreement with the exact results. Recalling the discussion on orientational structure of molecular fluids, we see that RAMG(1) fails to reproduce numerically the asymptotic result at closest approach, (especially in the case of g_{200}) whereas RAMY(1) is only slightly in error at small r in all cases. We conclude from these results for HOHD that RAMY(1) is the superior form of the theory for g^*_{klm}.

In Figures 5, 9 and 10, we show selected $g^*_{klm}(r)$ for other model fluids. The agreement of RAMY(1) with the exact results is seen to be excellent, even, for example, in the case of HSC at $\gamma = 3$. As for the case of HOHD, RAMY(1) is very accurate numerically at small r.

For the LJQQ and LJD models, no exact asymptotic results exist at small r, mainly because there is no "closest approach distance" for such models. For the LJQQ model, the T-shaped orientation is favored, although not heavily so, except at very low temperatures (*54*). The asymptotic result corresponding to Equation 58, if the T-shaped orientation were uniquely favored, would be

$$\lim_{r \downarrow 0} g_{klm}(r) = 4\pi Y^T_{k\bar{m}} Y^T_{lm} \tag{94}$$

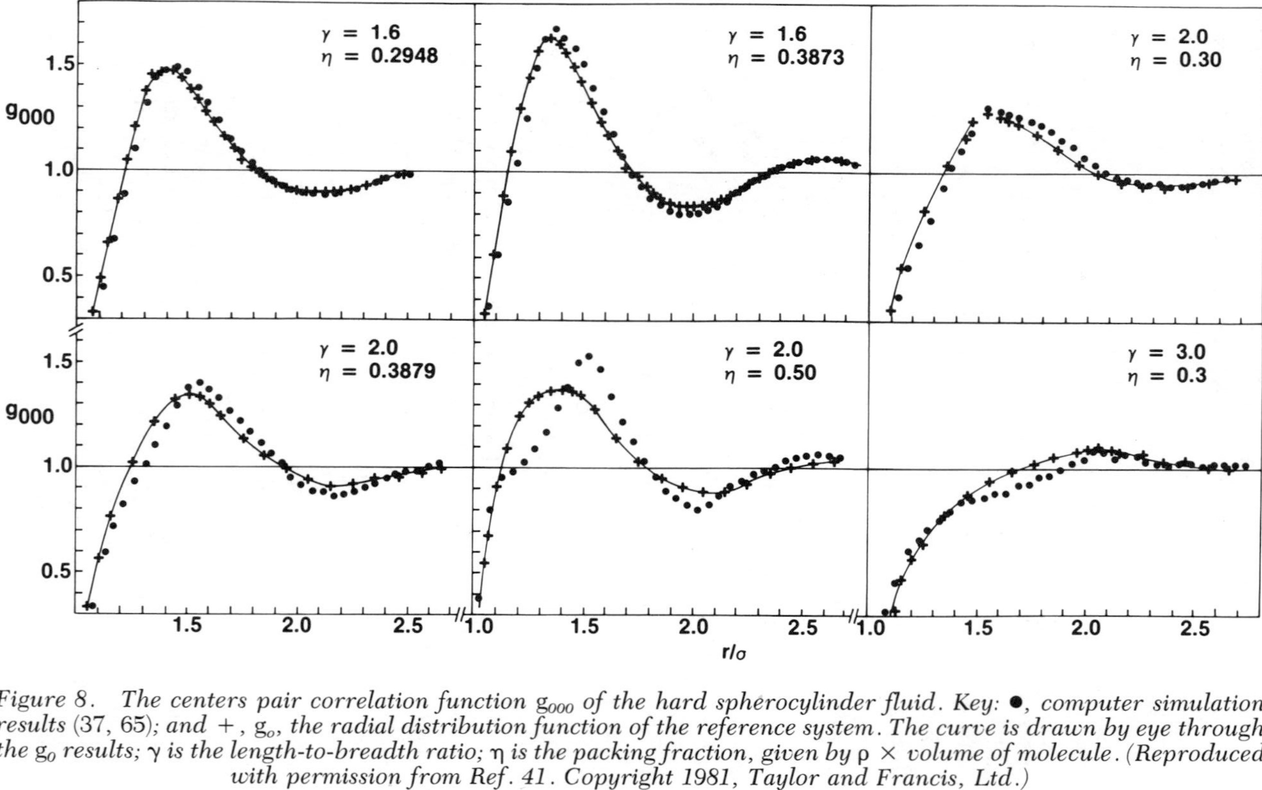

Figure 8. The centers pair correlation function g_{000} *of the hard spherocylinder fluid. Key:* ●, *computer simulation results* (37, 65); *and* +, g_o, *the radial distribution function of the reference system. The curve is drawn by eye through the* g_o *results;* γ *is the length-to-breadth ratio;* η *is the packing fraction, given by* ρ × *volume of molecule. (Reproduced with permission from Ref. 41. Copyright 1981, Taylor and Francis, Ltd.)*

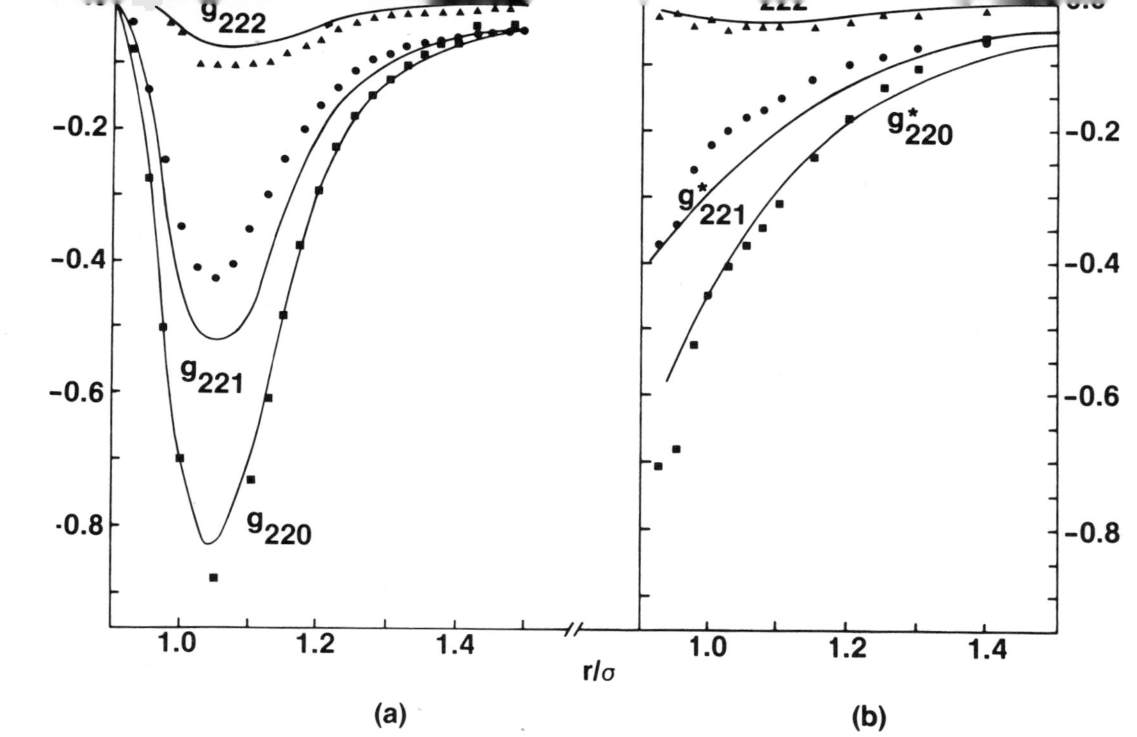

Figure 9. Selected spherical harmonic coefficients for the Lennard–Jones quadrupolar (LJQQ) fluid at $\rho^* = 0.75$, $T^* = kT/\varepsilon = 1.154$, *and reduced quadrupole moment* $Q^* = Q/(\varepsilon\sigma^5)^{1/2} = 1/\sqrt{2}$. *The points denote the computer simulation results of Haile (66). Left, the usual coefficients; right, the reduced spherical harmonics. The zeroth- and first-order RAMY results are indistinguishable. (Reproduced with permission from Ref. 54. Copyright 1982, Taylor and Francis, Ltd.)*

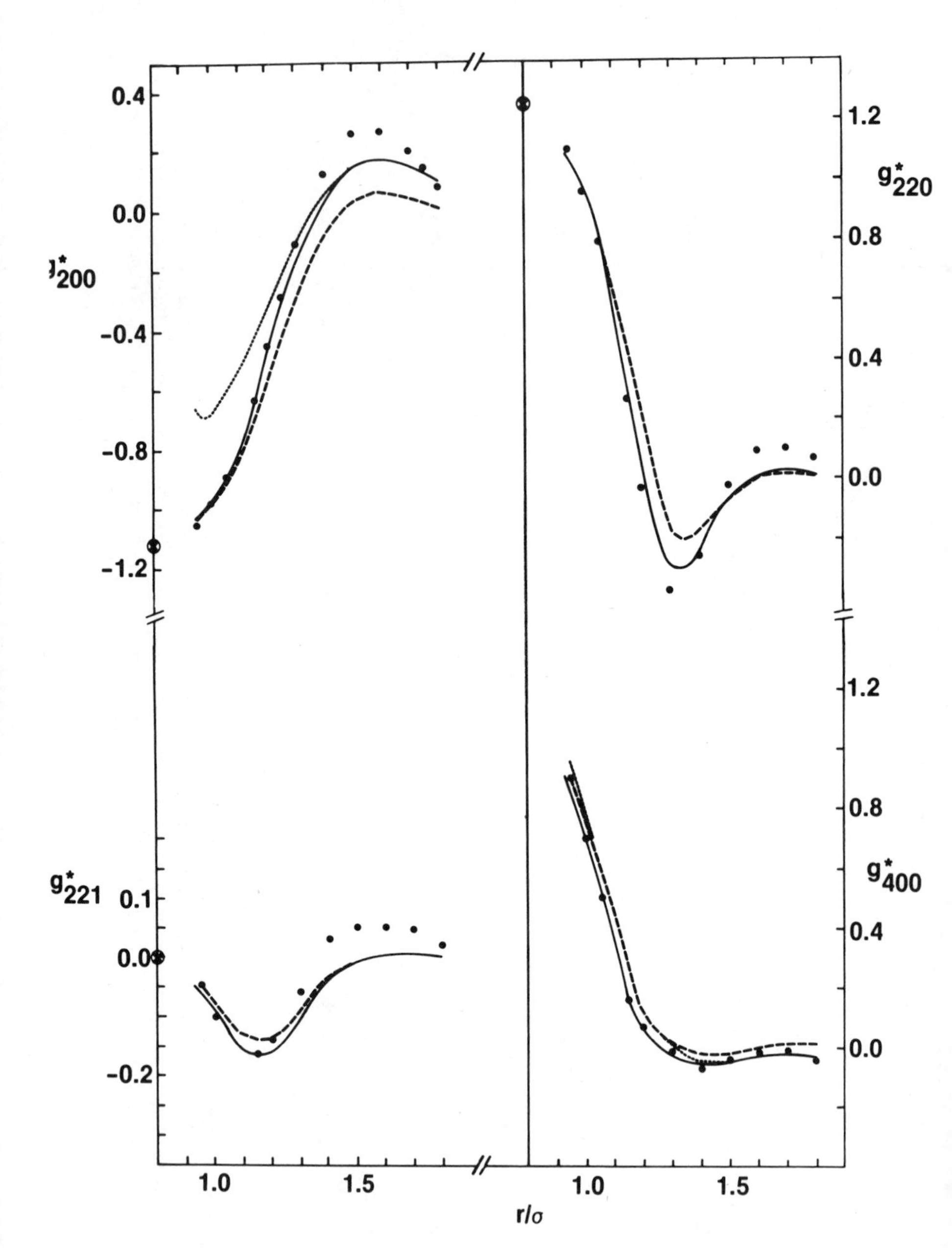

Figure 10. Spherical harmonic expansion coefficients for the Lennard–Jones diatomic fluid at ρ = 0.5, T* = 0.59, and L* = 0.5471. Points are the computer simulation results of Streett and Tildesley (31). Key: ---, zeroth-order RAMY results; ——, first-order RAMY results; and . . ., first-order RAMG result. (Reproduced with permission from Ref. 54. Copyright 1982, Taylor and Francis, Ltd.)*

The resulting values are zero for the 222 harmonic and −2.50 for the 220 harmonic. These would appear to be not inconsistent with the curves in Figure 9. However, the resulting value of zero for g_{221} would appear to be at variance with the corresponding curve in the figure. This may be because sufficiently small values of r are not displayed, or because the temperature is too high. For the LJD model, one would expect the crossed orientation to be heavily favored at sufficiently small separations, and hence that Equation 58 would hold in the limit $r \to 0$. In Figure 10, the limiting values are indicated. It would appear that the simulation and theoretical results are approaching these.

The main conclusion to be drawn is that g^*_{klm} is most accurately given by RAMY(1).

Results for g_Ω. Computer simulation results for g_Ω are available for the HOHD, HSC and LJQQ models. In Figures 4, 6, and 11, we show typical results. As can be seen, RAMY(1) is generally good in the case of the HSC models, although the agreement in the case of the T-shaped orientation begins to break down at large distances. For the LJQQ model, the agreement with computer simulation results is excellent everywhere.

The results of this and the preceding section suggest that $g^*(12)$ is given very accurately by RAMY(1). This suggests, in turn, that the approximation

$$g(12) = g_{000}(r_{12})g^*_{\mathrm{RAMY}}(12) \tag{95}$$

should be very accurate. We may combine this with Equation 93 to yield the further approximation

$$g(12) = g_0(r_{12})g^*_{\mathrm{RAMY}}(12) \tag{96}$$

where g_0 is the reference fluid pair-correlation function in the RAM theory. Equation 96 is to be preferred over Equation 95 if computer simulations are to be avoided and calculations are to be performed by using the theory from first principles.

Results for $G^{\alpha\beta}$. Computer simulation results are available for $G^{\alpha\beta}_{000}$ for the HOHD and LJD models considered previously, as well as for heteronuclear hard diatomics (HTHD) and symmetric hard triatomics (SHT), and for homonuclear diatomics whose atoms interact according to the purely repulsive part of the Lennard–Jones potential (RLJD). Results for $G^{\alpha\beta}_{100}$ and $G^{\alpha\beta}_{200}$ are available only in the case of LJD.

In view of the results discussed earlier for the center-of-mass coordinate system, we would expect $g_{\alpha\beta} \equiv G^{\alpha\beta}_{000}$ to be accurately given by RAMY(0). In Figures 12–14 are shown a selection of results.

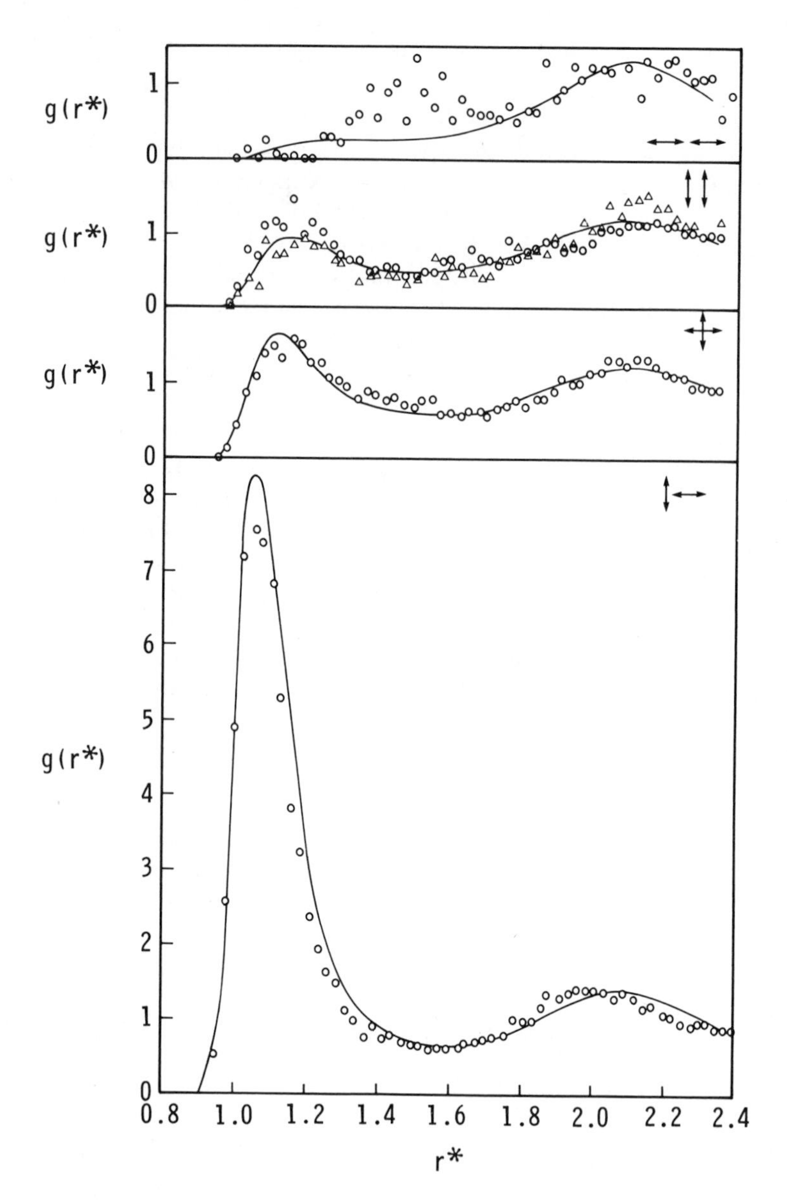

Figure 11. Radial slices through g(12) for the Lennard–Jones quadrupolar (LJQQ) fluid at Q = $1/\sqrt{2}$, T = 0.719, and ρ = 0.80. Key: ○, computer simulation results of Gubbins (67); and ——, the zeroth- and first-order RAMY results. (Reproduced with permission from Ref. 54. Copyright 1982, Taylor and Francis, Ltd.)

In the case of HOHD, shown in Figure 12, the results for $g_{\alpha\beta}$ are uniformly excellent, even at the large elongation and high density. The RISM results, shown for comparison, are everywhere inferior to RAMY(0). The cusps (slope discontinuities) in the computer simulation results for $g_{\alpha\beta}$ are reflected in similar discontinuities in $v_0(r)$, the reference system potential (shown in Figure 1). Apparently, v_0 is qualitatively similar to the actual force field experienced by the molecular sites, especially at small distances.

In Figure 13, we show $g_{\alpha\beta}$ for the two models based on the Lennard–Jones potential, LJD and RLJD. The results are again very good, similar to those for the HOHD model.

For the SHT model shown in Figure 14, RAMY(0) is good for both g_{BB} (which is identical to g_{000} in this case) and g_{AB}, and poor for g_{AA}. RISM is of similar accuracy in the former two cases, and much better in the latter.

The only models for which spherical harmonic coefficients are available from computer simulations are several LJD models, for which $G_{100}^{\alpha\beta}$ and $G_{200}^{\alpha\beta}$ have been computed. In Figure 15 some results are shown for RAMY(0). As might be expected from the results for g_{klm}^{*}, this is fairly accurate.

Equation of State. The equation of state may be calculated by any of Equations 78, 79, 82, or 85. Equation 78 is not useful, since it requires all spherical harmonic coefficients of g and of the potential.

For HOHD, Kohler et al. (*17*) used Equation 82 only. They used the Percus-Yevick theory (*8*) for the reference system and concluded that the RAM theory gave poor results. However, Smith and Nezbeda (*41*) showed that this poor performance of the RAM theory was likely due to using it directly instead of using it in reduced form. When Equation 85 is used, the theory produces much better results, which are in very good agreement with the simulation data over a wide range of anisotropy and density, as shown in Table I.

The most straightforward way to calculate the equation of state seems to be via Equation 79, or its simplified form, Equation 81. Nezbeda and Smith (*49*) considered Equation 81 for the equation of state, using the exact computer simulation results for $g_{\alpha\beta}(r)$, and these are displayed in Tables II and III.

The results are again seen to be excellent when the exact results for $g_{\alpha\beta}$ are used. This indicates that $G_{100}^{*\alpha\beta}$ at contact, which is required in the calculation of the equation of state, is very accurate. In view of the fact that the RAMY(0) results in the center-of-mass coordinate system are very accurate at short distances for the HOHD model, this accuracy in the site-centered coordinate system at short distances is to be expected. The results for the SHT model (*49*) lead to similar conclusions.

For the LJD and RLJD models, knowledge of $G_{100}^{\alpha\beta}$ is required over

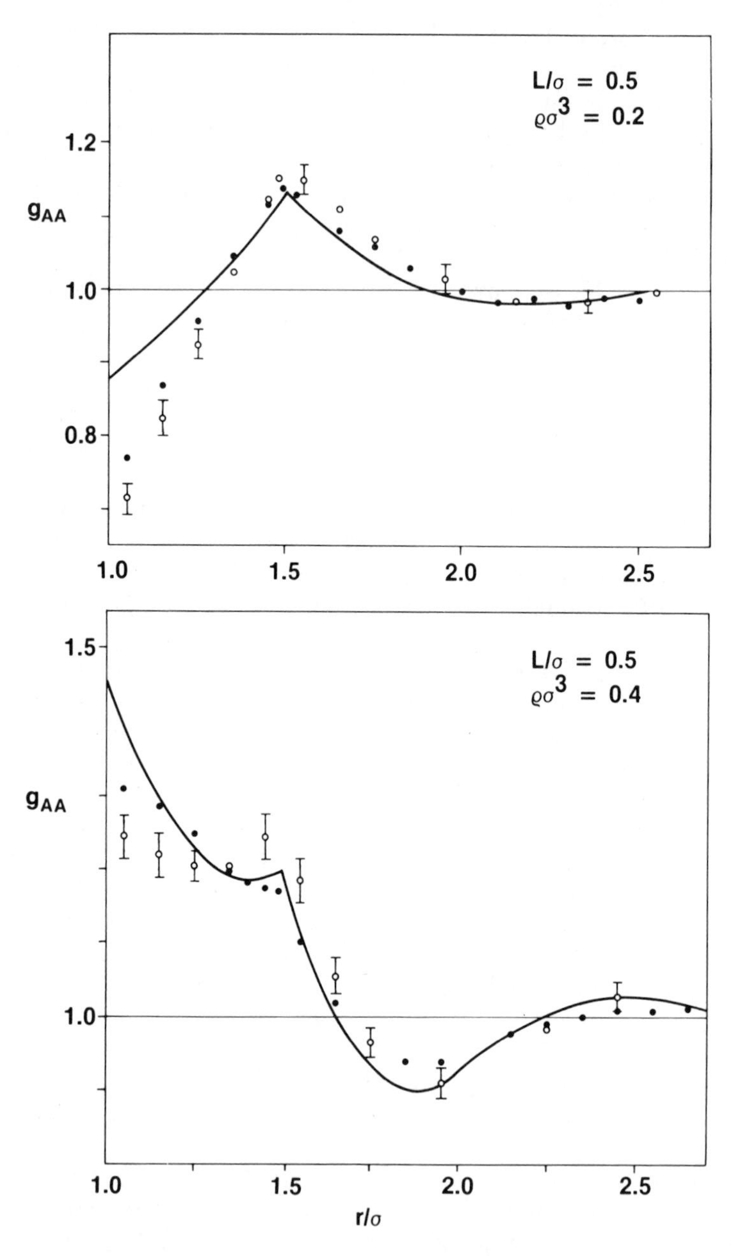

Figure 12. The site–site pair correlation function for homonuclear hard diatomics at the indicated state points. Key: ○, *computer simulation results of Morriss* (68) (L/σ = *0.5*), *Streett and Tildesley* (30) (L/σ = *0.4*) *and Chandler et al.* (15) (L/σ = *0.6*); ●, g^{0}_{AA}, *the simulation results for the RAM reference fluid; and* ——, *result of the RISM theory* (15). (*Reproduced with permission from Ref. 46. Copyright 1982, Taylor and Francis, Ltd.*) Continued.

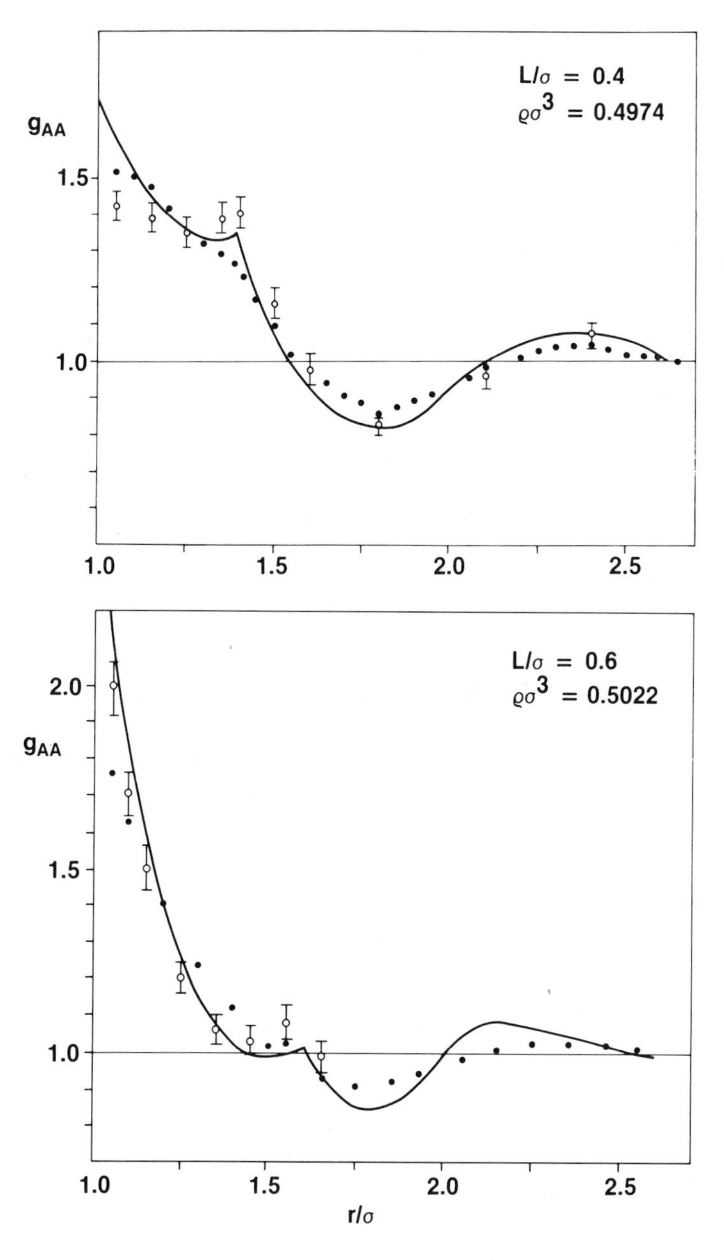

Figure 12. Continued.

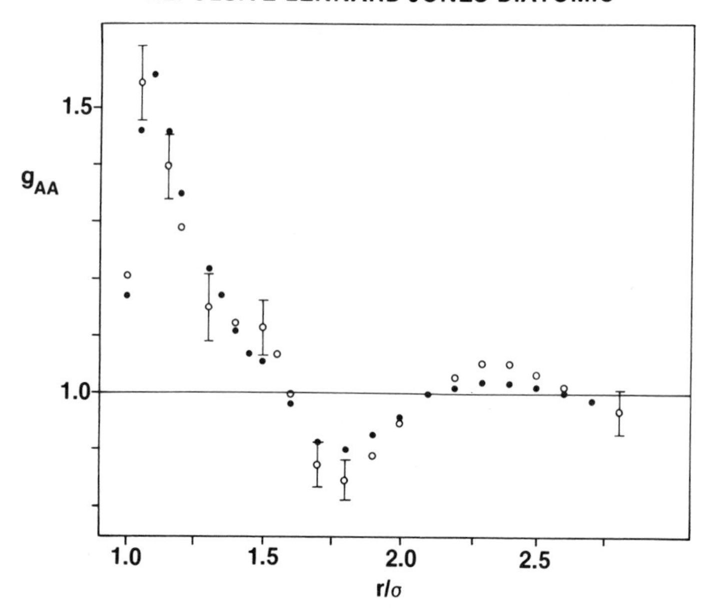

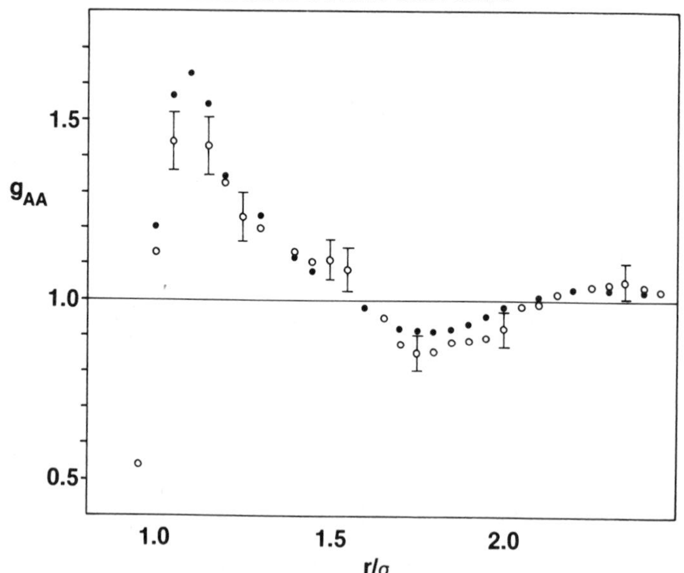

Figure 13. The site–site correlation functions of the RLJD and LJD fluids, at the state points (L^*, T^*, ρ^*) = *(0.5471, 0.524, 2.18) and (0.5471, 0.5, 2.36) respectively. Key:* ○, *computer simulation results of Streett and Tildesley (31); and* ●, *RAMY(0), obtained by means of the simulation results for the RAM reference fluid. (Reproduced with permission from Ref. 54, Copyright 1982, Taylor and Francis, Ltd.)*

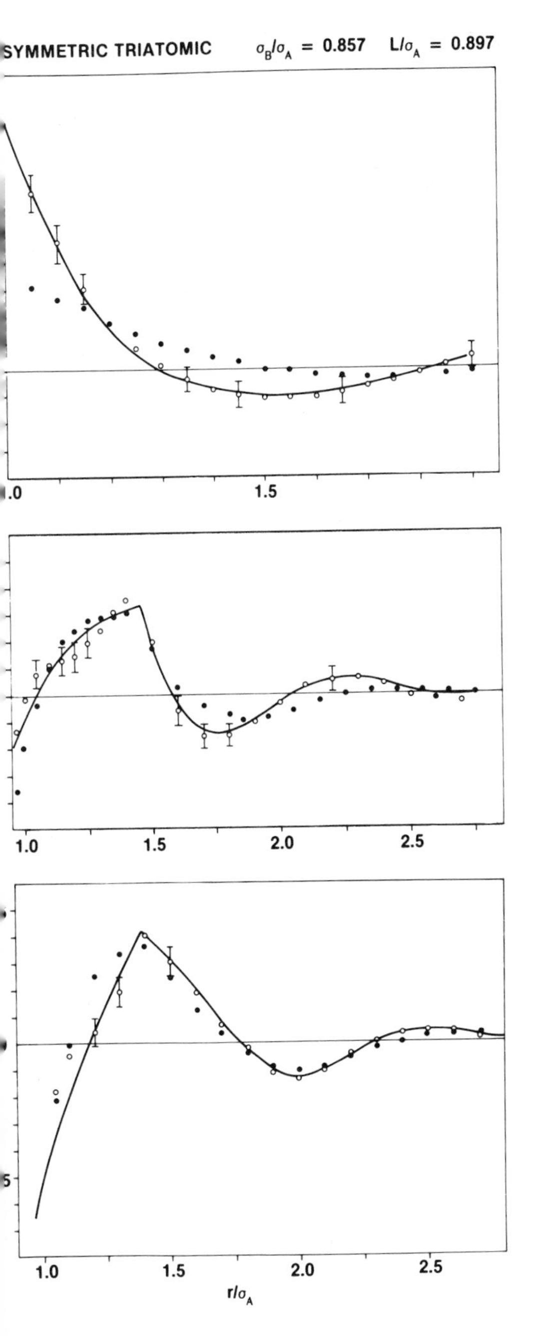

Figure 14. The site–site correlation functions for the linear symmetric hard triatomic fluid. Key: ○, simulation results of Streett and Tildesley (33); ●, RAMY(0) obtained by means of simulation results for the RAM reference fluid; and ——, result of the RISM theory. This figure is a corrected version of Figure 7 of Ref. 46, where the RISM results shown are incorrect. (Reproduced with permission from Ref. 46. Copyright 1982, Taylor and Francis, Ltd.)

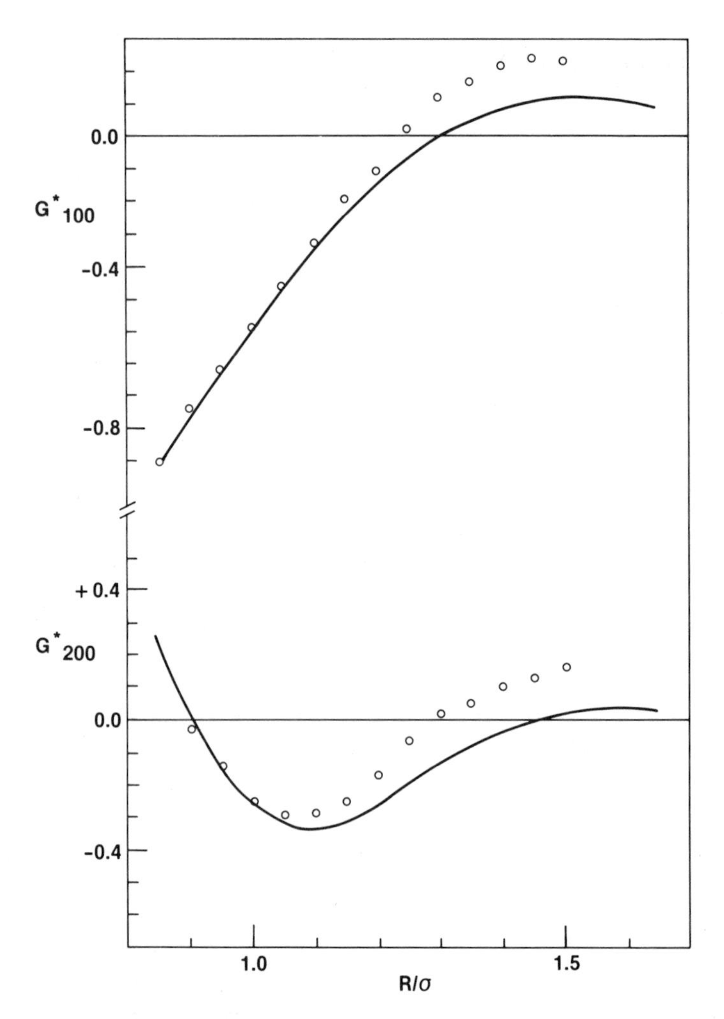

Figure 15. Selected reduced spherical harmonic expansion coefficients for G(12) in the site-centered coordinate system for the Lennard–Jones diatomic fluid at (ρ^*, T^*, L^*) = *(0.622, 2.90, 0.329). Key:* ○, *computer simulation results of Tildesley* (69)*; and* ——, *RAMY(0) result.*

the entire distance range, rather than just at contact, as for the hard-sphere ISM models. We show the results for the RLJD model in Table IV.

Discussion and Conclusions

We have seen that the RAM perturbation theory provides an accurate tool for calculating the structural and thermodynamic properties

Table I. Equation of State of the Hard Spherocylinder Fluid

		P/ρkT			
γ	η	*RAM*[a]	*Reduced RAM*[b]	*Unreduced RAM*[c]	*Zeroth-Order RAM*[d]
1.4	0.20	2.47	2.48	2.48	2.44
	0.25	3.17	3.19	3.18	3.10
	0.30	4.12	4.15	4.11	3.95
	0.35	5.42	5.43	5.33	5.03
	0.40	7.22	7.12	6.89	6.41
	0.45	9.80	9.35	8.86	8.15
1.6	0.15	1.98	1.99	1.98	1.95
	0.20	2.53	2.54	2.54	2.47
	0.25	3.26	3.29	3.26	3.11
	0.30	4.25	4.28	4.17	3.89
	0.35	5.60	5.57	5.32	4.84
	0.40	7.49	7.30	6.72	5.99
	0.45	10.17	9.64	8.39	7.36
2.0	0.15	2.07	2.04	2.05	1.99
	0.20	2.67	2.63	2.63	2.50
	0.25	3.47	3.43	3.35	3.10
	0.30	4.56	4.43	4.21	3.78
	0.35	6.04	5.64	5.20	4.55
	0.40	8.10	7.19	6.32	5.40
	0.45	11.02	9.34	7.55	6.33

[a] Calculated from Nezbeda equation of state (*70*).
[b] First-order reduced RAM theory, obtained using Equation 85.
[c] First-order unreduced RAM theory, obtained using Equation 82.
[d] Compressibility factor of the spherically symmetric reference fluid.

of a wide range of molecular fluid models. The theory is most accurate if reduced structural quantities and the first-order form based on the y expansion are used. Somewhat surprisingly, this holds for all quantities except for g_{000}, for which the zeroth-order form of the theory is the most accurate.

The least accurate part of the theory is apparently g_{000}. As noted in the discussion of this term, the first-order theory produces a deterioration compared to the zeroth-order results. For HOHD, however, the zeroth-order theory is extremely accurate. For heteronuclear molecules the accuracy deteriorates somewhat, as shown in Figure 14. For molecules with attractive forces, for example LJD, the accuracy of the approximation $g_{000} = g_0$ apparently deteriorates at large elongations (*59*). However, provided that an accurate g_{000} can be obtained from some other source,

Table II. Equation of State of Hard Homonuclear Diatomics

	$\beta P/\rho$				
ρd^3	$L^* = 0.2$	$L^* = 0.4$	$L^* = 0.6$	$L^* = 0.8$	$L^* = 1.0$
0.3	2.01	2.04	2.13	2.26	2.46
	2.04	2.05		2.24	2.49
0.4	2.59	2.64	2.78	3.01	3.36
	2.63	2.65	2.82	3.05	3.43
0.5	3.36	3.49	3.67	4.05	4.62
	3.40	3.51	3.71	4.12	4.80
0.6	4.45	4.59	4.95	5.48	6.40
	4.51	4.63	5.01	5.62	6.67
0.7	5.95	6.21	6.69	7.52	8.95
	6.02	6.27	6.84	7.77	9.47
0.8	8.02	8.42	9.23	10.54	12.64
	8.14	8.53	9.49	11.00	13.70
0.9	11.17	11.67	12.87	14.88	18.06
	11.20	11.77	13.21	15.49	19.80

Note: For each density, the first row is the Monte Carlo result of Streett and Tildesley (*30*). The second row is the zeroth-order RAM result, using Equation 81 and the computer simulation results for $g_{\alpha\beta}$. The term d is the diameter of a sphere having a volume equal to that of the molecule. At $L^* = 0.6$ and $\rho\sigma^3 = 0.3$, no result is shown since no $g_{\alpha\beta}(r)$ is available.

Table III. Equation of State of Hard Heteronuclear Diatomics

		$\beta P/\rho$		
L/σ_α	$\sigma_\beta/\sigma_\alpha$	*MC*[a]	*BN*[b]	*RAM*[c]
0.75	0.5	10.0 ± 0.7	9.74	10.38
0.625	0.5	8.9 ± 0.65	8.65	9.04
0.50	0.5	8.3 ± 0.6	7.92	8.45
0.375	0.5	7.8 ± 0.6	7.46	7.80
0.75	0.67	10.1 ± 0.7	9.88	10.52
0.75	0.84	9.9 ± 0.7	9.73	10.31

Note: All the compressibility factors are at the reduced density $\rho^* = \rho d^3 = 0.78$ where d is the diameter of a sphere having a volume equal to that of the diatomic.

[a] Monte Carlo data of Streett and Tildesley (*32*).

[b] From the Boublik–Nezbeda semiempirical equation of state (*71*).

[c] Zeroth-order RAM theory, using Equation 81 and the computer simulations results for $g_{\alpha\beta}$.

Table IV. Equation of State of the Repulsive Lennard–Jones Diatomic Liquid

Parameters			$P\sigma^3/\varepsilon$	
L/σ	$\rho\sigma^2$	$(\beta\varepsilon)^{-1}$	*MD*[a]	*RAM*[b]
0.5471	0.524	0.545	12.69	13.1
		0.610	13.48	13.7
0.6288	0.500	0.320	8.56	9.1

[a] Molecular dynamics data of Streett and Tildesley (*31*).
[b] Zeroth-order RAM theory, using Equation 81 and computer simulation results for $g_{\alpha\beta}$ (31).

the reduced form of the theory provides an accurate route to computing the angular-dependent properties. One possible approach might be to use the RAM theory for the repulsive part of the molecular interaction and then to treat the attractive part of the potential as a perturbation to the nonspherical reference system.

Since the reference fluid in the theory is spherically symmetric, its properties are readily computed in principle using relatively well-developed theories for such fluids. However, it is clear that these calculations must be very accurate in order to achieve good accuracy for the molecular fluid. To date, most of the calculations appearing in the literature and discussed in this review have used computer simulation results for g_0 for these reference systems, in order to provide unambiguous tests of the theory itself. This aspect of the RAM theory should serve to focus interest on the feasibility of rapidly and accurately computing g_0 for simple fluids. Provided this is possible, the RAM theory will provide a practical tool for computing molecular fluid properties.

The theory is currently being studied for molecular mixtures (*60*). Computer simulation results are available (*61, 62*) for such systems and the RAM theory provides one of the few theoretical means available for investigating such fluids.

Another application of the theory which has not been investigated is for molecular fluids whose molecules are nonlinear. The working equations of this paper will require some revision for such cases.

Finally, the theory is being investigated for the study of interfacial phenomena. Thompson et al. (*63*) have applied the theory to study vapor–liquid interfaces, and Nezbeda and Smith (*64*) have studied hard diatomics at a hard wall.

Literature Cited

1. Barker, J. A.; Henderson, D. *J. Chem. Phys.* **1967,** *47*, 2856.
2. Weeks, J. D.; Chandler, D.; Andersen, H. C. *J. Chem. Phys.* **1971,** *54*, 5237.

3. Smith, W. R. In "Statistical Mechanics"; Singer, K., Ed.; Special Periodical Reports; American Chemical Society: Washington, D. C., 1973; Vol. 1, Chap. 2.
4. Barker, J. A.; Henderson, D. *Rev. Mod. Phys.* **1976,** *48*, 587.
5. Andersen, H. C.; Chandler, D.; Weeks, J. D. *Adv. Chem. Phys.* **1976,** *14*, 105.
6. Green, H. S. "The Molecular Theory of Fluids"; Interscience: New York; 1952.
7. Van Leeuwen, J. M. J.; Groenveld, J., de Boer, J. *Physica* **1959,** *25*, 792.
8. Percus, J. H.; Yevick, G. J. *Phys. Rev.* **1958,** *110*, 1.
9. Lebowitz, J.; Percus, J. K. *Phys. Rev.* **1966,** *144*, 251.
10. Smith, W. R.; Henderson, D. *J. Chem. Phys.* **1978,** *69*, 319.
11. Gubbins, K. E.; Gray, C. G. *Mol. Phys.* **1972,** *23*, 187.
12. Pople, J. A. *Proc. Roy. Soc.* **1954,** *221A*, 498.
13. Morriss, G. P.; Perram, J. W.; Smith, E. R. *Mol. Phys.* **1979,** *38*, 465.
14. Chandler, D.; Andersen, H. C. *J. Chem. Phys.* **1972,** *57*, 1930.
15. Chandler, D.; Hsu, C. S.; Streett, W. B. *J. Chem. Phys.* **1977,** *66*, 5231.
16. Smith, W. R. *Can. J. Phys.* **1974,** *52*, 2022.
17. Koehler, F.; Marius, W.; Quirke, N.; Perram, J. W.; Hoheisel, C.; Breitenfelder-Manske, H. *Mol. Phys.* **1979,** *38*, 2057.
18. Mo, K. C.; Gubbins, K. E. *Chem. Phys. Lett.* **1974,** *27*, 144.
19. Tildesley, D. *Mol. Phys.* **1980,** *40*, 341.
20. Sandler, S. I. *Mol. Phys.* **1974,** *28*, 1207.
21. Chandler, D. *Annu. Rev. Phys. Chem.* **1978,** *29*, 441.
22. Sung, S.; Chandler, D. *J. Chem. Phys* **1972,** *56*, 4989.
23. Steele, W. A.; Sandler, S. I. *J. Chem. Phys.* **1974,** *61*, 1315.
24. Rushbrooke, G. S. *J. Chem. Soc., Trans. Far.* **1940,** *36A*, 1055.
25. Cook, D.; Rowlinson, J. S. *Proc. Roy. Soc.* **1953,** *219A*, 405.
26. Perram, J. W.; White, L. R. *Mol. Phys.* **1974,** *28*, 527.
27. Verlet, L.; Weiss, J. J. *Mol. Phys.* **1974,** *28*, 665.
28. Steinhauser, O.; Bertagnolli, H. *Zeit. Phys. Chem.* **1981,** *124*, 33.
29. Barker, J. A.; Henderson, D.; Smith, W. R. *Phys. Rev. Lett.* **1974,** *24*, 453.
30. Streett, W. B.; Tildesley, D. *Proc. Roy. Soc.* **1976,** *348A*, 485.
31. *Ibid.* **1977,** *355A*, 239.
32. Streett, W. B.; Tildesley, D. *J. Chem. Phys.* **1978,** *68*, 1275.
33. Streett, W. B.; Tildesley, D. *J. Chem. Soc., Faraday Disc.* **1978,** *66*, 27.
34. Streett, W. B.; Tildesley, D. *Mol. Phys.* **1980,** *41*, 85.
35. Tildesley, D.; Streett, W. B.; Wilson, D. S. *Chem. Phys.* **1979,** *36*, 63.
36. Tildesley, D.; Streett, W. B.; Steele, W. A. *Mol. Phys.* **1980,** *39*, 1169.
37. Nezbeda, I. *Czech. J. Phys.* **1980,** *30B*, 601.
38. Melnyk, T. W.; Smith, W. R. *Mol. Phys.* **1980,** *40*, 317.
39. Cummings, P.; Nezbeda, I.; Smith, W. R.; Morriss, G. *Mol. Phys.* **1981,** *43*, 1471.
40. Nezbeda, I.; Smith, W. R., unpublished data.
41. Smith, W. R.; Nezbeda, I. *Mol. Phys.* **1981,** *44*, 347.
42. Gaunt, J. A. *Trans. Roy. Soc.* **1928,** *228A*, 192.
43. Rose, M. E. "Elementary Theory of Angular Momentum"; Wiley: New York; 1957.
44. Nezbeda, I. *Mol. Phys.* **1977,** *33*, 1287.
45. Nezbeda, I.; Smith, W. R. *Chem. Phys. Lett.* **1981,** *81*, 79.
46. Nezbeda, I.; Smith, W. R. *Mol. Phys.* **1982,** *45*, 681.
47. Smith, W. R.; Nezbeda, I. *Chem. Phys. Lett.* **1981,** *82*, 96.
48. Steele, W. A. *J. Chem. Phys.* **1963,** *39*, 3197.

49. Nezbeda, I.; Smith, W. R. *J. Chem. Phys.* **1981,** *75*, 4060.
50. Smith, W. R.; Madden, W. G.; Fitts, D. D. *Chem. Phys. Lett.* **1975,** *36*, 195.
51. Smith, W. R. *Chem. Phys. Lett.* **1976,** *40*, 313.
52. Nezbeda, I.; Smith, W. R. *Chem. Phys. Lett.* **1979,** *64*, 146.
53. Labik, S.; Malijevsky, A.; Nezbeda, I. *Czech. J. Phys.* **1981,** *31B*, 8.
54. Melnyk, T. W.; Smith, W. R.; Nezbeda, I. *Mol. Phys.* **1982,** *46*, 629.
55. Quirke, N.; Perram, J. W., Jacucci, G. *Mol. Phys.* **1980,** *39*, 1311.
56. Oelschlaeger, H. Ref. 16 in Ref. 17.
57. Nezbeda, I., unpublished data.
58. Kohler, F.; Quirke, N.; Perram, J. W. *J. Chem. Phys.* **1979,** *71*, 4128.
59. Nezbeda, I.; Smith, W. R., unpublished data, 1982.
60. Smith, W. R.; Labik, S.; Nezbeda, I.; de Lonngi, D., in preparation.
61. Boublik, T.; Nezbeda, I. *Czech. J. Phys.* **1980,** *30B*, 121.
62. Nezbeda, I.; Boublik, T. *J. Phys.* **1980,** *30B*, 953.
63. Thompson, S. M.; Gubbins, K. E.; Haile, J. M. *J. Chem. Phys.* **1981,** *75*, 1325.
64. Nezbeda, I.; Smith, W. R., in preparation.
65. Monson, P. A.; Rigby, M. *Chem. Phys. Lett.* **1978,** *58*, *122*.
66. Haile, J. M., private communication, 1981.
67. Gubbins, K. E., private communication, 1978.
68. Morriss, G. Ph.D. Thesis, University of Melbourne, 1980.
69. Tildesley, D., private communication, 1981.
70. Nezbeda, I. *Chem. Phys. Lett.* **1976,** *41*, 55.
71. Nezbeda, I.; Boublik, T. *Czech. J. Phys* **1977,** *27B*, 1071.

RECEIVED for review January 27, 1982. ACCEPTED for publication October 18, 1982.

12

The Contribution of High Frequency Intermolecular Motions to the Structure of Liquid Water

PETER J. ROSSKY and FUMIO HIRATA[1]
The University of Texas, Department of Chemistry, Austin, TX 78712

A representation of liquid water structure obtained from a computer simulation in which the high frequency, hindered translational and rotational motions have been averaged out is analyzed with respect to geometrical and energetic characteristics of structure. The resulting average structure manifests a substantially higher degree of hydrogen bonding than is apparent in the vibrating structure, and it is shown that the observations are consistent with the removal of thermal excitation from an underlying distorted hydrogen-bonding network. Inferences with respect to the interpretation of structural variation in liquid water as a function of temperature and of perturbation by solutes are discussed, and it is suggested that such analyses can provide new insights into the molecular description of such systems.

THE MOLECULAR DESCRIPTION OF LIQUID WATER has been rapidly refined over the last decade with the ever increasing use of computer simulation as a standard research tool (*1*).

The structural and dynamical features of the liquid observed in simulation are in sufficiently good agreement with corresponding results from experimental studies that there is now little question that the simulation is rather faithful in its mimicry. Correspondingly, it has become possible to refine either or discard numerous structural inferences that had been made prior to the availability of this picture (*2*).

Among the earlier pictures that appear to have survived rather well is the continuum, distorted-hydrogen-bond view employed by Pople (*3*). According to this view, the structure is treated as fully hydrogen bonded but with substantial geometrical distortions present in the hydrogen bond network. A refined but corresponding picture, based on a random net-

[1] Current address: NASAC, Nishikanda 2–3–18, Chiyoda-Ku, Tokyo, Japan

0065-2393/83/0204-0281$06.00/0

work description of the liquid intermolecular hydrogen bonds, has been developed in some detail by Rice and coworkers (*4–7*).

The basic hydrogen bond networks characteristic of such descriptions can be thought of in dynamical terms as corresponding to the so-called V-structure of Eisenberg and Kauzmann (*8*). That is the structure that would be seen in the liquid if one averaged over the very rapid hindered local translational and librational motions, but not over the structural differences associated with net translational diffusion or reorientation. It is well established that the vibrational and net diffusional motions occur on well separated time scales (*1, 8*), and thus that such a picture is a physically interesting one.

One of the goals of the present consideration is to characterize quantitatively the V-structure of the liquid. However, the motivation for refining the available description of the pure liquid is not limited to an interest in pure water per se. With reasonable models for pure water available (*1*), detailed studies have grown naturally into the description of solvation and of solvent mediated solute–solute interaction (*1, 9–14*). In such solution studies, an essential element in the proper analysis of phenomena is an accurate and detailed description of the bulk solvent. It is the latter that must serve as the basis for comparison in any discussion of solute induced effects.

The purpose of the present chapter is to discuss the view of liquid water that one obtains from a realization of V-structure obtained from a computer simulation of the liquid (*15*) and to discuss some inferences that can be drawn from the results and that bear on solution structure. In the next section, we discuss the methodology employed in our study to obtain such a realization, and in the section on results we present representative results for the pure liquid structure. A further section presents a discussion of the implications regarding the study of both pure water and aqueous solutions. We include a consideration of hydrophobic and ionic hydration as well as comments on the interpretation of solvent isotope effects. The final section presents the conclusions.

Methodology

The numerical results presented here (*15*) were obtained from a 4-ps molecular dynamics computer simulation of bulk water at a temperature of 281 K and a density of 1 g/cm^3 using the ST2 potential (*1*) and a sample of 216 molecules.

The transformation from the sequence of instantaneous positions (I-structure) generated by the simulation to those representative of V-structure involves some subtleties, which are discussed elsewhere (*15*). The basic procedure involves dividing the dynamical history into a number of segments, each of a length τ_A of the order of an intermolecular vibrational period. The desired information consists of the sequence of mean molecular positions, each mean obtained from an average over the time τ_A of each segment. The averaging of coordinates was carried out in the six-dimensional space consisting of the center-of-mass position and the Euler angles (ϕ, θ, ψ) (*16*). From the mean center-of-mass positions and Euler angles obtained, we then recover a set of Cartesian coordinates for the sample of molecules. From the sequence of such sets, the analysis of the liquid structure proceeds in precisely the same manner as would analysis of the initial sequence of positions.

The choice of τ_A, the averaging time, is dictated up to a point by the dynamics inherent in the liquid, as reflected in the power spectrum for translational and rotational motion (*17*). The present result is shown in Figure 1. In the figure, the separate components associated with center-of-mass translational velocity and the rotational velocities of the principal axes are given. The librational motion is evident in the range from about 50–200 ps^{-1}, while the hindered translations span the range from nearly zero to about 70 ps^{-1}. The translational motions of highest frequency are apparently associated with local pairwise intermolecular vibrations (*8*, *17*, *18*), while the strong feature at approximately 10 ps^{-1} has been tentatively identified with three-body O–O–O bond-angle bending (*4*).

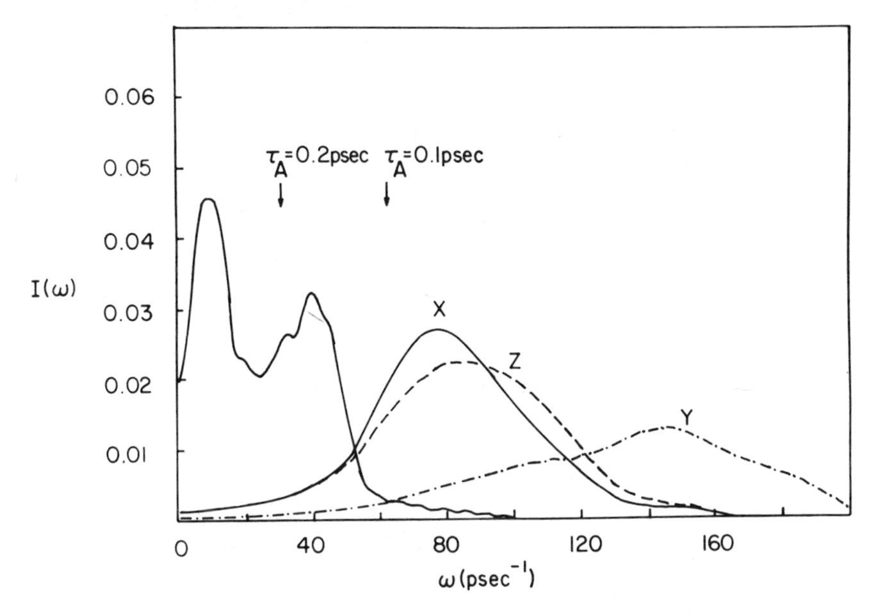

Figure 1. Velocity power spectra of single molecule for center-of-mass translational velocity (left-most curve) and principal axes rotational velocity (x, y, z)*. (Reproduced with permission from Ref. 15. Copyright 1981, American Institute of Physics.)*

If τ_A is chosen to be too small (say, $\tau_A \simeq 0.05$ ps) then it is clear that incomplete averaging of even the librational motion will occur. If τ_A is chosen to be large, say $\tau_A \geq 1$ ps, then the averaging will incorporate significant net reorientation. Further, to average effectively over the slower, many-molecule vibrations ($\omega \leq 10$ ps^{-1}, $\tau \geq 0.6$ ps^{-1}) would require an averaging time of at least 0.6 ps, and preferably longer. Since such times are not widely separated from the time over which net reorientation occurs, nor from estimates of hydrogen bond lifetimes ($\tau \simeq 0.6$ ps) (*19*), we have averaged only over the rapid librational motions associated with hindered rotation and, accordingly, the more rapid translational motions associated with pairwise vibrations. This suggests an averaging time τ_A in the range of about 0.1 to 0.2 ps, indicated in Figure 1; for $\tau_A \simeq 0.2$ ps, even the slowest librations experience roughly two periods. In the following analysis we have examined both $\tau_A = 0.1$ ps and $\tau_A = 0.2$ ps, with primary emphasis on the latter. For $\tau_A = 0.2$ ps, one obtains 20 V-structure configurations from the 4-ps simulation.

Results

In this section we describe the results obtained from an analysis of the generated V-structure and compare them with a corresponding analysis of the original simulation (I-structure). We consider both geometric and energetic measures of structure.

Geometric Analysis of V-Structure. We consider first intermolecular atomic radial pair correlation functions. In Figure 2 we show the results for $g_{OO}(r)$ and $g_{HH}(r)$. The solid line shows the result obtained from the original simulation and corresponds to that obtained in previous studies (*20*). The average structure obtained using $\tau_A = 0.2$ ps, and the corresponding result for $\tau_A = 0.1$ ps are both shown. Beyond $r_{OO} = 3.5$ Å and $r_{HH} = 4.5$ Å the three cases are indistinguishable, a result consistent with the longer time scale associated with the response of weaker, longer range correlations (*21*).

The averaging over rapid nearest neighbor vibrations leads to the expected predominant effect in these results, namely, a narrowing of the peaks associated with nearest neighbor molecules. This result is in agreement with earlier interpretations of O–D and O–H Raman bands (*18*). From a comparison of the results for $\tau_A = 0.1$ ps and $\tau_A = 0.2$ ps, it is also clear that the change associated with the vibrational averaging is basically developed after 0.1 ps, although some additional narrowing occurs for the larger value of τ_A. That some further narrowing occurs is not surprising, since (*see* Figure 1) the larger value incorporates a significantly larger portion of the hindered translation region of the power spectrum.

In accord with the averaging over librational motion, one expects

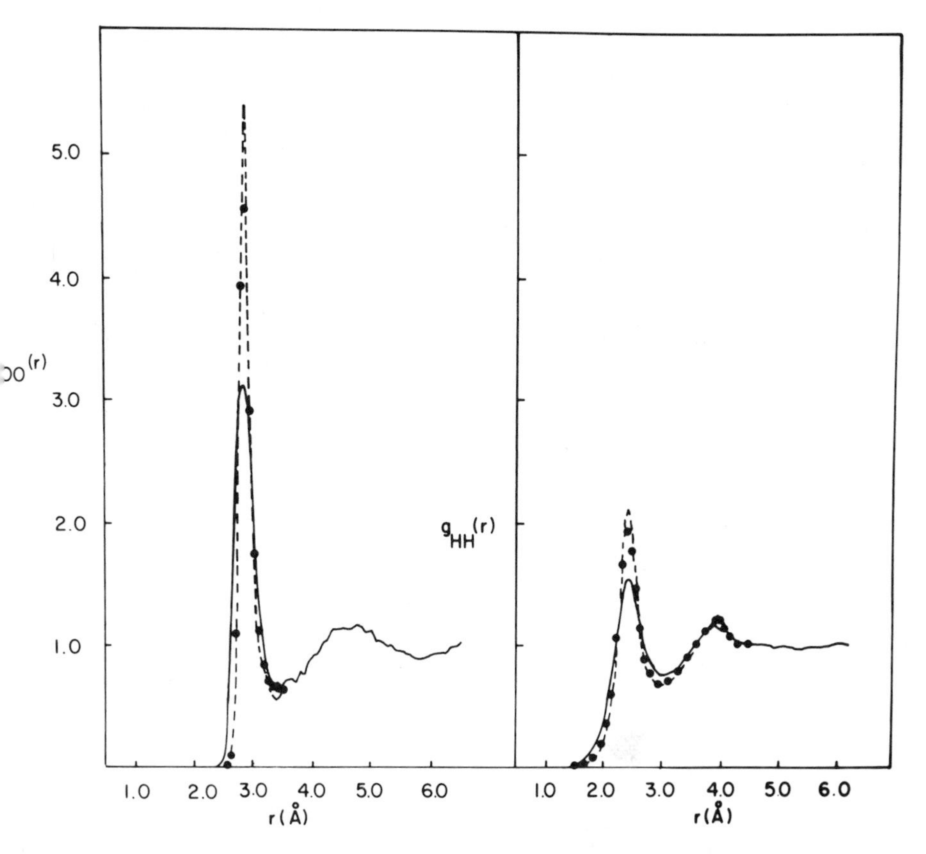

Figure 2. Intermolecular atomic radial pair correlation functions for O–O pairs (left) and H–H pairs (right). Key: —, I-structure; ----, V-structure, τ_A = 0.2 ps; and ●, V-structure, τ_A = 0.1 ps. Beyond r_{OO} = 3.5 Å and r_{HH} = 4.5 Å, the three cases are indistinguishable, and only one is shown. (Reproduced with permission from Ref. 15. Copyright 1981, American Institute of Physics.)

that the linearity of hydrogen bonds should be correspondingly enhanced. The relevant angle is θ_{OHO} formed by the pair of oxygen atoms and the shared proton. To avoid an a priori biasing of the results toward linear hydrogen bonds (e.g., by considering only pairs of molecules that are hydrogen bonded according to an energetic criterion), we consider all nearest neighbor pairs of molecules, with O–O separations less than 3.5 Å. For each pair, that proton (of the four associated with the molecular pair) which is nearest to the O–O line is located, and the angle θ_{OHO} is evaluated with that proton. This procedure does not discriminate against librationally distorted pairs. There is, however, a natural bias against acute angles ($\cos\theta > 0$), which is associated with the choice of the single proton nearest to the O–O line; hence we expect almost no density for positive values of cos θ. The calculated distribution is given in Figure 3.

The strongly hydrogen bonded pairs occur in the neighborhood of $\cos\theta = -1$ (linear hydrogen bonding), and as expected there is a substantial narrowing of the distribution. That is, the V-structure has significantly straighter hydrogen bonds; the occurrence probability of a linear hydrogen bond is enhanced by approximately 50%.

The shoulder in the distribution of Figure 3, centered at $\cos\theta \simeq 0.5$, is probably associated with the longest O–O distances included in this distribution, and correspondingly with nonbonded molecular pairs.

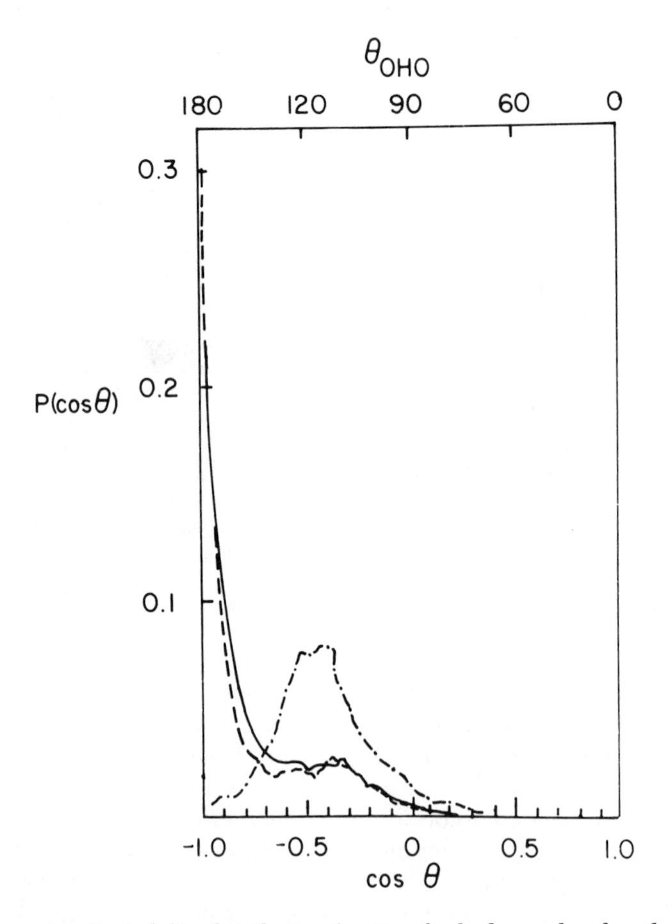

Figure 3. Probability distribution function for hydrogen bond angle θ_{OHO}. Key: —, I-structure, $r_{OO} < 3.5$ Å; ----, V-structure, $\tau_A = 0.2$ ps and $r_{OO} < 3.5$ Å; and -·-·-, V-structure, $\tau_A = 0.2$ ps and $3.5 \text{ Å} \leq r_{OO} \leq 4.2$ Å. (Reproduced with permission from Ref. 15. Copyright 1981, American Institute of Physics.)

To see this, we have evaluated the corresponding distribution for the region of O–O distances in the range 3.5 Å < r_{OO} ≤ 4.2 Å (cf. Figure 2). This result is also shown in Figure 3 and overlaps quite well with the shoulder in the original nearest-neighbor distribution.

Energetic Analysis of V-Structure. It is clear from the results given that the averaging procedure leads to a geometric structure with the anticipated sharpening of intermolecular bond length and bond angle distributions. However, although one expects correspondingly more negative hydrogen bond energies, it is not a priori clear how this will be quantitatively manifest in the structure. We therefore examine several energetic measures of the liquid structure.

We consider first the distribution of pair interaction energies. Figure 4 shows the probability of finding a pair of molecules interacting with a potential energy ε. The notation corresponds to that in Figure 2. As is clear from the figure, in the V-structure the pair energies are shifted to more negative values (the peak by approximately 0.7 kcal/mol) and the distribution is somewhat sharper. The observed shift is qualitatively very similar to that found to result from a decrease in temperature in bulk water (*20*). As for the geometric analysis, we find that the energetic structure is not sensitive to the choice of τ_A. Corresponding behavior has been found in the distribution of molecular binding energies (*15*).

A more graphic characteristic of the bonding structure is the proportion of molecules participating in a given number of hydrogen bonds. As has been discussed elsewhere (*17, 20, 22*), such a quantity is not an absolute but depends strongly on the assigned definition of an intact bond. This is true both of geometric (*6*) and energetic criteria. For geometric criteria, it appears possible to develop a less sensitive definition (*6*), which permits a bond to be defined in the Pople sense (*3*) even when intermolecular arrangements are highly distorted from the optimal energetic geometry. Such an approach is advantageous if one desires to construct a formal network theory (*5, 6*), but for the present purposes it does not offer any overriding advantages. Here we use the pair interaction energy as a criterion. However, we note that with a geometric definition, the numbers and locations of "intact" hydrogen bonds may vary substantially from that obtained using an energetic one, and these alternatives can provide complementary descriptions.

In Figure 5, we show, in the form of histograms, the results obtained for three different criteria for the most positive energy to be associated with an intact bond; the original simulation and V-structure are shown. Perhaps the most satisfactory energetic criterion is approximately −4 kcal/mol; it has been shown (*20*) that the population for more negative pair-energies decreases with increasing temperature, while the population for more positive pair-energies increases with increasing temperature. The right-hand distribution in Figure 5 corresponds closely to this value.

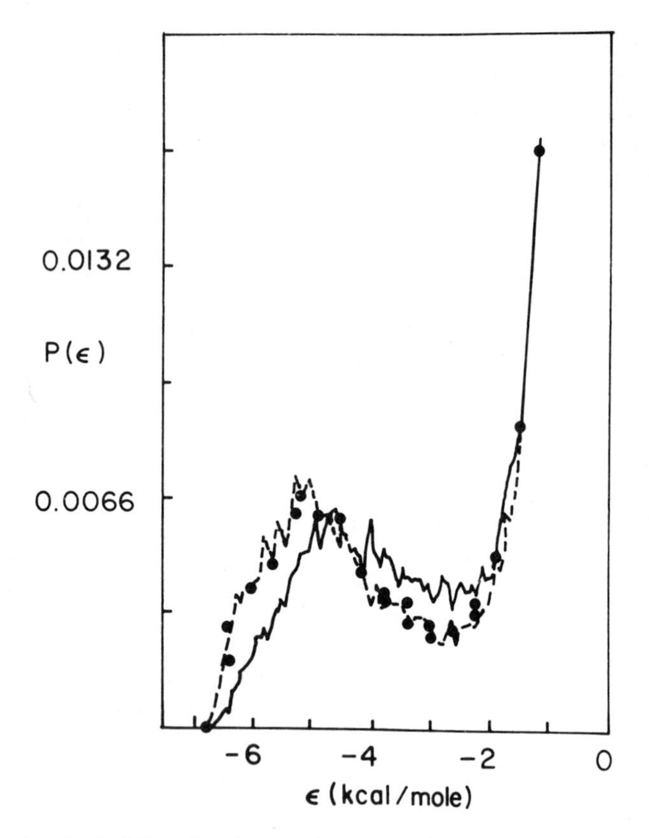

Figure 4. Probability distribution for intermolecular pair interaction energy ε. The key is the same as in Figure 2. (Reproduced with permission from Ref. 15. Copyright 1981, American Institute of Physics.)

The results in the figure show the enhancement in the degree of hydrogen bonding in the V-structure. With increasing stringency of the energetic criterion, the enhancement also increases. For the most negative criterion, which is a reasonable choice, the change is dramatic; the number of unbonded molecules ($n_{HB} = 0$) is reduced by a factor of four, and the number of four-bonded molecules is increased by a factor of three.

The enhancement in the degree of hydrogen bonding has a corresponding influence on the description of connectivity, via such bonds, in the liquid. As one measure of this, we have carred out an analysis of polygonal connectivity for the V-structure corresponding to that published earlier by Rahman and Stillinger (22). We have evaluated the number of non-cross-linked (non-short-circuited) polygons of different edge sizes ($n = 3$–11), with edges formed by intact intermolecular hydrogen bonds. The analysis is carried out in precisely the same manner as described in detail previously (22). An exhaustive search is made for polygons formed from sequentially connected molecules, and each polygon is tested to assure that no two members of the ring are short-circuited by a series of one of more hydrogen bonds that connect them by a shorter route.

A set of 20 configurations was used for the analysis, with resulting error estimates of about 10% for each population. The results are shown in Figure 6. The value N_p gives the number of polygons of edge size n per liquid molecule. The I-structure results agree with those published earlier (22) within our estimates of statistical uncertainty.

The connectivity analysis shows a dramatic increase in the frequency of ring structures in the V-structure, in parallel with the increase in the number of hydrogen bonds per molecule, shown above. For the most negative hydrogen bond criterion, the enhancement is more than 400%. It is therefore also fairly certain that other measures of connectivity, such as the percolation threshold (33), will manifest this change as well.

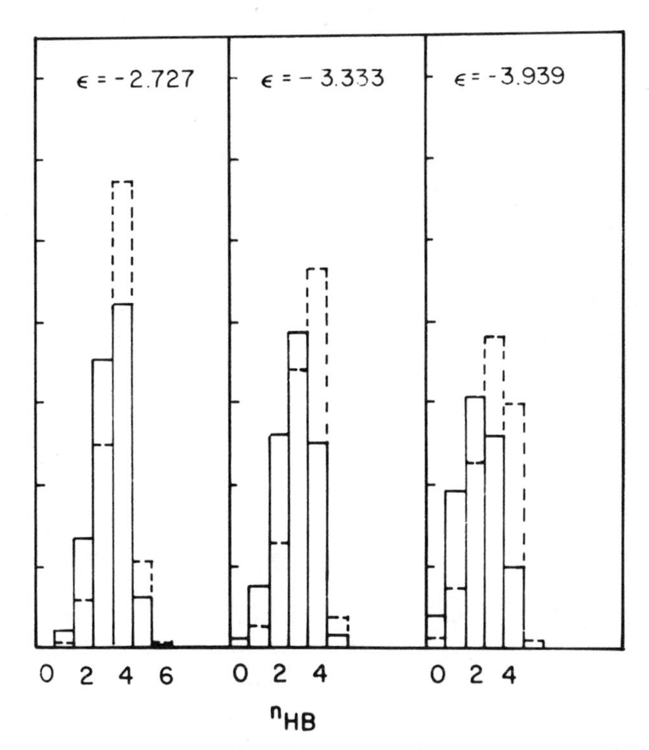

Figure 5. Fraction f of molecules participating in n_{HB} *hydrogen bonds, for various definitions of the maximum energy associated with an intact hydrogen bond* ε_{HB}. *Key: —, I-structure; and ----, V-structure,* $\tau_A = 0.2$ *ps. The hydrogen bond criteria are indicated above each histogram (kcal/mol).*

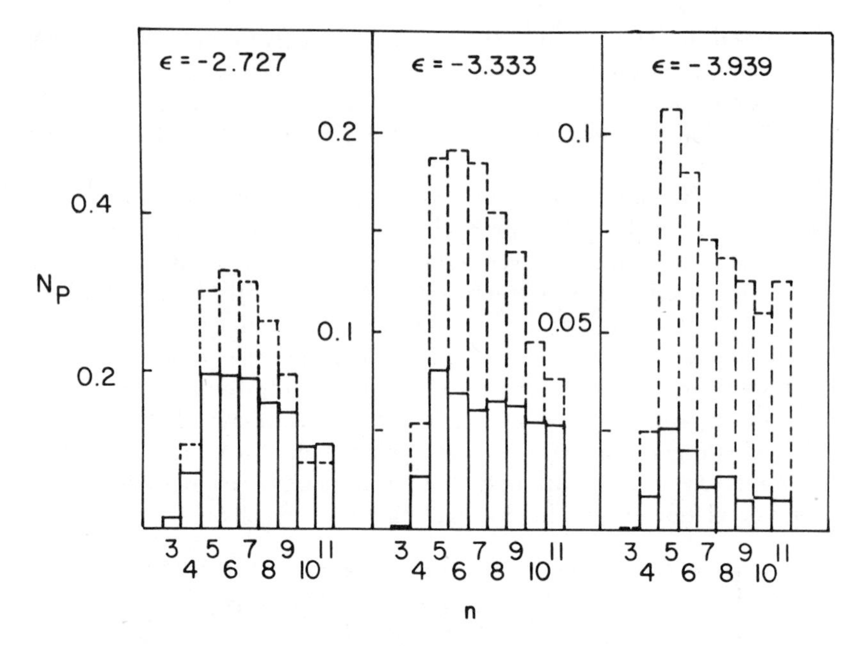

Figure 6. Number of non-short-circuited hydrogen bond polygons N_p *of edge length* n *(per liquid molecule) for various hydrogen bond energy criteria* ε_{HB}. *The key is the same as in Figure 5. (Reproduced with permission from Ref. 15. Copyright 1981, American Institute of Physics.)*

We note that a study using a related methodology but having a different main objective from the present work has recently appeared (7); the quantitative results obtained from structural analysis appear consistent with our own.

Discussion

In this section we consider the interpretation of the changes that are manifest after averaging out the high frequency intermolecular vibrations, and we consider the implications of this picture for solution phenomena.

Pure Liquid. We first address the question of how best to view the physical origin of the enhanced intermolecular interactions in the V-structure. It is clear that the existence of such enhancement is, at least, consistent with the averaging over rapid motion around a relatively stable point (a local minimum) on the many-body potential surface, so that the V-structure corresponds closely to that local minimum. If this is correct, it follows that the bonding characteristics observed for the V-structure with a given bond criterion ε_{HB} should correspond to that in the absence of averaging but with a criterion ε'_{HB} that is more lenient by an amount typical of the energy associated with the vibrational excitation.

To test this, we have reconsidered the bonding analysis presented above and, for representative cases, examined the dependence of the populations on ε_{HB}. The results are shown in Figure 7. The upper graph gives the fraction of molecules f participating in n hydrogen bonds for $n = 2$ and $n = 4$. The lower graph gives the polygon populations N_p for polygons of edge sizes $n = 5$ and $n = 9$. I-Structure results and corresponding V-structure results are shown. The solid lines are simply smooth curves drawn through the I-structure values; the dashed curve is equivalent, but it is shifted to more negative values of ε_{HB} by exactly k_BT (k_B is Boltzmann's constant). The latter value is an expected typical thermal excitation energy. Except for the most lenient choices of ε_{HB}, the V-structure results conform rather closely to the adjusted I-structure.

This comparison supports the view that the averaging procedure effectively averages out the local vibrational excitation and demonstrates that the V-structure is representative of the same hydrogen bond network structure as is the I-structure, despite the dramatic differences in the quantitative measures of the degree of bonding.

We have noted that the qualitative effect of the vibrational averaging is comparable to that of a substantial change in temperature, at least for energetic properties. This view is consistent with the analogy invoked elsewhere (*4, 8*) between the structure of the liquid and that of amorphous solid water. Our analysis has demonstrated that the observed enhancement in hydrogen bonding is, in fact, consistent with the removal of a typical thermal excitation energy from the intermolecular vibrational degrees of freedom. The residual disorder obtained here is, however, greater than that in the solid, particularly at longer range than nearest neighbors, a result of the time scale considered in the averaging (*see* the section on geometric analysis of V-structure).

In light of this analogy, it is interesting to ask about the temperature variation of V-structure. Although we have not examined this aspect in the present study, the discussion above suggests that the quantities examined (e.g., the distribution of bond energies and numbers of bonds) obtained from the V-structure should be significantly less temperature-dependent than are the corresponding features of I-structure. The residual dependence would reflect primarily the temperature dependence of the configuration of the underlying hydrogen bond network, rather than that of the degree of vibrational excitation present in that network. It is the former effect that leads, for example, to what Eisenberg and Kauzmann term the *configurational* contribution to the heat capacity of the liquid (*7*). Thus it appears that the present approach provides a fruitful avenue for the investigation of the structure of the pure liquid.

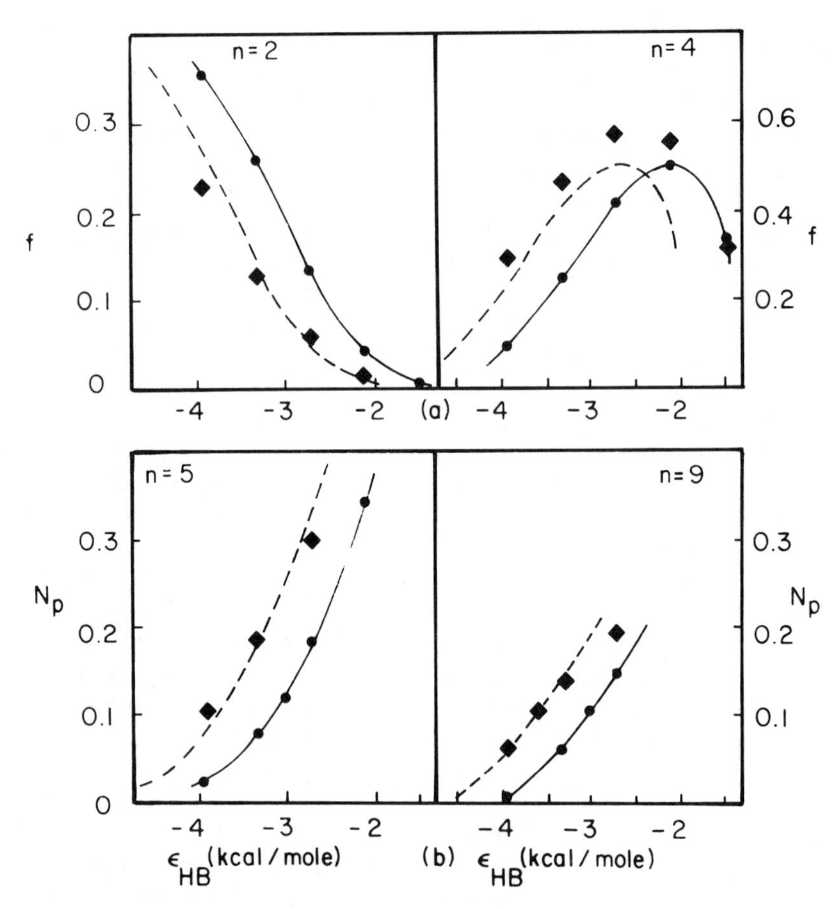

Figure 7. Hydrogen bonding description in V-structure and I-structure as a function of hydrogen bond energy criterion ε_{HB}. *Top, fraction* f *of molecules participating in* n *hydrogen bonds; left,* n = 2; *right,* n = 4. *Bottom, number* N_p *of non-short-circuited hydrogen bond polygons of edge length* n; *left,* n = 5, *right,* n = 9. *Key:* ●, *I-structure;* ♦, *V-structure,* τ_A = 0.2; ——, *smooth curve drawn through the I-structure values; and* ---, *equivalent curve, shifted by* k_BT. *(Reproduced with permission from Ref. 15. Copyright 1981, American Institute of Physics.)*

Aqueous Solutions. As indicated in the introduction to this chapter, one important motivation for the present study is to provide an information base needed to interpret aqueous solution phenomena. We comment here on several inferences that can be made at this point, although they are somewhat speculative.

Considering first apolar solutes, it is interesting to note a similarity between energetic quantities observed here in V-structure and those observed in the structure of water in some recent simulation studies of solutions of such solutes (*9–13*). In those studies, it has been generally observed that for water molecules vicinal to the solute, there is a small shift to more negative solvent–solvent pair interaction energies, and apparently a slight sharpening of the corresponding distribution functions. This behavior is apparently enhanced for water molecules located between a pair of nonpolar solute molecules (*9, 11*) and involved in a solvent-separated hydrophobic interaction (*24, 25*).

It has been suggested that the interaction between nonpolar solutes and water leads to a sharpening of the water–water pair correlation functions (*25*). The similarity between the V-structure energetics and that of the solvation water is consistent with this view. The clathratelike orientational structure associated with nonpolar hydration (*26*) arises from the inability of the solute to participate in hydrogen bonding. One correspondingly expects that the local orientational potential associated with librational deformations in which a solvent hydrogen bonding group becomes increasingly directed toward the apolar solute should be typically narrower than that for a corresponding deformation in bulk water. The implication is that one would expect the associated V-structure to be geometrically more similar to I-structure for the solvation region than it is for bulk water.

A corresponding argument can be made with regard to ionic solutes. That is, in solvation regions that are typically characterized as "structure broken" (*27*), because of balanced competition between solute–solvent and solvent–solvent interactions, one anticipates increased librational freedom, compared to the bulk. Correspondingly, we infer that the V-structure of such water regions should differ more from the I-structure than it does in bulk water.

If these inferences are correct, the comparison of the solvent V-structures manifest by pure water and by various solutions should produce significant new insight into the molecular structure of these systems that is not readily available from an analysis of I-structure.

A consideration of solvent isotope effects further supports the idea that V-structure may have substantial utility in describing solution structure. The difference between liquid H_2O and D_2O and between solution phenomena in the two solvents has been used by many authors as a probe of local molecular structure (*28–33*). It is important to note that, in the present context, the principal effect of the isotopic difference between the two arises from a difference in moments of inertia. The difference in total mass (affecting translational motion) is only about 10%, and differences in internal vibrational degrees of freedom affect solvent properties only indirectly. Further, one can account for the average differences in observed librational spectra between the two pure liquids simply in terms of the difference in the moments of inertia, without resort to changes in effective force constants (*29*). For lower frequencies (*see* Figure 1) one expects that classical behavior will be a rather good description. In accord with this view, we are led to the conclusion that the V-structure evaluated from the classical trajectory should be a characteristic of both D_2O and H_2O, while the classical I-structure is not truly characteristic of either.

It is reasonable to suppose that the same connection can be made for solutions; that is, that the V-structure is a characteristic of both H_2O and D_2O solutions. In fact, this hypothesis is supported by available experimental evidence (*28–33*). A description of transfer thermodynamics ($H_2O \rightarrow D_2O$) obtained by including librational frequency shifts due only to changes in the solvent moments of inertia, and ignoring any possible differences in the underlying liquid structure, are consistent with experimental results (*29*). That is, for "structure making" solutes (*27*) (e.g., Li^+, F^-, and apolar solutes), near room temperature, the enthalpy and entropy for transfer are both typically negative, while for "structure breaking" solutes (larger univalent ions) the sign is reversed (*28*). Using a simplified model, and the isotopic differences in moments of inertia, these trends have been shown to be consistent with an isotope-independent shift to larger effective librational force constants for the former case and smaller force constants in the latter (*29*).

Conclusions

We have presented results obtained from the analysis of a coarse-grained dynamical average of a classical trajectory of liquid water. The results manifest a consistent picture of the liquid V-structure, and this picture is not sensitive to the averaging time within reasonable limits. The energetic structural features, which reflect most sensitively the short range structure, manifest a dramatic enhancement in the degree of hydrogen bonding. This enhancement has a substantial influence, for example, on the degree of connectivity ascribed to the hydrogen bond network.

The results obtained, and a comparison with previous observed behavior in both experimental and simulated aqueous systems, have led us to suggest a number of avenues where the analysis of V-structure may well be very fruitful. Each of these is based on the ability of such an analysis to separate librational excitation from the structure of the underlying network. These areas include the structural changes in pure water associated with changes in pressure and temperature, the solvation structure present in aqueous solutions, and the interpretation of solvent isotope effects.

Correspondingly, we expect that the dynamics of net translation and reorientation in aqueous systems can be profitably studied by focusing on the rearrangement dynamics of the V-structure.

The extension of the analysis employed in the present study to the investigation of liquid water and of model aqueous solutions along these lines is presently underway in our laboratory.

Acknowledgments

Support of the research reported here by grants from the Robert A. Welch foundation and the Research Corporation is gratefully acknowledged.

Literature Cited

1. Wood, D. W. In "Water: A Comprehensive Treatise"; Franks, F. Ed.; Plenum Press: New York, 1979; Vol. 6, pp. 279–410.
2. Frank, H. S. In "Water: A Comprehensive Treatise"; Franks, F. Ed.; Plenum Press: New York, 1972; Vol. 1, pp. 515–544.
3. Pople, J. A. *Proc. Roy. Soc. (London)* **1951,** *205A*, 163.
4. Rice, S. A. *Top. Curr. Chem.* **1975,** *60*, 109.
5. Sceats, M. G.; Rice, S. A. *J. Chem. Phys.* **1980,** *72*, 6183.
6. Rice, S. A.; Belch, A. C.; Sceats, M. G. *Chem. Phys. Letters* **1981,** *84*, 245.
7. Belch, A. C.; Rice, S. A.; Sceats, M. G. *Chem. Phys. Lett.* **1981,** *77*, 455.
8. Eisenberg, D.; Kauzmann, W. "The Structure and Properties of Water"; Oxford University Press: New York, 1969.
9. Geiger, A.; Rahman, A.; Stillinger, F. H. *J. Chem. Phys.* **1979,** *70*, 263.
10. Swaminathan, S.; Harrison, S. W.; Beveridge, D. L. *J. Am. Chem. Soc.* **1978,** *100*, 5705.
11. Pangali, C.; Rao, M.; Berne, B. J. *J. Chem. Phys.* **1979,** *71*, 2982.
12. Clementi, E.; Barsotti, R.; *Chem. Phys. Lett.* **1978,** 59, 21, and references therein.
13. Rossky, P. J.; Karplus, M. *J. Am. Chem. Soc.* **1979,** *101*, 1913.
14. Palinkas, G.; Riede, W. C.; Heinzinger, K. Z. *Naturforsch. Teil A* **1977,** *32*, 1137, and references therein.
15. Hirata, F.; Rossky, P. J. *J. Chem. Phys.* **1981,** *74*, 6867.
16. Goldstein, H. "Classical Mechanics"; Addison-Wesley: Cambridge, MA, 1953; p. 107.
17. Rahman, A.; Stillinger, F. H. *J. Chem. Phys.* **1971,** *55*, 3336.
18. Falk, M. In "Proceedings of Electrochemical Society Symposium on Chemistry and Physics of Aqueous Gas Solutions"; Electrochemical Society: Toronto, 1975, pp. 19–41.
19. Montrose, C. J.; Bucaro, J. A.; Marshall-Coakley, J.; Litovitz, T. A. *J. Chem. Phys.* **1974,** *60*, 5025.
20. Stillinger, F. H.; Rahman, A. *J. Chem. Phys.* **1974,** *60*, 1545.
21. Streett, W. B.; Tildesley, P. J.; Saville, G. *Mol. Phys.* **1978,** *35*, 639.
22. Haan, S. W. *Phys. Rev.* **1979,** *20A*, 2516.
23. Rahman, A.; Stillinger, F. H. *J. Am. Chem. Soc.* **1973,** *95*, 7943.
24. Geiger, A.; Stillinger, F. H.; Rahman, A. *J. Chem. Phys.* **1979,** *70*, 4185.
25. Pratt, L. R.; Chandler, D. *J. Chem. Phys.* **1977,** *67*, 3683.
26. In "Water: A Comprehensive Treatise"; Franks, F. Ed.; Plenum Press: New York **1979,** Vol. 4, pp. 78–79.
27. Stillinger, F. H., *Science* **1980,** *209*, 451.

28. Frank, H. S.; Wen, W. Y. *J. Chem. Soc., Faraday Discuss.* **1957,** *24*, 133.
29. Arnett, E. M.; McKelvey, D. R. In "Solute–Solvent Interactions"; Coetzee, J. F.; Ritchie, C. D. Eds.; Marcel Dekker: New York, 1969, pp. 343–395.
30. Swain, C. G.; Bader, R. F. W. *Tetrahedron* **1960,** *10*, 182.
31. Dahlberg, D. B. *J. Phys. Chem.* **1972,** *76*, 2045.
32. Ben-Naim, A.; Wilf, J.; Yaacobi, M. *J. Phys. Chem.* **1973,** *77*, 95.
33. Jancso, G.; Van Hook, A. *Chem. Rev.* **1974,** *74*, 689.
34. Philip, P. R.; Desnoyers, J. E. *J. Solution Chem.* **1972,** *1*, 353.

RECEIVED for review January 26, 1982. ACCEPTED for publication September 9, 1982.

13

Monte Carlo Computer Simulation Studies of the Equilibrium Properties and Structure of Liquid Water

DAVID L. BEVERIDGE, MIHALY MEZEI, PREM K. MEHROTRA, FRANCIS T. MARCHESE, GANESAN RAVI-SHANKER, THIRUMALAI VASU, and S. SWAMINATHAN

Hunter College of the City University of New York, Chemistry Department, New York, NY 10021

This chapter reviews recent research studies of the structure of water at ordinary temperature and pressure and presents an opinion on the state of knowledge about this system. Current positions on the comparison between calculated and experimental properties of liquid water are presented, as well as current interpretation of these results in terms of structure at the molecular level. The application of large-memory, high-speed digital computers to the calculation of intermolecular interactions and the structure of liquids is discussed.

THE EXTENSIVE USE OF LIQUID WATER as a solvent and reagent in chemical reactions, the widespread occurrence of water on the planet Earth, and the unique role of water as a biological life-support system combine to make an understanding of the properties of liquid water in terms of structure a matter of central importance to chemistry, the earth sciences, and biology. The focus of this chapter is to review recent research studies of the structure of water at ordinary temperature and pressure, and to present an opinion on the state of knowledge about this system considered both as a structural problem in physical chemistry and as a methodological problem in computer simulation of the liquid state.

Studies of liquid water are among the earliest recorded research investigations in physical chemistry, and interest in this subject has remained strong throughout the past century. Diverse experimental data on the thermodynamic and physical properties of water have been obtained, but the unequivocal interpretation of these data in terms of structure remains today an area of active research. Now, as molecular detail in the liquid state becomes accessible to study at a high level of precision

0065-2393/83/0204-0297$14.25/0

via diffraction experiments and by computer simulation, water structure problems are being carried to a new level of understanding. One aim of this review is to present current positions on the comparison between calculated and experimental properties of liquid water and current interpretation of these results in terms of structure at the molecular level.

Liquid water is presently studied not only to gain new knowledge about the properties and structure of the system, but also to understand better the capabilities and limitations of the methods of study. Just as the Lennard–Jones fluid has become the prototype system of study for simple liquids, water has become the prototype system for the study of associated liquids. New advances in intermolecular potential functions and liquid-state simulation techniques are now invariably applied to water, since it is useless to proceed to the study of more complex chemical systems if this central problem cannot be treated reasonably well. Liquid water has thus become the testing ground for new methods of study for molecular liquids. The second aim of this review is to describe in detail computer simulation methodology, particularly Monte Carlo calculations, as currently applied to liquid water.

First in this chapter, the historical background for the problem is briefly reviewed. Next, the problem of describing intermolecular interactions among water molecules is considered, and the development of intermolecular potential functions for water molecules is reviewed. In the section entitled "Theory and Methodology," aspects of liquid state theory pertinent to computer simulation are reviewed. The emphasis in this chapter, as in our own research in this laboratory, is on the theoretical and computational approaches to the problem, particularly Monte Carlo computer simulation. Detailed description of the Monte Carlo method is given as presently implemented for studies on molecular fluids. Results from diverse studies on liquid water are then given in the next section, and are considered in the perspective of available experimental data. An analysis of simulation results based on quasi-component distribution functions is given in the section entitled "Analysis of Results." The final section gives a summary and conclusions.

Background

A vast amount of physicochemical experimentation and theoretical work has been carried out on the liquid water system. The multivolume series of review articles edited by Franks (*1*) is recommended as a point of departure for the original literature. A condensed, informative view is presented in the monograph by Eisenberg and Kauzmann (*2*). The collection of review articles on water and aqueous solutions edited by Horne (*3*) and the textbook by Ben-Naim (*4*) are other useful resources. We extract here selected points in the historical development pertinent to the relationship between properties and structure relevant to current liquid-state simulation studies.

The physical properties of liquid water have been studied in detail in diverse experimental investigations. Water gained quite early a reputation for anomalous behavior in its physical properties relative to simple liquids. The increase in density of water on warming from 0 to 4 °C and commensurate anomalous changes in the compressibility, thermal expansion, and viscosity were recognized by the turn of the 19th century. The 1912 monograph by Henderson (*5*) describes simply and elegantly how the anomalous physical properties of water are the basis for its unique suitability as a biological life support system. Measurements of the various thermodynamic and physical properties of liquid water are currently available (*6*). The experimental data most pertinent to the structural studies of liquid water are the intermolecular atom–atom radial distribution functions obtained from X-ray (*7, 8, 9*), electron (*10*), and neutron (*11–13*) diffraction studies and vibrational spectra, particularly from Raman measurements (*14*).

The earliest recorded ideas about the structure of liquid water developed as variations on ideas about the structure of ice. Ordinary ice has a lower density and a correspondingly larger void space than liquid water. This fact we know from X-ray crystallography (*15*) to be a consequence of the disposition of water molecules on a tetrahedral lattice in the solid, stabilized by the cohesive energy produced by intermolecular hydrogen bonds. In the most regular polymorphic form of the solid, ice Ic, the hydrogen bonds are essentially linear. Deviations from linearity in the hydrogen bonds are observed in ice II and other high pressure forms of the solid. Interactions other than those of hydrogen bonds are observed in ice VII, for example, which consists of interpenetrating networks of hydrogen-bonded water molecules. Clearly, considerable structural diversity is possible in interactions among water molecules.

The behavior of density as a function of temperature for water and the crystal structures of ice serve as a focus for introducing alternative conceptual models for the structure of the liquid. An early explanation of the anomalous behavior of density as a function of temperature is that of Roentgen in 1892 (*16*), who conceived of the system as a literal mixture of a low-density low-energy icelike form in equilibrium with a higher energy fluid polymorph. Interpretation of the density data follows from Le Chatelier's principle, with the density increase on melting seen as a consequence of the interconversion of the bulky icelike form to the denser fluid form. On further increase in temperature, the density decreases because of the thermal expansion of the fluid.

Roentgen's two-state view of the system is the simplest of the so-called mixture models of liquid water, and many subsequent quantitative theoretical studies of this problem are in one way or another elaborations of this idea. Ising models related to the system are due to Levine and Perram (*17*), Bell (*18*), and Angell (*19*), who developed an interesting two-state model based on configurational excitations. The extensive literature on mixture models of liquid water, reviewed by Davis and Jarzynski (*20*), shows how diverse definitions of "mixture" have been used, and how more than a few have been found to be partially successful for calculating liquid properties. This approach takes its most refined form in the five-state flickering cluster theory of Nemethy et al. (*21*). A mixture model was used in the interpretation of the IR spectrum of water by Luck (*22*) and of the Raman spectrum by Walrafen (*23*). The observation of isopiestic behavior in the temperature dependence of the Raman spectra of water was initially considered supportive of the mixture model. It was subsequently shown that this arises from possibly fortuitous circumstances and does not necessarily indicate a redistribution of associated species (*24*). A review (*25*) of vibrational spectra covers this and other detailed aspects of the vibrational spectra. Also, two-state models inevitably lead to contradictions with experimental data when diverse properties are considered (*26*).

Theories that hold that liquid structure involves contributions from structures in which voids in the icelike lattice are occupied lead to the so-called interstitial models for the system (*27–29*). Pauling (*30, 31*) used a modification of this idea in his clathrate model of water. Narten et al. (*8*) used the interstitial model for the interpretation of their early X-ray diffraction data on liquid water. The mathematical representation of the mixture and interstitial models in statistical mechanics have much in common.

An alternative explanation for the manner in which density increases as ice melts was initially put forward by Bernal and Fowler in 1933 (*32*), on the idea that the intermolecular hydrogen bond between water molecules could be bent without being broken. They interpreted the melting phenomenon in terms of a hydrogen-bonded lattice with bent hydrogen bonds. The distribution of hydrogen-bond angles could be altered by changes in pressure and temperature. A statistical mechanical representation of the idea was given by Pople (*33*) and gave rise to the present line of thought known as the *continuum* model for liquid water. "Continuum" is used here in the sense of an energetic continuum and not as a dielectric continuum representation of liquid water in the Kirkwood–Onsager sense (*34*). Continuum models for liquid water have been reviewed recently by Kell (*35*). Lentz et al. (*36*) showed that for a multistate model to have wide applicability, a broader distribution of cluster sizes was required, which, they noted, has much in common with the continuum model.

From a contemporary point of view, the mixture, continuum, and related approaches described above are considered to be ad hoc models of the system. In an ad hoc approach, a model (energetic or structural) is assumed that allows estimates of the energy states of the system. These energies depend parametrically on the quantitative characteristics of the assumptions. From the energies and the temperature of the system, the partition function is computed as a Boltzmann sum-over-states and, in the more eleaborate forms, involves the maximum term method. The thermodynamic indices follow from the partition function. At this point, the values of the disposable parameters are chosen to give best agreement between calculated and experimental values of observable quantities. However, agreement between observed results and those calculated from a theoretical model does not necessarily validate the model, and inferences are weak when the number of disposable parameters necessary to characterize the model is not sufficiently less than the number of experimental observations available. For these reasons, calculations based on ad hoc models, often involving very sophisticated statistical mechanics, still led to controversy about the nature of liquid water at the molecular level.

An obviously preferred approach to the theoretical computation of water structure would be to represent the system explicitly as a molecular assembly (a Hamiltonian model) and to develop the computation of the partition function, thermodynamic indices, and molecular distribution functions of the system in terms of intermolecular interactions. This can be considered to be an ab initio approach to the problem, noting that approximations may still be present, but that they enter at the level of intermolecular interactions, a more fundamental point in the theory. In the ab initio approach the structure of the fluid emerges as a result of the computation rather than entering as an assumption, as in the ad hoc methods, and diverse microscopic information on structure not accessible to direct experimental measurement can be computed. The ab initio alternative was not pursued in the early work since the intermolecular interactions were not known accurately, and precise calculation of the partition function for a molecular assembly of sufficient size to represent the liquid was essentially intractable.

The application of large-memory, high-speed digital computers to the calculation of intermolecular interactions and the structure of liquids now makes an ab initio, Hamiltonian approach to the problem feasible. The interaction of water molecules has been studied extensively, and analytical potential functions describing these interactions have been developed from both experimental data and quantum mechanical calculations. For the calculation of partition functions, even with high speed digital computers, the convergence of the sum-over-states was found to be too slow for molecular liquids, a consequence of the relatively small volume of configuration space that contributes significantly to the partition function integral. The calculation of thermodynamic and structural quantities has been accomplished successfully by statistical mechanics procedures developed in terms of intermolecular interactions but focused on the calculation of thermodynamic internal energy rather than the partition function. Configurational integration for internal energy and other average properties can be carried out numerically with reasonable precision, and calculations to a desired degree of accuracy on liquids have proved to be accessible to the present generation of digital computers. This numerical integration is the essence of liquid-state computer simulations.

Computer simulation of liquids, although based on statistical thermodynamics or kinetics, can be viewed in the framework of simulation theory in general. Procedures can be developed in either deterministic or probabilistic form. The deterministic approach reduces to Newtonian equations of motion for the particles of the system and is called *molecular dynamics*. Extensive studies on liquid water from this point of view have been carried out (*37*), and extensions of this work are being actively pursued (*38–45*). Molecular dynamical studies on liquid water have been reviewed by Stillinger (*46*) and by Wood (*47*), and methodological detail is particularly available in Wood's article. Analysis of molecular dynamics results on liquid water at the level of Eisenberg and Kauzmann's V-structure is given by Rossky and Hirato elsewhere in this volume. Results from molecular dynamics relevant to equilibrium properties and the water structure problem will be quoted at appropriate junctures in this review.

The probabilistic form of computer simulation in liquid-state theory is the Monte Carlo method, first applied to liquid water by Barker and Watts (*48*). An early application was reported by Sarkisov et al. (*49*). Subsequent studies of liquid water by Monte Carlo computer simulation were carried out by Clementi et al. (*50*), Owicki and Scheraga (*51*), and Swaminathan and Beveridge (*52*). Extensive further studies have been reported by Jorgensen (*53, 54*) and from our laboratory (*56–59*). Monte Carlo methodology as applied to the water structure problem was advanced by Rao et al. (*60, 61*) with the introduction of the force-bias method for convergence acceleration.

The theoretical study of the equilibrium properties and diffusionally averaged structure of a fluid can be approached by either Monte Carlo or molecular dynamics. Molecular dynamics gives dynamical as well as equilibrium properties of the system, but the calculations are usually done in the microcanonical ensemble, with the determination of temperature an empirical problem. Anderson (*55*) devised an approach for extending molecular dynamics to other ensembles; this has yet to be applied to liquid water. The Monte Carlo method is defined by convenient statistical thermodynamic ensembles and allows for various extensions necessary for the study of solvent effects on intermolecular and intramolecular interactions and for the calculation of the free energy, but it does not yield access to dynamical properties. Results on equilibrium properties of liquid water from Monte Carlo and molecular dynamics have been compared and found to agree closely (*56, 60–62*). The efficiencies of the two methods in computing equilibrium properties were also compared, and were similar (*60, 61*). Thus, for equilibrium properties and structure, either method of study is acceptable.

Research studies from our laboratory have used the Monte Carlo method. Convergence and convergence acceleration behavior in Monte Carlo calculation of configurational integrals have been studied (*56*) and procedures for the analysis of the results of Monte Carlo simulations as applied especially to water (*58*) and aqueous solutions (*63, 64*) have been developed and used. The emphasis in this review is on the Monte Carlo side of liquid-state computer simulation studies on liquid water.

Several significant contributions to the water structure problem have used methods other than computer simulation. In current work by Rice et al. (*65*), the amorphous ice system has been studied as a prototype for liquid water. In a series of articles they extended and refined the continuum model of liquid water. Their "random network model" treats diverse properties of the liquid by treating libration and hindered translation in a quasi-static network of hydrogen-bonded molecules, represented by a random network potential developed from spectroscopic data. Quantum dynamical effects are introduced. Good descriptions of the oxygen–oxygen radial distribution function, the dielectric constant, and bulk thermodynamic properties have been obtained. The results have been compared and are consistent with those obtained from computer simulation, providing a theoretical framework for quantitative extensions of the continuum model.

In another development, a generalized Ising model called polychromatic percolation theory was applied to the problem of water structure by Stanley et al. (*67*). A main motivation for this work was to further the understanding of supercooled water and the nature of cooperative effects in the pseudo-second-order phase transition in the system. Studies have now demonstrated that, for any reasonable definition of the hydrogen bond, at ordinary temperatures water is well above the percolation threshold. Stanley has shown that in spite of this, the main features of the hydrogen-bonded networks can be anticipated by percolation theory, which can provide a framework for further interpretation and extension of simulation results.

Several articles on water structure problems have been produced by integral equation methods. Rossky and Pettitt (*68*) reported a successful extension of the RISM method to waterlike systems. Patey reported studies of water based on the hypernetted-chain formalism (*69*). In general, the associated nature of waterlike liquids makes these systems relatively difficult to treat by integral equation methods (*4*) and this progress is quite significant.

Intermolecular Interactions

The individual characteristics of a molecular system are introduced into a computer simulation via the configurational potential energy. To define this quantity formally, let us specify an N-particle configuration of a molecular assembly by the configurational coordinate vector $\mathbf{X}^N$

$$\mathbf{X}^N = \{\mathbf{X}_1, \mathbf{X}_2, \mathbf{X}_3, \ldots, \mathbf{X}_N\} \tag{1}$$

where each $\mathbf{X}_i$ is a product of position $\boldsymbol{R}_i$ and orientation $\boldsymbol{\Omega}_i$

$$\mathbf{X}_i = \{\boldsymbol{R}_i, \boldsymbol{\Omega}_i\} \tag{2}$$

The configurational potential energy in this notation is $E(\mathbf{X}^N)$, and represents the energy of the N-molecule system in configuration $\mathbf{X}^N$ relative to the energy of N isolated molecules. The configurational energy may be expanded in terms of successive orders of interaction as

$$E(\mathbf{X}^N) = \sum_{i<j}^{N} E_2\,(\mathbf{X}_i, \mathbf{X}_j) + \sum_{i<j<k}^{N} \Delta E_3\,(\mathbf{X}_i, \mathbf{X}_j, \mathbf{X}_k) + \ldots \tag{3}$$

where E_2 is the energy of dimerization

$$E_2(\mathbf{X}_i, \mathbf{X}_j) = E(\mathbf{X}_i, \mathbf{X}_j) - 2E(\mathbf{X}_i) \tag{4}$$

and ΔE_3 is a correction term for three-body effects

$$\begin{aligned}\Delta E_3(\mathbf{X}_i, \mathbf{X}_j, \mathbf{X}_k) = {} & E(\mathbf{X}_i, \mathbf{X}_j, \mathbf{X}_k) - 3\,E(\mathbf{X}_i) \\ & - [E_2(\mathbf{X}_i, \mathbf{X}_j) + E_2(\mathbf{X}_i, \mathbf{X}_k) + E_2(\mathbf{X}_j, \mathbf{X}_k)]\end{aligned} \tag{5}$$

Analogous terms ΔE_4, ΔE_5, and so forth can be developed to represent even higher order effects. The terms ΔE_n for $n > 2$ introduce cooperative effects into the configurational potential. When all of these terms are neglected, the configurational energy is expressed as a sum of interaction energies for molecular pairs, an assumption referred to as "pairwise additivity."

Computer simulation requires rapid evaluation of the configurational energy for a large number of N-molecule complexions of the system. This task is accomplished by means of potential energy functions, simple analytical expressions for interaction energy as a function of configurational coordinates and a set of disposable parameters. Potential functions for intermolecular interaction energies can be grouped for purposes of discussion into three classes: model, empirical, and quantum mechanical. Typical model potential functions are the hard-sphere and Lennard–Jones potentials, both studied extensively in the formal development of liquid-state theory. Empirical potentials result when the disposable parameters of a function are selected on the basis of experimental data. In quantum mechanical potentials, disposable parameters are determined on a best fit criterion from a discrete data base of quantum mechanically calculated interaction energies.

Both empirical and quantum mechanical approaches have been used for the determination of potential functions describing the interaction energy of water molecules, and there are advantages and disadvantages to both. The construction of functions based on experimental data has the decided advantage of building all possible observed information about the intermolecular interactions into the function. In addition, the experimental nature of the data partially compensates for the assumption of pairwise additivity in the functional form, leading to so-called effective pair potentials that include higher order effects in some averaged form. There are extensive experimental data to draw upon in this approach, including electric moments, vibrational frequencies, lattice constants, and so forth. However, the available experimental information corresponds only to certain limited regions of intermolecular configuration space, and a function determined from experimental data only is not necessarily accurate in regions not represented in the data base; i.e., the behavior of the interaction energy in the configuration space is considerably underdetermined by the available data. In practice, a sensible functional form partially compensates for this problem.

On the quantum mechanical side, the nonempirical calculation of intermolecular interactions for small and modestly sized systems using molecular quantum mechanics is now feasible. The main advantages here are that interaction energies can be determined at rigorously defined levels of approximation, and that any possible geometrical arrangement of molecules can be considered. Various reasonable approaches to sampling configuration space have been suggested and used effectively, and fitting functional forms to data can be accomplished with reasonable precision. However, the task of generating the data base of quantum mechanically calculated interaction energies becomes prohibitively expensive as the size of the system under consideration increases, when larger basis sets are necessary, or when electron correlation must be included. Compromises in the quality of the quantum mechanical calculations used for the data base remain, of course, inherent in the resulting intermolecular potential function. A particular problem in quantum mechanical calculations of intermolecular interaction energies is the basis set superposition error, whereby small basis sets result in spuriously inflated interaction energies and commensurate errors in other properties, particularly a foreshortening of the calculated equilibrium intermolecular separation.

A succession of potential functions has been developed for the pairwise interaction energy of water molecules, from which configurational energy can be computed assuming pairwise additivity. The first empirical potential functions developed for water–water interactions were based on a point-charge model for the attractive part of the function joined smoothly to a repulsive core term representing Pauli forces. The basic idea is expressed in the function of Bernal and Fowler (*31*) in their early work on water. Geometrical aspects of the intermolecular interaction energy are very easily dealt with in point-charge models, and generalizations of the model to systems other than water are usually straightforward. An early significant contribution in this vein was that of Rowlinson (*70*). His RLS2 potential described water interactions by four point charges and a Lennard–Jones term, parameterized on virial coefficient data and the molecular dipole. An adaption of Bjerrum's four-point charge model for water (*71*) was used by Ben-Naim and Stillinger (*72*) in the development of the BNS potential (*73*), first used in molecular dynamics simulation studies. Initial results prompted a minor revision of the function into what is now known as the ST2 potential (*74*), which, at this time, is the most widely used empirical potential for water interactions in computer simulation work. Transferable empirical potentials based on point charges for electrons and nuclei (EPEN) were developed by Scheraga and workers (*75*). Results have been obtained with three-point-charge models which significantly reduce the computation time required for energy evaluations. Jorgensen (*54*) introduced a water potential of this type in his set of transferable intermolecular potentials (TIPS), and he developed parameters for treating interactions of organic molecules in the liquid state. Berendsen et al. (*40*) developed a three-point simple point charge (SPC) potential for computer simulation studies on biological water. Jorgensen (*76*) proposed a very promising extension of his TIPS model called TIPS2, where, analogously to the MCY potential discussed later, the negative charge is displaced from the oxygen site along the HOH bisector towards the hydrogens, and the oxygen site is retained for the Lennard–Jones interaction. The similarity of MCY and TIPS2 functional forms to that of Bernal and Fowler has been pointed out by Klein (*75*).

Stockmayer's potential for water (*78*), used in virial coefficient calculations, was the first of a genre of functions based on the multipolar expansion of the molecular interaction energy, and involves point dipole terms. An ab initio approach using high-order multipole expansion of the quantum mechanically obtained charge density of water coupled with the representation of cooperativity by induced dipoles was developed by Campbell and Mezei and applied to small clusters (*79*) and various ice forms (*80*). Barnes et al. developed the polarized electropole (PE) potential (*81*), which includes the permanent dipole and quadrupole interactions as well as the induced dipole interactions in the energy terms, parameterized from experimental data, and applied to small clusters and several ice forms as well as to the liquid state.

Another class of empirical potentials incorporates intramolecular as well as intermolecular effects. The central force model introduced by Rahman et al. (*82*) was the first of this type and has gone through several modifications (*83*). Watts developed an interesting potential (*84*, *85*) coupling the intermolecular force field of the water molecule to a description of intermolecular interactions. Further modifications of this potential, the RWK1 and the RWK2, have appeared (*86*). The current sequence of polarization models (*87*, *88*) is designed to accommodate both the cooperativity via the polarization and the molecular dissociation of H_2O smoothly into H^+ and OH^-, and thus provide an entry into reaction chemistry.

The development of quantum mechanical potentials describing the interaction of water molecules has been pursued mainly by Clementi et al. (*89*). Functions describing pairwise interactions over all configuration space were successfully developed from data bases of O(100) quantum mechanical calculations and produced functions representative of molecular orbital calculations near the Hartree–Fock (HF) limit and for several levels of electron correlation via electronic configuration interaction (CI). These models place positive charges on the hydrogen atoms and a negative charge at a point on the bisector of the HOH angle. The repulsive core is represented by exponential terms centered on the individual atoms, making the repulsion anisotropic. The quantum mechanical potential due to Matsuoka et al., [MCY–CI(1)] (*90*), based on intermolecular CI calculations, is the most widely used of the quantum mechanical potentials in computer simulation studies of liquid water. A potential based on intramolecular configuration interactions, MCY–CI(2) was also determined. In the following discussions, the MCY label will denote the MCY–CI(1) potential exclusively. Considerable work on potential functions representative of molecular orbital calculations has been reported by Jorgensen et al. (*91*), and references therein. The EPEN functional form was successfully fitted, in our laboratory (*57*), to the ab initio water dimer energies obtained by Clementi et al. (*90*) and is labeled by QPEN. Table I contains the information required to compare the relative computational expense in evaluating interaction energies from the various potential functions.

The experimental results from microwave spectroscopy of water dimer (*93*) and from second virial coefficients serve as useful points of reference for considering the calculated properties of the various intermolecular potential functions for water. The equilibrium geometry found experimentally for the water dimer, Figure 1, features a linear hydrogen bond involving a proton of the donor molecule and the oxygen atom of the acceptor, with R_{eq} = 2.98 ± 0.04 Å and ϕ_{eq} = 60° ± 10°. The binding energy of the water dimer has been estimated to be 5.44 ± 0.07 kcal/mol (92).

Before further discussion of the intermolecular potential function for liquid water, we consider some relevant calculations on the water dimer based on ab initio molecular quantum mechanics. Quantum mechanical calculations at the near Hartree–Fock level (*see* Ref. 94 for a review) of the equilibrium separation R_{eq} in $(H_2O)_2$ are found to be in the range 2.98–3.02 Å, in close agreement with the microwave values. Improving the wavefunction calculation by including electron correlation as described by Matsuoka et al. (*90*) and Diercksen et al. (*95*) produced a significantly smaller value for R_{eq} (2.86–2.92 Å) and an anomalously larger discrepancy with experiment. Newton and Kestner studied this discrepancy vis-a-vis the basis set superposition error (*94*). They found the superposition error to be significant, and that correcting for it results in R_{eq} values in the range of 2.98–3.0 Å, in close agreement with the observed dimer value. The superposition error in energy was found to be in the range of 1.0–1.8 kcal/mol, and correcting for it produced correspondingly lower dimerization energies, closer to the experimental value.

The potential energies of the linear water dimer calculated from the potential functions just described are shown as a function of the R-coordinate (with optimized ϕ-coordinates) in Figure 2 and as a function of the ϕ-coordinate (with optimized R-coordinates) in Figure 3. The ST2 potential with a binding energy of −6.84 kcal/mol at R = 2.85 Å and ϕ = 54° is seen to have the strongest emphasis on tetrahedral character of the potentials in the set. The MCY potential has a binding energy of −5.67 kcal/mol at R = 2.78 Å and ϕ = 37°, and is quite close to the

Table I. Factors in Computational Expense for Various Water–Water Potentials

Factor	*ST2*	*MCY*	*TIPS*	*SPC*	*EPEN*	*TIPS2*
Charge centers	4	3	3	3	7	3
van der Waals centers	1	3	1	1	4	1
Distances	17	9,16[a]	9	9	49	10
Exponentials	0	0,9[a]	0	0	0	0

[a] The two numbers given for MCY water reflect the difference in the cutoff used for the Coulombic and exponential terms in the potential.

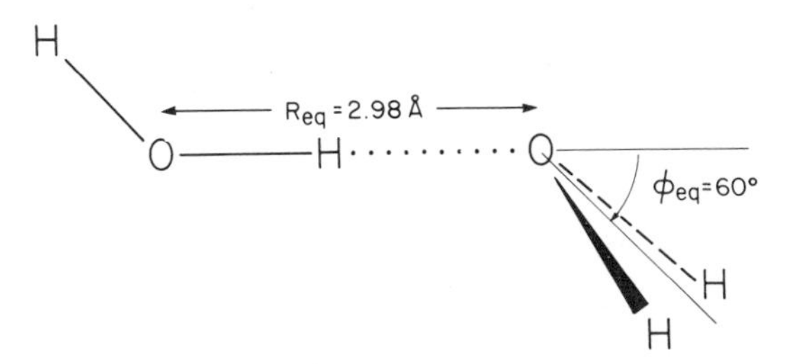

Figure 1. Structure of the linear water dimer as determined from microwave spectra (93), and definition of the structural parameters R *and* ϕ.

very large CI results reported by Diercksen et al. (95). The corresponding characteristics of the HF potential show a binding energy of -4.60 kcal/mol at $R = 3.00$ Å and $\phi = 30$–45°. Comparison of the CI and HF results shows the importance of correlation effects on the water–water hydrogen bond to be 1 kcal/mol, but Newton and Kestner (94) point out that correcting for the superposition error reduces the calculated correlation effect to 0.3–0.4 kcal/mol. The TIPS potential ($E_2 = -5.70$ kcal/mol at $R = 2.78$ Å and $\phi = 27°$) and SPC potential ($E_2 = -6.59$ kcal/mol at $R = 2.75$ Å and $\phi = 27°$) are essentially two different parameterizations of the same functional form. The TIPS2 potential (not shown

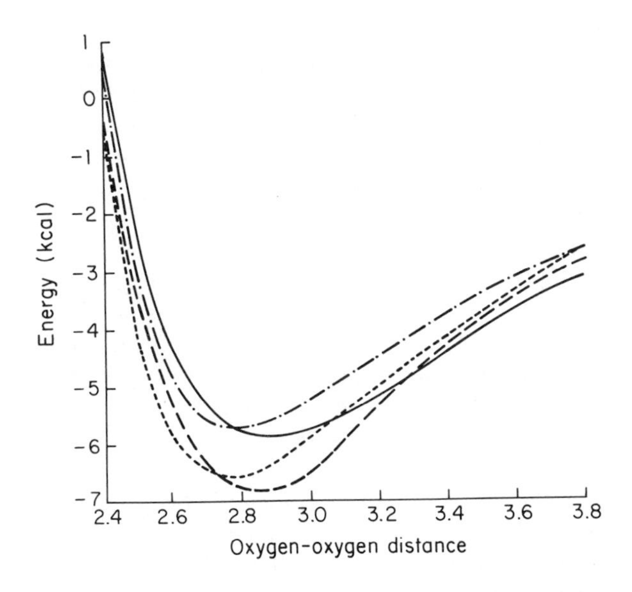

Figure 2. Plots of calculated E_2(R) *vs.* R *as determined from the ST2 (– –), MCY (—), TIPS (• –), and SPC (- - -) potentials.*

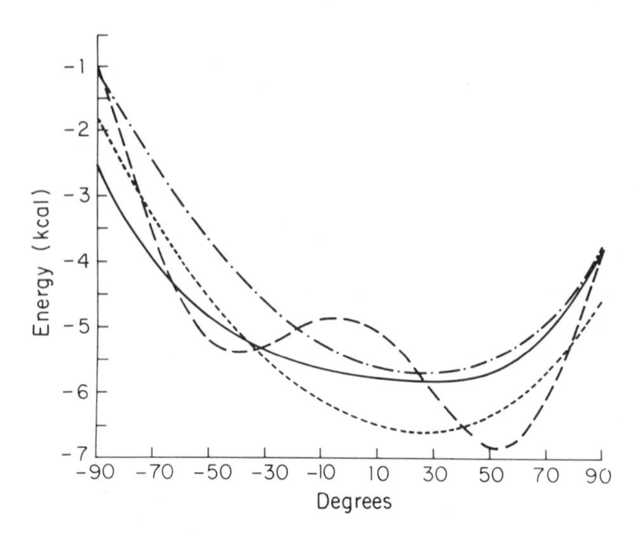

Figure 3. Plots of calculated $E_2(\phi)$ *vs.* ϕ *as determined from the ST2 (--), MCY(—), TIPS(·–), and SPC (---) potentials.*

in the figures) has $E_2 = -6.20$ kcal/mol at $R = 2.79$ Å and $\phi = 47°$. The PE potential (not shown) has $E_2 = -5.00$ kcal/mol at $R = 3.0$ Å and $\phi = 52°$. The ϕ-dependence of E_2 is similar for MCY, TIPS, and SPC potentials, and shows a single broad minimum. The ST2 potential exhibits a double minimum in $E_2(\phi)$.

The second virial coefficient, $B(T)$, of steam is related via statistical mechanics to the pairwise interaction potential for liquid water. Virial coefficient data were used to parametrize the Watts potential and to check the MCY, ST2 and PE potentials. The PE potential performs well for $B(T)$. Both the MCY and ST2 deviate significantly in $B(100°\ C)$, giving values of -827 and -1023 mL/mol, respectively, to be compared with the experimental estimates ranging from -580 to -450 mL/mol (*89, 90*). Although improved agreement is obtained for higher temperatures, some question of the reliability of these functions as pure pair potentials is indicated. An extensive comparison of virial coefficient performance on various potentials has been carried out by Reimers et al. (*86*). The best pure dimer potential is not necessarily optimum for describing the condensed phases under the assumption of pairwise additivity, if cooperative effects are significant.

Morse and Rice (*96*) tested the accuracy of most of the popular water–water pair potentials by comparing calculated and observed crystal structures of four polymorphic forms of ice. The MCY potential consistently gave the most accurate predictions for calculated and observed O–O–O angles and hydrogen-bond network topologies. The MCY potential tested there was the potential fitted to intramolecular CI results [MCY-CI(2)], not the one used successfully in liquid water computer simulations. The densities of the ice crystals were uniformly too low by ~20%, and the O–O separations were too large by ~7%. The ST2 potential performed well on ice Ih and II but was found to have hidden flaws, which are exhibited when it is used to predict the structure and density of ice IX and ice VII. Serious discrepancies were found in all the central-force potentials. The Watts and Lemberg–Stillinger potentials did not produce an arrangement of hydrogens that satisfies the Pauling rules for ice Ih, while the RLS2 potential led to spurious bifurcated hydrogen bonds in ice VIII. The various pair potentials are thus found to perform inconsistently with experiment for calculations on the ices and similar performance can be expected in simulations on liquid water. Reimers et al. (*83*) reported a comparison of a wide range of gas, solid, and liquid properties using several potential functions; they found that no one model gives a satisfactory account of all three phases.

The study of cooperative effects among water interactions and their effect on liquid state structure is a matter of considerable interest. The importance of the cooperative effects in condensed phases of H_2O is clearly indicated by comparing R observed in the water dimer, 2.98 Å, with that for ice, 2.76 Å. Two basic approaches to developing potential functions including cooperative effects are the n-body expansion method and the polarization model. In the n-body potential, functions for ΔE_3, ΔE_4, . . . are constructed, most likely from quantum mechanical calculations. The basic features of three-body cooperativity were described by Hankins et al. (*97*). They considered hydrogen-bonded water trimers where the central water is double donor, double acceptor, or a donor–acceptor. In the Hartree–Fock approximation, the three-body term was found to stabilize donor–acceptor trimers by 0.94 kcal/mol at R_{OO} = 2.76 Å. Both the double donor and the double acceptors are destabilized by the three-body contributions by 1.30 and 0.77 kcal/mol, respectively. The most recent studies have been reported by Clementi et al. (*98*). Considering 29 trimer configurations, the nonadditive component of the interaction energy was quantified at ~1 kcal/mol and attributed to dipole-induced dipole interactions. Problems with basis set superposition errors and compensations using the counterpoise (ghost orbital) method were also discussed. Considerable interesting literature exists on quantum mechanical calculations of water clusters (*99–102*), but is beyond the scope of this review.

The n-body method can be rigorously extended to include all significant ΔE_n, at least in principle. No higher order ab initio potential functions have been reported as yet, although work is in progress in this area. The limiting factors in the development of cooperative potentials include the size of the quantum-mechanical calculations required and details of the form of the analytical fitting function. The four-body contribution to clusters of small molecules was estimated by Murrell (*103*) and was an order of magnitude smaller than the three-body contributions for most cases studied.

Polarization functions introduce cooperative effects via the induced electric moments in the multipolar expansion of the interaction energy, and include contributions from all ΔE_n for $n \leq N$. Cooperative contributions to the repulsive contributions to the potential are neglected. Updating the induced moments at each Monte Carlo step significantly increases the computational overhead in calculation of the configurational energy using a polarization function. Barnes et al. showed in their N-body PE potential that the three-body terms are the dominant nonadditive contribution. For the polarization model of Campbell and Mezei (*79*), Campbell and Belford (*104*) found that for both small clusters and ice Ih the four-body contributions amount to 10–15% of the three-body contributions. Also, by optimizing the geometry of various trimers, tetramers, and hexamers, they found that the induced dipole approximation to the cooperative effects can account for the decrease in the O–O distance found in liquid water and ice.

Theory and Methodology

A theoretical description of liquid state properties and compositional characteristics of a system at a temperature T and density N/V (number of particles, N, per unit volume, V) follows from the semiclassical canonical ensemble partition function

$$Q(T, V, N) = (q^N/(8\pi^2)\Lambda^{3N}N!)\, Z(T, V, N) \tag{6}$$

$$Z(T, V, N) = \int \cdots \int \exp[-E(\mathbf{X}^N)/kT]d\mathbf{X}^N \tag{7}$$

Here q is the partition function for internal degrees of freedom and Λ is the one-dimensional translational partition function for each particle. The quantity $E(\mathbf{X}^N)$ is the configurational energy of the system defined in the preceding section by Equation 3. The integration ranges over the configurational coordinates $\mathbf{X}^N$ of the N particles of the system. The formalism is developed here in the context of the (T, V, N) ensemble. Parallel developments can be given for the (T, P, N) and (T, V, μ) ensembles (*4, 105*).

The thermodynamical properties of the system follow from the statistical thermodynamic definition of the Helmholtz free energy

$$\mathcal{A} = -kT \ln Q(T, V, N) \tag{8}$$

and its derivatives; the thermodynamic internal energy $\mathcal{U}$ (much discussed herein) is given by $-T(\partial\mathcal{A}/\partial T)_{V,N}$. On substitution, differentiation, and management of terms, $\mathcal{U}$ can be partitioned into "ideal" and "excess" contributions, with the latter expressed as the configurational average of $E(\mathbf{X}^N)$

$$U = \int \cdots \int E(\mathbf{X}^N)\, P(\mathbf{X}^N)\, d\mathbf{X}^N = \langle E(\mathbf{X}^N) \rangle \tag{9}$$

where $P(\mathbf{X}^N)$ is the Boltzmann probability for the system to be found in configuration $\mathbf{X}^N$

$$P(\mathbf{X}^N) = \exp[-E(\mathbf{X}^N)/kT]/Z(T, V, N) \tag{10}$$

Other thermodynamic properties can be expressed in an analogous manner. Of particular interest here will be the constant volume excess heat capacity, given by

$$C_v = (\langle E(\mathbf{X}^N)^2 \rangle - \langle E(\mathbf{X}^N) \rangle^2)/kT^2 \tag{11}$$

pressure, given by

$$P = (kT/V)\,(N - \langle \sum_{i=1}^{N} (\mathbf{R}_i \cdot \partial E(\mathbf{X}^N)/\partial \mathbf{R}_i) \rangle) \tag{12}$$

and atom–atom spatial distribution functions

$$g_{\alpha\beta}(R) = N_{\alpha\beta}(R)/(\rho 4\pi R^2 \Delta R) \tag{13}$$

where R is the interatomic separation, $N_{\alpha\beta}(R)$ is the average number of β neighbors of an α atom in a spherical shell between R and $R + \Delta R$ and ρ is the bulk density.

An alternative expression for the Helmholtz excess free energy, advantageous for computer simulation, is

$$A = \int_0^1 U(\xi)\, d\xi \tag{14}$$

Here the integrand $U(\xi)$ can be expressed as an ensemble average

$$U(\xi) = \int \cdots \int E(\mathbf{X}^N) P(\mathbf{X}^N/\xi) d\mathbf{X}^N \tag{15}$$

where $P(\mathbf{X}^N/\xi)$ is the probability of observing the system in configuration $\mathbf{X}^N$, conditional upon the value of the parameter ξ

$$P(\mathbf{X}^N/\xi) = \exp[-E(\mathbf{X}^N/\xi)/kT] / \int \cdots \int \exp[-E(\mathbf{X}^N/\xi)/kT] d\mathbf{X}^N \quad (16)$$

when the system is coupled by the auxiliary parameter through the expression

$$E(\mathbf{X}^N/\xi) = \xi E(\mathbf{X}^N) \quad (17)$$

the free energy is defined with respect to an ideal gas reference state of liquid density (*106, 107*). Mezei, following Torrie and Valleau (*108*), has studied the use of other reference states for tractability and computational efficiency (*109*). Results from this work are quoted below. Methods for calculating free energy in computer simulation remain the subject of active research (*110–112*). With the internal energy and free energy available, the excess entropy can be easily obtained from the expression

$$S = (A - U)/T \quad (18)$$

The additional equilibrium properties of the system accessible to calculation in the (T, P, N) ensemble simulation are the constant-pressure heat capacity:

$$C_p = (\langle H^2 \rangle - \langle H \rangle^2)/kT^2 \quad (19)$$

isothermal compressibility

$$\kappa = (\langle V^2 \rangle - \langle V \rangle^2)/(kT^2 \langle V \rangle) \quad (20)$$

and coefficient of thermal expansion

$$\alpha = (\langle VH \rangle - \langle V \rangle \langle H \rangle)/(kT^2 \langle V \rangle) \quad (21)$$

where H is the enthalpy of the system.

Liquid-state computer simulation is based on the simultaneous numerical integration of integrands analogous to the internal energy expression, Equation 9. This integral is well-known to be ill-conditioned for direct numerical integration, since the integrand $E(\mathbf{X}^N)\ P(\mathbf{X}^N)$ is large in only a very restricted region of configuration space. The integration can be successfully carried out as suggested by Metropolis et al. (*113*) by sampling from the Boltzmann distribution in order to automatically concentrate the effort in the important regions of configuration space. In the Metropolis method, the N-molecule configurations of the problem are taken to be the states of an irreducible Markov chain. The one-step transitions between any two states k and l of the chain have the probability

$$p_{kl} = Pr[\mathbf{X}_{t+1}^N = l \mid \mathbf{X}_t^N = k] \tag{22}$$

The p_{kl} terms can be collected in array form as a stochastic transition probability matrix, $\mathbf{P} = [p_{kl}]$. Integration in the Metropolis method is carried out by means of a stochastic walk through configuration space, generating a realization of an irreducible Markov chain whose unique limiting stationary distribution $\boldsymbol{\pi}$ is the Boltzmann distribution, i.e.

$$\boldsymbol{\pi} = \boldsymbol{\pi}\mathbf{P} \tag{23}$$

where $\boldsymbol{\pi}_i = P(\mathbf{X}_i^N)$. In a realization of this process, the configurations, $\mathbf{X}^N$, are then sampled with a frequency proportional to $P(\mathbf{X}^N)$, and the determination of average properties reduces to a simple summation over the property of the individual configurations $\mathbf{X}_i^N$, for example

$$\overline{E(\mathbf{X}^N)} = (1/M) \sum_{t=1}^{M} E[\mathbf{X}_t^N] \tag{24}$$

Provided the system is ergodic, as $M \rightarrow \infty$, $\overline{E(\mathbf{X}^N)} \rightarrow \langle E(\mathbf{X}^N) \rangle$, the cumulative average energy becomes an increasingly good estimator of the energy expectation value. The computation of heat capacity and other configurational properties of the system take analogous form.

To implement this procedure in practice, a particle is selected, and a trial move is attempted from configuration $\mathbf{X}_k^N$ to $\mathbf{X}_l^N$. The change in energy of the configurational energy of the system is used to calculate the quantity

$$R = (q_{lk}/q_{kl}) \exp[-(E(\mathbf{X}_l^N) - E(\mathbf{X}_k^N))/kT] \tag{25}$$

If R is greater than or equal to one, the move is accepted. If R is less than one, then the move is accepted with the probability R and rejected with the probability $1 - R$. To do this, one compares R with a random number on the interval (0, 1). If R is greater than the random number, the move is accepted; otherwise the move is rejected. Repeated application of this process forms a sequence of configurations that is a realization of the desired Markov process, and any configurational average property of the system may, in principle, be calculated by averaging over these configurations. Optimum sampling for a property other than energy may require sampling from a modified Boltzmann or even non-Boltzmann distribution (*108, 114*).

The convergence and statistical error bounds of Metropolis Monte Carlo calculations are generally monitored according to the method of block averages (also known as the method of batch means) (*106, 107*). Here the Monte Carlo realization is partitioned into several nonoverlapping blocks of equal lengths, and the averages of the property under consideration (e.g., mean energy) are computed over each block. Let $\overline{f}_i$ denote the average of the property f computed over the block i. Under the assumptions that the $\overline{f}_i$'s are independent and normally distributed, and that the Markov chain is ergodic, the error bounds for the property f at a ~95% confidence level are 2σ, where

$$\sigma^2 = (1/B(B - 1)) \sum_{i=1}^{B} [\overline{f}_i^2 - (\overline{f}_i)^2] \tag{26}$$

and the summation runs over the B blocks. In computer simulations of small lengths, the assumptions are honored more in the breach than in observance, and thus computed error bounds by the method of batch means are to be taken with caution. To ensure the validity of the estimate by the batch means method, the block size has to be increased until reliable statistical tests show that the batch means are indeed independent. Other methods have also been proposed to estimate the confidence intervals of Monte Carlo estimates of this type. Good reviews can be found in Refs. 118–120.

The details of the Metropolis method and subsequent elaborations thereof can be specified in the following general notation: The elements of the one-step transition probability matrix of the Markov chain, p_{kl}, are written as a product of two terms,

$$p_{kl} = q_{kl}\alpha_{kl} \tag{27}$$

The first factor, q_{kl}, is dependent on the method of generating the state l from the state k in a single-step transition. The second factor, α_{kl}, depends also on the way in which state l is accepted. The Metropolis choice

$$\alpha_{kl} = \min(1, P(\mathbf{X}_l^N)\, q_{lk}/\, P(\mathbf{X}_k^N)\, q_{kl}) \tag{28}$$

has been shown by Peskun (*121*) to be asymptotically optimum. The elements of the one-step transition probability matrix for the Markov chain can be rewritten into a more popular notation

$$p_{kl} = q_{kl} \min(1, p_l q_{lk}/p_k q_{kl}),\ k \neq l \tag{29}$$

and

$$p_{kk} = 1 - \sum_{k \neq l} p_{kl} \tag{30}$$

The various sampling methods discussed herein differ essentially in the definition of q_{kl}, i.e., the way in which state l is generated from state k in a single-step transition. In principle, all the sampling schemes allow for more than one-particle moves. However, in practice, for convergence efficiencies, the moves are restricted to a single particle. Thus, the configurational coordinates of the state l are related to the configurational coordinates of state k by

$$\mathbf{X}_l^N = \mathbf{X}_k^N + \boldsymbol{\delta}^N \tag{31}$$

where

$$\boldsymbol{\delta}^N = \{\mathbf{0}, \mathbf{0}, \ldots, \boldsymbol{\delta}(\mathbf{X}_m), \mathbf{0}, \ldots\} \tag{32}$$

and $\boldsymbol{\delta}(\mathbf{X}_m)$ is a displacement vector for the molecule m selected for the move. For rigid polyatomic molecules

$$\boldsymbol{\delta}(\mathbf{X}_m) = \{\delta x_{cm}, \delta y_{cm}, \delta z_{cm}, \delta\omega, \boldsymbol{\eta}\} \tag{33}$$

where δx_{cm}, δy_{cm}, δz_{cm} are the displacements for the center of mass and $\delta\omega$ is the rotation around a chosen axis, $\boldsymbol{\eta}$, passing through the center of mass of the molecule m. The magnitudes for the center-of-mass displacement and for the rotation angle are further restricted by certain step-size parameters Δr and $\Delta\omega$, which are optimized in the initial stages of the simulation. In Metropolis sampling, the components of the displacement vector $\boldsymbol{\delta}(\mathbf{X}_m)$ are obtained by uniformly sampling from the domain D located at the center of mass of the molecule m in the state k, and defined by the step-size parameters Δr and $\Delta\omega$. The elements q_{kl} of the transition probability matrix $\mathbf{Q}$ are then

$$\begin{aligned} q_{kl} &= \text{a constant} \qquad && \boldsymbol{\delta}(\mathbf{X}_m) \in D \\ q_{kl} &= 0 && \boldsymbol{\delta}(\mathbf{X}_m) \notin D \end{aligned} \tag{34}$$

It follows that, $\mathbf{Q}$ is a symmetric matrix.

In a typical Monte Carlo computer simulation on a molecular liquid in the (T, V, N) ensemble, the system consists of a simulation cell containing N molecules in a volume V determined by N/ρ, where ρ is the experimental density at the system temperature T. The configurational energy of the system is computed by means of analytical potential functions. The system is presented with a condensed phase environment by means of periodic boundary conditions, with the central cell surrounded at each face, edge, and vertex by a self-image. Calculations from this laboratory use mainly simple cubic or face-centered-cubic boundary conditions. To reduce the effect of the periodic images, most calculations include only interactions between the nearest images of each pair (minimum-image cutoff). Quite often, an additional cutoff criterion is applied to the nearest pair to decrease the computational effort (spherical cutoff). Calculations on liquid water reported from our laboratory used the spherical cutoff criterion. The initial segment of the calculation is an equilibration phase, and it is discarded in the formation of ensemble averages.

An example of a Monte Carlo computer simulation is the realization of 4400×10^3 configurations on liquid water carried out in this laboratory (*56*). The standard Metropolis method was used, with a potential function representative of quantum mechanical calculations of the intermolecular interaction energy, developed by Matsuoka et al. (*80*). The calculation was carried out on 125 water molecules under simple cubic periodic boundary conditions at T = 25 °C and at a density of 0.997 g/mL.

The convergence profile from this study is shown in Figure 4. An expanded scale was chosen for the energy ordinate here and in subsequent analogous figures; the convergence characteristics here are discussed in tenths of kilocalories per mole. The calculation achieves a mean

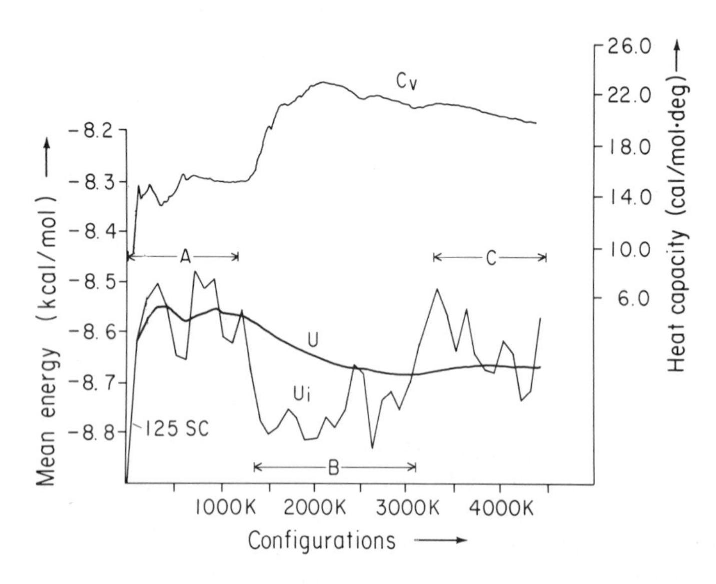

Figure 4. Convergence profile for Metropolis Monte Carlo computer simulation of $[H_2O]_l$ at 25 °C; K denotes 10^3 configurations.

energy of -8.57 kcal/mol after 200×10^3 configurations. The mean energy oscillates within 0.1 kcal/mol of this value for the next 1000×10^3 configurations. The convergence profile indicates that a mean energy value of -8.56 ± 0.03 kcal/mol was reached for this section of the run. At $N = 1200 \times 10^3$ to 1400×10^3, the control function shows a sharp decline of 0.13 kcal/mol in energy, to -8.78 kcal/mol at $N = 1400 \times 10^3$, and the onset of a region of 1600×10^3 configurations with a mean energy of -8.75 ± 0.02 kcal/mol. Concomitant with this decline is a sharp increase in heat capacity. At $N = 3000 \times 10^3$, the control function rises again, and at termination is oscillating about -8.64 ± 0.03 kcal/mol. The heat capacity is relatively constant from $N = 2000 \times 10^3$ on. The general appearance of the control function suggests that the high frequency oscillations in the control function are superimposed on a grand oscillatory cycle with an amplitude of 0.2 kcal in the realization, of which this calculation covers one and one-half cycles. Similar behavior in simulations based on the ST2 potential has also been discussed by Pangali et al. (*61*). The cumulative mean energy is -8.65 kcal/mol, with a heat capacity of 14.1 cal/mol °C, which, after the kinetic energy correction, results in -6.87 kcal/mol, and 20.1 cal/mol °C for mean energy and heat capacity, respectively.

The above results can be used to give an idea of the computer time required for simulation studies. Our sampling rate is 100×10^3 configurations per hour on an Amdahl 470/V6 using the Metropolis method on a system of 216 MCY water molecules. The calculation of stable values for internal energy and radial distribution functions requires $\sim 1500 \times 10^3$ configurations or 22 hours of computer time. Fluctuation properties such as heat capacity require roughly twice as much sampling. Properties such as the dipole correlation function required for the calculation of dielectric properties are even more slowly convergent, in part because of the statistical factor lost when less than full orientational averaging is involved. Overall, with this magnitude of computer time involved, the acceleration and improvement of convergence is a matter of continuous concern.

Following the work of Geperley et al. (*122*) on quantum liquids, several methods have been proposed for acceleration of convergence in Monte Carlo calculations, with development and testing carried out on the liquid water system. These procedures involve appending an additional importance-sampling criterion to the Metropolis method. The principal procedures currently under consideration are the force-bias procedure developed for liquid water by Pangali et al. (*60*) and an alternative force-bias scheme based on Brownian dynamics proposed by Rossky et al. (*123*). In the force-bias sampling, the particle moves are biased in the direction of forces and torques on the molecule selected for the move. The elements q_{kl} of the transition probability matrix for force-bias sampling are given by the expressions

$$q_{kl} = \mathcal{N}(\mathbf{X}_k^N)\exp\{\lambda(\boldsymbol{F}_m(\mathbf{X}_k^N)\cdot\delta\boldsymbol{r} + \boldsymbol{N}_m(\mathbf{X}_k^N)\cdot\delta\boldsymbol{\omega})/kT\} \qquad (35)$$

$$\boldsymbol{\delta}(\mathbf{X}_m) \in D$$

and

$$q_{kl} = 0,\ \boldsymbol{\delta}(X_m) \notin D \qquad (36)$$

Here $\delta\boldsymbol{r} = \{\delta x_{cm}, \delta y_{cm}, \delta z_{cm}\}$, $\delta\boldsymbol{\omega} = \boldsymbol{\zeta}\delta\omega$, $\mathcal{N}(\mathbf{X}_k^N)$ is a normalization constant and λ is a parameter to be optimized. The quantities $\boldsymbol{F}_m(\mathbf{X}_k^N)$ and $\boldsymbol{N}_m(\mathbf{X}_k^N)$ are the forces and torques, respectively, in the state k on the particle m to be moved. Note that $q_{kl} \neq q_{lk}$. In our force-bias calculations, λ is set to 0.5, following Rossky et al. (*123*).

For an absolute comparison between Metropolis sampling and force-bias sampling, the magnitudes of the step-size parameters Δr and $\Delta\omega$ and other set-up characteristics must be optimized as fully as possible in an initial short segment of the realization. One useful criterion for this optimization, based on translational and angular diffusion of particles, was reported by Pangali et al. (*60*). A different criterion related to particle diffusion was also considered. Using the analogy of random walks, Kalos (*124*) has proposed the quantity $\Delta r\,\langle A\rangle^2$ as an index of sampling efficiency, where A denotes the acceptance rate. To include the cage effect to first order, we have modified this index to $\Delta r\,\langle A\rangle^2(1 + \langle\cos(\theta)\rangle)$ (*125*), where θ is the angle between the successive accepted moves of a molecule. The best positional and angular displacement for the force-bias method, according to the new criterion, were found to be nearly double the respective displacement in the standard Metropolis method. This finding is in agreement wth the results of Kincaid and Scheraga, who showed that significant improvement can be obtained with the regular Metropolis method when larger than usual step sizes are used (*126*).

Convergence acceleration studies using force bias were carried out on MCY water at 25 °C (*127*) and analyzed in terms of particle diffusion, "rate" of equilibration, and the evolution of internal energy and heat capacity during the Monte Carlo realization. The particle diffusion rate was found to be four times greater in the force-bias computation, indicative of the increased sampling of the configuration space for individual particles. The equilibration of the computation in the initial segment of the realization was found to be two to three times faster with force bias. The convergence profile, Figure 5, shows that both internal energy and heat capacity are well settled down after 1000×10^3 configurations, an improvement by a factor of two to three over the standard Metropolis results shown in Figure 5. Preliminary studies of the energy autocor-

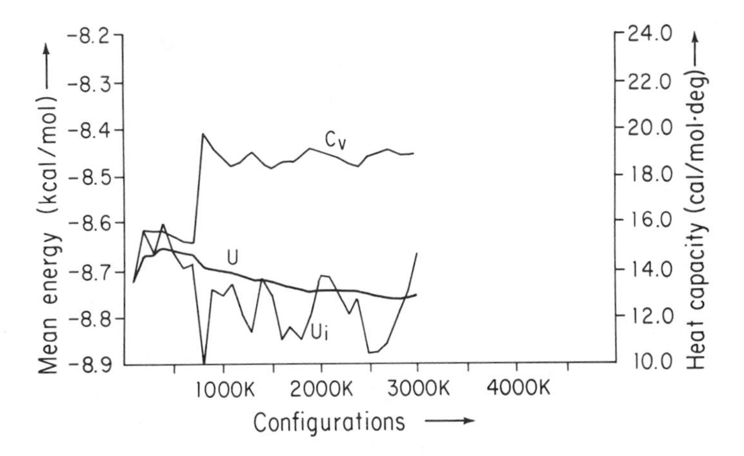

Figure 5. Convergence profile for force-bias Monte Carlo computer simulation of $[H_2O]_l$ at 25 °C; K denotes 10^3 configurations.

relation functions for the force-bias and standard Metropolis method show that in the latter, significant correlations persist even after some 2000 × 10^3 configurations of sampling. In the force-bias simulations, the energy correlations are greatly reduced after a mere 500 × 10^3 configurations. For an overall evaluation, however, the extra cost involved in the calculation of forces is also to be considered. Depending on the system size and set-up characteristics, the extra computational expense is in the range of 70–100%. Preliminary results using the Brownian force-bias technique produced similar results (*125*). For $(H_2O)_l$, these results indicate that these gradient-bias samplings are a definite improvement over the standard Metropolis method constitute and are the preferred procedure for doing the simulations.

Monte Carlo computer simulation can also be configured for the (T, P, N) ensemble (*117*), with the volume and thus the density of the system computed as an ensemble average. In practice, this requires about 10–50% additional computational effort in the Monte Carlo procedure, depending on the frequency of volume perturbations. Mezei recently showed (*128*) that the efficiency of the volume-change sampling can be improved significantly if the volume change is biased in the direction of the volume derivative of the pseudo-Boltzmann factor used in computing the acceptance probabilities.

Results

In this section the results of computer simulations of liquid water, based on the most widely used and promising potential functions, are collected. A critical comparison of these results with available experimental data is presented. Thermodynamic properties are considered first, followed by radial distribution functions. Finally, the temperature dependence of the results is considered. We address two main questions: (1) the accuracy with which experimental observables for liquid water can be calculated by computer simulation, and (2) the sensitivity of these results to the choice of potential function, the main variable in the calculations.

Most of the potentials discussed in the section entitled "Background" have been used in computer simulation of liquid water. The ST2 and MCY potentials have proved to be of continuing interest and serve well to illustrate broadly the capabilities and limitations of potential functions in liquid-water simulations. The three most recently proposed potentials, TIPS, SPC, and TIPS2, are especially interesting from the point of view of computational economies, and of demonstrating the effect of small parametrization changes on the calculated liquid structure. The QPEN potential is of importance for the development of potentials for large solutes, because of the transferability of the parameters of this functional form (*129*). The PE function points up problems in introducing cooperative effects. Thus, the results of computer simulations are described here based on the ST2, MCY, TIPS, SPC, TIPS2, RWK2, QPEN, and PE functions.

The MCY results are taken from the standard Metropolis Monte Carlo calculation described in the preceding section and elsewhere (*56*, *58*, *59*), from the (T, P, N) simulation studies of Owicki and Scheraga (*51*), and from the molecular dynamics calculations of Rapaport and Scheraga (*44*). The ST2 results derive from a Metropolis Monte Carlo simulation of 4700×10^3 configurations on 216 molecules at a density of 1.0 g/mL at 10 °C (*56*)–conditions chosen to coincide directly with the previous ST2 liquid water calculations by Rahman and Stillinger and from results obtained by Pangali et al. (*60*, *61*). The TIPS, SPC, and QPEN results are based on (T, V, N) ensemble force-bias Monte Carlo simulations on 125 molecules under simple cubic periodic boundary conditions for a temperature of 25 °C and a density of 0.997 g/mL, recently carried out in our laboratory by Mezei (*130*). These simulations consist of 720 $\times$ 10^3, 1500×10^3, and 1020×10^3 configurations, respectively. The TIPS2 results are the work of Jorgensen (*76*). The PE results are based on a 40×10^3 simulation by the standard Metropolis method for ambient temperature by Barnes et al. (*81*).

The thermodynamic properties calculated for liquid water from these various Monte Carlo computer simulations are collected along with the corresponding experimental data in Table II. Internal energy is the configurationally averaged energy of the system, and good agreement between calculated and observed values implies that the hydrogen-bond well depth and the number and energy of thermally accessible configurations is in order. The calculated internal energy is seen to be within 15% of the observed value for all of the potentials considered, which represents reasonable agreement, considering that cooperative effects are either neglected entirely or included only implicitly. Comparing the results from the various potentials, the calculated magnitude of the internal energy is directly correlated with the energy of the linear hydrogen bond in the corresponding water dimer. The ST2, SPC, and TIPS2 potentials, with dimerization energies of 6.84, 6.79, and 6.20 kcal/mol, respectively, are seen to overestimate slightly both the hydrogen-bond energy of the dimer and the internal energy of the liquid, whereas the MCY and TIPS potentials underestimate the internal energy.

In comparing calculated and observed values for internal energy and other properties, quantum corrections (*105, 131*) are not included. Owicki and Scheraga estimate the quantum corrections to the internal energy to be only 0.2 kcal/mol (*51*). The importance of the quantum effect has also been stressed by Rice et al. (*66*). The subject is under detailed study by Wilson et al. (*132*), and indications are that corrections can be quite significant.

The configurational excess heat capacity of the system is computed from the fluctuations in internal energy in the course of the simulation and, as a second moment, reflects the width of the distribution of binding energies for the molecules of the system. The observed heat capacity of liquid water is roughly double that of ice or steam, because of the additional modes of energy uptake in the hindered translational and rotational motions in the liquid. A large calculated heat capacity is given by all of the potentials for which results are available. The magnitude is, however, overestimated by 100% by the ST2 potential. The MCY, TIPS, SPC, RWK2, and QPEN potentials are all within 15% of the observed value.

The calculated pressure of the liquid should be identically equal to atmospheric pressure for a liquid at equilibrium under ordinary conditions. In liquid-state computer simulation, this quantity depends on the shape of the hydrogen-bond potential well, and the magnitude of the calculated pressure is thus expected to be highly sensitive to fine detail in the potential. Consideration of the experimental value of the compressibility or results from (T, P, N) simulation studies (*51*) show that a 2–3% difference in the density can change the calculated pressure by ~500 atm; thus agreement between calculated and observed values to within O(100) atm. can be considered reasonable performance for a potential. Considering the (T, V, N) results for pressure in Table II, the ST2, TIPS, and SPC functions, and the QPEN function to a somewhat lesser degree, are seen to be acceptable for this criterion. The pressure is seriously overestimated by the MCY potential.

In (T, P, N) ensemble simulation, external pressure is assigned to

Table II. Calculated and Observed Values

Property	*ST2*	*MCY*	*TIPS*	*SPC*
ρ (g/mL)	1.000^a	0.997^a	0.997^a	0.997^a
T (K)	283^a	298^a	298^a	298^a
$\langle U \rangle$ (kcal/mol)	-10.7 ± 0.2	-8.7 ± 0.1	-8.17 ± 0.2	-10.18 ± 0.2
$C_V{}^c$ (cal/mol°C)	20.6 ± 2	11.5 ± 2	13.6	11.4
$P_o{}^d$	1232^e	7552 ± 238	1807 ± 270	269 ± 226
P	543	6760 ± 311	716 ± 305	-886 ± 227
A^f (kcal/mol)	-5.40	-3.96	—	-4.41
S^f (cal/°C)	-17.2	-15.6	—	-19.4

Note: Experimental values refer to 298 K.

[a] Assumed values.

[b] This value, widely quoted, assumes ideality in water vapor. Jorgensen points out (*74*) that consideration of nonideality of water vapor leads to a modified value of −10.08 kcal/mol.

[c] Corrected by $6R$ for the kinetic contribution.

[d] Obtained without the Dirac δ contribution from the cutoff in potential.

[e] From Ref. 61.

[f] From Ref. 109.

be 1 atmosphere and the density is calculated from the configurationally averaged volume. Owicki and Scheraga found the density produced by the MCY potential to be 24% lower than the experimental value, a consequence of the same features that led to the pressure error noted previously. The TIPS2 potential of Jorgensen was parametrized to give the correct density at 25 °C in (T, P, N) simulation; at two other temperatures generally good calculated values for the density are produced. Under the considerably larger error bounds expected for fluctuation properties, the calculations are in reasonable accord with experiment.

Estimates of the excess free energy and entropy have been calculated for the MCY, ST2, and SPC potentials (*125*) by using thermodynamic integration methods (*106*). The free energy is best reproduced with the ST2 potential, and the entropy with the MYC potential. Both the ST2 and the SPC predict entropies lower than those observed, indicative of excessive structure in the model liquid.

The major point of contact between experimental data and computer simulation of liquid water is the comparison between observed and calculated atom–atom radial distribution functions. The experimental data on these quantities are obtained from X-ray and neutron diffraction experiments. The dominant contribution to the X-ray diffraction of liquid water is caused by scattering from oxygen atoms, but the neutron diffraction comes mainly from hydrogen scattering with a smaller contribution from the oxygen atoms. A combination of X-ray and neutron diffraction experiments is required for complete determination of struc-

for Thermodynamic Properties of Liquid Water

TIPS2	*QPEN*	*RWK2*	*EXP*	*MCY*	*MCY*
0.994 ± 0.007	0.997[a]	0.997[a]	0.997	0.993[a]	0.988[a]
298[a]	298[a]	298[a]	298	210[a]	323[a]
−10.08 ± 0.04	−9.41 ± 0.1	−8.24	−9.9[b]	−8.57 ± 0.1	−8.40 ± 0.1
—	11.7	10.5	12.0	14.0 ± 1	13.0 ± 2
—	1911	—	—	—	—
1350	1403 ± 200	—	1.0	6922 ± 299	7178 ± 193
—	—	—	−5.74	—	—
—	—	—	−14.0	—	—

ture factors for the system. The quantities $g_{OO}(R)$, $g_{OH}(R)$, and $g_{HH}(R)$, the oxygen–oxygen, oxygen–hydrogen, and hydrogen–hydrogen radial distribution functions, respectively, are obtained from the structure factors by Fourier inversion, with intramolecular components subtracted out.

Diffraction experiments on liquids are difficult both to carry out and to analyze, and progress in this area has been a major accomplishment in modern chemical physics. There are particular obstacles to overcome in the neutron diffraction studies, for the large cross section for incoherent scattering from hydrogen relative to coherent scattering results in a low signal-to-noise ratio. The high neutron flux required to attain an acceptable level of statistical accuracy is available only at a few large-scale reactor sites. Also, inelastic and recoil effects complicate the determination of the static, coherent scattering function from the measured effective cross sections. The Fourier transformation is susceptible to errors caused by truncation, which affect the analysis of both X-ray and neutron diffraction experiments.

X-Ray diffraction patterns of liquid water were discussed initially in the 1933 Bernal and Fowler paper and early measurements were reported by Morgan and Warren (*7*). Narten et al. (*8*, *9*) obtained high-resolution data from X-ray diffraction data and published widely cited results on $g_{OO}(R)$ as a function of temperature in the ambient region. Palinkas et al. (*10*) used a combination of X-ray, early neutron, and electron diffraction data to determine $g_{OO}(R)$, $g_{OH}(R)$, and $g_{HH}(R)$. Another report of neutron diffraction experiments on liquid water is by Narten et al. (*11*) with a corresponding determination of $g_{OH}(R)$ and $g_{HH}(R)$ and a redetermination of $g_{OO}(R)$. A neutron diffraction measurement of $g_{HH}(R)$ was reported by Soper and Silver (*12*). The collected experimental measurements of the atom–atom distribution functions for liquid water are discussed in the following paragraph, along with a consideration of corresponding results for these quantities obtained from computer simulation.

The oxygen–oxygen radial distribution functions from the reported diffraction studies are collected in Figure 6, presented in typical form, as relative fluctuations of local density in the liquid relative to bulk density. The $g_{OO}(R)$ carries information on the relative positional disposition of molecules in the liquid, and the successive peaks in the distribution can be identified with the distinct hydration shells of a reference molecule. The three reported measurements differ slightly in quantitative detail, but at the qualitative level they are in substantial accord. The assignment of the peaks based on oxygen–oxygen separations in the linear water tetramer is indicated on an inset to the figure.

The first peak in $g_{OO}(R)$, centered at ~2.8 Å, gives the mean value and distribution of nearest neighbor separations in the liquid. Comparison of the position of this peak and the corresponding value of the interoxygen separation in ice Ih (2.76 Å) with the value observed for the equilibrium separation in the water dimer (2.98 Å) indicates a significant contribution from cooperative effects to the structure in the liquid and

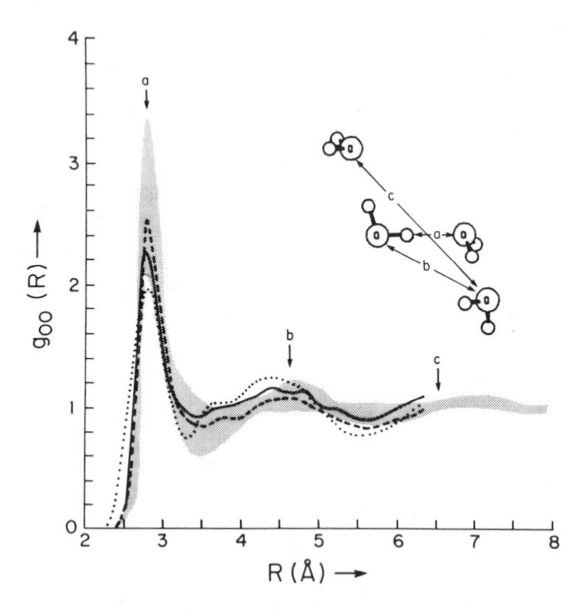

Figure 6. Observed g_{OO}(R) *vs* R *from diffraction experiments (—: Narten et al.* [8], *···: Palinkas et al.* [10], *---: Narten et al.* [11]*). Collective results on* g_{OO}(R) *calculated by computer simulations fall within the shaded area. The structure insert depicts a sequential tetramer based on* R = 2.82 Å *and* ϕ = 60°*; the various interatomic separations fall as indicated on the figure.*

solid phases. The effective diameter of a water molecule in the liquid is equivalent to the mean first-neighbor separation of 2.8 Å.

The second peak in $g_{OO}(R)$ for liquid water appears with a broader distribution centered at ~4.5 Å. In a simple liquid composed of effectively spherical particles of diameter σ, successive peaks in the distribution are expected at separations σ, 2σ, 3σ, etc. The position of the second peak in $g_{OO}(R)$ for liquid water at ~4.5 Å rather than ~5.6 Å indicates that specific orientational correlations are present, and these distinguish water as an associated liquid rather than a simple liquid. The specific value of 4.5 Å is characteristic of second-neighbor separations in tetrahedral coordination, as indicated in the structure insert to Figure 6. On this basis, a third peak would be predicted at 6.5–7 Å. The experimental data of Narten et al. and Palinkas et al. are in close accord on the general shape of the oxygen–oxygen distribution but differ in quantitative detail on peak intensities and on structure in the region of the second shell.

Reported experimental data on the oxygen–hydrogen distribution function for liquid water are shown in Figure 7. These functions carry distance information on hydrogen bond lengths and describe further the orientational correlations in the liquid. Palinkas et al. report peaks in

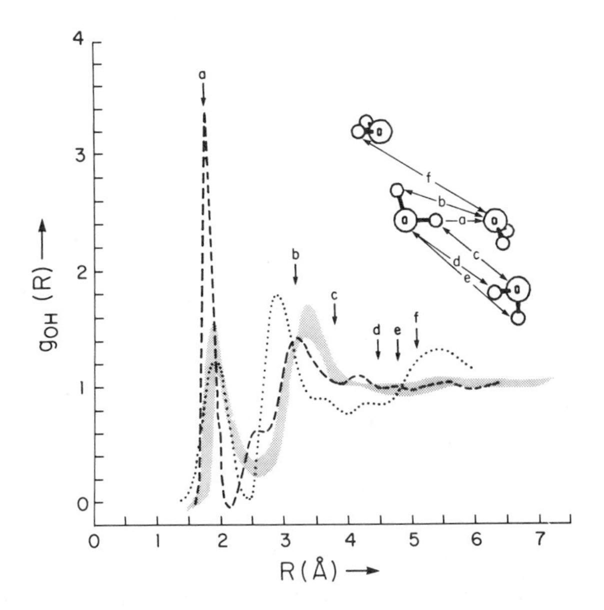

Figure 7. Observed g_{OH}(R) *vs* R *from diffraction experiments* (···: *Palinkas et al.* [10], ---: *Narten et al.* [11]). *Collective results on* g_{OH}(R) *calculated by computer simulations fall within the shaded area. The structure insert depicts a sequential tetramer based on* R = 2.82 Å *and* ϕ = 60°; *the various interatomic separations fall as indicated on the figure.*

$q_{OH}(R)$ at 1.95 and 2.95 Å, the former corresponding to the $O\cdots H$ hydrogen bond and the latter to a longer range oxygen–hydrogen correlation. These results are consistent with a decidedly bent hydrogen-bond structure for the liquid. The results of Narten et al. show a narrow, intense first peak at 1.86 Å and a second peak at 3.2 Å for the longer range correlations. These results are consistent with a structure featuring nearly linear hydrogen bonds. Thus significant discrepancies are found between the two presently available sets of experimental results for $g_{OH}(R)$.

The results of the three separate experimental measurements reported on $g_{HH}(R)$ for liquid water are shown in Figure 8. The first peak in $g_{HH}(R)$ is quite pronounced and higher than the second peak in the results of Narten et al., but the first peak is slightly lower than the second peak in the results of Palinkas et al. There are also slight differences in peak positions in these two sets of results. The $g_{HH}(R)$ reported by Soper and Silver differs significantly from both earlier reports, although their results agree closely with the first peak of Palinkas et al. and with the second peak of Narten et al. Overall, significant discrepancies among the experimental results are again also quite evident for $g_{HH}(R)$. The apparent disagreement among the experimentalists in meas-

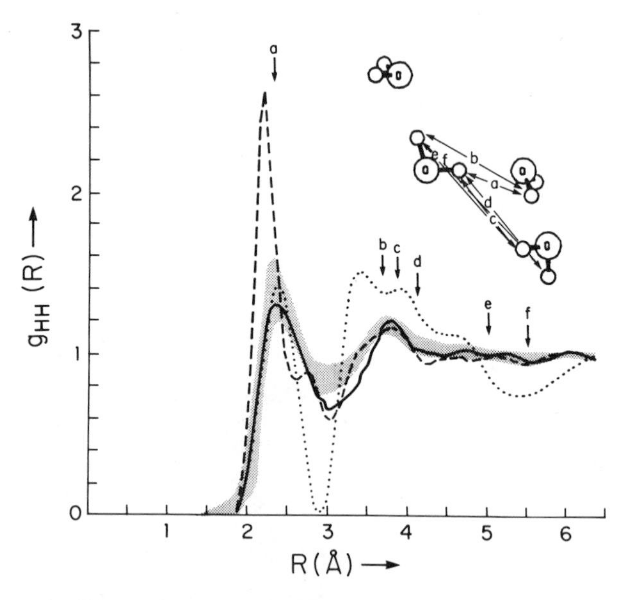

Figure 8. Observed g_{HH}(R) *vs* R *from diffraction experiments (···: Palinkas et al.* [10], *—: Soper and Silver* [12], *---: Narten et al.* [11]*). Collective results on* g_{HH}(R) *calculated by computer simulations fall within the shaded area. The structure insert depicts a sequential tetramer based on* R = 2.82 Å *and* ϕ = 60°*; the various interatomic separations fall as indicated on the figure.*

urements based on neutron diffraction must be considered in the context of the difficulty of the measurement and the corresponding data reduction problems, as well as apparently significant differences in sample configuration in the various experiments. In any case, the reconciliation of the diverse experimental measurements of $g_{OH}(R)$ and $g_{HH}(R)$ is clearly an important next step in understanding detailed aspects of liquid water structure at the molecular level.

The atom–atom radial distribution functions obtained from Monte Carlo computer simulation of liquid water are shown in Figures 9–11. Considering first the results for $g_{OO}(R)$, for the first peak we find general agreement among the calculated results on position but some significant diversity in performance on amplitude. There is ~35% variation in peak height in passing from results of simulations based on the TIPS and MCY potentials to those based on the ST2 potential. There is considerable variation among the different potentials in the description of the second peak in $g_{OO}(R)$, ranging from well-defined maxima from the ST2, MCY, QPEN and TIPS2 functions to weakly defined or nonexistent peaks from the SPC and TIPS functions. The RWK2 potential (86) also lacks a second peak in $g_{OO}(R)$. For the potentials producing a second peak, the position varies in the range between 4 and 4.5 Å, but in any case the essential associated nature of the liquid is evidently well described. The shoulder on the inside of the second peak, discussed particularly by Rice et al.

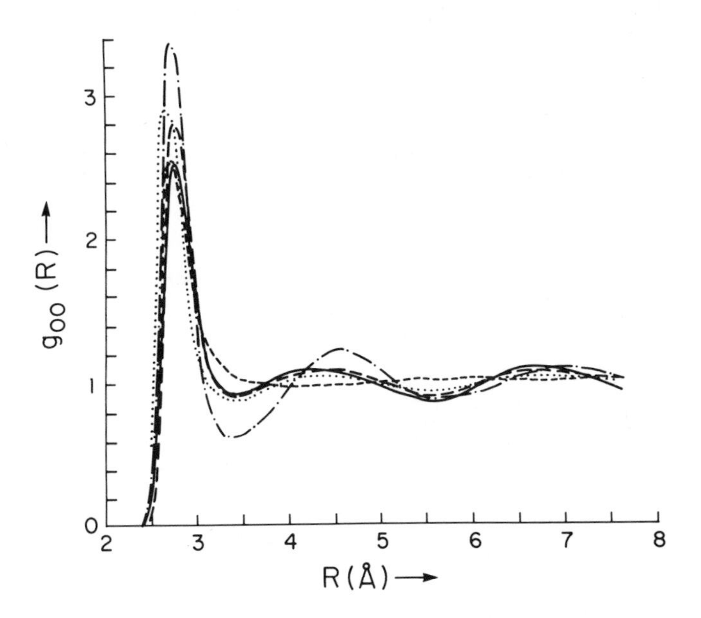

Figure 9. Calculated $g_{OO}(R)$ *vs* R *from liquid water simulation (—: MCY, -·-·-: ST2, ——: TIPS2, ···: SPC, ---: TIPS).*

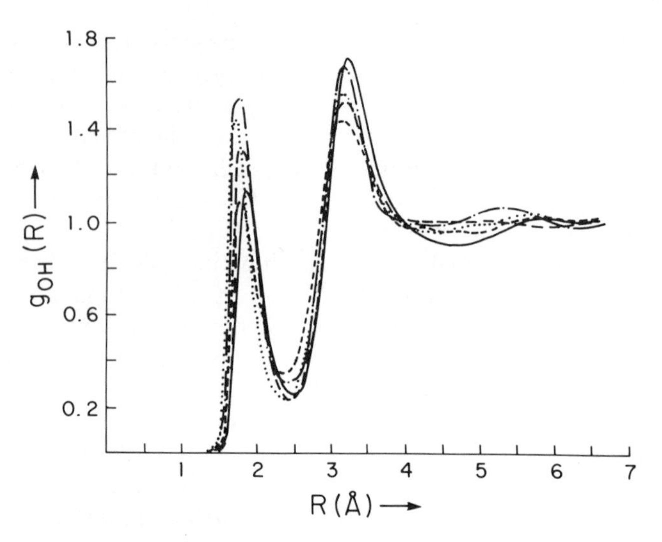

Figure 10. Calculated g_{OH}(R) *vs* R *from liquid water simulation (——: MCY, -·-·-: ST2, ––––: TIPS2, ···: SPC, ---: TIPS).*

(65) in studies of amorphous ice relative to liquid water, has not been reproducibly seen in computer simulation. All of these potentials perform similarly in the region of the third peak in $g_{OO}(R)$.

Results from the PE potential are not included here. An early report of $g_{OO}(R)$ based on this function (79) showed notable discrepancies with

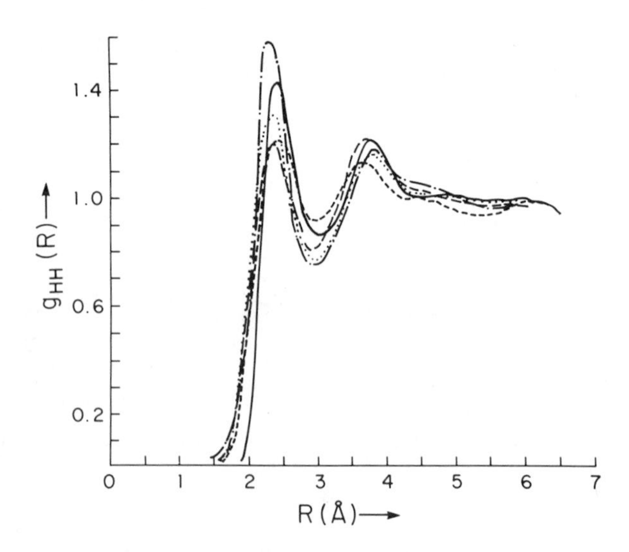

Figure 11. Calculated g_{HH}(R) *vs* R *from liquid water simulation (——: MCY, -·-·-: ST2, ––––: TIPS2, ···: SPC, ---: TIPS).*

experiment, on very limited ensemble averaging. This function is being modified at present, and clarification of the role of cooperative effects in water structure thus remains to be established. Also not included is a simulation on the original Bernal–Fowler potential, carried out by Jorgensen (*133*), which shows $g_{OO}(R)$ in serious disagreement with experiment; here, the second peak occurs at ~5.8 Å rather than at 4.5 Å.

For $g_{OH}(R)$ and $g_{HH}(R)$, there is general agreement among the potentials on the peak positions but, once again, considerable variation in peak heights. However, a quite general accord on the shape of the O–O and O–H distribution functions is clearly evident, indicating that the essential features of the structure described by the various potentials are similar.

Considering now the comparison of observed and calculated results for atom–atom radial distribution functions, in both domains a range of results has been reported for each quantity. To compare results with these variations in perspective, we have taken the calculated results collectively for each type of distribution and formed plots of the maximum and minimum extrema. These plots are superimposed on the plots of experimentally observed distribution functions in Figures 6–8, with the region between the extrema shaded. Simply speaking, all calculated results fall within the shaded region regardless of choice of potential function.

Discrepancies between the calculated and observed results are quite evident. Considering first the comparison of observed and calculated results for $g_{OO}(R)$, Figure 9, one observes a general tendency in the water potentials to overestimate intensity and to underestimate slightly the position for the first peak in computer simulation. There is a general qualititative accord between calculated and observed values on the structure beyond the first peak, with the results of Narten et al. within the range of calculated results and the results of Palinkas et al. slightly outside this range. Results from particular simulation studies provide a calculated $g_{OO}(R)$ in close agreement with the experimental data of Narten et al. As pointed out by Jorgensen, there is a strong correlation between the calculated intensities of the first and second peaks in $g_{OO}(R)$, and it has proved impossible to date to produce a potential function that gives in computer simulation a good description of the second peak without overestimating the intensity of the first peak.

For $g_{OH}(R)$, referring back to Figure 10, the collective calculated results are in serious disagreement with both sets of experimental measurements reported to date. Particularly, the intensity of the first peak in all calculated results is far below that reported by Narten et al. and significantly above that of Palinkas et al. The shape of the calculated distribution differs with both sets of experimental results in the region of the first minimum and lacks the shoulder at 2.5 Å reported by Narten et al. The shape of the remainder of the calculated distribution agrees fairly well with the Narten et al. measurement. The calculations show a serious discrepancy with the Palinkas et al. data on the position of the second peak and on the existence of a third peak in the distribution.

For $g_{HH}(R)$, referring back to Figure 11, the collective calculations are in general accord on the shape of the distribution seen in the Narten et al. results, although the calculated intensity of the first peak is lower and the shoulder is not reproduced. Serious discrepancies are evident between the calculated results and the Palinkas et al. measurements. However, the measurement of $g_{HH}(R)$ reported by Soper and Silver is in close qualitative accord and shows good semiquantitative agreement with $g_{HH}(R)$ predicted by computer simulation. In view of the correspondence between the collective simulation results for $g_{OO}(R)$ and X-ray diffraction results, and between the simulation results for $g_{HH}(R)$ and Soper and Silver's data, we feel there is a strong chance that the simulation results for $g_{OH}(R)$ are more reliable than either the Narten et al. or Palinkas et al. measurements, and that computational theory is actually predicting accurately the correct $g_{OH}(R)$ and $g_{HH}(R)$ ahead of the experimental measurements. Impey et al. (*134*) point out various problems with the interpretation of scattering experiments and suggest that comparisons be made by transforming the computer simulation data into k-space.

Comparing now the relative performance of the various intermolecular potential functions with respect to the experimental data and with each other, we focus on results for internal energy, heat capacity, density, and $g_{OO}(R)$. The calculated percentage errors for simulation results based on the different potentials for each of the properties under consideration are shown in Figure 12. For internal energy, heat capacity, and density, calculated values are generally within 15% of experiment, with the TIPS2 potential performing best. For $g_{OO}(R)$, a general tendency of peak position to be computed accurately is clearly evident, with the percentage error generally less than 5%. Calculated peak heights, however, are seen to vary widely. The MCY potential shows the best overall performance on the oxygen–oxygen radial distribution function, in spite of its well-known problematic performance on density or pressure.

The temperature dependence of the internal energy is implicitly reflected in the heat capacity, and agreement between calculated and observed values for this quantity have been discussed earlier. The temperature dependence of compressibility and of the thermal expansion coefficient has been treated by computer simulation (*76*), and the basic qualitative features of the anomalous behavior of water properties with respect to temperature have been accounted for. A temperature of maximum density has been found in both molecular dynamics (*74*) and (T, P, N) ensemble Monte Carlo work (*76*) for the ST2 and TIPS2 potentials, respectively.

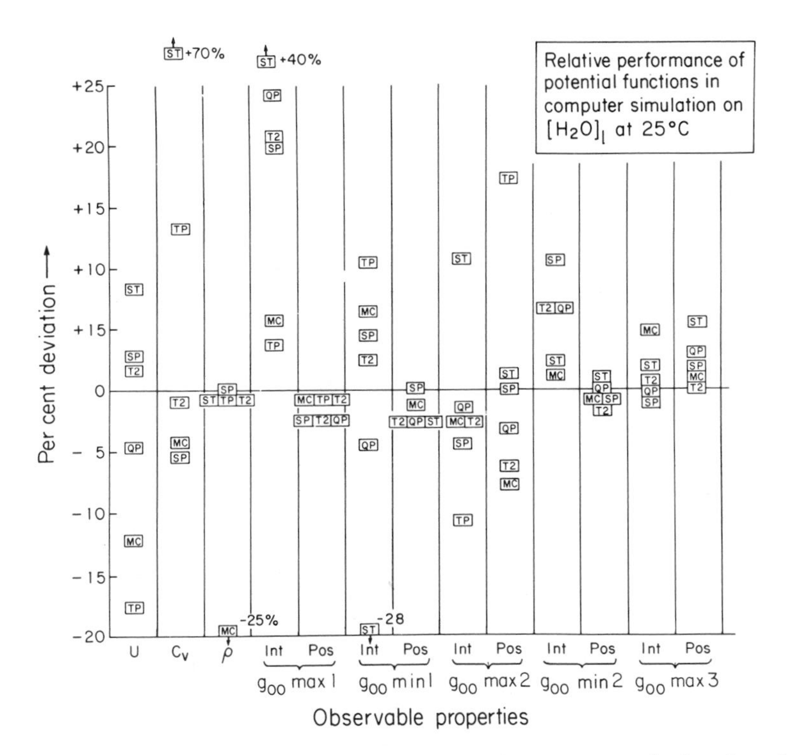

Figure 12. Comparison of the percentage deviation in calculated and observed results on thermodynamic properties and the X-ray g_{OO}(R) *for liquid water. Water models are abbreviated as follows: MCY:MC, ST2:ST, TIPS2:T2, SPC:SP, TIPS:TP, QPEN:QP.*

The water–water radial distribution functions have been determined experimentally as a function of temperature by Narten. These data have been compared with molecular dynamics calculations using the MCY potential by Impey et al. (*42*) and by Rapaport and Scheraga (*44*). Results for $g_{OO}(R)$ for several temperatures are shown in Figure 13. The calculated and observed radial distribution functions are in good agreement. Monte Carlo results and detailed QCDF analysis of the temperature dependence of the MCY water have been described by Mezei and Beveridge (*58*), and for the TIPS2 water by Jorgensen (*76*). Thus temperature effects on liquid water structure are seen to be qualitatively and semi-quantitatively accounted for by liquid-state computer simulation. A highly quantitative study of temperature effects by a new approach, the isochoric temperature differential of structure factors, is currently emerging (135, 136). Egelstaff and Root took up a characterization of the many-body effects on water structure using this approach, and indicate that these effects are possibly the cause of the larger temperature effects in OH and HH correlations than in the OO correlation.

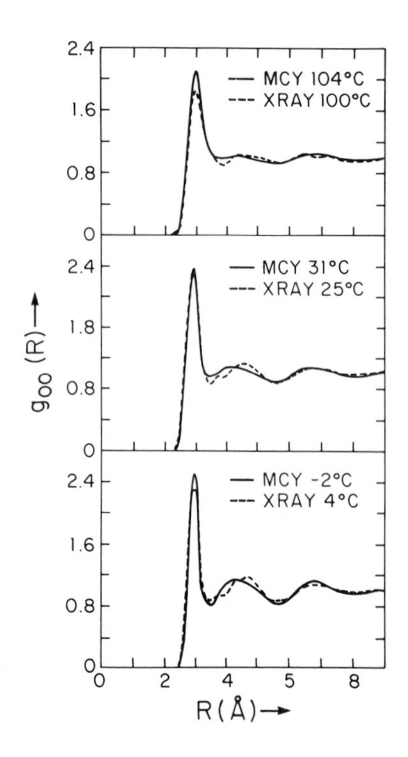

Figure 13. Calculated $g_{OO}(R)$ *vs.* R *from liquid-water simulations at various temperatures based on the MCY potential.*

Analysis of Results

The subject of this section is interpretation of the results obtained in liquid-state computer simulations of water and the determination of the structural and energetic composition of the system. The composition of a liquid must be defined with regard to the statistical state of the system, and compositional indices must be defined in terms of statistically weighted structural alternatives rather than any single supermolecular structure.

A general theoretical framework for such analyses was mapped out by Ben-Naim (*4*) in terms of "generalized molecular distribution functions" and the closely related quasi-component distribution functions (QCDF). In this approach, one obtains the concentration in mole fraction $x_Q(p)$ of particles with certain well-defined values p of a compositional characteristic Q. Specifically, we consider the QCDF for the following compositional characteristics: coordination number K, binding energy ν, near-neighbor pair energy ε, near-neighbor dipole angle θ, cavities of radius R_c (58) and hydrogen-bond parameters R_{OO}, θ_H, θ_{LP} and δ_D (59). In the following discussion, the QCDF's are presented either in graphical form or in tabular form and are characterized by their salient features as follows: the location of the maximum, p^{max}, the value at the maximum,

$x_Q(p^{max})$, the beginning and end point of the smallest interval that contains 99.9% of the distribution, $p_<$ and $p_>$, and the smallest and largest p values such that $x_Q(p) = x_Q(p^{max})/2$, $p_<^{1/2}$ and $p_>^{1/2}$.

The calculated results for atom–atom radial distribution functions for liquid water showed a considerable degree of similarity regardless of the potential function used. As we turn now to further structural analysis, the results at the qualitative and semiquantitative level are found to be substantially similar for the various models of water–water interactions. We arbitrarily focus our discussion in this section on results based on the MCY potential, studied in detail in this laboratory, and quote illustrative results based on other potentials for comparison. We expect most, if not all, of the conclusions discussed here to be completely general.

The coordination number of a given water molecule is defined here as the number of neighbors whose center of mass lies within a sphere of radius R_M around the center of mass of the given molecule. The value of R_M is chosen to be the location of the first minimum in the center-of-mass radial distribution function, $R_M = 3.3$ Å. The QCDF for coordination number K can be defined as the mole fraction

$$x_C(K) = \langle \sum_i^N \delta[C_i(\mathbf{X}^N) - K] \rangle / N \tag{37}$$

where the summation involves the Dirac delta counting function $\delta[C_i(\mathbf{X}^N) - K]$ for the number of particles with coordination number K in configuration $\mathbf{X}^N$. The average coordination number, $\overline{K}$, is given by the expression

$$\overline{K} = \sum_{K=0}^{\infty} K x_C(K) = \rho \int_0^{R_M} g(R)\, 4\pi R^2 dR \tag{38}$$

The $x_C(K)$ obtained for the MCY water at 25 °C is presented in Figure 14. The $x_C(K)$ values are given in Table III for the water models considered. The four-coordinate nature of the liquid water is clearly exhibited by all these systems. The TIPS and SPC results are rather similar and resemble those of the MCY water. The value of $x_C(0)$ is nonzero for all of the water models studied, implying the existence of significant density fluctuations in the liquid.

The binding energy of a molecule in the liquid is the quantity

$$B_i(\mathbf{X}^N) = E(\mathbf{X}^N) - E(\mathbf{X}_1, \ldots, \mathbf{X}_{i-1}, \mathbf{X}_{i+1}, \ldots, \mathbf{X}_N) \tag{39}$$

i.e., the negative of the vertical dissociation energy of the ith molecule.

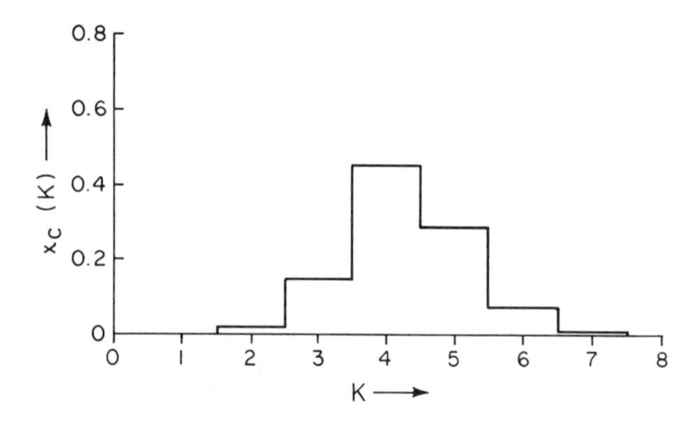

Figure 14. Calculated QCDF for coordination number $x_C(K)$ *vs.* K *for MCY water at 25 °C.*

The QCDF for binding energy ν follows as the mole fraction of particles with binding energy between ν and $\nu + d\nu$:

$$x_B(\nu) = \langle \sum_{i=1}^{N} \delta[B_i(\mathbf{X}^N) - \nu] \rangle / N \quad (40)$$

The distribution of binding energies $x_B(\nu)$, as pointed out by Ben-Naim, is a diagnostic index of the mixture model versus the continuum model; $x_B(\nu)$ is expected to be polymodal for a system to be viewed in terms of a mixture model and unimodal for a continuum model.

The $x_B(\nu)$ for the MCY water at 25 °C is given in Figure 15. The characteristics of the $x_B(\nu)$ values for the MCY, ST2, TIPS, SPC, and

Table III. Comparison of the QCDF of Coordination Number for the Various Water Models

	ST2	*MCY*	*TIPS*	*SPC*	*QPEN*	*MCY*	*MCY*
T (K)	283	298	298	298	298	310	323
$x_C(0)$	0.00001	0.00002	0.00012	0.00003	0.00003	0.00009	0.00010
$x_C(1)$	0.00006	0.0013	0.0048	0.0022	0.0016	0.0039	0.0044
$x_C(2)$	0.0036	0.0224	0.0492	0.0324	0.0250	0.0449	0.0505
$x_C(3)$	0.0524	0.1516	0.2068	0.1846	0.1574	0.2170	0.2300
$x_C(4)$	0.3927	0.4561	0.3902	0.4408	0.4529	0.4589	0.4472
$x_C(5)$	0.3417	0.2901	0.2603	0.2664	0.2739	0.2242	0.2170
$x_C(6)$	0.1619	0.0766	0.0759	0.0650	0.0766	0.0468	0.0458
$x_C(7)$	0.0442	0.0094	0.0117	0.0080	0.0115	0.0041	0.0046
$x_C(8)$	0.0076	0.0006	0.0009	0.0006	0.0010	0.0002	0.0002
$\overline{K}$	4.807	4.306	4.131	4.167	4.254	4.012	3.979

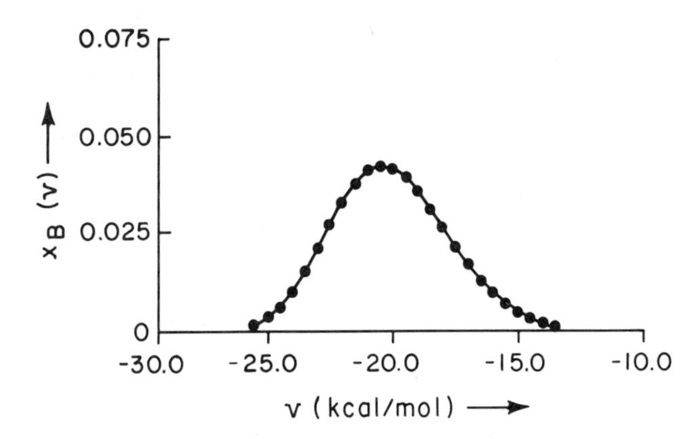

Figure 15. Calculated QCDF for binding energy $x_B(\nu)$ *vs.* ν *for MCY water at 25 °C.*

QPEN methods are collected in Table IV. The function $x_B(\nu)$ is unimodal for all potentials studied, and thus the computer simulation results for liquid water are uniformly supportive of the continuum model. The unimodal nature found for the MCY water at 25 °C can also be found at higher temperatures.

The QCDF for the near-neighbor pair interaction energy, $x_P(\varepsilon)$ is defined as

$$x_P(\varepsilon) = \langle \sum_{i<j}^{N} \delta[\, \varepsilon_{ij}(X^N) - \varepsilon] \, C_{ij}(X^N) \,\rangle / \langle \sum_{i<j}^{N} C_{ij}(X^N) \rangle \tag{41}$$

where $\varepsilon_{ij}(\mathbf{X}^N)$ is the value of the pair-energy between the molecules i and j, C_{ij} $(\mathbf{X}^N)$ is unity if the distance between i and j is less than R_M, and zero otherwise. Generally, pair-energy curves are published for all pairs in the system, resulting in a very large peak at $\varepsilon = 0$ and a shoulder or a second peak at the location of the peak of our $x_P(\varepsilon)$. Restriction to near-neighbors eliminates this large but trivial peak.

The $x_P(\varepsilon)$ values for both the MCY and ST2 waters are given in Figure 16. The characteristics of the MCY, ST2, TIPS, SPC, and QPEN $x_P(\varepsilon)$ distributions are given in Table V. These curves are also unimodal, supporting the conclusion drawn from the unimodality of $x_B(\nu)$. The locations of the peaks vary, in accordance with the depth of the pair-potential well. For all water models, it can be seen that the perfect linear hydrogen bond is a very rare occurrence, because the most probable hydrogen-bond energies are 2–3 kcal/mol higher than the minimum for all functions studied. Thus the hydrogen bonds in liquid water are weaker, on the average, than the hydrogen bond in the water dimer. The volume of the configuration space with hydrogen-bond energies near the mini-

Table IV. Characterization of the QCDF's $x_B(\nu)$ and $x_o(R_C)$ for the Various Water Models

	ST2	*MCY*	*TIPS*	*SPC*	*QPEN*	*MCY*	*MCY*
T (K)	283	298	298	298	298	310	323
$\nu^{\max}$	−21.60	−17.40	−16.50	−20.75	−20.00	−17.75	−17.50
$x_B(\nu^{\max})$	0.0388	0.0431	0.0309	0.0277	0.0278	0.0296	0.0287
$\nu_<$	−28.95	−26.15	−25.50	−30.75	−28.50	−27.25	−27.25
$\nu_>$	−8.30	−5.15	−7.50	−7.50	−7.75	−6.00	−5.50
$\nu_<^{1/2}$	−25.10	−20.55	−20.25	−30.75	−23.50	−21.25	−21.25
$\nu_>^{1/2}$	−16.35	−12.85	−12.50	−15.00	−16.25	−13.50	−13.00
$\Delta\nu$	0.35	0.35	0.25	0.25	0.25	0.25	0.25
$R_C^{\max}$	1.60	1.60	1.60	1.55	1.50	1.60	1.60
$x_o(R_C^{\max})$	0.0849	0.0893	0.0450	0.0432	0.0432	0.0896	0.0883
ΔR_C	0.10	0.10	0.05	0.05	0.05	0.10	0.10
$P\ (R_C > 2.4)$	0.0291	0.0141	0.0204	0.0255	0.0245	0.0145	0.0146
$P\ (R_C > 2.6)$	0.0084	0.0027	0.0053	0.0075	0.0069	0.0029	0.0027
$P\ (R_C > 2.8)$	0.0018	0.0003	0.0011	0.0017	0.0015	0.0004	0.0003

Note: The Δp values represent the grid size used to compute the QCDF $x_Q(p)$. The $x_Q(p)$ values given represent the right end point of the grid interval.

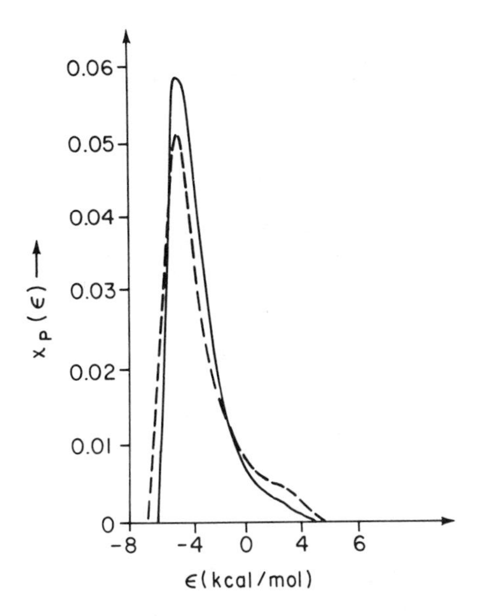

Figure 16. Calculated QCDF for near-neighbor pair-energies $x_P(\varepsilon)$ *versus* ε *for both MCY water at* $T = 25$ *°C (solid line) and ST2 water at 10 °C (broken line).*

mum is very small, and thus linear structures, which are energetically favorable, contribute little to the statistical state of the system.

The near-neighbor dipole correlation function $x_D(\theta)$ is computed for water pairs with center-of-mass distance less than or equal to R_M, the radius of the first solvation shell. This function is defined as

$$x_D(\theta) = \langle \sum_{i<j}^{N} \delta[\theta_{ij}(\mathbf{X}^N) - \theta]\, C_{ij}(\mathbf{X}^N) \rangle / \langle \sum_{i<j}^{N} C_{ij}(\mathbf{X}^N) \rangle \tag{42}$$

where θ is the angle between the HOH bisectors, $\theta_{ij}(\mathbf{X}^N)$ is the angle between the dipoles of the molecules i and j. Furthermore, in studying structural parameters in statistical mechanical context, there are both probabilistic and energetic factors to consider, and the most favorable parameter value energetically may not be the most probable, particularly when it is associated with a relatively small region of configuration space. This circumstance is expressed quantitatively by a comparison of $x_D(\theta)$ and $x_D(\theta)/\sin(\theta)$, the latter quantity being normalized by the volume element of the configuration space with respect to θ. The normalized distribution is thus proportional to the frequency of θ per unit volume of configuration space.

Plots of $x_D(\theta)$ and $x_D(\theta)/\sin(\theta)$ are shown for the MCY water at 25 °C

Table V. Characterization of the QCDF's $x_P(\varepsilon)$ and $x_D(\theta)$ for the Various Water Models

	ST2	*MCY*	*TIPS*	*SPC*	*QPEN*	*MCY*	*MCY*
T (K)	283	298	298	298	298	310	323
ε^{max}	−4.60	−4.40	−4.10	−5.10	−5.00	−4.60	−4.60
$x_P(\varepsilon^{max})$	0.0531	0.0581	0.0522	0.0514	0.0581	0.0581	0.0561
$\varepsilon_<$	−6.60	−5.60	−5.60	−6.50	−6.40	−5.80	−5.80
$\varepsilon_>$	5.40	5.20	$>$2.90	$>$2.90	4.40	5.00	5.20
$\varepsilon_<^{1/2}$	−5.60	−5.20	−5.20	−6.10	−5.80	−5.40	−5.40
$\varepsilon_>^{1/2}$	−2.60	−2.20	−1.80	−2.80	−3.00	−2.40	−2.20
θ^{max}	63	71	61	64	71	81	78
$x_D(\theta^{max})$	0.0103	0.0101	0.0108	0.0108	0.0103	0.0099	0.0101
$\theta_<$	0	1	1	1	1	1	1
$\theta_>$	170	174	172	173	173	173	174
$\theta_<^{1/2}$	14	24	22	22	19	22	23
$\theta_>^{1/2}$	156	125	113	115	117	126	125

Note: The grid sizes used were 0.2 kcal/mol and 1.0° for $x_P(\varepsilon)$ and $x_D(\theta)$, respectively. The $x_Q(p)$ values given represent the right end point of the grid interval.

in Figure 17. The characteristics of the MCY, ST2, TIPS, SPC, and QPEN $x_D(\theta)$ values are also given in Table V. The $x_D(\theta)$ values of the MCY, SPC, and TIPS are very similar, and all are unimodal. The quantity $x_d(\theta)$ for ST2 water was found to be bimodal. This finding is in accord with the appearance of the double minimum found for the ST2 $E_2(\phi)$, a consequence of the tetrahedrally located lone-pair electrons.

The interior of ice Ih contains large void spaces. The extent to which these persist in the liquid can be characterized by means of the QCDF for cavity size, obtained by generating uniformly distributed test points and finding the distance of the closest water molecule to each test point. To avoid arbitrary definition of the molecular radius, the distribution of the distance to the nearest water center of mass is given. Thus the actual cavity size is obtained by deducting the assumed molecular radius of water from the distances shown here. The distribution function $x_o(R_C)$ is defined as

$$x_o(R_c) = \int_V C(\mathbf{X}, R_c, \Delta R)/\Delta R \, d\mathbf{X}/V \tag{43}$$

$$C(\mathbf{X}, R, \Delta R) = \begin{cases} 1, \text{ if } R < \min_i |\mathbf{R}_i - \mathbf{X}| \leq R + \Delta R \\ 0, \text{ otherwise} \end{cases} \tag{44}$$

The $x_o(R_C)$ values for both the MCY and ST2 waters are displayed in Figure 18. The values of $x_o(R_C)$ at selected large R_C values are listed

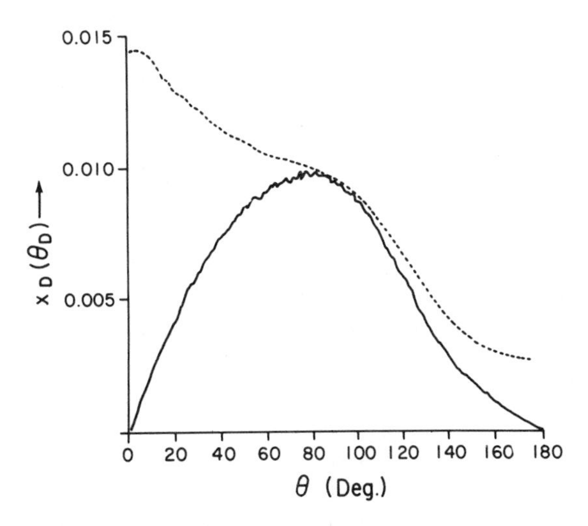

Figure 17. Calculated QCDF for near-neighbor dipole-correlation functions $x_D(\theta)$ *vs.* θ *for MCY water at 25 °C. Key: solid line,* $x_D(\theta)$*; dotted line,* $x_D(\theta)/sin(\theta)$.

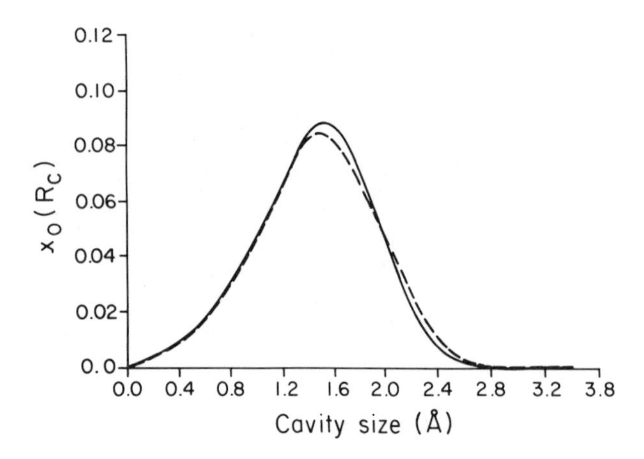

Figure 18. Probability of finding a cavity of radius R_C *for both the MCY water at 25 °C (solid line) and the ST2 water at 10 °C (broken line).*

in Table IV for the MCY, ST2, TIPS, SPC, and QPEN waters. The more structured nature of the ST2 water can be seen from the much higher probability of finding a cavity of molecular size. The $x_o(R_C)$ terms of the SPC and TIPS potentials are rather similar to the ST2 $x_o(R_C)$, showing a frequency of large cavities higher than that from the MCY $x_o(R_C)$. This comes as a consequence of the anomalously high pressure of the MCY water.

The four internal coordinates of the water dimer that are relevant for the description of hydrogen bonding are defined in Figure 19. Here R_{OO} is the inter-oxygen separation, the angle θ_H is the angle between the H–O and O–O bonds and θ_{LP} is the angle between the LP–O and O–O bonds. The angle δ_D is the dihedral angle between the planes H–O–O and LP–O–O. In these definitions, LP is a suitably located "pseudo-atom" on the water molecule, corresponding to the qualitative idea of tetrahedrally oriented lone-pair (LP) orbitals. They were placed in such a way that the LP–O–LP triangle is of the same dimensions as the H–O–H triangle and oriented perpendicular to it. For each water, the atom or pseudo-atom participating in a hydrogen bond with another water was taken as the atom on the donor water closest to the oxygen atom of the acceptor water.

A quantitative geometric definition of the hydrogen bond further requires the specification of cutoff values for each of these parameters. Qualitative notions concerning the hydrogen bond place an upper bound on θ_H and θ_{LP} because it is natural to require that the atoms on one molecule proximal to the oxygen of the other molecule should be H and LP, respectively. The tetrahedral character of the interaction leads to a

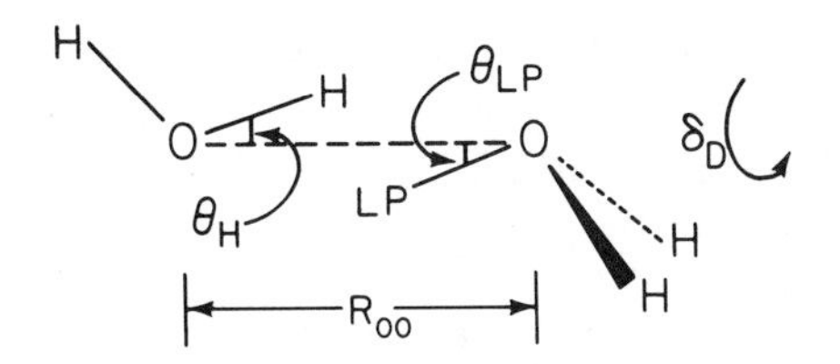

Figure 19. Definitions of the hydrogen-bonding parameters R_{OO}, θ_H, θ_{LP}, *and* δ_D.

"minimal" definition of the hydrogen bond as

$$\begin{aligned} R_{OO} &\leqslant R_{max} \\ \theta_H &\leqslant 70.53° \\ \theta_{LP} &\leqslant 70.53° \\ \delta_D &\leqslant 180.0° \end{aligned} \tag{45}$$

A natural choice for R_{max} is the cutoff value R_M for the previously determined coordination number distribution function, 3.3 Å.

The four parameters described above give rise to the following four hydrogen-bonding QCDF terms:

$$x_H(p) = \langle \sum_{i<j}^{N} \delta[p_{ij}(\mathbf{X}^N) - p] C_{ij}^H(\mathbf{X}^N) \rangle \,/\, \langle \sum_{i<j}^{N} C_{ij}^H(\mathbf{X}^N) \rangle \tag{46}$$

where the four possible choices of p are R_{OO}, θ_H, θ_{LP} and δ_D, and $p_{ij}(\mathbf{X}^N)$ is the value of the parameter p for the pair (i, j). The quantity $C_{ij}^H(\mathbf{X}^N)$ is a counting function for the hydrogen bond; it equals one if the pair (i, j) is hydrogen bonded, and zero otherwise. The volume element of the configuration space associated with the four hydrogen-bond parameters are

$$\begin{aligned} v(R_{OO}) &= 4\pi R_{OO}^2 \\ v(\theta_H) &= \sin(\theta_H) \\ v(\theta_{LP}) &= \sin(\theta_{LP}) \\ v(\delta_D) &= 1 \end{aligned} \tag{47}$$

and are used to obtain the normalized QCDF values $x_H^n(p) = x_H(p)/v(p)$.

Results for the MCY water are presented in Figure 20, using a strong and a weak hydrogen-bond definition. The strong hydrogen bond employs cutoffs of (3.3 Å, 45°, 45°, 90°) while the weak hydrogen bond is defined by the cutoffs (4.0 Å, 70.53°, 70.53°, 90°). The hydrogen-bond QCDF terms using the strong hydrogen-bond definition for the ST2, TIPS, and SPC waters and for waters at higher temperatures are characterized in Table VI. The values quoted for the MCY and ST2 waters differ slightly from the values in Ref. 59, because of minor program modifications. Also, the functions $x_{\mathrm{H}}^{n}(p)$ were normalized to unity.

The peak of $x_{\mathrm{H}}(R_{\mathrm{OO}})$ coincides with the peak of $g_{\mathrm{OO}}(R)$ for all functions studied, as expected, because the dominant intermolecular interaction in liquid water is the hydrogen bonding. The QCDF values $x_{\mathrm{H}}(\theta_{\mathrm{H}})$ and $x_{\mathrm{H}}(\theta_{\mathrm{LP}})$ show the prevalence of bent hydrogen bonds for all functions, but the average degree of bending varies with potential function from 12.5° (ST2) to 37.5° (TIPS). Sceats and Rice estimate the mean square average of θ_{H} and θ_{LP} from the vibrational spectra to be ~20°. The MCY results are quite close to the experimental estimate, while the ST2 pre-

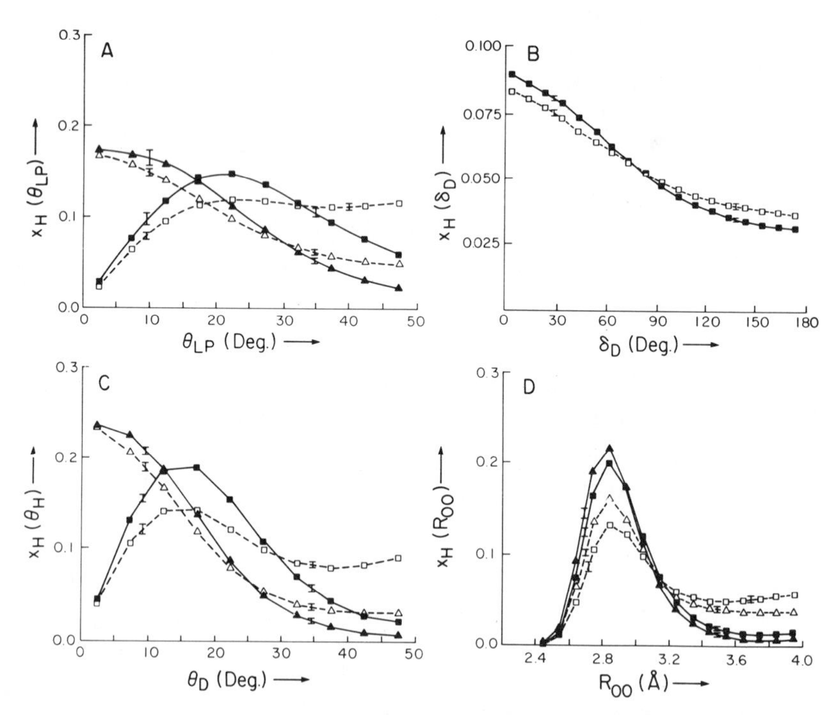

Figure 20. Hydrogen-bonding QCDF values for the parameters R_{OO}, θ_H, θ_{LP}, *and* δ_D, *MCY water at 25 °C. Key:* ■, $x_H(p)$, *strong hydrogen bond;* ▲, $x_H^n(p)$, *strong hydrogen bond;* □, $x_H(p)$, *weak hydrogen bond;* △, $x_H^n(p)$, *weak hydrogen bond.*

Table VI. Hydrogen-Bond QCDF Values for the Various Water Models, Computed by Using the Strong Hydrogen-Bond Definition

	ST2	*MCY*	*TIPS*	*SPC*	*QPEN*	*MCY*	*MCY*
T (K)	283	298	298	298	298	310	323
R_{OO}^{max}	2.85	2.85	2.85	2.75	2.75	2.85	2.85
$x_H(R_{OO}^{max})$	0.264	0.231	0.221	0.240	0.250	0.226	0.222
θ_H^{max}	12.5	17.5	12.5	12.5	12.5	17.5	17.5
$x_H(\theta_H^{max})$	0.213	0.187	0.189	0.220	0.232	0.182	0.179
θ_H^{max}	2.5	2.5	2.5	2.5	2.5	2.5	2.5
$x_H^n(\theta_H^{max})$	0.294	0.229	0.293	0.344	0.329	0.233	0.228
θ_{LP}^{max}	12.5	22.5	32.5	27.5	22.5	22.5	22.5
$x_H(\theta_{LP}^{max})$	0.176	0.138	0.113	0.116	0.131	0.134	0.133
θ_{LP}^{max}	2.5	2.5	2.5	2.5	7.5	2.5	2.5
$x_H^n(\theta_{LP}^{max})$	0.245	0.163	0.125	0.132	0.152	0.164	0.159
δ_D^{max}	5.0	5.0	5.0	5.0	5.0	5.0	5.0
$x_H(\delta_D^{max})$	0.090	0.090	0.069	0.071	0.094	0.088	0.078

Notes: Superscript n represents the normalized QCDF $x_H(p)/v(p)$. Grid intervals are 0.1 Å, 5°, 5°, and 10° for the variables R_{OO}, θ_H, θ_{LP} and δ_D, respectively. Variable values refer to the midpoints of the grid interval. The symbol p^{max} denotes the location of the maximum of both $x_H(p)$ and $x_H^n(p)$.

dicts less bending and the TIPS and SPC functions predict more bending. The normalized curves $x_H^n(\theta_H)$ and $x_H^n(\theta_{LP})$ indicate that the prevalence of bending is a primary consequence of geometric rather than energetic factors. The large values for the location of the maximum of $x_H(\theta_{LP})$ found for the TIPS and SPC potentials, considered together with the difficulty of reproducing the second peak in $g_{OO}(R)$, lead to the conclusion that three-center models provide a significantly weaker representation for the angular correlations than do models with more than three centers.

In conclusion, we present in Figures 21 and 22 two specific water structures arbitrarily chosen from a Monte Carlo calculation on MCY water (*138*). No one structure in a simulation is necessarily representative of the statistical state of the system. The urge to look at structures is nevertheless irresistible, and we present them here along with the appropriate warning of "caveat emptor." There is a notable prevalence of bent hydrogen bonds throughout both structures.

Summary

Computer simulation now has been applied to the calculation of thermodynamic properties and molecular distribution functions for liquid

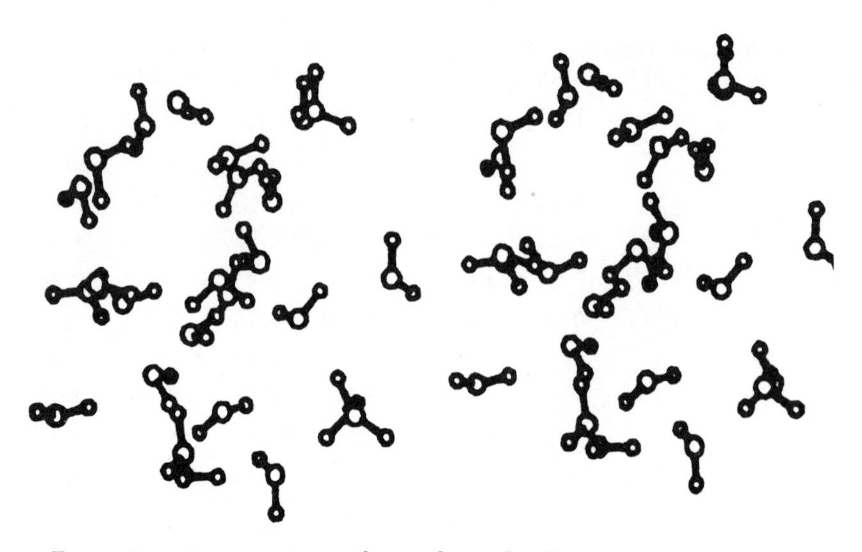

Figure 21. Stereo picture of an arbitrarily chosen water cluster from MCY water simulation.

water, and to the interpretation of the experimental results. The applicability of computer simulation procedures to problems in associated liquids has been demonstrated. System size of O(100) particles under periodic boundary conditions has been found to be sufficient for the representation of the pure liquid case. Stable results can be obtained for O(100) molecules with O(1000 × 10^3) configurations of sampling. Thus

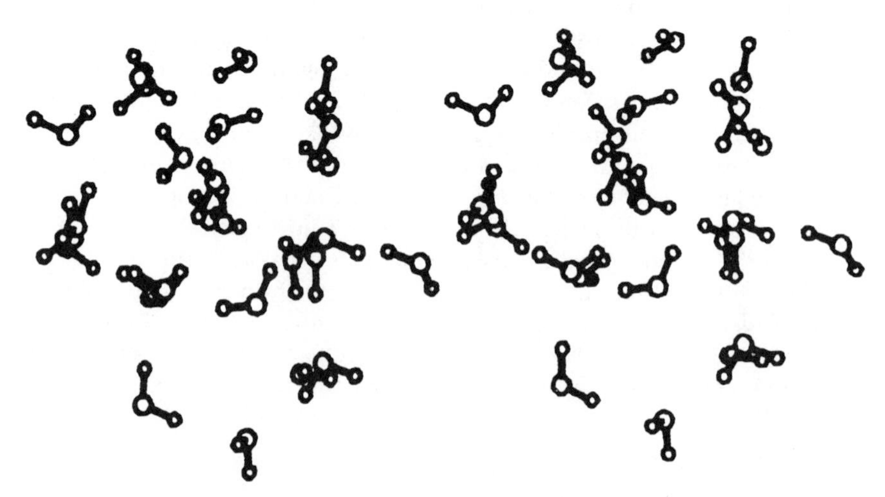

Figure 22. Stereo picture of a second arbitrarily chosen water cluster from MCY water simulation.

the quality of the results obtained depends essentially on the description of configurational energy in the calculation.

Configurational energy in computer simulations of liquid water has been computed for the most part under the assumption of pairwise additivity. The general features of internal energy, heat capacity, and intermolecular separation are accommodated by the better potentials and the results of all agree in the essential description of liquid water structure under ordinary conditions as a dynamically organized fully developed network of molecules interacting via bent hydrogen bonds. The temperature dependence of the liquid structure between 0 and 100 °C is well described at the qualitative level.

The quantitative agreement between calculated and observed results is sensitive to the details of the potential function, and no single potential is found to give uniformly good agreement for all properties. The simulation results for $g_{OO}(R)$ are in reasonable accord with experiment, but significant unanswered questions remain concerning the relationship of calculated and observed $g_{OH}(R)$ and $g_{HH}(R)$ data. Quantum effects are currently an active area of study. Recent extension of simulation techniques to the calculation of larger areas of the phase diagram for water are also under way by Yamamoto and coworkers (*139*) and by Bol (*140*).

Overall, we are encouraged that theoretical calculations on the properties and structure of liquid water have become accessible to study at the ab initio level and that the results, considered in perspective of the known approximations in configurational energy, have created significant new knowledge about the structure. Monte Carlo simulation studies of liquid water have also resulted in significant methodological developments, particularly the introduction of the force-bias method for convergence acceleration of the Metropolis Monte Carlo method. Within the foreseeable future, we feel that the remaining discrepancies between experiment and theory in the problem of liquid water structure can be resolved.

Note Added in Proof

Professor Jorgensen informs us of results based on the potential functions TIPS3 and TIPS4 that supersede the potentials TIPS and TIPS2.

Acknowledgments

This research was supported by National Institutes of Health Grant 5-R01-GM-24914, National Science Foundation Grant CHE-8203501, and Faculty Research Awards of the City University of New York.

In preparing this review we have had numerous interesting and educational discussions with many of the scientists cited herein. We hereby acknowledge our appreciation and gratitude to each of them. Of course, this does not imply that there is unanimous agreement on all or even any of the views expressed herein. We all do agree, however, on the interest and importance of liquid water as a system for physicochemical investigation.

Literature Cited

1. Franks, F., Ed., "Water: A Comprehensive Treatise"; Plenum: New York 1972–1980; Vol. 1–6.
2. Eisenberg, D.; Kauzmann, W. "The Structure and Properties of Water"; Oxford Univ. Press: New York, 1969.
3. Horne, R. A., Ed., "Water and Aqueous Solutions"; Wiley: New York, 1972.
4. Ben-Naim, A. "Water and Aqueous Solutions"; Plenum: New York, 1974.
5. Henderson, L. J. "The Fitness of the Environment"; Macmillan: New York, 1913, xii + 317pp.
6. Kell, G. S. In "Water: A Comprehensive Treatise"; Franks, F., Ed., Plenum: New York; Vol. 1, Chap. 10.
7. Morgan, J.; Warren, B. E. *J. Chem. Phys.* **1938,** *6*, 666.
8. Narten, A. H.; Danford, M. D.; Levy, H. A. *J. Chem. Soc. Disc. Faraday* **1967,** *43*, 97.
9. Narten, A. H.; Levy, H. A. *J. Chem. Phys.* **1971,** *55*, 2263.
10. Palinkas, G.; Kalman, E.; Kovacs, P. *Mol. Phys.* **1977,** *34*, 525.
11. Thiessen, W. E.; Blum, L.; Narten, A. H. *Science* **1982,** *217*, 1033.
12. Soper, A. K.; Silver, R. N. *Phys. Rev. Lett.* **1982,** *49*, 471.
13. Dore, J. C.; *Faraday Disc.* **1978,** *66*, 82.
14. Sceats, M.; Rice, S. A.; Butler, J. E. *J. Chem. Phys.* **1976,** *63*, 5390.
15. Kamb, B. *Science* **1965,** *150*, 205.
16. Roentgen, W. *Ann. Phys.* **1892,** *45*, 91.
17. Levine, S.; Perram, J. W. In "Hydrogen-Bonded Systems," Covington, A. K.; Jones, P., Eds., Taylor and Francis: London, 1968; p.79.
18. Bell, G. M.; *Solid State Phys.* **1972,** *5*, 889.
19. Angell, C. A.; Sare, E. J. *Science* **1970,** *168*, 280.
20. Davis, C. M.; Jarzynski, J. In "Water and Aqueous Solutions"; Horne, R. A., Ed., Wiley: New York, 1972.
21. Hagler, A. T.; Scheraga, H. A.; Nemethy, G. *Ann. NY Acad. Sci.* **1973,** *204*, 51.
22. Luck, W. A. P. In "Water: A Comprehensive Treatise"; Franks, F., Ed., Plenum: New York, Vol. 2, Chap. 4.
23. Walrafen, G. E. In "Water: A Comprehensive Treatise"; Franks, F., Ed., Plenum: New York, Vol. 1, Chap. 5.
24. Scherer, J. R.; Go, M. K.; Kint, S. *J. Phys. Chem.* **1974,** *78*, 1304.
25. Gorbunov, B. Z.; Naberukhin, Yu. I. *J. Struct. Chem.* **1975,** *16*, 703.
26. Kauzmann, W. *Colloques Internationaux du CNRS* **1976,** *246*, 63.
27. Samoilov, O. Ya. *Zh. Fiz. Khim.* **1946,** *20*, 1411.
28. Forslind, E. *Acta Polytech., Scand.* **1952,** *115*, 9.
29. Danford, M. D.; Levy, H. A. *J. Am. Chem. Soc.* **1962,** *84*, 3965.
30. Pauling, L. *J. Am. Chem. Soc.* **1935,** *57*, 2680.
31. Pauling, L. *Science*, **1961,** *134*, 15.
32. Bernal, J. D.; Fowler, R. H. *J. Chem. Phys.* **1933,** *1*, 515.
33. Pople, J. A. *Proc. Roy. Soc., London* **1951,** *205A*, 163.

34. Kirkwood, J. G.; Buff, F. P. *J. Chem. Phys.* **1951,** *19*, 774.
35. Kell, G. S. "Water and Aqueous Solutions"; Horne, R. A., Ed., Plenum: New York, 1974; Chap. 9.
36. Lentz, B. R.; Hagler, A. T.; Scheraga, H. A. *J. Phys. Chem.* **1974,** *78*, 1531.
37. Rahman, A.; Stillinger, F. H. *J. Chem. Phys.* **1971,** *55*, 336.
38. Geiger, A.; Stillinger, F. H.; Rahman, A. *J. Chem. Phys.* **1979,** *70*, 4185.
39. Berendsen, H. J. C. In "Molecular Dynamics and Monte Carlo Calculations on Water"; Report CECAM Workshop: Orsay, France, 1972; p. 63.
40. Berendsen, H. J. C.; Postma, J. P. M.; van Gunsteren, W. F.; Hermans, J. In "Jerusalem Symposia on Quantum Chemistry and Biochemistry"; Pullman, B., Ed., Reidel: Dordrecht, Holland, 1981.
41. Klein, M. L.; McDonald, I. R. *J. Chem. Phys.* **1979,** *71*, 298.
42. Impey, R. W.; Klein, M. L.; McDonald, I. R.; *J. Chem. Phys.* **1981,** *74*, 647.
43. Impey, R. W.; Madden, P. A.; McDonald, I. R. *Mol. Phys.* **1982,** *46*, 513.
44. Rapaport, D. C.; Scheraga, H. A. *Chem. Phys. Lett.* **1981,** *78*, 491.
45. Rossky, P. J.; Hirata, F. Chap. 12 in this book.
46. Stillinger, F. H. *Adv. Chem. Phys.* **1975,** *31*, 1.
47. Wood, D. W. "Water: A Comprehensive Treatise", Franks, E., Ed., Plenum: New York, Vol. 6.
48. Barker, J. A.; Watts, R. O. *Chem. Phys. Lett.* **1969,** *3*, 144.
49. Sarkisov, G. N.; Dashevsky; Malenkov, G. G. *Mol. Phys.* **1974,** *27*, 1249.
50. Clementi, E. "Liquid Water Structure"; Springer Verlag: New York, 1976; and references therein.
51. Owicki, J. Ç.; Scheraga, H. A. *J. Am. Chem. Soc.* **1977,** *99*, 8382.
52. Swaminathan, S.; Beveridge, D. L. *J. Am. Chem. Soc.* **1977,** *99*, 8392.
53. Jorgensen, W. L. *J. Amer. Chem. Soc.* **1979,** *101*, 2016.
54. Jorgensen, W. L. *J. Am. Chem. Soc.* **1981,** *103*, 335.
55. Anderson, H. C. *J. Chem. Phys.* **1980,** *72*, 2384.
56. Mezei, M.; Swaminathan, S.; Beveridge, D. L. *J. Chem. Phys.* **1979,** *71*, 3366.
57. Marchese, F. T.; Mehrotra, P. K.; Beveridge, D. L. *J. Phys. Chem.* **1981,** *85*, 1.
58. Mezei, M.; Beveridge, D. L. *J. Chem. Phys.* **1982,** *76*, 593.
59. Mezei, M.; Beveridge, D. L. *J. Chem. Phys.* **1981,** *74*, 622.
60. Pangali, C. S.; Rao, M.; Berne, B. J. *Chem. Phys. Lett.* **1979,** *55*, 413.
61. Rao, M.; Pangali, C. S.; Berne, B. J. *Mol. Phys.* **1979,** *37*, 1773.
62. Mezei, M. *Chem. Phys. Lett.* **1980,** *74*, 105.
63. Mehrotra, P. K.; Beveridge, D. L. *J. Am. Chem. Soc.* **1978,** *100*, 5705.
64. Beveridge, D. L.; Mezei, M.; Mehrotra, P. K.; Marchese, F. T.; Vasu, T. R.; Ravi-Shanker, G. *Ann. NY Acad. Sci.* **1981,** *367*, 108.
65. Sceats, M. G.; Rice, S. A. *J. Chem. Phys.* **1980,** *72*, 6183.
66. Rice, S. A.; Sceats, M. G. *J. Phys. Chem.* **1981,** *85*, 1108.
67. Stanley, H. E.; Teixeira, J. *J. Chem. Phys.* **1980,** *73*, 3404.
68. Pettitt, M.; Rossky, P. unpublished data.
69. Patey, G. N. unpublished data.
70. Rowlinson, J. S.; *J. Chem. Soc., Trans. Faraday* **1951,** *47*, 120.
71. Bjerrum, N.; *Kgl. Danske Videnskab. Selksab, Skr.* **1951,** *27*, 1.
72. Ben-Naim, A.; Stillinger, F. H. In "Water and Aqueous Solutions"; Horne, R. A.; Ed., Wiley: New York, 1972; p. 295.
73. Ben-Naim, A.; Stillinger, F. H. In "Water and Aqueous Solutions"; Horne, R. A., Ed., Wiley: New York, 1972; p. 295.
74. Stillinger, F. H.; Rahman, A. *J. Chem. Phys.* **1974,** *60*, 1545.

75. Snir, J.; Nemenoff, R. A.; Scheraga, H. A. *J. Phys. Chem.* **1979,** *82*, 2497.
76. Jorgensen, W. L. *J. Chem. Phys.* **1982,** *77*, 4156.
77. Klein, M. unpublished data.
78. Stockmayer, W. H. *J. Chem. Phys.* **1941,** *9*, 398.
79. Campbell, E. S.; Mezei, M. *J. Chem. Phys.* **1977,** *67*, 2338.
80. Campbell, E. S.; Mezei, M. *Mol. Phys.* **1980,** *41*, 883.
81. Barnes, P.; Finney, J. L.; Nicholas, J. D.; Quinn, J. E. *Nature* **1979,** *282*, 459.
82. Rahman, A.; Stillinger, F. H.; Lemberg, H. L. *J. Chem. Phys.* **1975,** *63*, 5223.
83. Stillinger, F. H.; Rahman, A. *J. Chem. Phys.* **1968,** *68*, 666.
84. Watts, R. O. *J. Chem. Phys.* **1983,** *94,* 198.
85. McDonald, I. R.; Klein, M. L. *J. Chem. Phys.* **1978,** *68*, 4865.
86. Reimers, J. R.; Watts, R. O.; Klein, M. L. *Chem. Phys.* **1982,** *64*, 95.
87. Stillinger, F. H.; David, C. W. *J. Chem. Phys.* **1978,** *69*, 1473.
88. Stillinger, F. H.; Weber, T.; David, C. W. private communications.
89. Clementi, E. "Computational Aspects for Large Chemical Systems"; Springer-Verlag: New York, 1980; and references therein.
90. Matsuoka, O.; Clementi, E.; Yoshimine, M. *J. Chem. Phys.* **1976,** *64*, 1351.
91. Jorgensen, W. L. *J. Am. Chem. Soc.* **1979,** *101*, 2011.
92. Curtis, L. A.; Frurip, D. J.; Blander, M. *J. Chem. Phys* **1979,** *71*, 2703.
93. Dyke, T. R.; Muenter, J. S. *J. Chem. Phys.* **1974,** *60*, 2929.
94. Newton, M. D.; Kestner, N. R. *Chem. Phys. Letts.* in print.
95. Diercksen, G. H. F.; Kraemer, W. P.; Roos, B. O. *Theo. Chim. Acta* **1975,** *36*, 249.
96. Morse, M. D.; Rice, S. A. *J. Chem. Phys.* **1982,** *76*, 650.
97. Hankins, D.; Moskowitz, J. W.; Stillinger, F. H.; *J. Chem. Phys.* **1970,** *53*, 4544.
98. Kolos, W.; Ranghino, G.; Novaro, O.; Clementi, E. *Int. J. Quant. Chem.* **1980,** *17*, 429.
99. Del Bene, J.; Pople, J. A. *J. Chem. Phys.* **1970,** *52*, 4858.
100. Lentz, B. R.; Scheraga, H. A. *J. Chem. Phys.* **1973,** *58*, 5296.
101. Newton, M. *J. Am. Chem. Soc.* **1977,** *99*, 611.
102. Kistenmacher, H.; Lie, G. C.; Popkie, H.; Clementi, E. *J. Chem. Phys.* **1974,** *59*, 546.
103. Murrell, J. N. CCP5 Meeting, London, 1981.
104. Campbell, E. S.; Belford, D. unpublished data.
105. Barker, J. A.; Henderson, D. *Rev. Mod. Phys.* **1976,** *48*, 587.
106. Kirkwood, J. G. In "Theory of Liquids"; Alder, B. J., Ed., Gordon and Breach: New York, 1968.
107. Mezei, M.; Swaminathan, S; Beveridge, D. L. *J. Am. Chem. Soc.* **1978,** *100*, 3255.
108. Torrie, G.; Valleau, J. P. *J. Comput. Phys.* **1977,** *23*, 187.
109. Mezei, M. *Mol. Phys.* **1982,** *47*, 1307.
110. Romano, S.; Singer, K. *Mol. Phys.* **1979,** *37*, 1765.
111. Jacucci, G.; Quirke, N. *Mol. Phys.* **1980,** *40*, 1005.
112. Quirke, N.; Jacucci, G. *Mol. Phys.* **1982,** *45*, 823.
113. Metropolis, N.; Rosenbluth, A. W.; Rosenbluth, M. N.; Teller, A. H.; Teller, E. *J. Chem. Phys.* **1953,** *21*, 1087.
114. Patey, G. N.; Valleau, J. P. *J. Chem. Phys.* **1974,** *61*, 534.
115. Blackman, R. B.; Tuckey, J. W. "The Measurement of Power Spectra"; Dover: New York, 1958.

116. Blackman, R. B. "Data Smoothing and Prediction"; Addison-Wesley: Reading, MA, 1965.
117. Wood, W. W. In "Physics of Simple Liquids"; Temperley, H. N. V.; Rowlinson, J. S.; Rushbrooke, G. S., Eds., North-Holland: Amsterdam, 1968.
118. Fishman, G. S. In "Principles of Discrete Event Simulation"; Wiley-Interscience: New York, 1978.
119. Law, A. M.; Kelton, W. D. Tech. Rep. No. 78-56, Univ. WI, 1978.
120. Heidelberger, P.; Welch, P. D. *Comm. of ACM* **1981,** *24*, 233.
121. Peskun, P. H. *Biometrika* **1973,** *60*, 3.
122. Ceperley, D.; Chester, G. V.; Kalos, M. H. *Phys. Rev.* **1977,** *16B*, 3081.
123. Rossky, P. J.; Doll, J. D.; Friedman, H. L. *J. Chem. Phys.* **1978,** *69*, 4628.
124. Kalos, M. H. private communication.
125. Mezei, M.; Mehrotra, P. K.; Beveridge, D. L. unpublished data.
126. Kincaid, R.; Scheraga, H. A. *J. Comp. Chem.* in press.
127. Mehrotra, P. K.; Mezei, M.; Beveridge, D. L. *J. Chem. Phys.* accepted for publication.
128. Mezei, M. *Mol. Phys.*, in press.
129. Marchese, F. T.; Mehrotra, P. K.; Beveridge, D. L. *J. Phys. Chem.* **1982,** *86*, 2592.
130. Mezei, M.; Beveridge, D. L. unpublished data.
131. Powles, J. G.; Rickayzen, G. *Mol. Phys.* **1979,** *38*, 1875.
132. Berens, P. H.; Mackay, D. H. J.; White, G. M.; Wilson, K. R. *J. Chem. Phys.* submitted for publication.
133. Jorgensen, W. L. unpublished data.
134. Impey, R. W.; Madden, P. A.; McDonald, I. R. *Chem. Phys. Lett.* **1982,** *88*, 589.
135. Egelstaff, P. A.; Polo, J. A.; Root, J. H.; Hahn, L. J.; Chen, S. H. *Phys. Rev. Lett.* **1981,** *47*, 1733.
136. Bosio, L.; Chen, S. H.; Teixeira, J. submitted for publication.
137. Egelstaff, P. A.; Root, J. H. *Chem. Phys. Lett.* **1982,** *91*, 96.
138. Beveridge, D. L.; Mezei, M.; Swaminathan, S.; Harrison, S. In "Computer Modeling of Matter"; Lykos, P. G., Ed., ACS SYMPOSIUM SERIES 86, American Chemical Society: Washington, D.C., 1978.
139. Okazaki, K.; Nose, S.; Kataoka, Y.; Yamamoto, T. *J. Chem. Phys.* **1982,** *77*, 5699.
140. Bol, W. *Mol. Phys.* **1982,** *45*, 605.

RECEIVED for review March 15, 1982. ACCEPTED for publication November 18, 1982.

14

Fluid Phase Equilibria at High Pressures: Correlations and Predictions

ULRICH K. DEITERS

University of Bochum, Department of Chemistry, Bochum, Federal Republic of Germany

In fluid mixtures the limits between liquid–gas, liquid–liquid, and gas–gas equilibria are not clearly defined, and transitions occur at high pressures. This is demonstrated for binary mixtures of hydrocarbons with carbon tetrafluoride and for some inert gas mixtures. The experimental results are compared with calculations. These calculations make use of three mathematical relations:

1. *an equation of state, from which the thermodynamic stability criteria are derived*
2. *mixing rules, which refer the characteristic parameters of a mixture to those of the pure substances*
3. *combining rules, which estimate binary interaction parameters from pure substance parameters*

These three relations are discussed; the reliability of the equation of state approach is demonstrated for several equations of state. The correlation of high pressure phase equilibria is shown to be a severe test for the quality of the mixing rules as well as for the usefulness of an equation of state.

As TEMPERATURE AFFECTS THE MOTION OF MOLECULES, pressure affects the average distances of molecules and therefore their average potential energy. Varying the pressure, in addition to varying the temperature, is therefore a second way to control the balance of kinetic and potential energies in a fluid system. This balance is of central importance for static as well as dynamic and transport properties. By varying the

0065-2393/83/0204-0353$06.00/0

pressure in supercritical fluid chromatography (SFC) (*1*) it is possible to affect activity and diffusion coefficients to obtain any intermediate state between gas chromatography and high pressure liquid chromatography. High pressure fluid extraction techniques permit the extraction of delicate organic substances without the need for high temperatures or toxic solvents (*2, 3*). Modern production of oil or natural gas is closely tied up with the understanding of high pressure phase equilibria.

It has long been known that a rigid discrimination between vapor–liquid equilibria and liquid–liquid equilibria cannot be maintained; investigations using high pressure techniques show continuous transitions between these two types of equilibria and eventually to a third type of fluid phase equilibrium, the so-called gas–gas equilibrium (*4*). A typical example of this class is shown in Figure 1. In the phase diagram of the system neon–krypton the critical line originating from the critical point of krypton shows a temperature minimum; for temperatures above this minimum phase separations can be achieved by raising the pressure of the system.

In order to demonstrate the transitions between the three types of fluid phase equilibria, and in order to find correlations between equilibrium type and molecular parameters, several series of fluid systems have been investigated. Examples of this systematic research are studies of noble gas mixtures (*5–7*), methane–alkane mixtures (*4*), and carbon tetrafluoride–alkane mixtures (*8*). The critical lines of the latter systems are shown in Figure 2. With increasing chain length of the alkane component the liquid–liquid equilibrium critical line shifts more and more to higher temperatures, until it "overlaps" with the vapor–liquid equilibrium domain, thus giving rise to gas–gas equilibrium-like phase diagrams.

Thermodynamic Conditions

Most methods for calculating phase equilibria are characterized by the use of activity coefficients by which the properties of a mixture are related to those of a perfect mixture or a perfect gas. These methods, while working very well and efficiently for low pressure vapor–liquid equilibria, are difficult to apply to high pressure phase equilibria because the Poynting corrections become very large and because—with supercritical components—no reference states of the pure component are available. In addition, critical coalescence of phases has to be accounted for. It is therefore advantageous to use one equation of state for the description of all phases of a fluid mixture, thereby assuming that the concept called the continuity of phases (*9*) holds. Because most equations of state are written as functions of molar volume and temperature, it is useful to regard the Helmholtz energy A as the central property of a mixture, from which all other properties may be derived. The Helmholtz energy

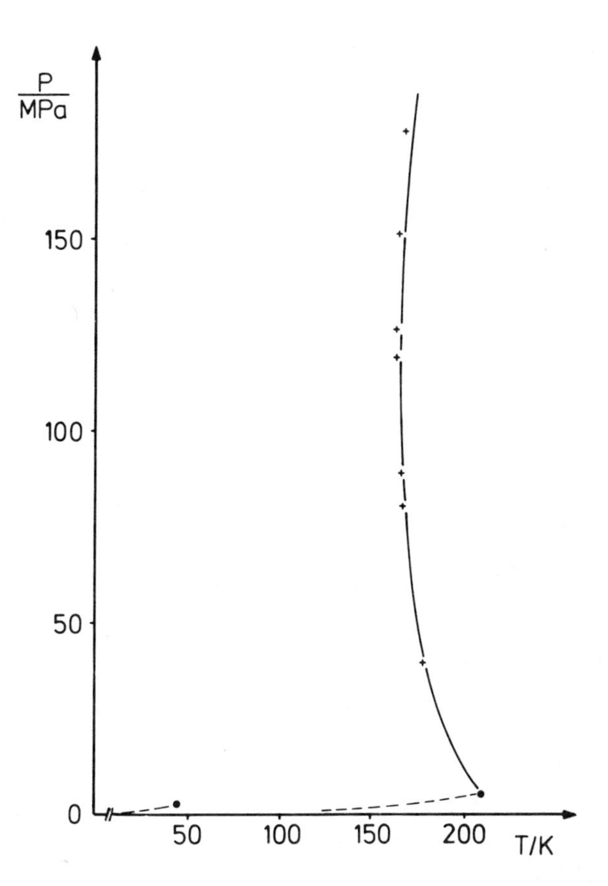

Figure 1. P–T *diagram of the neon–krypton system. Key:* —, *critical line;* ----, *vapor pressure curves;* ●, *critical points of the pure substances; and* +, *experimental binary critical points. (Reproduced with permission from Ref. 18, Copyright 1982, Pergamon Press Ltd.)*

of a binary mixture is given by the following equation (a detailed derivation is given elsewhere (*10–12*))

$$A = n_1 A_1^+(V_m^+, T) + n_2 A_2^+(V_m^+, T) - \int_{V^+}^{V} P\, dV + RT(n_1 \ln x_1 + n_2 \ln x_2) \quad (1)$$

The A_i^+ terms denote the molar Helmholtz energies of the pure substances in the perfect gas state at temperature T and the very large volume V^+. The pressure P is given by the equation of state.

The conditions of phase equilibrium are then represented by the

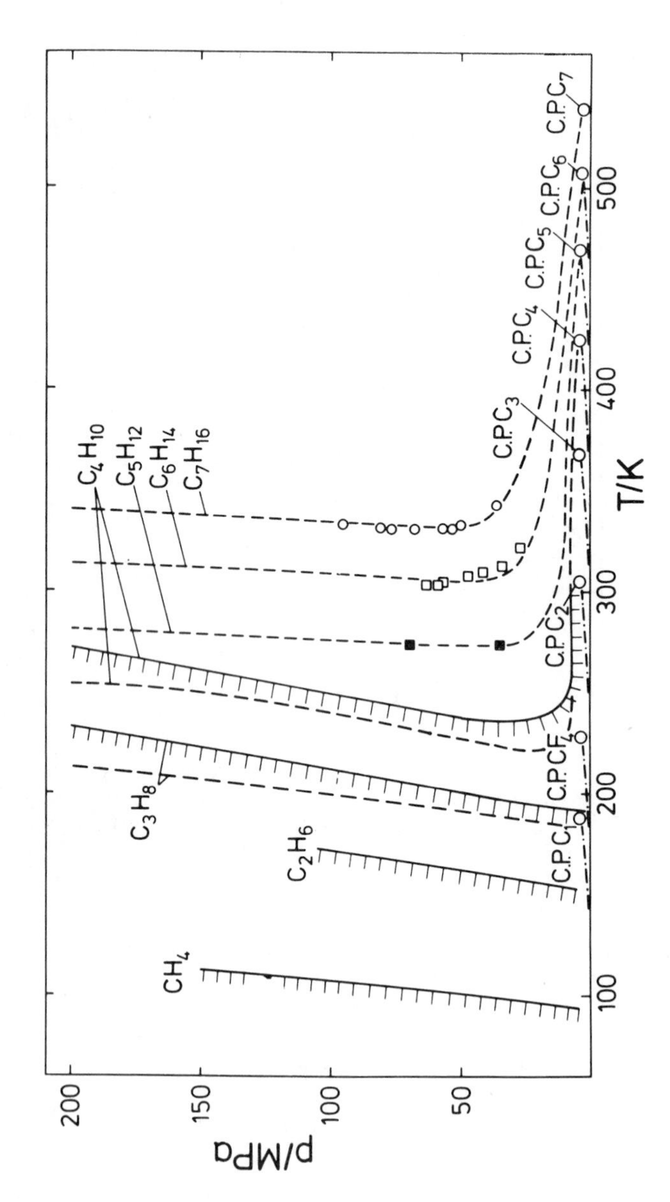

Figure 2. P–T *diagram of carbon tetrafluoride–alkane mixtures. Key:* —, *experimental critical lines;* ·-·-·, *vapor pressure lines;* ○, *critical points of the pure substances; and* ----, □, ○, *experiments and calculations of Mendonça* (35). *(Reproduced with permission from Ref.* 8. *Copyright 1982, Academic Press Inc. (London) Ltd.)*

following system of equations (different phases are denoted by ′ and ″)

$$
\begin{aligned}
T' &= T'' \\
P' &= P'' \\
\mu_i' &= \mu_i'' \qquad i = 1, 2
\end{aligned}
\tag{2}
$$

The chemical potentials, μ_i, are obtained from the thermodynamic relation

$$
\mu_i = \left(\frac{\partial A}{\partial n_i}\right)_{T,V,n_{j\neq i}}
\tag{3}
$$

A binary critical point is defined by

$$
\left(\frac{\partial^2 G_m}{\partial x_1^2}\right)_{T,P} = 0 \qquad \left(\frac{\partial^3 G_m}{\partial x_1^3}\right)_{T,P} = 0
\tag{4}
$$

This can be expressed in terms of A as

$$
\begin{aligned}
&A_{2x}A_{2v} - A_{vx}^2 = 0 \\
&A_{3x}A_{2v}^2 - 3A_{v2x}A_{vx}A_{2v} + 3A_{2vx}A_{vx}^2 - A_{3v}A_{2x}A_{vx} = 0
\end{aligned}
\tag{5}
$$

Here each subscript x or v indicates a partial differentiation of the molar Helmholtz energy A_m with respect to V_m or x_1. Once the dependence of the pressure on temperature, density, and composition is known, the determination of fluid phase equilibrium states or critical properties is accomplished by solving systems of nonlinear equations. Computer algorithms and the conditions under which Equations 5 hold and criteria for the elimination of physically unreasonable solutions are discussed elsewhere (*10*, *13*).

The Equation of State

Any relation that permits the calculation of the pressure from density and temperature may serve as an equation of state. There are, however, several requirements that different *PVT* relations will meet to a different degree. Fluid mixtures are sometimes stable under conditions that cannot be approached by pure fluid substances. Calculation procedures based on the corresponding states principle or on a strictly empirical equation of state may be beyond their working ranges in such cases. Examples are mixtures of noble gases under high pressure. These mixtures can be in a fluid state, while a pure noble gas with the same reduced temperature and density would be a solid (*14*).

Purely empirical equations of state are seldom valid beyond the range of density and temperature to which they have originally been fitted, and therefore cannot always be considered safe for high pressure calculations. On the other hand, there is as yet no purely theoretical equation of state with sufficient precision to cover wide ranges of temperature and density. It is possible, however, to combine the advantages of these two kinds of equations of state in so-called semiempirical equations.

Even the simplest representative of this class, the van der Waals equation, is able to reproduce all of the kinds of fluid phase equilibria (*15*), although agreement with experimental data is qualitative only. Now, a large number of equations of state for the quantitative treatment of phase equilibria has become available, ranging from the rather simple Redlich–Kwong equation to the very sophisticated perturbed chain equation by Beret and Prausnitz (*16*). It is impossible to give here a complete list and evaluation of all equations of state; recently a comparison of cubic equations has been compiled by Peneloux (*17*), and of several noncubic equations by Vera and Prausnitz (*18*).

The following equations of state have been used by us:

1. The equation of Redlich and Kwong (*19*)

$$P = \frac{RT}{V_m - b} - \frac{aT^{-0.5}}{V_m(V_m + b)} \tag{6}$$

2. The equation of Peng and Robinson (*20*)

$$P = \frac{RT}{V_m - b} - \frac{a(T)}{V_m(V_m + b) + b(V_m - b)} \tag{7}$$

3. A new three-parameter equation of state derived from a square-well model of intermolecular interaction (*12, 21*)

$$P = \frac{RT}{V_m}\left[1 + cc_0\frac{4\xi - 2\xi^2}{(1 - \xi)^3}\right] - \frac{Rab}{V_m^2}\tilde{T}_{\text{eff}}(\exp(\tilde{T}_{\text{eff}}^{-1}) - 1)I_1 \tag{8}$$

In this equation ξ denotes a reduced density, $\tilde{T}_{\text{eff}}$ a reduced effective temperature, and I_1 a polynomial representing a first-order perturbation contribution. The repulsive part of Equation 8 is a Carnahan–Starling function with two modifications, which have been discussed in detail

(*21*); c_0 is a constant accounting for deviations from rigid core repulsion, and c is a shape parameter for nonspherical molecules. Equation 8 is shown to be valid even for pressures beyond 100 MPa for several simple molecules and to yield good vapor pressure data. In addition, it can be fitted to real critical compressibility factors (*22*).

For calculations of properties of mixtures we assume that a mixture may be considered as a hypothetical pure substance, with the same equation of state as a pure substance, but with the parameters a and b (and c in Equation 8) depending on composition. The following mixing rules are used:

For the Redlich–Kwong and Peng–Robinson equations

$$\begin{aligned} a &= a_{11}x_1^2 + 2a_{12}x_1x_2 + a_{22}x_2^2 \\ b &= b_{11}x_1^2 + 2b_{12}x_1x_2 + b_{22}x_2^2 \end{aligned} \tag{9}$$

For Equation 8

$$\begin{aligned} a &= x_1a_{11} + x_2a_{22} \\ &\quad + \frac{2x_1s_1q_2\Delta a}{1 + \sqrt{1 + 4q_1q_2\left(\exp\left(-\dfrac{\Delta a s_{12}}{kT}\right) - 1\right)}} \\ b &= b_{11}x_1^2 + 2b_{12}x_1x_2 + b_{22}x_2^2 \\ c &= c_1x_1 + c_2x_2 \end{aligned} \tag{10}$$

In this equation the s_i terms denote contact numbers per molecule, and the q_i, contact fractions. The harmonic mean of the s_i is s_{12}. The formula for a is an extension of the mixing functions for a strictly regular solution according to Guggenheim (*23*); it can be applied to mixtures of spherical molecules of different size.

The contact numbers per molecule are not proportional to the "surface area" of a molecule, but roughly to the power 2.4 of the diameter. This is shown by studies of the maximum number of molecules that can be grouped around a central molecule of given size (*24*).

The mixing rules, Equations 9 and 10, enable us to calculate the parameters of the equations of state for any composition of the mixture under consideration, provided that parameters for the pure substances and for unlike interaction (a_{12}, b_{12}) are available. Pure substance parameters are calculated from the critical data or from vapor pressure data.

The unlike interaction parameters are linked to the pure substance parameters by combining rules:

$$b_{12} = (1 - \zeta) \frac{1}{2} (b_{11} + b_{22}) \tag{11}$$

$$\frac{a_{12}}{s_{12}} = (1 - \vartheta) \sqrt{\frac{a_{11}}{s_1} \frac{a_{22}}{s_2}} \tag{12}$$

(For the Redlich–Kwong and Peng–Robinson equations, $s_1 = s_2 = s_{12} = 1$.)

Parameters ϑ and ζ are adjustable. Their values are calculated from one equilibrium state of the mixture under consideration. If ζ is set to zero, b_{12} becomes the arithmetic mean of b_{11} and b_{22}, and the b-mixing rules in Equation 9 and 10 degenerate to linear mixing rules. Linear mixing rules for b are widely adopted in literature, and are usually sufficient for vapor–liquid equilibrium calculations. The influence of deviations from linearity increases with density, however, and therefore a quadratic mixing rule is useful for calculations of high pressure fluid phase equilibria. It has been shown that the introduction of ζ greatly improves the representation of critical curves using the Redlich–Kwong equation (*10, 11*).

The mixing rules (Equation 10) have been specifically designed for spherical molecules. Mixing theories for more complicated systems have been discussed elsewhere (*25, 26*).

Application to Mixtures

When several equations of state are to be compared, one must realize that most of the modern equations of state are of nearly equal precision when it comes to the calculation of vapor–liquid equilibria for mixtures of simple molecules. As examples we quote an experimental and computational investigation of the systems carbon dioxide–dimethyl ether (*27*) and methane–krypton (*28*). In both cases the Redlich–Kwong and Peng–Robinson equations and Equation 8 have been used to correlate the experimental vapor–liquid equilibria data (up to 5.2 MPa). The interaction parameters had been fitted to one isotherm. All three equations of state are able to represent this isotherm with 0.3% deviation in pressure, which is comparable to the scatter of the experimental data. The other isotherms could be predicted within 1% deviation in pressure by all equations of state; however, Equation 8 is shown to be superior for the prediction of supercritical phase equilibria.

The real test for equations of state is the correlation of high pressure phase equilibria. Figure 3 shows three experimental isotherms of the hydrogen–methane system together with calculated curves (29). Again, all equations of state have been fitted to the middle isotherm (for the Peng–Robinson equations and Equation 8 only ϑ has been adjusted), and the same set of parameters has been used to predict the other isotherms. Hydrogen causes special problems in calculations; because of quantum effects its pure substance parameters had to be extracted from *PVT* data rather than calculated from critical or vapor pressure data.

Again it is evident from Figure 3 that with all equations of state a similar agreement between experimental and computed data is achieved

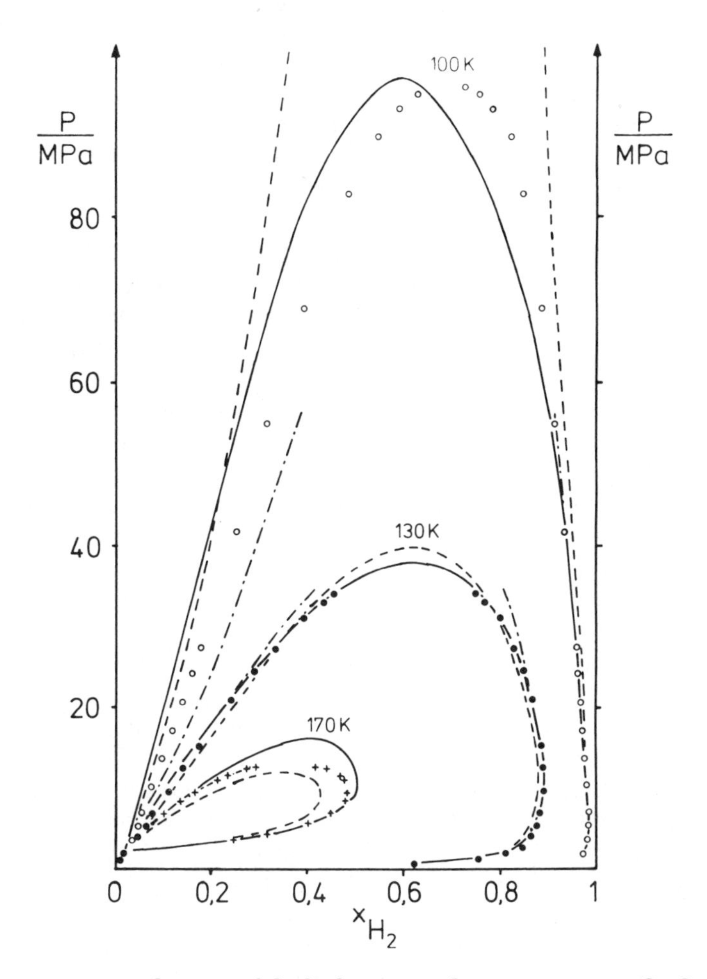

Figure 3. P–x *diagram of the hydrogen–methane system. Key:* ●, ○, +, *experimental data;* —, *calculated with Equation 8;* ----, *calculated with the Redlich–Kwong equation; and* -··-·-, *calculated with the Peng–Robinson equation.*

at moderate pressures, although the Peng–Robinson equation proves to be superior to the Redlich–Kwong equation. At high pressures, only Equation 8 leads to an agreement with the experimental data. The Redlich–Kwong equation produces very large errors in pressure, and the Peng–Robinson equation has no solutions beyond 60 MPa.

Similar results have been obtained for the systems hydrogen–carbon monoxide and hydrogen–carbon dioxide (*30, 31*).

The gas–gas equilibrium in the system neon–krypton (*5*) is represented quite well by Equation 8. Calculated and experimental critical points agree very well (Figure 1). Although the interaction parameters for the calculation had been fitted to an equilibrium state at 178.15 K and 20 MPa, the critical double point is predicted correctly within 3 K. Figure 4 shows three isotherms of this system. The agreement with the experimental data is very good (*18*). The dashed curve in this diagram had been calculated without size and nonrandomness corrections to the mixing rule; the importance of these refinements to the mixing rule is evident.

An especially interesting way of checking the validity of the equation of state approach is the study of series of mixtures, e.g., carbon tetrafluoride with a series of homologous alkanes (Figure 2). In this case, one expects a correlation for the interaction parameters with the chain length of the alkane. The first four critical curves in Figure 2 have therefore also been calculated with the Redlich–Kwong equation (Equation 6). The agreement of calculated and experimental data is very good; the curves virtually coincide for large pressure ranges (*8, 11*). It must be noted, however, that the ϑ parameter for the RK calculation varies from 0.05 to the rather large value of 0.21 in the carbon tetrafluoride–alkane series, whereas it varies for Equation 8 only from 0.02 to 0.05.

In the mixtures of carbon tetrafluoride with alkanes, a continuous transition from liquid–liquid equilibrium to a gas–gas equilibrium-like phase diagram takes place. A similar transition is found for carbon dioxide–alkane mixtures (*4*). Transitions from gas–gas equilibrium of the second kind to gas–gas equilibrium of the first kind have been reported for water–alkane mixtures (*32*) or for helium–noble gas mixtures (*33, 34*).

From a theoretical point of view, the use of adjustable binary interaction parameters might be considered as a weak point of the equation of state approach. There is indeed the danger that the adjustable parameters will not only take care of deviations from the Lorentz–Berthelot combining rules (Equations 11 and 12), but also absorb inadequacies of the mixing rules, the equation of state, and the thermodynamic assumptions inherent in Equation 1. This compensation effect, however, makes itself felt in physically unreasonable values of the adjustable parameters. In addition, the investigation of series of mixtures (as men-

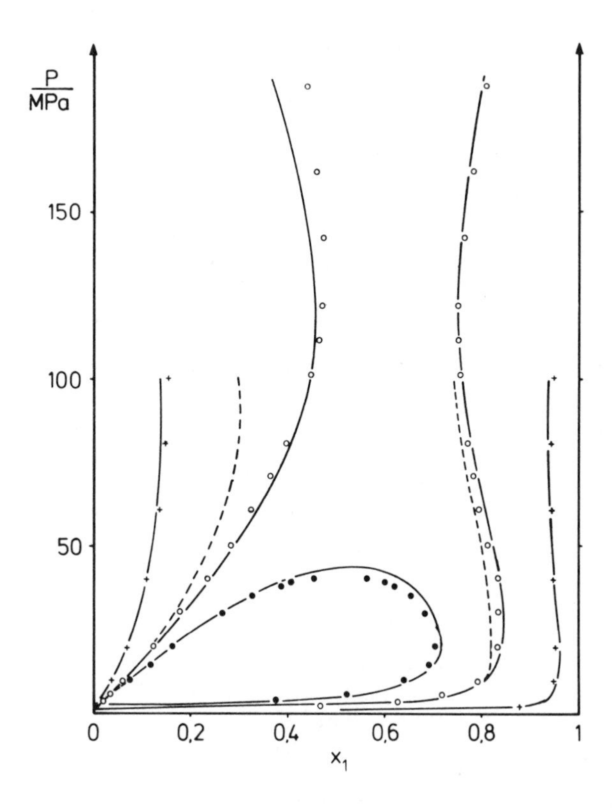

Figure 4. P–x *diagram of the neon–krypton system. Key:* —, *calculated with nonrandom mixing rules;* ----, *calculated with random mixing rules;* +, *experimental data at 133.16 K;* ○, *at 163.15 K; and* ●, *at 178.15 K. (Reproduced with permission from Ref. 18. Copyright 1982, Pergamon Press Ltd.)*

tioned earlier) will not lead to useful correlations of these parameters. It is one of the advantages of Equation 8 over the Redlich–Kwong equation that its parameters are less prone to unrealistic variations. Furthermore, the adjustable parameters of Equation 8 are less temperature-dependent than those of the Redlich–Kwong equation.

Equations of state must be regarded as useful tools for the calculation of high pressure phase equilibria. In spite of many improvements of the experimental techniques, calculations with equations of state have kept up with the precision of the experimental data and with the recent efforts to extend our knowledge of intermolecular interactions and statistical mechanics. It is to be hoped that they will keep up in the future.

Acknowledgment

The author expresses his gratitude to G. M. Schneider for friendly support of this work and many helpful discussions.

Literature Cited

1. van Wasen, U.; Swaid, I.; Schneider, G. M. *Angew. Chem. Int. Ed. Engl.* **1980,** *19*, 575.
2. Schneider, G. M. *Angew. Chem. Int. Ed. Engl.* **1980,** *17*, 701.
3. Schneider, G. M.; Stahl, E.; Wilke, G. "Extraction with Supercritical Gases"; Verlag Chemie: Weinheim, 1980;
4. Schneider, G. M. In "A Specialist Periodical Report: Chemical Thermodynamics"; McGlashan, M. L. Ed., Chemical Society: London, 1978; Vol. 2, Chap. 4.
5. Trappeniers, N. J.; Schouten, J. A. *Physica* **1974,** *73*, 527.
6. De Swaan Arons, J.; Diepen, G. A. M. *J. Chem. Phys.* **1966,** *44*, 23.
7. Streett, W. B.; Erickson, A. L. *Phys. Earth Planet. Inter.* **1972,** *5*, 357.
8. Jeschke, P.; Schneider, G. M. *J. Chem. Thermodyn.* **1982,** *14*, 547.
9. van der Waals, J. D.; Kohnstamm, P. "Lehrbuch der Thermodynamik"; Barth, Leipzig 1912 Vol. 2, 10ff.
10. Deiters, U. Thesis, Bochum, 1976.
11. Deiters, U.; Schneider, G. M. *Ber. Bunsenges. Phys. Chem.* **1976,** *80*, 1316.
12. Deiters, U. Ph.D. Thesis, Bochum, 1979.
13. Hicks, C. P.; Young, C. L. *J. Chem. Soc., Faraday Trans. II* **1977,** *73*, 597.
14. Deiters, U. *Chem. Eng. Sci.* **1982,** *37*, 855.
15. van Konynenburg, P. H. Ph.D. Thesis, Univ: CA, Los Angeles, 1968.
16. Beret, S.; Prausnitz, J. M. *AIChE J.* **1975,** *21*, 1123.
17. Péneloux, A.; Rauzy, E. *Fluid Phase Equil.* **1982,** *8*, 3.
18. Vera, J. H.; Prausnitz, J. M. *Chem. Eng. J.* **1972,** *3*, 1.
19. Redlich, O.; Kwong, J. N. S. *J. Chem. Rev.* **1949,** *44*, 233.
20. Peng, D.-Y.; Robinson, D. B. *Ind. Eng. Chem. Fund.* **1976,** *15*, 59.
21. Deiters, U. *Chem. Eng. Sci.* **1981,** *36*, 1139.
22. Deiters, U. *Chem. Eng. Sci.* **1981,** *36*, 1147.
23. Guggenheim, E. A. "Mixtures"; Clarendon Press: Oxford, 1952; Chap. 4,
24. Deiters, U. K. *Fluid Phase Equil.* **1982,** *8*, 123.
25. Donohue, M. D.; Prausnitz, J. M. *AIChE J.* **1978,** *24*, 849.
26. Gmehling, J.; Liu, D. D.; Prausnitz, J. M. *Chem. Eng. Sci.* **1979,** *34*, 951.
27. Tsang, C. Y.; Streett, W. B. *J. Chem. Eng. Data* **1981,** *26*, 155.
28. Calado, J. C. G.; Deiters, U.; Streett, W. B. *J. Chem. Soc., Faraday Trans. I* **1981,** *77*, 2503.
29. Tsang, C. Y.; Clancy, P.; Calado, J. C. G.; Streett, W. B. *Chem. Eng. Commun.* **1980,** *6*, 365.
30. Tsang, C. Y.; Streett, W. B. *Chem. Eng. Sci.* **1981,** *36*, 993.
31. Tsang, C. Y.; Streett, W. B. *Fluid Phase Equil.* **1981,** *6*, 261.
32. de Loos, Th. W.; Wijen, A. J. M.; Diepen, G. A. M. *J. Chem. Thermodyn.* **1980,** *12*, 193.
33. Streett, W. B. *J. Chem. Soc., Faraday Trans.* **1969,** *65*, 696.
34. Tsiklis, D. S.; Rott, L. A. *Usp. Khim.* **1967,** *36*, 869.
35. Mendonça, J. M. M. Ph.D. Thesis, London, 1979.

RECEIVED for review January 27, 1982. ACCEPTED for publication September 28, 1982.

15

Thermodynamics of Molecular Fluids and Their Mixtures

Y. SINGH[1] and K. P. SHUKLA[2]
Banaras Hindu University, Department of Physics, Varanasi-221005, India

The problem of calculating the equilibrium properties of fluids having nonspherical rigid molecules of arbitrary symmetry and their mixtures is studied. A perturbation expansion in which all tensor interactions (anisotropic pair and three-body nonadditive interactions) are taken as a perturbation of the central pair potential is discussed. Theoretical expressions are given and calculations made for the virial coefficients of the equation of state, Helmholtz free energy, configurational energy, entropy, and pressure. These are compared with experimental data for nitrogen, oxygen, carbon monoxide, carbon dioxide, and methane. The theoretical predictions for binary mixtures are compared with experimental results for argon–nitrogen, argon–oxygen, argon–carbon monoxide, nitrogen–oxygen, and nitrogen–carbon monoxide at the zero pressure isobar and temperature equal to 83.82 K. Agreement with experiment is very satisfactory for all of these systems.

THE POTENTIAL ENERGY of N interacting molecules has nonadditive interaction terms in addition to the sum of pair potentials. A substantial improvement in the quantitative understanding of the behavior of real, dense fluids can be made if the nonadditive interactions are included in theoretical calculations. For atomic fluids, it has been shown that the long-range triple-dipole three-body dispersion (Axilrod–Teller) interaction (*1*) contributes substantially to the thermodynamic properties of fluids (*2*, *3*) and to low-order cluster integrals appearing in the density expansion of the radial distribution function (*4–6*). Barker, Henderson, and Smith (*3*) have found that the calculated pressure, internal energies, and critical constants of dense gaseous argon are in reasonable agreement

[1] Current address: University of Illinois, School of Chemical Sciences, Urbana, IL 61801.
[2] Current address: University of Duisburg, Fachgebiet Thermodynamik, Duisberg, Federal Republic of Germany.

0065-2393/83/0204-0365$08.25/0

with experiment provided that the three-body contributions are included. Goldman (*7*) has found that the effect of the triple-dipole three-body dispersion interaction on the Henry's law constant for krypton in argon is so large that it cannot possibly be treated accurately by any effective pair potential.

The effect of the three-body nonadditive dispersion interactions on the structure factor of liquid rare gases has, however, been found negligibly small (*8*, *9*).

The extent to which the three-body nonadditive interaction and the anisotropy in pair interaction contribute to the equilibrium properties of dense polyatomic fluids has been a subject of considerable interest in recent years (*10–12*). Singh and Singh (*13*, *14*) have found that both the dielectric and equation-of-state virial coefficients and the dilute-gas viscosity of polyatomic fluids can be explained satisfactorily with one set of force parameters, provided that the molecular asymmetry and the nonadditivity of the (three-body) interactions are explicitly taken into account in calculating the virial coefficients.

The calculation of thermodynamic properties and correlation functions of molecular fluids in the presence of three-body forces is relatively difficult. The solutions of integral equations such as the hypernetted-chain (HNC) equation, the Percus–Yevick (PY) equation, the mean-spherical approximation (MSA), or the optimized random phase approximation (ORPA) are difficult to obtain even in the absence of three-body forces. This is because the solution of these equations involves, even for axially symmetric rigid molecules, repetitive sixfold, numerical integrations (more in the presence of three-body forces) and requires the calculation of the full anisotropic pair correlation function, a procedure that is numerically very complicated but that can be accomplished by a spherical harmonic expansion (*see*, e.g., *15*). Another method that can be applied to molecular fluids with relative ease is a use of a perturbation scheme (*11*, *12*, *16–26*) in which quantities of interest are obtained by applying a perturbation correction to the corresponding quantities of some reference system.

In this chapter we describe a method for computing the thermodynamic properties of fluids and their mixtures, assuming rigid nonspherical molecules of arbitrary symmetry. The procedure is based upon a perturbation expansion (*11*, *12*, *25*, *26*) in which all tensor interactions (angle-dependent pair and triplet potentials) are taken as a perturbation of the central pair potential. The procedure applies most directly to molecules with electric multipoles embedded in a core that deviates only marginally from spherical symmetry, so that its nonsphericity can be treated as a perturbation of the spherical core. In this type of perturbation expansion, in which a spherically symmetric reference potential (SSRP) is used, the series is summed up by using the Padé approximant of Stell et al. (*24*).

The Padé approximant has been found to yield results for systems with dipolar and/or quadrupolar forces that are in good agreement with machine simulation results even when the electric multipole moments are very large (*27*, *28*). Even for anisotropic overlap forces, the Padé approximant has been found to work well for not too elongated molecules (*29*).

The details of the pair and triplet (three-body nonadditive) interactions used in the calculation are given in the first section below. In the following section we review a thermodynamic perturbation theory of a multi-component mixture (*12*). Molecular asymmetry of very general types arising from permanent electric multipole moments, induced dipole moments, and anisotropic dispersion forces are considered. Permanent moment interactions involving dipoles and quadrupoles are treated through the third-order of the perturbation, while all other anisotropic interactions including three-body nonadditive interactions are treated to the second-order term only. The contribution of the higher-order terms in the expansion, arising because of the first few permanent moment interactions, are approximated by means of a simple [1, 0] Padé extrapolation procedure.

An application of the theory to the description of the equilibrium properties of some nonpolar fluids is then presented. In the last section we apply the theory to predict the properties of binary mixtures.

Molecular Interactions

We consider a fluid mixture consisting of t components contained in a volume V at temperature T. The system is described by a potential function that depends on the orientation and the relative center of mass coordinates of molecules, but that is independent of rotational momentums and internal vibrational states. We approximate the total potential energy of interaction of the system as a sum of the interaction energies of isolated pairs and triplets.

The pair potential energy, $\Phi_{ab}(\mathbf{X}_{ai}, \mathbf{X}_{bj})$, is assumed to consist of a spherically symmetric component $U_{ab}(\boldsymbol{r}_{ai}, \boldsymbol{r}_{bj})$ plus a contribution $V_{ab}(\mathbf{X}_{ai}, \mathbf{X}_{bj})$ due to the nonsphericity of the molecular charge distribution. That is

$$\Phi_{ab}(\mathbf{X}_{ai}, \mathbf{X}_{bj}) = U_{ab}(\boldsymbol{r}_{ai}, \boldsymbol{r}_{bj}) + V_{ab}(\mathbf{X}_{ai}, \mathbf{X}_{bj}) \qquad (1)$$

where $\mathbf{X}_{ai} \equiv (\boldsymbol{r}_{ai}, \omega_{ai})$ is a vector describing both the location $\boldsymbol{r}_{ai}$ of the center of mass and the orientation ω_{ai} of the ith molecule of species a. For neutral molecules, V_{ab} is conveniently divided into terms representing the classical electrostatic (permanent and induction) interaction, the anisotropy of the quantum mechanical dispersion, and overlap forces.

Thus

$$V_{ab} = V_{ab}^{perm} + V_{ab}^{in} + V_{ab}^{dis} + V_{ab}^{sh} \tag{2}$$

where (*10, 30, 31, 32*)

$$V_{ab}^{perm} = \sum_{s=1}^{\infty} \sum_{l=1}^{\infty} a_{sl} \mathbf{M}_{ai}^{(s)}[s] \mathbf{T}_{ai\,bj}^{(s+l)}[l] \mathbf{M}_{bj}^{(l)} \tag{3}$$

$$\begin{aligned} V_{ab}^{in} = & -\frac{1}{2} \sum_{s=1}^{\infty} \sum_{l=1}^{\infty} a_{sl} [\mathbf{M}_{ai}^{(s)}[s] \mathbf{T}_{ai\,bj}^{(s+1)}[1] \hat{\boldsymbol{\alpha}}_{bj}[1] \\ & \cdot \mathbf{T}_{bj\,ai}^{(l+1)}[l] \mathbf{M}_{ai}^{(l)} \\ & + \mathbf{M}_{bj}^{(s)}[s] \mathbf{T}_{bj\,ai}^{(s+1)}[1] \hat{\boldsymbol{\alpha}}_{ai}[1] \mathbf{T}_{ai\,bj}^{(l+1)}[l] \mathbf{M}_{bj}^{(l)}] \end{aligned} \tag{4}$$

and

$$V_{ab}^{dis} = -\frac{2}{3} A_{ab} [(\hat{\boldsymbol{\alpha}}_{ai} \cdot \mathbf{T}_{ai\,bj}^{(2)}) : (\hat{\boldsymbol{\alpha}}_{bj} \cdot \mathbf{T}_{ai\,bj}^{(2)}) - \bar{\alpha}_a \bar{\alpha}_b \mathbf{T}_{ai\,bj}^{(2)} : \mathbf{T}_{ai\,bj}^{(2)}] \tag{5}$$

with

$$\begin{aligned} a_{sl} &= \frac{(-1)^{s+1}\, 2^{s+l}\, s!\, l!}{(2s)!(2l)!} \\ \bar{\alpha}_a &= \frac{1}{3}\, Tr\, \hat{\alpha}_{ai} \end{aligned} \tag{6}$$

and

$$A_{ab} = \frac{\varepsilon_{ab}\, \sigma_{ab}^6}{\bar{\alpha}_a\, \bar{\alpha}_b}$$

$\hat{\boldsymbol{\alpha}}_{ai}$ is electric dipole polarizability tensor of molecule ai, $\mathbf{M}_{ai}^{(s)}$ is the s rank molecular electric multipole tensor, and $\mathbf{T}_{ai\,bj}^{(s+l)}$ is the $s+l$ rank interaction tensor. The notations $[s]$ and $[l]$ indicate, respectively, s-fold and l-fold contraction of the tensors. In Equations 4 and 5, the contributions arising from the tensor describing induced moments higher than dipole moments and hyperpolarizabilities are ignored. Although the neglect of higher order induction terms may lead to errors in some cases (*33*) for the systems considered in the present chapter, this is an excellent approximation.

We use the Pople expression (*16*) for V^{sh}. This is an r^{-12} form with the first nonvanishing spherical harmonic as angular dependence. For interaction between symmetric linear molecules (e.g., nitrogen and carbon dioxide) this takes the form

$$V_{ab}^{sh}(\mathbf{X}_{ai}, \mathbf{X}_{bj}) = 4D_{ab}\varepsilon_{ab}\left(\frac{\sigma_{ab}}{r_{ai\ bj}}\right)^{12} [\,3\cos^2\theta_{ai} + 3\cos^2\theta_{bj} - 2] \quad (7)$$

where D_{ab} is a dimensionless shape parameter, and θ_{ai} and θ_{bj} are the angles that determine the orientations of molecules *ai* and *bj* with respect to the line joining the centers of the molecules. Positive D_{ab} values correspond to prolate molecules, while negative values correspond to oblate molecules.

In an assembly of molecules, the three-body nonadditive interactions arise from two sources. First, there is potential nonadditivity arising from the dispersion and short-range three-body exchange forces. Second, the potential nonadditivity arises from the classical electric induction interaction between asymmetric molecules. From the theoretical point of view, very little is known with certainty about short-range overlap many-body interactions, except for the work of O'Shea and Meath (*34*) on hydrogen atoms. However, there is very strong experimental evidence (*see*, e.g. *35–37*) that their effects on thermodynamic properties of solid and liquid rare gases are very small, at least up to pressure of about 20 kbar. We expect similar behavior for the molecular liquids of interest here, and therefore we neglect the three-body overlap interactions in the present treatment. Thus (*10, 32*)

$$V_{abc} = V_{abc}^{in} + V_{abc}^{dis} \quad (8)$$

where

$$V_{abc}^{in} = -\frac{1}{2}\sum_{i_1 i_2 i_3}\sum_{s=1}^{\infty}\sum_{l=1}^{\infty} a_{sl}\, \mathbf{M}_{i_1}^{(s)}[s]\mathbf{T}_{i_1 i_2}^{(s+1)}[1]\hat{\boldsymbol{\alpha}}_{i_2}[1]\mathbf{T}_{i_1 i_3}^{(l+1)}[l]M_{i_3}^{(l)} \quad (9)$$

where ($i_1 \neq i_2 \neq i_3$) and

$$V_{abc}^{dis} = \frac{\nu_{abc}}{3\bar{\alpha}_a\bar{\alpha}_b\bar{\alpha}_c}\,(\hat{\boldsymbol{\alpha}}_{ai}[1]\mathbf{T}_{ai\ bj}^{(2)}[1]\hat{\boldsymbol{\alpha}}_{bj})[2](\mathbf{T}_{bj\ ck}^{(2)}[1]\hat{\boldsymbol{\alpha}}_{ck}[1]\mathbf{T}_{ck\ ai}^{(2)}) \quad (10)$$

In Equation 9, each index in the three-fold sum takes the three values *ai*, *bj*, and *ck*, and there is the restriction that $i_1 \neq i_2 \neq i_3$. Thus there are only six terms in the sum corresponding to interactions due to the simultaneous presence of all the three molecules *ai*, *bj*, and *ck*. The term

ν_{abc} appearing in Equation 10 can be written as

$$\nu_{abc} = \frac{2R_aR_bR_c(R_a + R_b + R_c)}{(R_a + R_b)(R_b + R_c)(R_c + R_a)} \tag{11}$$

with

$$R_a^{-1} = (\psi_{ab}\bar{\alpha}_c)^{-1} + (\psi_{ac}\bar{\alpha}_b)^{-1} - (\psi_{bc}\bar{\alpha}_a)^{-1}$$

where ψ's are defined as $\psi_{ab} = 4\varepsilon_{ab}\sigma_{ab}^6$ for Lennard–Jones 6–12 potential. In the special case of a one-component fluid $\nu = \frac{3}{4}\bar{\alpha}\,(4\varepsilon\sigma^6)$.

Perturbation Expansion

The configurational Helmholtz free energy per particle is given as

$$f = -\frac{kT}{N} \ln Z_N(V, T) \tag{12a}$$

where

$$Z_N(V, T) = \int\int \exp[-\beta U_N(\mathbf{X}_1, \ldots, \mathbf{X}_N)] \prod_{i=1}^{N} d\mathbf{X}_i \tag{12b}$$

and $\beta = (kT)^{-1}$. The volume element $d\mathbf{X}_i$ in Equation 12b is equivalent to $dr_i\, d\omega_i$ where $dr = dr_x dr_y dr_z$ and $d\omega = (8\pi^2)^{-1} \sin\theta d\theta\, d\phi dX$. Each integration is extended to the sample volume V for positions and to the usual domains $0 \leq \theta \leq \pi$, $0 \leq \phi \leq 2\pi$, and $0 \leq X \leq 2\pi$ for angles.

In developing a perturbation theory for simple molecular liquids and their mixtures, we begin by writing the total potential energy of interactions as a sum of two parts

$$U_N(\mathbf{X}_1, \ldots, \mathbf{X}_N) = U_N^{(0)}(\mathbf{X}_1, \ldots, \mathbf{X}_N) + \lambda U_N^{(p)}(\mathbf{X}_1, \ldots, \mathbf{X}_N) \tag{13}$$

where λ is a perturbation parameter. For a mixture

$$U_N^{(0)}(\mathbf{X}_1, \ldots, \mathbf{X}_N) = \frac{1}{2} \sum_{a,b}^{t} \sum_{i}^{N_a} \sum_{\neq j}^{N_b} U_{ab}(\boldsymbol{r}_{ai}, \boldsymbol{r}_{bj}) \tag{14a}$$

and

$$U_N^{(p)}(\mathbf{X}_1, \ldots, \mathbf{X}_N) = \frac{1}{2} \sum_{a,b}^{t} \sum_{i}^{N_a} \sum_{\neq j}^{N_b} [V_{ab}(\mathbf{X}_{ai}, \mathbf{X}_{bj}) + \frac{1}{3} \sum_{c}^{t} \sum_{\substack{k \\ (\neq i, j)}}^{N_c} V_{abc}(\mathbf{X}_{ai}, \mathbf{X}_{bj}, \mathbf{X}_{ck})] \tag{14b}$$

When Equations 13 and 14 are substituted in Equation 12, and the right hand side is expanded in ascending powers of λ, one gets (*12*, *25*)

$$f = f^{(0)} + \sum_{n=1}^{\infty} f^{(n)} \tag{15}$$

where

$$f^{(n)} = \frac{\rho}{2n} \sum_{a,b}^{t} x_a x_b f_{ab}^{(n)} + \frac{\rho^2}{6n} \sum_{a,b,c}^{t} x_a x_b x_c f_{abc}^{(n)} \tag{16}$$

Here $x_i = N_i/N$, $\rho = N/V$, and

$$f_{ab}^{(n)} = \int \langle V_{ab}(\mathbf{r}_{a1\ b2};\ \omega_{a1},\ \omega_{b2}) g_{ab}^{(n-1)}(\mathbf{r}_{a1\ b2};\ \omega_{a1},\ \omega_{b2})\rangle_{\omega_{a1}\omega_b} d\mathbf{r}_{a1\ b2} \tag{17}$$

$$f_{abc}^{(n)} = \iint \langle V_{abc}(\mathbf{r}_{a1\ b2},\ \mathbf{r}_{b2\ c3},\ \mathbf{r}_{c3\ a1};\ \omega_{a1},\ \omega_{b2},\ \omega_{c3}) g_{abc}^{(n-1)}$$
$$(\mathbf{r}_{a1\ b2},\ \mathbf{r}_{b2\ c3},\ \mathbf{r}_{c3\ a1};\ \omega_{a1},\ \omega_{b2},\ \omega_{c3})\rangle_{\omega_{a1}\omega_{b2}\omega_{c3}} d\mathbf{r}_{b2} d\mathbf{r}_{c3} \tag{18}$$

Here and elsewhere in this chapter, the notation $\langle \ldots \rangle_{\omega_i}$ indicates an unweighted averaging over the orientations of molecule i; the terms $g_{ab}^{(n)}$ and $g_{abc}^{(n)}$ are nth perturbation terms in the expansion of pair and triplet correlation functions, respectively. For pure fluids, the first few terms in the expansion of pair and triplet correlation functions in the presence of three-body forces are given by Singh (25). These expressions have been generalized for mixtures by Shukla and Singh (*12*). Gray, Gubbins, and Twu (*38*) have given expressions for $g_{ab}^{(1)}$ and $g_{ab}^{(2)}$ in the absence of three-body forces.

It may be noted that while Equation 18 contains the contributions arising exclusively from triplet potential, Equation 17 contains contributions arising from both pair and triplet potentials. The latter contribution arises because of the effect of triplet potentials on the pair correlation function and appears for $n \geqslant 2$.

First-Order Terms. In the first-order of perturbation, the terms with V_{ab}^{perm}, V_{ab}^{dis}, V_{ab}^{sh}, and V_{abc}^{in} as integrands give zero upon integration over the Euler angles. But the integrals of V_{ab}^{in} and V_{abc}^{dis} are not, in general, zero. We find

$$f_{ab}^{(1)} = f_{ab}^{(1,ad)}(in) \tag{19}$$
$$= -\sum_{s=1}^{\infty} \frac{2^{(s-2)}(2s+2)(s!)^2}{(2s)!} (\overline{\alpha}_b N_a^{(s)} + \overline{\alpha}_a N_b^{(s)}) I_{ab}^{(2s+4)}$$

$$f_{abc}^{(1)} = f_{abc}^{(1,non)}\ (dis) \tag{20}$$
$$= \nu_{abc} K_{ab,bc,ca}^{(1;\ 3,3,3)}$$

where the superscripts *ad* and *non* designate the contributions to $f^{(1)}$ arising, respectively, from pair and triplet interactions and the notation in parentheses indicates the type of interaction that contributes to the property under calculation

$$N_l^{(s)} = \mathbf{M}_l^{(s)}\,[s]\,\mathbf{M}_l^{(s)}$$

$$I_{ab}^{(2s)} = \int g_{ab}^{(0)}\,(\boldsymbol{r}_{a1\ b2})r_{a1\ b2}^{-2s}d\boldsymbol{r}_{a1\ b2} \tag{21}$$

and

$$K_{ab,bc,ca}^{(1;\ 3,3,3)} = \int d\boldsymbol{r}_{b2}\int d\boldsymbol{r}_{c3}\ g_{abc}^{(0)}\,(\boldsymbol{r}_{a1\ b2}, \boldsymbol{r}_{b2\ c3}, \boldsymbol{r}_{c3\ a1}) \times F_{ab,bc,ca}\,(1)\,(\boldsymbol{r}_{a1\ b2}\boldsymbol{r}_{b2\ c3}\boldsymbol{r}_{c3\ a1})^{-3} \tag{22}$$

with

$$F_{ab,bc,ca}(1) = 3\overline{C}_{12}\overline{C}_{23}\overline{C}_{31} + 1 \tag{23}$$

Here $\overline{C}_{12}$, etc., means the cosine of the angle opposite the side $a1b2$, etc., of the triangle formed by molecules $a1$, $b2$ and $c3$.

The molecular symmetries are used in evaluating $N^{(s)}{}_l$. For example, for linear molecules exhibiting axial symmetry one gets (39)

$$N_a^{(s)} = \mathbf{M}_a^{(s)}\,[s]\,\mathbf{M}_a^{(s)} = \frac{(2s)!}{2^s(s!)^2}\,\{\mathbf{M}_a^{(s)}\}^2 \tag{24}$$

Second-Order Terms. In the second-order of perturbation one finds it convenient to rewrite Equation 16 in the following form

$$f^{(2)} = \frac{1}{4}\rho\sum_{ab}^{t} x_a x_b\, f_{ab}^{(2,ad)} + \frac{1}{12}\rho^2\sum_{abc}^{t} x_a x_b x_c\, f_{abc}^{(2,non)} \tag{25}$$

where the superscripts *ad* and *non* have the meaning defined earlier. In the expression $f_{ab}^{(2,ad)}$, there are squares of the permanent moments, induction, dispersion, and anisotropic shape terms as well as cross terms such as product of permanent moments with induced moments, and so forth. In evaluating the integrals over Euler angles, it is convenient to ignore powers of V_{ab}^{in} and V_{abc}^{in} greater than the first because Equations 4 and 9 are correct only to the first power in the polarizability. The final results, which we get after performing the integrations over angles appearing in expressions for $f_{ab}^{(2,ad)}$ and $f_{abc}^{(2,non)}$, can be written in a convenient way as

$$\begin{aligned} f_{ab}^{(2,ad)} &= f_{ab}^{(2,ad)}\,(perm) + f_{ab}^{(2,ad)}\,(anis - dis) + f_{ab}^{(2,ad)}\,(sh) \\ &+ f_{ab}^{(2,ad)}\,(perm \times in) + f_{ab}^{(2,ad)}\,(perm \times dis) + f_{ab}^{(2,ad)}\,(in \times dis) \\ &+ f_{ab}^{(2,ad)}\,(sh \times in) + f_{ab}^{(2,ad)}\,(sh \times dis) \end{aligned} \quad (26)$$

$$\begin{aligned} f_{abc}^{(2,non)} &= f_{abc}^{(2,non)}\,(perm \times in) + f_{abc}^{(2,non)}\,(perm \times dis) \\ &+ f_{abc}^{(2,non)}\,(in \times dis) + f_{abc}^{(2,non)}\,(dis) + f_{abc}^{(2,non)}\,(dis \times dis) \end{aligned} \quad (27)$$

where

$$f_{ab}^{(2,ad)}\,(perm) = -\beta \sum_{s=1}^{\infty} \sum_{l=1}^{\infty} \frac{2^{(s+l)}\,(2s + 2l)!(s!\,l!)^2}{(2s)!(2l)!(2s+1)!(2l+1)!} N_a^{(s)} N_b^{(l)} I_{ab}^{(2s+2l+2)} \quad (28)$$

$$\begin{aligned} f_{ab}^{(2,ad)}\,(anis - dis) &= -\frac{4\beta}{225} A_{ab}^2 [19 P_a^{(0)} P_b^{(0)} - 27 \bar{\alpha}_a^2 P_b^{(0)} \\ &- 27 \bar{\alpha}_b^2 P_a^{(0)} - 9 \bar{\alpha}_a^2 \bar{\alpha}_b^2] \times I_{ab}^{(12)} - \frac{8}{15} \beta\rho \sum_c x_c \,\{A_{ab} A_{ac} [\bar{\alpha}_a \bar{\alpha}_b (P_a^{(0)} \\ &- 3\bar{\alpha}_c^2)]\, J_{abc}^{(6,0,6)} + A_{ab} A_{bc} [\bar{\alpha}_a \bar{\alpha}_b (P_b^{(0)} - 3\bar{\alpha}_c^2)]\, J_{bac}^{(6,6,0)}\} \end{aligned} \quad (29)$$

$$f_{ab}^{(2,ad)}\,(sh) = -\frac{128}{5} \beta D_{ab}^2 I_{ab}^{(24)}\, \varepsilon_{ab}^2 (\sigma_{ab})^{24} \quad (30)$$

$$f_{ab}^{(2,ad)}\,(perm \times in) = \beta \left\{ \frac{8}{5} [P_a^{(2)} Z_b^{(1)} + P_b^{(2)} Z_a^{(1)}] I_{ab}^{(11)} + \frac{32}{35} [P_a^{(2)} Z_b^{(2)} + P_b^{(2)} Z_a^{(2)}] I_{ab}^{(13)} \right\} \quad (31)$$

$$f_{ab}^{(2,ad)}\,(perm \times dis) = \frac{32}{25} \beta A_{ab} P_a^{(2)} P_b^{(2)} I_{ab}^{(11)} \quad (32)$$

$$f_{ab}^{(2,ad)}\,(in \times dis) = \frac{2}{75}\,\beta A_{ab}[3\overline{\alpha}_a(\overline{\alpha}_b^2 + 3P_b^{(0)})N_a^{(1)}$$

$$+ (27\overline{\alpha}_b^2 - 19P_b^{(0)}) \times Z_a^{(0)} + 3\overline{\alpha}_b(\overline{\alpha}_a^2 + 3P_a^{(0)})N_b^{(1)}$$

$$+ (27\overline{\alpha}_a^2 - 19P_a^{(0)}]Z_b^{(0)}]I_{ab}^{(12)} - \frac{1}{15}\,\rho \sum_c^t x_c\{A_{ac}\overline{\alpha}_c[3\overline{\alpha}_b(Z_a^{(0)} - \overline{\alpha}_a N_a^{(1)})$$

$$+ (P_a^{(0)} - 3\overline{\alpha}_a^2)N_b^{(1)}] \times J_{abc}^{(6,0,6)} + A_{bc}\overline{\alpha}_c[3\overline{\alpha}_a(Z_b^{(0)} - \overline{\alpha}_b N_b^{(1)})$$

$$+ (P_b^{(0)} - 3\overline{\alpha}_b^2)N_a^{(1)}]\,J_{bac}^{(6,6,0)}\} \tag{33}$$

$$f_{ab}^{(2,ad)}\,(sh \times in) = \frac{16}{5}\,\beta D_{ab}[\overline{\alpha}_a N_b^{(1)} + \overline{\alpha}_b N_a^{(1)}]I_{ab}^{(18)}\,\varepsilon_{ab}(\sigma_{ab})^{12}$$

$$+ \frac{128}{35}\,\beta D_{ab}[\overline{\alpha}_a N_b^{(2)} + \overline{\alpha}_b N_a^{(2)}]I_{ab}^{(20)}\,\varepsilon_{ab}(\sigma_{ab})^{12} \tag{34}$$

$$f_{ab}^{(2,ad)}\,(sh \times dis) = \frac{64}{5}\,\beta D_{ab}[k_a + k_b]I_{ab}^{(18)}\,(\varepsilon_{ab})^2\,(\sigma_{ab})^{18} \tag{35}$$

$$f_{abc}^{(2,non)}\,(perm \times in) = 2\beta\Bigg\{\frac{1}{3}\,\overline{\alpha}_c N_a^{(1)} N_b^{(1)} K_{ab,bc,ca}^{(1,3,3,3)}$$

$$+ \frac{1}{15}\,[N_c^{(2)}\overline{\alpha}_a N_b^{(1)} + \overline{\alpha}_b N_a^{(1)}]K_{ab,bc,ca}^{(2;3,4,4)} + \frac{1}{15}\,\overline{\alpha}_c N_a^{(2)} N_b^{(2)} K_{ab,bc,ca}^{(3;5,4,4)} + \ldots$$

$$+ \text{permutations}\Bigg\} \tag{36}$$

$$f_{abc}^{(2,non)}\,(in \times dis)$$

$$= -\frac{4}{75}\,\beta\{A_{ab}\overline{\alpha}_c P_a^{(2)} P_b^{(2)} K_{ab,bc,ca}^{(5,6,4,4)} + \text{permutations}\} \tag{37}$$

$$f_{abc}^{(2,non)}\,(perm \times dis)$$

$$= -\frac{8}{25}\,\beta\,\frac{\nu_{abc}}{\overline{\alpha}_a\overline{\alpha}_b\overline{\alpha}_c}\,\{\overline{\alpha}_c P_a^{(2)} P_b^{(2)} K_{ab,bc,ca}^{(1;8,3,3)} + \text{permutations}\} \tag{38}$$

$$f_{abc}^{(2,non)}\,(dis \times dis) = \frac{2}{225}\,\beta\nu_{abc}\left\{\frac{1}{\overline{\alpha}_a^2\overline{\alpha}_b^2}\,[19P_a^{(0)}P_b^{(0)} - 27\overline{\alpha}_a^2P_b^{(0)}\right.$$

$$- 27\overline{\alpha}_b^2P_a^{(2)} - 9\overline{\alpha}_a^2\overline{\alpha}_b^2]K_{ab,bc,ca}^{(1,9,3,3)} + \frac{4}{\overline{\alpha}_a^2\overline{\alpha}_b^2}\,[P_a^{(0)}P_b^{(0)} - 33\overline{\alpha}_a^2P_b^{(0)}$$

$$\left.- 33\overline{\alpha}_b^2P_a^{(0)} + 189\overline{\alpha}_a^2\overline{\alpha}_b^2]\,J_{abc}^{(9,3,3)} + \text{permutations}\right\} \qquad (39)$$

$$f_{abc}^{(2,non)}\,(dis) = -\frac{1}{3}\,\beta\nu_{abc}^2L_{ab,bc,ca}^{(1,3,3,3)} + \text{permutations} \qquad (40)$$

with

$$J_{abc}^{(p,q,r)} = \iint d\boldsymbol{r}_{b2}\,d\boldsymbol{r}_{c3}\,g_{abc}^{(0)}\,(\boldsymbol{r}_{a1\ b2}, \boldsymbol{r}_{b2\ c3}, \boldsymbol{r}_{c3\ a1})$$
$$\times\ \boldsymbol{r}_{a1\ b2}^{-p}\,\boldsymbol{r}_{b2\ c3}^{-q}\,\boldsymbol{r}_{c3\ a1}^{-r}\,P_2\left(\frac{\boldsymbol{r}_{a1\ b2}\cdot\boldsymbol{r}_{b2\ c3}}{\boldsymbol{r}_{a1\ b2}\,\boldsymbol{r}_{b2\ c3}}\right) \qquad (41)$$

$$K_{ab,bc,ca}^{(i,p,q,r)} = \iint d\boldsymbol{r}_{b2}\,d\boldsymbol{r}_{c3}\,g_{abc}^{(0)}\,(\boldsymbol{r}_{a1\ b2}, \boldsymbol{r}_{b2\ c3}, \boldsymbol{r}_{c3\ a1})$$
$$\times\ \boldsymbol{r}_{a1\ b2}^{-p}\,\boldsymbol{r}_{b2\ c3}^{-q}\,\boldsymbol{r}_{c3\ a1}^{-r}\,F_{ab,bc,ca}^{(i)} \qquad (42)$$

and

$$L_{ab,bc,ca}^{(1,3,3,3)} = \iint d\boldsymbol{r}_{b2}\,d\boldsymbol{r}_{c3}\,g_{abc}^{(0)}\,(\boldsymbol{r}_{a1\ b2}, \boldsymbol{r}_{b2\ c3}, \boldsymbol{r}_{c3\ a1})[(\boldsymbol{r}_{a1\ b2}\,\boldsymbol{r}_{b2\ c3}$$
$$\times\ \boldsymbol{r}_{c3\ a1})^{-3}\,F_{ab,bc,ca}(1)]^2 \qquad (43)$$

Here $P_l^{(0)} = \hat{\boldsymbol{\alpha}}_l : \hat{\boldsymbol{\alpha}}_l, P_l^{(2)} = \hat{\boldsymbol{\alpha}}_l : \mathbf{M}_l^{(2)}$, $Z_l^{(0)} = \mathbf{M}_l^{(1)} \cdot \hat{\boldsymbol{\alpha}}_l \cdot \mathbf{M}_l^{(1)}$, $Z_l^{(1)} = \mathbf{M}_l^{(1)} \cdot \mathbf{M}_l^{(2)} \cdot \mathbf{M}_l^{(1)}$, $Z_l^{(2)} = \mathbf{M}_l^{(2)} : (\mathbf{M}_l^{(1)} \cdot \mathbf{M}_l^{(1)})$, and κ is anisotropy in the polarizability of an isolated molecule.

$F_{ab,bc,ca}(1)$ is already defined in Equation 23 and other values are given below:

$$F_{ab,bc,ca}(2) = 15\overline{C}_{12}^2\overline{C}_{23}\overline{C}_{31} - 5\overline{C}_{12}^3 - 3\overline{C}_{13}\overline{C}_{23} + 9\overline{C}_{12} \qquad (44)$$

$$F_{ab,bc,ca}(3) = 35\overline{C}_{12}\overline{C}_{13}^2\overline{C}_{23}^2 + 20\overline{C}_{12}^2\overline{C}_{13}\overline{C}_{23}$$
$$+ 5\overline{C}_{12}^2 + 10\overline{C}_{13}\overline{C}_{23} - 2\overline{C}_{12} \qquad (45)$$

$$F_{ab,bc,ca}(5) = 45\overline{C}_{12}\overline{C}^2_{23}\overline{C}^2_{31} + 30\overline{C}^2_{12}\overline{C}_{23}\overline{C}_{31} + 10\overline{C}^3_{12} + 12\overline{C}_{13}\overline{C}_{23} - 3\overline{C}_{12} \quad (46)$$

The notation *permutations* indicates two more sets of terms like those explicitly written out, but with cyclic permutations of the three pairs of indices *ab*, *bc* and *ca*. While all expressions given above are valid for molecules of any symmetry, Equations 30, 34, and 35 are valid strictly speaking only for symmetric linear molecules.

Third-Order Terms. In the third-order perturbation term the only contribution that is of some significance is the one that arises from the permanent moment interaction branch of the pair potential. The contribution of all other branches of pair and triplet potentials to thermodynamic properties of pure, simple molecular fluids such as nitrogen (or even carbon dioxide) and their mixtures is negligible. Thus

$$f^{(3)}_{ab} \simeq f^{(3,ad)}_{ab}\ (perm) \quad (47)$$

$$f^{(3)}_{ab} = \frac{1}{2}\beta^2\left\{\frac{24}{5}Z^{(1)}_a Z^{\omega}_b I^{(11)}_{ab} + \frac{96}{35}\left(Z^{(1)}_a Z^{(2)}_b + Z^{(1)}_b Z^{(2)}_a\right) I^{(13)}_{ab} + \frac{768}{245}\left[Z^{(2)}_a Z^{(2)}_b I^{(15)}_{ab}\right]\right\} + \frac{1}{36}\rho\beta^2 \sum_c x_c\left\{\frac{2}{9}N^{(1)}_a N^{(1)}_b N^{(1)}_c \times K^{(1,3,3,3)}_{ab,bc,ca} + \frac{2}{15}N^{(1)}_a N^{(1)}_b N^{(2)}_c K^{(2;\ 3,4,4)}_{ab,bc,ca} + \frac{2}{15}N^{(2)}_a N^{(2)}_b N^{(1)}_c \times K^{(3,5,4,4)}_{ab,bc,ca} + \frac{2}{225}N^{(2)}_a N^{(2)}_b N^{(2)}_c K^{(4,5,5,5)}_{ab,bc,ca} + \cdots + \text{permutations}\right\} \quad (48)$$

where "permutation" has the same meaning as noted earlier, and

$$K^{(4;\ 5,5,5)}_{ab,bc,ca} = \int d\boldsymbol{r}_{b2}\int d\boldsymbol{r}_{c3}\, g^{(0)}_{abc}\left(\boldsymbol{r}_{a1\ b2}, \boldsymbol{r}_{b2\ c3}, \boldsymbol{r}_{c3\ a1}\right) \cdot \left(r_{a1\ b2}\, r_{b2\ c3}\, r_{c3\ a1}\right)^{-5} F_{ab,bc,ca}(4) \quad (49)$$

with

$$F_{ab,bc,ca}(4) = 245\,\overline{C}^2_{12}\overline{C}^2_{23}\overline{C}^2_{31} + 110\overline{C}_{12}\overline{C}_{13}\overline{C}_{23} - 35\left[\overline{C}^2_{12}\overline{C}^2_{13} + \overline{C}^2_{12}\overline{C}^2_{23} + \overline{C}^2_{13}\overline{C}^2_{23}\right] + 18 \quad (50)$$

The contribution arising from terms beyond third-order can be approximated by using the [1,0] Padé approximant of Stell et al. (*24*).

That is

$$\beta f_{ab}^{(ad)}\,(perm) = \beta f_{ab}^{(2,ad)}\,(perm)\left[1 + \left|\beta f_{ab}^{(3,ad)}\,(perm)/\beta f_{ab}^{(2,ad)}\,(perm)\right|\right] \quad (51)$$

The first-, second-, and third-order perturbation contributions to $\beta P/\rho$, $\beta U/N$ and S/Nk (where P, U, and S are the configurational contributions to the pressure, internal energy, and entropy, respectively) are found by applying the usual thermodynamic relationships to the corresponding terms of the free energy, i.e.

$$\beta P^{(n)}/\rho = \rho\frac{\partial(\beta f^{(n)})}{\partial\rho}; \quad \frac{\beta U^{(n)}}{N} = -T\frac{\partial(\beta f^{(n)})}{\partial T}$$

and

$$\frac{S^{(n)}}{Nk} = \frac{\beta U^{(n)}}{N} - \beta f^{(n)} \quad (52)$$

where superscript n indicates the order of perturbation.

For theoretical developments similar to our treatment but without three-body nonadditive interactions we refer the reader to References *21*, *22*, *24*, *38*, Gubbins et al. (*40*), and Shing and Gubbins (*41*). Another perturbation method, which uses nonspherical reference potential and is in principle more accurate in dealing with strongly anisotropic forces, is given by References *18–20* and Kohler and Quirke (*42*).

Thermodynamic Properties of Pure Fluids

The theory described above has been applied by Shukla et al. (*11*) to calculate the equilibrium properties of nitrogen, oxygen, carbon monoxide, carbon dioxide, and methane. Of these five nonpolar polyatomic fluids, the first four have an axis of at least threefold symmetry and therefore have quadrupole moments described by a single scalar. Methane is tetrahedrally symmetric and has an octupole moment as the first nonvanishing electric multipole moment.

The core of the methane molecule is very nearly spherically symmetric (spherical top) while all of the other molecules have rodlike cores. For nitrogen, Weis and Levesque (*43*) have found from molecular dynamic simulation that the anisotropic contribution, which arises from the nonspherical repulsive core, to the static structure factor is quite small. The structure factor associated with the center of mass motion bears a close resemblance to that of the monatomic liquid. Wang et al. (*44*) have found from Monte Carlo studies of $g(\mathbf{r}_1, \mathbf{r}_2; \omega_1, \omega_2)$ that the anisotropic overlap interaction V_{12}^{sh} may be taken as a perturbation of the Lennard–

Jones 6-12 potential as long as $D \leqslant 0.15$. In the presence of dispersion and induction interactions, this limit on D may be still higher (*26*). Thus the use of the spherically symmetric reference potential (SSRP) for perturbation expansion for the above systems is justified.

The properties of the reference system, which in our case is the Lennard–Jones 6-12 system, are calculated using the Verlet–Weis (*45*) version of the Weeks–Chandler–Andersen (*46*) perturbation theory. Two- and three-body integrals, $I_x^{(n)}$ and $K_x^{(i,n',n'',n''')}$ that appear in the expansion of additive and nonadditive contributions to the thermodynamic properties can be evaluated from the expressions given by Gubbins and Twu (*47*), Shukla et al. (*11, 48*), and Larsen et al. (*49*).

It has been found that at liquid densities the contributions of the three-body nonadditive interactions and anisotropy in pair interactions to the Helmholtz free energy are separately substantial. However, the net effect of these contributions is, in general, small because the two contributions are in opposite directions (exceptions are oxygen and methane). The other thermodynamic properties vary in their degree of sensitivity to the anisotropic forces. The contribution of anisotropic forces is relatively small for the configurational energy, whereas it is of similar order of magnitude to the isotropic contribution in the case of pressure. In particular, the pressure is more sensitive to short range anisotropic forces than are the other properties.

Singh and Singh (*13, 14*) have calculated the second and third virial coefficients of nitrogen, carbon monoxide, carbon dioxide, carbon disulfide, ethylene, and benzene and have calculated the dielectric second virial coefficient of nitrogen and carbon dioxide for a set of force parameters that has been found suitable for explaining the dilute gas viscosity data. The agreement between theory and experiment is very satisfactory for all of these systems. In Figures 1 and 2 we give the calculated and experimental third virial coefficients for nitrogen and carbon dioxide. Because the experimental values of the third virial coefficient are derived from gas compressibility measurements by fitting an isotherm with a polynomial in the density, the values are subject to uncertainties in the isotherm and the degree of polynomial used, in addition to experimental inaccuracies. Moreover, the values reported by different authors differ from each other considerably. In view of these uncertainties, the agreement found by Singh and Singh (*13*) is very satisfactory.

In Tables I and II we compare at selected temperatures and densities the calculated values of the configurational energy, entropy, and pressure with the experimental data for nitrogen, oxygen, and methane, and carbon monoxide and carbon dioxide. Details are given by Shukla et al. (*11*). Even for carbon dioxide, for which we expect the theory to be less accurate because its "rod-likeness" is too large compared to that of nitrogen, the agreement found between theory and experiment is very

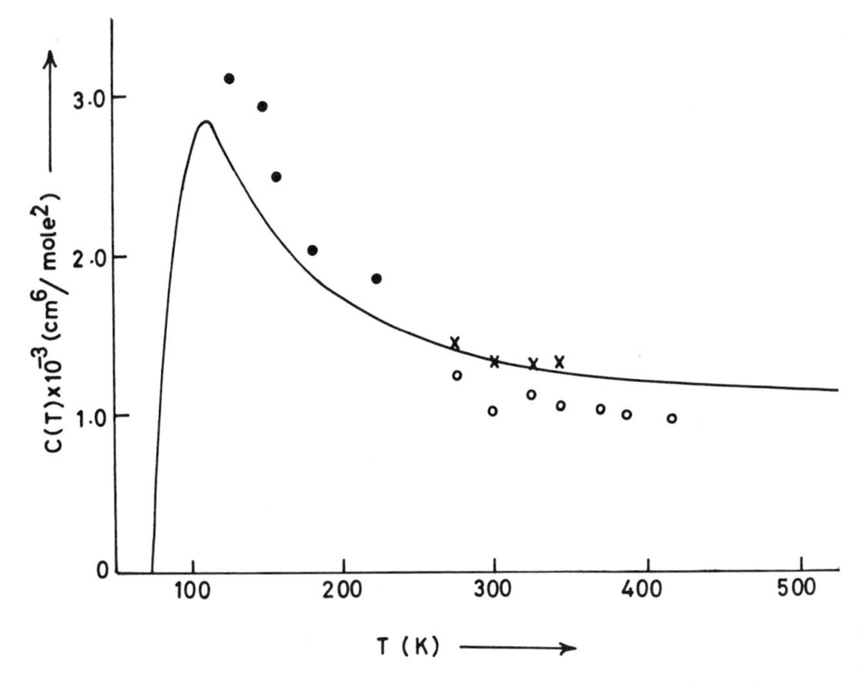

Figure 1. Comparison of calculated and experimental third virial coefficients for nitrogen. Key: —, calculated; ●, data of Hoover et al. (64); *○, data of Michels et al.* (65); *and ×, data of Michels et al.* (66). *(Reproduced with permission from Ref. 13. Copyright 1976,* Physica (Ultrecht).)

good. From this we may probably conclude that the thermodynamic properties of molecular liquids are not very sensitive to the nonsphericity in the shape of molecules.

The force parameters found suitable for calculating the equilibrium properties at liquid densities are given in Table III. The central force parameters ε and σ are obtained from the saturated liquid density data, and D is obtained from the vapor pressure data near the boiling point. Multipole moments and anisotropic polarizabilities are available from independent experimental measurements. It may be noted that the central force parameters given in Table III give a deeper potential well (on the average about 7% for nitrogen, carbon monoxide and carbon dioxide) and smaller molecular "diameter" (about 2%) than those determined by Singh and Singh from gas data. This difference may be attributed to one or both of the following reasons: (1) The Lennard–Jones 6-12 potential model is probably inadequate to represent the central forces of molecules; and (2) the four-body and higher nonadditive interactions are probably not negligible at liquid densities.

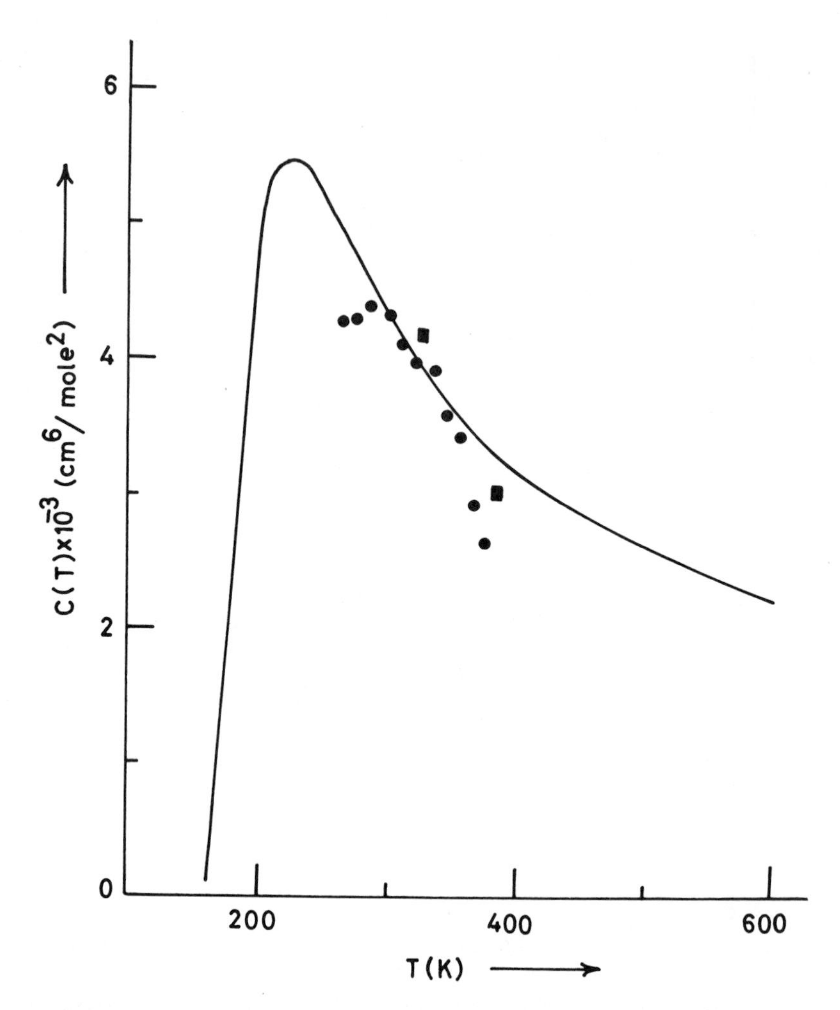

Figure 2. Comparison of calculated and experimental third virial coefficients for carbon dioxide. Key: —, calculated; ●, *Butcher and Dadson* (67); *and* ■, *McCormack and Schneider* (68). *(Reproduced with permission from Ref. 13. Copyright 1976,* Physica (Ultrecht).)

Ananth et al. (*21*) have calculated the thermodynamic properties of nitrogen, oxygen, and methane. Their calculation, however, differs from ours in two respects: (1) they have neglected both the pair induction and three-body nonadditive interactions (arising from induction and dispersion forces), and (2) they have neglected the contribution arising from third- and higher-order terms in the perturbation series of permanent multipole interactions. The induction and nonadditive terms make significant contributions and improve the agreement for nitrogen, oxygen, and methane. The extension to higher order perturbation terms makes

Table I. Comparison of Calculated and Experimental Thermodynamic Properties of Nitrogen, Oxygen, and Methane

			$-U^c/Nk_BT$			$-S^r/Nk_B$			P (*atm*)		
System	T (°K)	ρ (*mol/L*)	*Expt.*	*Shukla et al.*[a]	*Ananth et al.*[b]	*Expt.*	*Shukla et al.*[a]	*Ananth et al.*[b]	*Expt.*	*Shukla et al.*[a]	*Ananth et al.*[b]
N_2	100	24.57	5.060	5.071	5.010	2.630	2.643	2.480	7.72	7.68	7.13
		25.01	5.166	5.134	5.099	2.698	2.795	2.544	30	30.53	29.12
		26.14	5.402	5.378	5.323	2.888	2.909	2.748	100	100.56	102.40
		27.29	5.648	5.583	5.547	3.093	3.060	2.959	200	201.76	206.90
		28.88	5.953	5.890	5.842	3.363	3.374	3.272	400	402.26	413.30
		30.04	6.151	5.995	6.039	3.393	3.607	3.512	600	604.91	615.40
O_2	110	32.360	5.970	5.954	5.990	2.850	2.832	2.830	5.36	5.40	5.05
		33.356	6.141	6.117	6.191	2.993	2.985	3.004	98.70	99.45	114.90
		34.200	6.284	6.239	6.339	3.124	3.107	3.141	197.40	199.69	220.00
		34.916	6.402	6.436	6.463	3.237	3.209	3.262	296.10	306.29	326.20
CH_4	230	16.200	2.073	2.097	2.253	1.248	1.138	1.378	200	201.9	191.4
		18.560	2.367	2.318	2.557	1.494	1.433	1.644	300	301.8	297.9
		21.020	2.708	2.798	2.871	1.833	1.796	1.965	500	498.3	573.9
		34.160	3.080	2.893	3.231	2.328	2.359	2.437	1000	955.7	1058.3

Note: Experimental data from Refs. 17, 62, 63.
[a] From Ref. 11.
[b] From Ref. 21.
Source: Reproduced with permission from Ref. 11. Copyright 1979, American Institute of Physics.

Table II. Comparison of Calculated and Experimental Thermodynamic Properties of Carbon Monoxide and Carbon Dioxide

System	T (°K)	ρ (mol/L)	$-U^c/NkT$ Expt.	$-U^c/NkT$ Calculated	$-S^r/Nk$ Expt.	$-S^r/Nk$ Calculated	P (atm) Expt.	P (atm) Calculated
CO	68.2	30.211	9.313	8.707	5.562	5.376	0.152	0.186
	74.4	29.235	7.931	7.931	5.159	5.123	0.400	0.436
	88.3	27.322	5.628	5.723	4.331	4.363	2.000	1.899
	101.5	24.873	3.201	3.293	3.568	3.732	6.000	6.071
	114.8	21.929	2.341	2.181	2.929	3.180	14.000	16.983
CO_2	223.2	26.209	7.465	7.426	4.069	4.196	6.7	7.8
	243.2	24.408	6.293	6.158	3.356	3.353	14.1	14.3
	263.2	22.373	5.256	5.168	2.781	2.729	26.1	28.4
	283.2	19.614	4.264	4.159	2.072	2.067	44.4	44.9
	293.2	17.609	3.731	3.697	1.634	1.696	56.5	58.5
	298.2	16.207	3.394	3.376	1.323	1.377	63.5	65.4

Note: Experimental data from Refs. 17, 62, 63.
Source: Reproduced with permission from Ref. 11. Copyright 1979, American Institute of Physics.

Table III. Force Parameters Used in Calculations of This Chapter

System	ε/k (K)	σ (Å)	D	$\bar{\alpha} \times 10^{24}$ (cm^3)	κ	$\theta \times 10^{26}$ (e.s.u. cm^2)	$\Omega \times 10^{34}$ (e.s.u.)
Ar	119.8	3.405	—	—	—	—	—
	(119.8)	(3.405)					
N_2	103.29	3.615	0.08	1.730	0.176	−1.40	0.0
	92.50	3.650	0.02				
	(100.15)	(3.619)					
O_2	120.97	3.406	0.12	1.600	0.239	−3.9	—
	110.90	3.431	0.10				
	(122.44)	(3.388)					
CO	103.23	3.624	0.10	1.970	0.168	−2.20	3.1
	95.50	3.672	0.10				
	(105.54)	(3.643)					
CO_2	203.50	3.781	0.16	2.925	0.257	−4.30	0.0
	197.50	3.932	0.16				
CH_4	137.80	3.826	0.00	2.60	0.00	0.00	2.04
	137.00	3.882	0.00				

Note: Upper lines, the central force parameters found by Shukla et al. (*11*) from liquid state data. Lower lines, those found by Singh and Singh (*13*) from gas data. The force parameters given in parentheses are those used by Leonard et al. (*60*) to study the excess properties of binary mixtures of these systems.

possible comparison with experimental data for carbon monoxide and carbon dioxide.

Numerical calculations carried out by Gubbins et al. (*40*) for thermodynamic properties of fluids of nonaxial quadrupolar molecules show that, in general, the nonaxial part of the quadrupole moment has a large influence and must be properly accounted for. If the quadrupole moment is weak, it is possible to use the axial expression with the effective axial approximation. However, this will be a poor approximation for many nonaxial fluids (e.g., H_2O) and for binary mixtures of a nonaxial molecule (e.g., C_2H_4) with an axial molecule (e.g., N_2O).

Binary Mixtures

The perturbation theory based on SSRP and the Padé approximant has been used to study a system of hard spheres mixed with hard spheres that have embedded point dipoles by Melnyk and Smith (*50*). Chambers and McDonald (*23*) have used this theory to study a nonpolar–polar liquid mixture, using a Lennard–Jones potential model for the nonpolar constituent and a Stockmayer potential for the polar constituent. Gubbins and Gray and co-workers (*22, 38, 47, 51, 52, 53, 54*) have used this theory to evaluate the effect of polar, quadrupolar, octopolar, dispersion, and anisotropic overlap forces under the assumption of pairwise additivity of the potential on the excess properties of binary liquid mixtures (Xe–HCl, Xe–HBr, HCl–HBr, Xe–N_2O, Xe–C_2H_4, Kr–C_2H_4, Xe–CF_4) at low and moderate pressures. They found that the anisotropic forces have a large effect on excess properties, producing a positive deviation from Raoult's law. They have also found that the strong polar and quadrupolar forces may cause liquid–liquid immiscibility to occur. In their calculations, they have, however, ignored the contributions arising due to three-body non-additive interactions.

Based on the work of Shukla and Singh (*12*) we discuss here the relative importance of the pair and three-body nonadditive interactions on the excess and total thermodynamic properties of binary fluid mixtures.

For a binary fluid mixture, Equation 16 can be rewritten as

$$f^{(n)} = \frac{\rho}{2n}[x_a^2 f_{aa}^{(n,ad)} + 2x_a x_b f_{ab}^{(n,ad)} + x_b^2 f_{bb}^{(n,ad)}] + \frac{\rho^2}{6n}[x_a^3 f_{aaa}^{(n,non)} + 3x_a^2 x_b f_{aab}^{(n,non)} + 3x_a x_b^2 f_{abb}^{(n,non)} + x_b^3 f_{bbb}^{(n,non)}] \quad (53)$$

where again n indicates the order of perturbation and superscripts *ad* and *non* represent the contributions that arise from pairwise additive and three-body nonadditive interactions, respectively.

For a binary mixture consisting of a component a of spherically symmetric molecules and another component b of axially symmetric molecules, $f_{aa}^{(n,ad)} = 0$ for $n \geq 1$, and the contribution to $f_{ab}^{(n,ad)}$ arises only from induction interaction which, as discussed earlier, is considered to first-order term only. Thus, $f_{ab}^{(1,ad)} = f_{ab}^{(1,ad)}(in)$, and $f_{ab}^{(n,ad)} = 0$ for $n \geq 2$. Since the molecules of species b are axially symmetric and have a quadrupole moment as the first nonvanishing electric pole, we consider $f_{bb}^{(n,ad)}$ up to third-order terms and apply the Padé approximant to estimate the contribution of higher order terms. Because the molecules of species a are spherically symmetric, it is sufficient to consider $f_{aaa}^{(n,non)}$ and $f_{aab}^{(n,non)}$ to the first-order perturbation term only, whereas $f_{abb}^{(n,non)}$ and $f_{bbb}^{(n,non)}$ are considered up to second-order terms. On the other hand, if we consider a binary mixture in which both components are axially symmetric molecules, we must consider all of the contributions discussed earlier. The expressions in terms of molecular parameters for the above terms are obtained from the section giving the perturbation expansion by introducing the symmetry of molecules.

The properties of the reference system are evaluated using van der

Waals one-fluid theory of mixtures (55). This theory approximates the properties of a mixture by those of a fictitious pure fluid with the interaction parameters

$$\begin{aligned} \sigma_x^3 &= \sum_{a,b} x_a x_b \, \sigma_{ab}^3 \\ \varepsilon_x &= \sigma_x^{-3} \sum_{a,b} x_a x_b \, \varepsilon_{ab} \, \sigma_{ab}^3 \end{aligned} \tag{54}$$

The unlike force parameters ε_{ab} and σ_{ab} of the potential model are approximated from the following combination rule

$$\begin{aligned} \sigma_{ab} &= \frac{1}{2}(\sigma_{aa} + \sigma_{bb}) \\ \varepsilon_{ab} &= \xi_{ab}(\varepsilon_{aa} \, \varepsilon_{bb})^{1/2} \end{aligned} \tag{55}$$

where ξ_{ab} is an adjustable parameter with the requirement that $\xi_{ab} \leqslant 1$. The excess properties are found to be very sensitive to the value of ξ_{ab}.

In the van der Waals one-fluid theory of mixtures, the free energy, enthalpy, and pressure of the mixture are written as

$$\begin{aligned} A_0 &= A_x + NkT \sum_a x_a \ln x_a + \text{second order terms} \\ H_0 &= H_x + \text{second order terms} \\ P_0 &= P_x + \text{second order terms} \end{aligned} \tag{56}$$

where A_x, H_x, and P_x are values for a pure fluid containing N molecules in volume V at temperature T, the molecules interacting with a Lennard–Jones 6-12 potential having parameters σ_x and ε_x given by Equation 54. Values of A_x, H_x, and P_x are calculated using curves fitted either to Monte Carlo data of McDonald and Singer (56) or pure fluid Percus–Yevick data of Grundke et al. (57). The theory is known to give very good results for the argon–krypton mixture and compares very well with machine simulation results.

The two-body integrals $I_{ab}^{(2s)}$ that appear in perturbation terms are evaluated using the values of the mixture radial distribution function obtained from the zeroth order term in the conformal solution expansion of Mo et al. (58). In this approximation, the mixture pair distribution function $g_{ab}^{(0)}$ is equated to that of a pure fluid at reduced temperature T_x^* and reduced density ρ_x^*. That is

$$g_{ab}^{(0)}\left(\frac{r_{a1\,b2}}{\sigma_{ab}}, \frac{kT}{\varepsilon_{ab}}, \rho\sigma_{ab}^3\right) \simeq g^{(0)}\left(\frac{r_{a1\,b2}}{\sigma_{ab}}, T_x^*, \rho_x^*\right) \tag{57}$$

Thus

$$I_{ab}^{(n)}(\rho_{ab}^*,\ T_{ab}^*) = \left(\frac{\sigma_{ab}}{\sigma_x}\right)^{n-3} I_x^{(n)}(\rho_x^*,\ T_x^*) \tag{58}$$

where $I_x^{(n)}$ is the pure fluid integral at reduced density ρ_x^* and temperature T_x^*.

To evaluate the three-body integrals $K_{abc}^{(i,p,q,r)}$ we make use of the superposition approximation for the triplet correlation function of the reference mixture system, i.e.

$$g_{abc}^{(0)}(\boldsymbol{r}_{12},\ \boldsymbol{r}_{13},\ \boldsymbol{r}_{23}) \simeq g_{ab}^{(0)}(\boldsymbol{r}_{12})g_{bc}^{(0)}(\boldsymbol{r}_{23})g_{ac}^{(0)}(\boldsymbol{r}_{13}) \tag{59}$$

This approximation gives negligible error in pure fluid integrals and is expected to hold equally well in the case of mixtures of molecules of approximately the same size. For each of $g_{lm}^{(0)}$ in Equation 59 we use the value obtained from Equation 57. With this, the three-body integral of mixtures is converted into the three-body pure fluid integrals in the following way

$$K_{abc}^{(i,\ n',\ n'',\ n''')} \simeq \left(\frac{\sigma_{ab}}{\sigma_x}\right)^{n'-2}\left(\frac{\sigma_{ac}}{\sigma_x}\right)^{n''-2}\left(\frac{\sigma_{bc}}{\sigma_x}\right)^{n'''-2} K_x^{(i,\ n',\ n'',\ n''')} \tag{60}$$

where $K_x^{(i,\ n',\ n'',\ n''')}$ is the pure fluid integral at the reduced state conditions T_x^* and ρ_x^*. The value of integral $K_x^{(i,\ n',\ n'',\ n''')}$ is obtained in terms of the reduced density and reduced effective hard sphere diameter by approximants of the Padé type discussed by Shukla et al. (*11*).

The magnitude of the contributions of various branches of pair and three-body nonadditive interactions to the Helmholtz free energy (*see* Table IV) and internal energy per particle in an equimolar binary mixture have been evaluated by Shukla and Singh (*12*) for five binary mixtures, $Ar–N_2$, $Ar–O_2$, Ar–CO, $N_2–O_2$ and N_2–CO, three of which belong to the category of spherically and axially symmetric molecules and the other two to the category of axially symmetric molecules. Among the three-body nonadditive interaction branches, the most significant contribution arises from dispersion interaction in the first-order perturbation term for all of the systems. The contributions of the induction and dispersion branches in the second order term are small. This justifies the truncation of the perturbation series to the second-order term in case of three-body nonadditive interactions.

The contribution arising from pair induction interaction in first-order perturbation is substantial for N_2–CO, small for Ar–CO, $N_2–O_2$, and $Ar–N_2$, and negligible for $Ar–O_2$ mixtures. In the second-order pertur-

Table IV. Contribution of Various Branches of Pair and Triplet Potentials to Free Energy for Binary Liquid Mixtures (T = 83.82 K, P = 0.0, x_1 = x_2 = 0.5)

Contributions	*Ar–N_2*	*Ar–O_2*	*Ar–CO*	*N_2–O_2*	*N_2–CO*
$\beta f^{(1,ad)}(in)$	−0.0156	−0.0019	−0.0430	−0.0383	−0.1083
$\beta f^{(2,ad)}(perm)$	−0.0436	−0.0005	−0.2561	−0.0549	−0.4847
$\beta f^{(2,ad)}(anis\text{–}dis)$	−0.1160	−0.1545	−0.1090	−0.3966	−0.2047
$\beta f^{(2,ad)}(shape)$	−0.0401	−0.1184	−0.0640	−0.1936	−0.1235
$\beta f^{(2,ad)}(shape \times in)$	0.0013	0.0002	0.0044	0.0036	0.0104
$\beta f^{(2,ad)}(shape \times dis)$	0.0688	0.1966	0.0852	0.4847	0.2853
$\beta f^{(2,ad)}(perm \times in)$	0.0003	0.0000	0.0021	0.0005	0.0039
$\beta f^{(2,ad)}(perm \times dis)$	0.0229	0.0055	0.0521	0.0512	0.1355
$\beta f^{(3,ad)}(perm)$	0.0030	0.0000	0.0436	0.0050	0.0751
$\beta f^{(1,non)}(dis)$	0.4998	0.6415	0.5412	0.6773	0.5662
$\beta f^{(2,non)}(dis^2)$	−0.0113	−0.0182	−0.0133	−0.0201	−0.0146
$\beta f^{(2,non)}(perm \times in)$	0.0009	0.0000	0.0053	0.0029	0.0183
$\beta f^{(2,non)}(perm \times dis)$	−0.0009	−0.0000	−0.0023	−0.0017	−0.0059
$\beta f^{Padé}(perm)$	−0.0408	−0.0000	−0.2188	−0.0503	−0.4158
$\beta f^{(L\text{–}J)}$	−4.5565	−5.1843	−4.6676	−4.5788	−4.1398
$\beta f^{(ad)}(anis)$	−0.1190	−0.0730	−0.2847	−0.1384	−0.4110
$\beta f^{(non)}$	0.4885	0.6233	0.5309	0.6584	0.5640
$\beta f(total)$	−4.1870	−4.6340	−4.4214	−4.0588	−3.9868

Source: Reproduced with permission from Ref. 12. Copyright 1980, American Institute of Physics.

bation term, the contribution arising from the permanent moment (quadrupole moment) interaction branch of the pair potential to free energy and internal energy is significant for N_2–CO and Ar–CO, small for Ar–N_2 and N_2–O_2, and negligible for Ar–O_2 mixtures. This contribution, in the case of free energy in the third-order perturbation term, is found to be 7%, 9%, 15%, and 17% of the corresponding contribution arising in the second-order term for Ar–N_2, N_2–O_2, N_2–CO, and Ar–CO, respectively. In the case of internal energy, the third-order term contributes about 12%, 14%, 24%, and 27% to that of the contribution in second-order term for these systems. The contributions arising from higher-order terms are approximated using the [1,0] Padé approximant. The other branches of pair interaction that are of significance in the second-order term are anisotropic dispersion, shape, and the cross terms, shape × dispersion and permanent moment × dispersion interactions. The contributions of these branches of interactions to the third-order term are negligibly small, and therefore the truncation of the perturbation series at the second-order term is justified.

In Table V we compare the excess properties (Gibbs free energy, G^E, and enthalpy, H^E) at zero pressure calculated using Monte Carlo and Percus–Yevick values for the reference system (since these methods give almost identical values, they are not reported separately) and assuming ξ_{12} to be an adjustable parameter with the experimental (*59*) properties of several equimolar mixtures. In adjusting ξ_{12}, Shukla and Singh (*12*) used the data of G^E instead of H^E, which is more sensitive to changes in ξ_{12} (by roughly twice). This is done because values of H^E are not always available, and it is preferable to use the same procedure for all the mixtures.

Leonard et al. (*60*) have studied the excess properties of these mixtures by representing the interaction between like and unlike molecules by the Lennard–Jones 6-12 potential, and neglecting completely the

Table V. Excess Properties: Comparison of Theory and Experiment (P = 0, T = 83.82 K, $x_1 = x_2$ = 1/2)

				Calculated from L–J (6-12) Potential[b]	
System	*Property*	*Expt.*[a]	*Present Work*	*vdWl*	*B–H*
$Ar–N_2$	ξ_{12}	—	0.9995	1.002	0.999
	G^E	34	34	34	34
	H^E	51	44	33	30
$Ar–O_2$	ξ_{12}	0.982	0.9832	0.988	0.987
	G^E	37	37	37	37
	H^E	60	60	52	51
Ar–CO	ξ_{12}	—	0.994	0.988	0.985
	G^E	57	57	57	57
	H^E	—	90	78	71
N_2–CO	ξ_{12}	0.986	0.9975	0.991	0.990
	G^E	23	23	23	23
	H^E	—	32	35	33
$N_2–O_2$	ξ_{12}	—	0.9983	1.003[c]	0.999[c]
	G^E	39	39	42[c]	42[c]
	H^E	—	60	41[c]	36[c]

Note: G^E and H^E in joules/mol; ξ_{12} assumed to be an adjustable parameter.
[a] From Ref. 59.
[b] From Ref. 40.
[c] These values refer to T = 78 K.
Source: Reproduced with permission from Ref. 12. Copyright 1980, American Institute of Physics.

contributions arising from pair isotropic and three-body nonadditive interactions. The values they have found from van der Waals one-fluid theory of mixtures (vdW1) and from the generalized Barker–Henderson (*61*) perturbation theory (B–H) are given in columns 5 and 6 of Table V. We immediately observe that the agreement found by Shukla and Singh

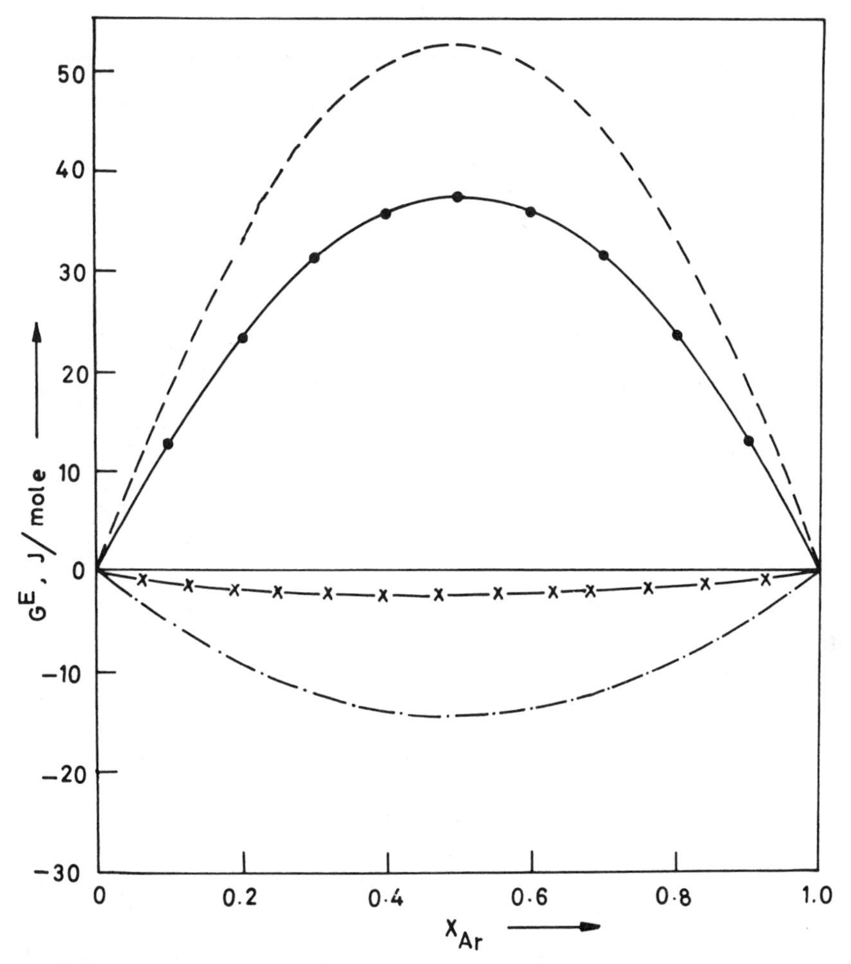

Figure 3. Excess Gibbs free energy, G^E, as a function of composition of argon–oxygen binary mixtures at T = 83.82 K and P = 0. Key: ●●●, experimental points; ——, theoretical curve (total contribution); − − −, total contribution arising from pair central potential (reference system), -×-×-, total contribution arising from pair anisotropic interaction; and -·-·-·-, total contribution arising from three-body nonadditive interactions. (Reproduced with permission from Ref. 12. Copyright 1980, American Institute of Physics.)

(*12*) between theory and experiment is good and is better than that found by Leonard et al. (*60*).

Shukla and Singh (*12*) have also studied the composition dependence of G^E and H^E for the binary mixtures Ar–N_2, Ar–O_2, Ar–CO, N_2–O_2 and N_2–CO. We plot the results in Figures 3–5 for only two systems, Ar–O_2 and N_2–O_2. In these figures we also plot separately the total contributions

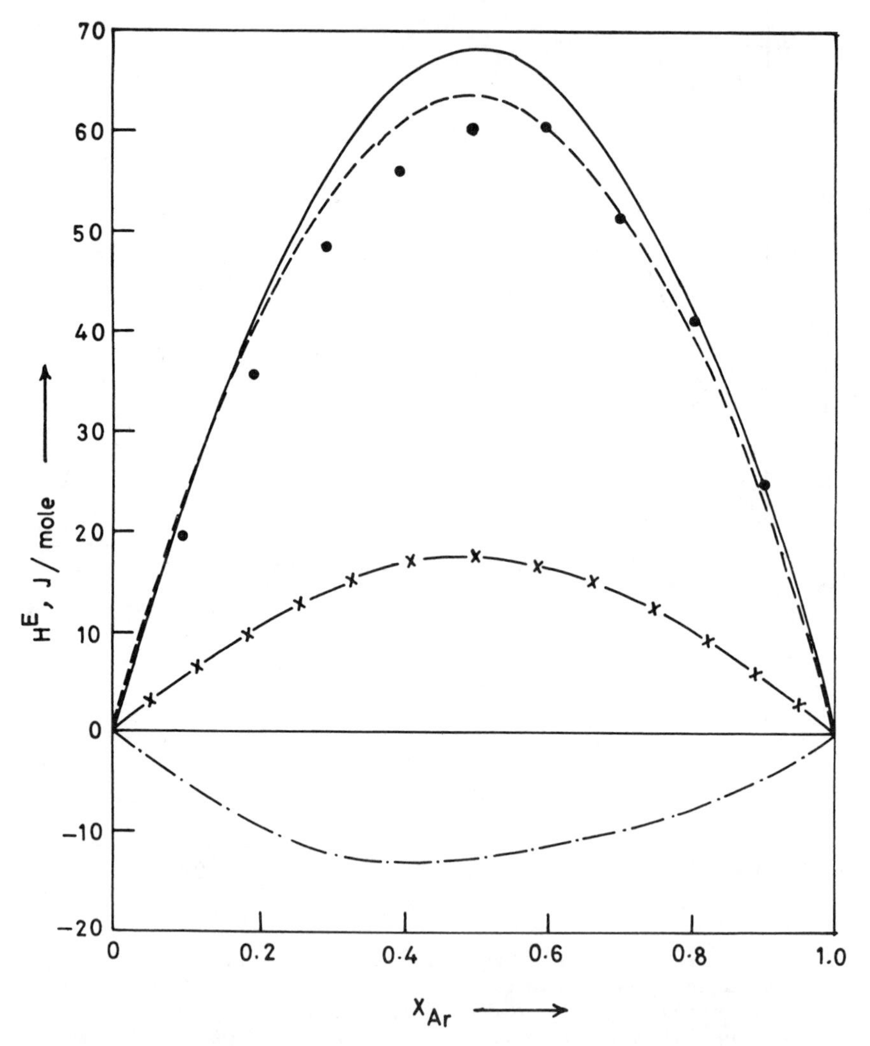

Figure 4. The excess enthalpy, H^E, *as a function of composition of argon–oxygen binary mixtures at* T = *83.82 K and* P = *0. Key as in Figure 3. (Reproduced with permission from Ref. 12. Copyright 1980, American Institute of Physics.)*

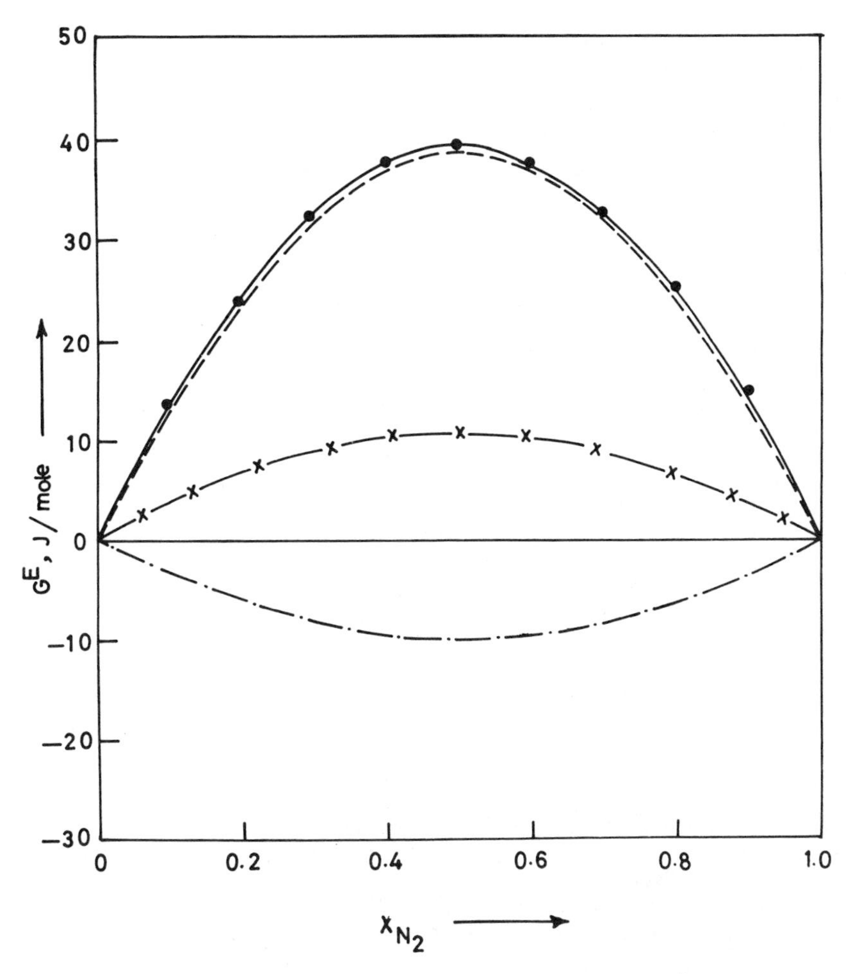

Figure 5. The excess Gibbs free energy, G^E, *as a function of composition of nitrogen–oxygen binary mixtures at* T = *83.82 K and* P = *0. Key as in Figure 3. (Reproduced with permission from Ref. 12. Copyright 1980, American Institute of Physics.)*

arising from the pair central potential, pair anisotropic interactions, and three-body nonadditive interactions. We note that the heat of mixing is more sensitive to pair anisotropic and three-body nonadditive interactions than it is to G^E. Similar conclusions hold for other systems.

We thus find that the predictions of the perturbation theory are in generally good agreement with experiment. It is shown that a better agreement with experiment is obtained when the anisotropy of pair interaction and the three-body nonadditivity are taken into account even for such simple systems as methane, oxygen, and nitrogen and such mixtures as Ar–O_2 and Ar–N_2. In conclusion, we therefore feel that the perturbation theory in which all the tensor interactions are taken as perturbations of the isotropic pair interaction is appropriate for describing the thermodynamic functions of simple molecular fluids and their mixtures.

Literature Cited

1. Axilrod, B. M.; Teller, E. *J. Chem. Phys.* **1943,** *11*, 299.
2. Simanoglu, O. *Adv. Chem. Phys.* **1967,** *12*, 283.
3. Barker, J. A.; Henderson, D.; Smith, W. R. *Mol. Phys.* **1968,** *17*, 579, and references therein.
4. Copeland, D. A.; Kestner, N. R. *J. Chem. Phys.* **1966,** *49*, 5214.
5. Chen, C. T.; Present, R. D. *J. Chem. Phys.* **1972,** *53*, 585.
6. Dulla, R. J.; Rowlinsonn, J. S.; Smith, W. R. *Mol. Phys.* **1971,** *21*, 299.
7. Goldman, S. *Acc. of Chem. Res.* **1979,** *11*, 409.
8. Sinha, S. K.; Ram, J.; Singh, Y. *J. Chem. Phys.* **1977,** *66*, 5013.
9. Ram, J.; Sinha, S. K.; Singh, Y. submitted for publication.
10. Stogryn, D. E. *J. Chem. Phys.* **1969,** *50*, 4967.
11. Shukla, K. P.; Singh, S.; Singh, Y. *J. Chem. Phys.* **1979,** *70*, 3086.
12. Shukla, K. P.; Singh, Y. *J. Chem. Phys* **1980,** *72*, 2719.
13. Singh, S.; Singh, Y. *Physica, Ultrecht* **1976,** *83A*, 339.
14. Singh, S.; Singh, Y. *Physica, Ultrecht* **1977,** *87A*, 207.
15. Gray, C. G.; Henderson, R. J. *Can. J. Phys.* **1979,** *57*, 1605, and references therein.
16. Pople, J. A. *Proc. Roy. Soc.* **1954,** *221A*, 498, 507.
17. Rowlinson, J. S. "Liquids and Liquid Mixtures"; Butterworths: London, 1969.
18. Sung, S.; Chandler, D. *J. Chem. Phys.* **1972,** *56*, 4989.
19. Mo, K. C.; Gubbins, K. E. *J. Chem. Phys.* **1975,** *63*, 1490.
20. Sandler, S. *Mol. Phys.* **1974,** *28*, 1207.
21. Ananth, M. S.; Gubbins, K. E.; Gray, C. G. *Mol. Phys.* **1974,** *28*, 1005.
22. Flytzani-Stephanopoulos, M.; Gubbins, K. E.; Gray, C. G. *Mol. Phys.* **1975,** *30*, 1649.
23. Chambens, M. V.; McDonald, K. R. *Mol. Phys.* **1975,** *29*, 1053.
24. Stell, G.; Rasaiah, J. C.; Narang, H. *Mol. Phys.* **1974,** *27*, 1393.
25. Singh, Y. *Mol. Phys.* **1975,** *29*, 155.
26. Shukla, K. P.; Pandey, L.; Singh, Y. *J. Phys. C., Solid State Phys.* **1979,** *12*, 4151.
27. Verlet, L.; Weis, J. J. *Mol. Phys.* **1974,** *28*, 1207.
28. Stell, G.; Weis, J. J. *Phys. Rev.* **1977,** *16A*, 757.
29. Gray, C. G.; Gubbins, K. E.; Twu, C. H. *J. Chem. Phys.* **1978,** *69*, 182.
30. Kielich, S. *Acta Phys. Polon.* **1965,** *28*, 459.
31. Buckingham, A. D. *Adv. Chem. Phys.* **1967,** *12*, 107.
32. Stogryn, D. E. *Mol. Phys.* **1971,** *22*, 81.
33. Wertheim, M. S. *Mol. Phys.* **1979,** *37*, 83.
34. O'Shea, S. F.; Meath, W. J. *Mol. Phys.* **1974,** *28*, 1431.
35. Barker, J. A. In "Rare Gas Solids"; Klein, M. L.; Venables, J. A. Eds., Academic Press: New York, 1976; Vol. 1, Chap. 4.
36. Barker, J. A.; Fisher, R. A.; Walts, R. O. *Mol. Phys.* **1971,** *21*, 657.
37. Klein, M. L.; Kochler, T. R. In "Rare Gas Solids"; Klein, M. L.; Venables, J. A. Eds., Academic Press: New York, 1976; Vol. 1, Chap. 6.

38. Gray, C. G.; Gubbins, K. E.; Twu, C. H. *J. Chem. Phys.* **1978,** *69*, 182.
39. Kielich, S. *Physica, Ultrecht* **1965,** *31*, 444.
40. Gubbins, K. E.; Gray, C. G.; Machado, J. R. S. *Mol. Phys.* **1981,** *42*, 817.
41. Shing, K.; Gubbins, K. E., Chap. 4 in this book.
42. Kohler, F.; Quirke, N., Chap. 10 in this book.
43. Weis, J. J.; Levesque, D. *Phys. Rev.* **1976,** *13A*, 450.
44. Wang, S. S.; Gray, C. G.; Egelstaff, P. A.; Gubbins, K. E. *Chem. Phys. Lett.* **1974,** *24*, 453.
45. Verlet, L.; Weis, J. J. *Phys. Rev.* **1972,** *5A*, 939.
46. Weeks, J. D.; Chandler, D.; Andersen, H. C. *J. Chem. Phys.* **1971,** *54*, 5237.
47. Gubbins, K. E.; Twu, C. H. *Chem. Eng. Sci.* **1978,** *33*, 863.
48. Shukla, K. P.; Ram, J.; Singh, Y. *Mol. Phys.* **1976,** *31*, 873.
49. Larsen, B.; Rasaiah, J. C.; Stell, G. *Mol. Phys.* **1977,** *33*, 987.
50. Melnyk, T. W.; Smith, W. R. *Chem. Phys. Lett.* **1974,** *28*, 313.
51. Twu, C. H.; Gubbins, K. E.; Gray, C. G. *Mol. Phys.* **1975,** *29*, 713.
52. Twu, C. H.; Gubbins, K. E.; Gray, C. G. *J. Chem. Phys.* **1976,** *64*, 5186.
53. Clancy, P.; Gubbins, K. E.; Gray, C. G. *J. Chem. Soc., Faraday Disc.* **1979,** *66*, 116.
54. Lobo, L. Q.; McClure, D. W.; Staneley, L. A. K.; Clancy, P.; Gubbins, K. E.; Gray, C. G. *J. Chem. Soc., Faraday Trans. II* **1981,** *77*, 425.
55. Leland, T. W. Jr.; Rowlinson, J. S.; Sather, G. A. *J. Chem. Soc., Trans. Faraday* **1968,** *64*, 1447.
56. McDonald, I. R.; Singer, K. *Mol. Phys.* **1972,** *23*, 29.
57. Grundke, E. W.; Henderson, D.; Murphy, R. D. *Can. J. Phys.* **1973,** *51*, 1216.
58. Mo, K. C.; Gubbins, K. E.; Jacucci, G.; McDonald, I. R. *Mol. Phys.* **1972,** *23*, 29.
59. Pool, R.; Saville, G.; Herrington, T. M.; Shields, B. D. C.; Stavaley, L. A. K. *J. Chem. Soc., Trans. Faraday* **1962,** *58*, 1692.
60. Leonard, P. J.; Henderson, D.; Barker, J. A. *J. Chem. Soc., Trans. Faraday* **1970,** *66*, 2439.
61. Barker, J. A.; Henderson, D. *Rev. Mod. Phys.* **1976,** *48*, 587.
62. Din, F., "Thermodynamic Functions for Gases"; Butterworths: London, 1963, Vol. 1–3.
63. Streett, W. B.; Staveley, L. A. K. *Adv. Cryog. Eng.* **1963,** *13*, 363.
64. Hoover, A. E.; Canfield, F. B.; Kobayashi, R.; Leland, T. W. *J. Chem. Eng. Data* **1964,** *9*, 568.
65. Michels, A.; Lunbeck, R. J.; Wolkers, G. J. *Physica* **1951,** *17*, 801.
66. Michels, A.; Wouters, H.; deBoer, J. *Physica, Ultrecht* **1934,** *1*, 587.
67. Butcher, E. G.; Dadson, R. S. *Proc. Roy. Soc., London* **1964,** *277A*, 448.
68. MacCormack, K. E.; Schneider, W. G. *J. Chem. Phys.* **1950,** *18*, 1269.

RECEIVED for review January 27, 1982. ACCEPTED for publication September 9, 1982.

16

Investigation of Binary Liquid Mixtures via the Study of Infinitely Dilute Solutions

D. A. JONAH

University of Sierra Leone, Department of Mathematics, Fourah Bay College, Freetown, Sierra Leone, West Africa

The prediction and correlation of activity coefficients and their derivatives for binary liquid mixtures at infinite dilution have been considered on both the thermodynamic and molecular levels. The related considerations of gas solubilities in liquids and the solubilities of solids in supercritical fluids have also been studied. Based on a method of numerical differentiation using intercepts rather than slopes, it has been demonstrated that the above limiting thermodynamic functions can be estimated with sufficient accuracy from either excess Gibbs free energy or vapor pressure data available near both ends of the composition range. Limiting activity coefficients and their derivatives so estimated have been used in three empirical expressions for excess Gibbs free energy, to extrapolate from the dilute portion to the other parts of the composition range; such extrapolations have been found to be in good agreement with experimental measurements. Using rigorous statistical mechanics, combined with certain semiempirical arguments, a number of tested correlations of the limiting residual chemical potential (and also solubilities) with pure solvent properties are presented; these are found to be in good agreement with experimental data.

THE STUDY OF VERY DILUTE SOLUTIONS deserves serious attention for at least three reasons:

1. Such a study has important practical applications in extractive and azeotropic distillations where important components often occur in very low concentrations.

0065-2393/83/0204-0395$08.00/0

2. The constants in certain empirical equations for excess functions are related to thermodynamic functions at infinite dilution (e.g., the limiting activity coefficients and their first derivatives); further, these constants are needed for the prediction of multicomponent vapor–liquid equilibria.
3. Very dilute solutions are of considerable interest to the theoretician for at least two reasons. First, the absence of solute–solute interactions helps to isolate the unlike solute–solvent interactions against a background of solvent–solvent interactions, so that thermodynamic properties at infinite dilution provide a convenient source of information about these unlike pair interactions so necessary in any attempt to relate these to the like interactions. Second, the dominance of the solvent–solvent interactions makes the contribution from the solute–solvent interactions to the total energy a veritable small perturbation. Therefore, an infinitely dilute system is ideal for the application of a perturbation about the pure solvent, as one possible route in the prediction of mixture properties from those of pure components.

It is therefore surprising that more attention has not been devoted to this important aspect of mixture theory.

As is evident from Reason 2, the study of infinitely dilute solutions can form the basis for the study of binary mixtures at finite concentrations. This appears to be a viable alternative to the usual approach, which focuses attention on the mixture at finite concentration. Then, at some stage in the statistical thermodynamical development, information about the composition-dependence of the principal thermodynamic functions is injected into the theory, through assumptions expressing the composition-dependence of mixture molecular parameters in their relationship to corresponding parameters of the pure components.

In this alternative approach, however, the form of composition-dependence of a basic thermodynamic function, such as the excess Gibbs free energy, is assumed beforehand through the use of one of the empirical equations with constants related to limiting thermodynamic quantities. Attention is then focused on the evaluation or prediction of these constants with the help of molecular thermodynamics.

In this chapter, we shall apply this alternative approach to the study of binary nonelectrolyte mixtures. Infinitely dilute solutions will, of course, also be studied in their own right in attempts to predict such properties as Henry constants, solubilities of liquids in gases, and the solubilities of liquids and solids in supercritical gases. There will be two broad sections, one based on purely thermodynamic considerations, and the other on molecular thermodynamics.

The "Thermodynamic Considerations" section briefly reviews some empirical equations whose constants are related to limiting quantities; special attention is given to a modified version of the two-constant Van Laar equation, called the Conic equation (*1, 2*). Examination of this equation as a correlating and extrapolative device—that is, a means for the extrapolation of excess functions measured for very dilute solutions into the rest of the composition range—is next considered; then follows an examination of the evaluation of the constants in this equation from vapor pressure–composition curves. Comparison is made with experimental data.

The "Molecular Thermodynamics" section addresses the problem of predicting the constants in the empirical equations or correlating them with pure component properties. My approach combines rigorous statistical thermodynamics with empirical and semiempirical arguments. Also considered are the solubilities of liquids and solids in supercritical gases. Comparison is made between my equations and experimental data.

Thermodynamic Considerations

Some Empirical Equations. Although several useful empirical and semiempirical equations express the composition-dependence of the excess Gibbs free energy, G^E, few of these have the useful and desirable property of their constants being simply related to the limiting derivatives of the excess functions. The more familiar equations in this desirable class are the well-known two-constant Van Laar and Margules equations and the one-constant regular solution equation. In each of these the constants are related to the limiting activity coefficients (or the limiting first derivatives of the excess function, G^E).

However, these equations seldom do well when the constants are identified as just stated; invariably, these constants have to be evaluated by some method (e.g., by least squares) using data over the whole concentration range. However, in not identifying the constants with limiting properties, we are sacrificing a very valuable feature of these equations that makes them useful not only as correlating equations, but also as extrapolative equations for deducing the excess function over the rest of the composition range, given a few measurements near either end of this range. Hence, these equations need to be modified to make it possible to use limiting thermodynamic quantities for the constants and achieve, at the same time, a close fit to the experimental data.

Such modifications of the Van Laar and Margules equations have been suggested by Jonah and Ellis (*1, 2*) and by Abbott and Van Ness (*3*). The modified Van Laar form, the Conic equation, is given by

$$q^2 + a_{12}x_1x_2 + q(\alpha_1x_1 + \alpha_2x_2) = 0 \tag{1}$$

where $q \equiv G^E/RT$.

The modified Margules forms are

$$M_1:\ q = x_1x_2\{A_{21}x_1 + A_{12}x_2 - (\lambda_{21}x_1 + \lambda_{12}x_2)x_1x_2\} \tag{2}$$

and

$$M_2:\ q = x_1x_2\left(A_{21}x_1 + A_{12}x_2 - \frac{\lambda_{12}\,\lambda_{21}x_1x_2}{\lambda_{12}x_1 + \lambda_{21}x_2}\right) \tag{3}$$

The constants in Equations 1, 2, and 3 are all related to the limiting activity coefficients and their derivatives as follows

$$\begin{aligned}\lambda_{12} &= A_{21} - 2A_{12} - \frac{1}{2}\left(\frac{\partial \ln \gamma_1}{\partial x_1}\right)^{\infty}\\ \lambda_{21} &= A_{12} - 2A_{21} - \frac{1}{2}\left(\frac{\partial \ln \gamma_2}{\partial x_2}\right)^{\infty}\end{aligned} \tag{4}$$

$$a_{12}^{-1} = \begin{cases} \dfrac{1}{2A_{12}^3}\left[\left(\dfrac{\partial \ln \gamma_1}{\partial x_1}\right)^{\infty} + \dfrac{2A_{12}^2}{A_{21}}\right] \\[2ex] \dfrac{1}{2A_{21}^3}\left[\left(\dfrac{\partial \ln \gamma_2}{\partial x_2}\right)^{\infty} + \dfrac{2A_{21}^2}{A_{12}}\right] \end{cases} \tag{5}$$

$$\alpha_1 = \frac{-a_{12}}{A_{21}} \quad \text{and} \quad \alpha_2 = \frac{-a_{12}}{A_{12}}$$

$$A_{12} = \ln \gamma_1^{\infty} \quad \text{and} \quad A_{21} = \ln \gamma_2^{\infty}$$

To see that Equation 1 is a modified form of the Van Laar equation, we rewrite it in the form

$$b_{12}q^2 + x_1x_2 - q\left(\frac{x_1}{A_{21}} + \frac{x_2}{A_{12}}\right) = 0 \tag{6}$$

where $b_{12} \equiv a_{12}^{-1}$. When $b_{12} = 0$, Equation 1 reduces to the two-constant Van Laar equation; if $A_{12} = A_{21}$, the regular solution equation is recovered.

In these modifications of the two-constant Van Laar and Margules equations the additional constants merely take account of the curvature of the excess Gibbs free energy curves at the ends of the composition range. If the curvature is negligible, then we expect to find the two-constant equations adequate for correlation purposes. Also, whereas the modified Margules equations, Equations 2 and 3, require limiting derivatives at both ends of the composition range, the Conic equation, Equation 1, requires just one of these same quantities. As will become evident later, this is a very useful feature. Often, in practice, both limiting derivatives cannot be evaluated with the same degree of certainty; in such cases we choose the more reliable of the two limiting derivatives.

The introduction of curvature parameters into the two-constant Van Laar and Margules equations has the effect of constraining the fitted excess curve to pass through the maximum (or minimum) point. It is therefore not surprising to find that a_{12} in the Conic equation can be expressed in terms of the maximum (or minimum) excess function

$$-a_{12}^{-1} = \frac{\{1 - e(A_{21}^{-1} - A_{12}^{-1})\}^2 - 4eA_{12}^{-1}}{4e^2} \tag{7}$$

where $e \equiv G_{\max}^{E}$. Using Equation 5 in Equation 7, we have the alternative expressions for the limiting derivatives of the activity coefficients:

$$\left(\frac{\partial \ln \gamma_1}{\partial x_1}\right)^{\infty} = -\frac{A_{12}^3\{1 - e(A_{21}^{-1} - A_{12}^{-1})\}^2 - 4eA_{12}^2}{2e^2} - \frac{2A_{12}^2}{A_{21}}$$

and (7a)

$$\left(\frac{\partial \ln \gamma_2}{\partial x_2}\right)^{\infty} = -\frac{A_{21}^3\{1 - e(A_{12}^{-1} - A_{21}^{-1})\}^2 - 4eA_{21}^2}{2e^2} - \frac{2A_{21}^2}{A_{12}}$$

For correlation purposes, when we have data over the whole composition range, Equation 7a is obviously preferable to a direct method of evaluating the limiting derivatives for data satisfying the Conic equation. On the other hand, Equation 7a is not useful for extrapolation of data from the very dilute regions of the concentration range into other regions of this range. However, for data satisfying the Conic equation, Equation 7a provides a standard by which to judge the effectiveness of a direct method, such as the method of intercepts, for evaluating derivatives, and it can serve as a check on the estimates of the derivatives used to evaluate this same constant by Equation 5.

The Conic equation does surprisingly well in correlating the excess functions of a wide class of binary systems, in spite of its unattractiveness in that it gives q as an implicit function of composition. The criterion for applicability is simple: the excess function has to satisfy a "rule of rectilinear diameters" (*see* Figure 1). I have yet to come across excess Gibbs free energy values for binary systems that depart significantly from this rule. The slope of the straight line joining the midpoints of the diameters can be simply related to the limiting first derivatives of the excess function.

Evaluation of Limiting Derivatives. The usefulness of Equations 1–3 as extrapolating equations depends very much on the accurate evaluation of the limiting first and second derivatives of the excess functions. A method for the evaluation, by the method of intercepts, of the first and second derivatives of experimental data has already been developed and described elsewhere (7); the method has been tested against several sets of experimental data, and found to yield reliable estimates. I shall merely quote the principal equations here and give the results of applying them to the excess Gibbs free energy data for three binary systems. Let $\phi(x)$ be some function of x whose first and second derivatives are required at $x = b$. Then these are given by

$$\left(\frac{d\phi}{dx}\right)_{x=b} = \left(\frac{d\phi_T}{dx}\right)_{x=b} + (b - a)L_b^{(1)} \tag{8}$$

where

$$L_b^{(1)} \equiv \lim_{x \to b} \frac{\phi - \phi_T}{(x - a)(x - b)} \tag{8a}$$

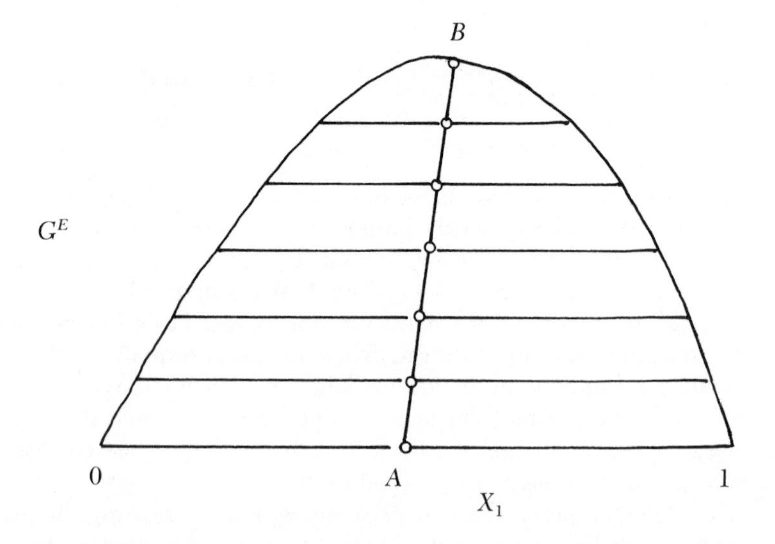

Figure 1. Rectilinear diameter criterion that must be satisfied by an excess function in order to be closely fitted by Equation 1. The slope s *of the locus* AB *of the midpoints of diameters is given by* s $= 2A_{12}A_{21}/(A_{21} - A_{12})$.

and

$$\left(\frac{d^2\phi}{dx^2}\right)_{x=b} = 2(b - a)\left\{\left(\frac{dq_1}{dx}\right)_{x=b} + (b - a)L_b^{(2)}\right\} + 2L_b^{(1)} + 2\left(\frac{d\phi_T}{dx}\right)_{x=b} \quad (9)$$

where

$$L_b^{(2)} \equiv \lim_{x \to b} \frac{\Phi - q_1}{(x - a)(x - b)} \quad (9a)$$

$$\Phi \equiv (\phi - \phi_T)/(x - a)(x - b) \quad (10)$$

$\phi_T(x)$ and $q_1(x)$ are certain trial functions, which may be linear, quadratic, or otherwise. The linear forms (which we employ in this chapter) are given by

$$\phi_T = \phi(a) + \frac{\phi(b) - \phi(a)}{b - a}(x - a) \quad (11)$$

$$q_1 = L_a^{(1)} + \frac{L_b^{(1)} - L_a^{(1)}}{b - a}(x - a) \quad (12)$$

The value of a in the above equations is chosen according to convenience.

The important feature in the above formulas is that they make use of intercepts, rather than slopes, for the evaluation of derivatives. They have been applied in the next section in evaluating the first and second limiting pressure derivatives (Equations 15 and 16), and also the limiting activity coefficients and their derivatives from excess Gibbs free energy–composition curves, for use in Equations 1–3. Three binary systems have been considered:

ethanol–*n*-heptane (30 °C) (*5*)
acetone–carbon tetrachloride (45 °C) (*6*)
nitromethane–carbon tetrachloride (45 °C) (*6*)

The results are summarized in Table I.

In Table I, "$g^E(1)$ calc" indicates excess free energies obtained from the Conic equation and the modified versions of Margules equations, M_1 and M_2, given in Equations 1–3. The constants A_{12} and A_{21} have been chosen to obtain the best possible fit with the Conic equation, using the values obtained by the method of intercepts as first approximations. The limiting derivatives of the activity coefficients are evaluated according to Equation 7a and should be close to the true values, considering the very

Table I. Excess Gibbs Free Energy From Equations 1–3 with Constants Identified with Limiting Properties Evaluated by Equations 8–12, 13a, and 14a

x_1	g^E *Expt.*	$g^E(1)$ *Calc.*			$g^E(2)$ *Calc.*	g^E (*From* P–x *Data*)		
		C	M_1	M_2	C	C	M_1	M_2
			Ethanol (1) and *n*-Heptane (2) (30 °C) (*19*)					
0.1	0.262	0.262	0.242	0.246	0.262	0.266	0.275	0.276
0.3	0.508	0.511	0.365	0.407	0.511	0.508	0.449	0.456
0.5	0.570	0.568	0.380	0.440	0.566	0.559	0.444	0.454
0.7	0.478	0.476	0.378	0.409	0.475	0.469	0.393	0.399
0.9	0.219	0.219	0.211	0.214	0.219	0.218	0.210	0.211
Constants						*Jonah*		*Klaus et al.*
A_{12}	=		3.52		3.55	3.75		4.09
A_{21}	=		2.65		2.65	2.68		2.69
$\left(\frac{\partial \ln \gamma_1}{\partial x_1}\right)^\infty$	=		−26.2		−21.8	−23.11		−40.62
$\left(\frac{\partial \ln \gamma_2}{\partial x_2}\right)^\infty$	=		−11.18		−11.3	−12.81		−12.42
λ_{12}	=		8.71		6.45	6.74		—
λ_{21}	=		3.81		3.90	4.80		—
			Nitromethane (1) and Carbon Tetrachloride (2) (45 °C) (*23*)					
0.0459	0.110	0.108	0.108	0.108	0.108			
0.0918	0.199	0.198	0.196	0.197	0.199			
0.1954	0.350	0.350	0.340	0.343	0.354			
0.2829	0.432	0.434	0.415	0.421	0.441			
0.3656	0.482	0.482	0.456	0.465	0.491			
0.4659	0.504	0.504	0.476	0.485	0.514			
0.5366	0.501	0.498	0.471	0.480	0.507			
0.6065	0.476	0.475	0.452	0.459	0.483			
0.6835	0.432	0.429	0.412	0.418	0.435			
0.8043	0.317	0.313	0.307	0.309	0.317			
0.9039	0.176	0.176	0.175	0.175	0.176			
0.9488	0.099	0.099	0.099	0.099	0.099			
Constants								
A_{12}	=		2.578		2.547			
A_{21}	=		2.072		2.056			
$\left(\frac{\partial \ln \gamma_1}{\partial x_1}\right)^\infty$	=		−10.59		−9.37			
$\left(\frac{\partial \ln \gamma_2}{\partial x_2}\right)^\infty$	=		−5.50		—			
λ_{12}	=		2.211		—			
λ_{21}	=		1.184		—			

Table I. Excess Gibbs Free Energy From Equations 1–3 with Constants Identified with Limiting Properties Evaluated by Equations 8–12, 13a, and 14a (Continued)

x_1	g^E *Expt.*	$g^E(1)$ *Calc.* C	M$_1$	M$_2$	$g^E(2)$ *Calc.* C	g^E (*From* P–x *Data*) C	M$_1$	M$_2$
			Acetone (1) and Carbon Tetrachloride (2) (45°C) (*6*)					
0.0556	0.057	0.057	0.057	0.057	0.057	0.055	0.055	0.056
0.0903	0.087	0.086	0.086	0.086	0.086	0.084	0.084	0.086
0.2152	0.161	0.161	0.157	0.161	0.162	0.160	0.159	0.168
0.2929	0.188	0.189	0.182	0.188	0.189	0.188	0.187	0.199
0.3970	0.209	0.207	0.198	0.206	0.208	0.207	0.206	0.222
0.4769	0.208	0.208	0.199	0.207	0.210	0.208	0.209	0.224
0.5300	0.203	0.204	0.195	0.203	0.205	0.204	0.205	0.219
0.6047	0.191	0.191	0.183	0.190	0.192	0.191	0.193	0.204
0.7128	0.158	0.158	0.154	0.158	0.160	0.158	0.161	0.167
0.8088	0.117	0.117	0.115	0.117	0.118	0.116	0.119	0.121
0.9090	0.061	0.061	0.061	0.061	0.062	0.061	0.062	0.062
0.9636	0.026	0.026	0.026	0.026	0.026	0.025	0.026	0.026
Constants								
A_{12}	=		1.147		1.146		1.102	
A_{21}	=		0.730		0.739		0.718	
$\left(\frac{\partial \ln \gamma_1}{\partial x_1}\right)^\infty$	=		−4.869		−4.70		−4.14	
$\left(\frac{\partial \ln \gamma_2}{\partial x_2}\right)^\infty$	=		−1.258		—		−0.756	
λ_{12}	=		0.8705		—		0.584	
λ_{21}	=		0.316		—		0.044	

Note: $g^E \equiv G^E/RT$

close fits by the Conic equations. The modified Margules equations M_1 and M_2 do not give as close fits as the Conic equations, except for the system acetone–carbon tetrachloride, for which M_2 does as well as the Conic equation. It is, however, probable that M_1 and M_2 may give closer fits, by not identifying the constants with limiting thermodynamic properties. The close agreement between $g^E(1)$ and experimental values demonstrates the effectiveness of the above equations for purposes of correlation.

In Table I, "$g^E(2)$ calc" indicates values of the excess free energy as obtained from the Conic equation using constants that have been calculated entirely from limiting activity coefficients and their derivatives, as estimated by method of intercepts. Corresponding values for M_1 and M_2 are not shown, as these equations require both limiting derivatives, which in general are not available. The good agreement between $g^E(2)$ and the experimental values demonstrates the potentiality of extrapolating excess free energy from the dilute ends of the concentration range to other parts of the range.

We next examine the possibility of calculating excess free energies over the whole composition range, given a few total vapor pressure measurements near both ends of the range.

Excess Gibbs Free Energy from Vapor Pressure Data. The limiting activity coefficients and their derivatives can be expressed in terms of the limiting first and second pressure derivatives with respect to composition; corresponding expressions in terms of limiting first and second temperature derivatives with respect to composition for isobaric data are also available. Gautreax and Coates (8) were the first to derive such relationships but only for the limiting activity coefficients within the symmetric standard states convention. Jonah has rederived these relations for both the symmetric and unsymmetric choice of standard states; analogous relations for the limiting first derivatives of the activity coefficients within the symmetric standard states convention, have also been derived (9). Only the isothermal relations, which have been applied to two binary systems, acetone–carbon tetrachloride (45 °C), and ethanol–n-heptane (30 °C) are given here:

$$\gamma_1^\infty = \left(\frac{\phi_1}{f_1{}^*}\right)^\infty P_2^* \left\{1 + \frac{v_2^{*g} - v_2^*}{RT}\left(\frac{\partial P}{\partial x_1}\right)^\infty\right\} \tag{13}$$

$$\left(\frac{\partial \ln \gamma_1}{\partial x_1}\right)^\infty = \frac{\dfrac{\partial^2 P}{\partial x_1^2} - 2\gamma_1 \dfrac{\partial}{\partial x_1}\left(\dfrac{f_1{}^*}{\varphi_1}\right) + 2\dfrac{\partial P}{\partial x_1}\left\{(1 - z_2^*) - P_2^* \dfrac{\partial}{\partial x_1}\left(\dfrac{v^E}{RT}\right)\right\} - \dfrac{\partial^2}{\partial x_1^2}\left(\dfrac{f_2{}^*}{\varphi_2}\right)}{2\gamma_1\left(\dfrac{f_1{}^*}{\varphi_1}\right) - P_2^*} \tag{14}$$

All quantities on the right-hand sides of Equations 13 and 14 have been evaluated in the state of infinite dilution, the mole fraction x_1 tending to zero; in particular all derivatives are limiting derivatives. In Equations 13 and 14, P_i^* is the saturated vapor pressure of the pure component i; v_i^{*g} and v_i^{*l} are the molar volumes of the pure component i in the vapor and liquid phases, respectively; v^E is the excess volume of mixing; f_i^* is the fugacity of pure component i; φ_i is the fugacity coefficient of component i in the binary mixture; and z_i^* is the compressibility factor of pure component i.

In applying Equations 13 and 14, we shall neglect vapor-phase non-ideality; the liquid phase molar volume v_1^{*l} and the derivative of the excess volume of mixing will also be neglected. Under these conditions, Equations 13 and 14 take the simpler forms given as Equations 13a and 14a, respectively

$$\gamma_1^\infty = \frac{1}{P_1^*}\left\{P_2^* + \left(\frac{\partial P}{\partial x_1}\right)^\infty\right\} \tag{13a}$$

$$\left(\frac{\partial \ln \gamma_1}{\partial x_1}\right)^\infty = \frac{\left(\dfrac{\partial^2 P}{\partial x_1^2}\right)^\infty}{2\left(\dfrac{\partial P}{\partial x_1}\right)^\infty + P_2^*} \tag{14a}$$

Analogous relations to Equations 13 and 14 and Equations 13a and 14a can be readily written down from symmetry for the other activity coefficient. Equation 14a has been quoted earlier by Abbott and Van Ness (2) but without details of its derivation.

We now apply Equations 13a and 14a to the two binary systems ethanol–*n*-heptane (30 °C), and acetone–carbon tetrachloride (45 °C). Equations 8–12 have been used to evaluate the limiting pressure derivatives, the linear forms having been chosen for ϕ_T and q_1, viz

$$\begin{aligned}\phi_T &= P_1^* x_1 + P_2^* x_2 \\ q_1 &= L_1^{(1)} x_1 + L_0^{(1)} x_2\end{aligned} \tag{15}$$

where

$$L_0^{(1)} \equiv \lim_{x_1 \to 0} \frac{\Delta P}{x_1(x_1 - 1)} \quad \text{and} \quad L_1^{(1)} \equiv \lim_{x_1 \to 1} \frac{\Delta P}{x_1(x_1 - 1)}$$

ΔP is the excess pressure defined by

$$\Delta P = P - x_1 P_1^* - x_2 P_2^*$$

When the above expressions for ϕ_T and q_1 are used in Equations 8 and 9, we have for the limiting pressure derivatives

$$\begin{aligned}\left(\frac{\partial P}{\partial x_1}\right)^\infty &= -L_0^{(1)} + P_1^* - P_2^* \\ \left(\frac{\partial^2 P}{\partial x_1^2}\right)^\infty &= 2(L_0^{(2)} + 2L_0^{(1)} - L_1^{(1)})\end{aligned} \tag{16}$$

We recall that the intercept $L_0^{(2)}$ is given by

$$L_0^{(2)} \equiv \lim_{x_1 \to 0} \frac{\{\Delta P/x_1(x_1 - 1)\} - q_1}{x_1(x_1 - 1)}$$

Analogous expressions for the limiting pressure derivatives at $x_1 = 1$ can be readily written by symmetry.

We have applied the above equations to two of the three binary systems considered in the previous section, namely, ethanol–*n*-heptane (30 °C) (5) and acetone–carbon tetrachloride (45 °C) (*6*). In the case of the system ethanol–*n*-heptane, the vapor pressure–composition data have been differentiated throughout the composition range (including $x_1 = 0$ and $x_1 = 1$) by Klaus and Van Ness (*10*), using an extended version of the rigorous spline technique (*4*). Thus in Table I, which summarizes the results of applying Equations 13a and 14a to this system, we have compared the values of the limiting derivatives of the excess function G^E, as obtained from their limiting pressure derivatives, with that based on our estimate of the limiting pressure derivatives according to the method of intercepts. It is readily seen that there is good agreement at the ethanol end of the composition range, but marked divergences at the other end. At the *n*-heptane end of the composition range, our estimate of the limiting first pressure derivative is 3320 (± 100) as compared with 4660 obtained by Klaus and Van Ness; while our estimate for the limiting second pressure derivative is −155,000 (± 18,000) as compared with −381,000 obtained by Klaus and Van Ness.

The limiting activity coefficients and their derivatives, based on the two sets of estimates for the limiting pressure derivatives, have been used in Equations 1–3 for evaluating the excess Gibbs free energy over the whole of the composition range; the results are shown in the column headed g^E (from P–x data).

For the system acetone–carbon tetrachloride, all three equations provide good predictions for the excess free energy over the whole composition range. For the system ethanol–*n*-heptane however, the Conic equation does a much better job then either the M_1 or M_2 equation. Further, the close agreement between g^E from P–x data (as obtained from the Conic equation) and the experimental values suggest that the estimates of the limiting pressure derivatives as obtained by our method are closer to the true values at the *n*-heptane end of the composition range than are those of Klaus and Van Ness.

Our method of estimating the limiting pressure derivatives employs the raw, unsmoothed experimental data near both ends of the composition range; no fitting of data to any curve is involved.

Molecular Thermodynamic Considerations

The preceding sections demonstrated that knowledge of the limiting activity coefficients and their first derivatives for binary liquid systems enables us to extrapolate fairly accurately data for dilute solutions into the regions of finite composition. Further, we have demonstrated that these limiting quantities can be reliably estimated by a method of numerical differentiation based on the measurement of intercepts, rather than slopes.

This section addresses the problem of predicting these limiting thermodynamic properties of interest, through molecular thermodynamic considerations. We shall also be concerned with very dilute systems in their own right, in trying to predict and correlate with pure solvent properties such quantities as Henry's constants, solubilities of liquids in gases, and the solubilities of solids and liquids in supercritical gases. Most of the relevant results have already been discussed elsewhere (*11*), so, for the most part, only the main results will be reviewed.

Exact Expressions for Limiting Residual Chemical Potentials and Derivatives. As reported elsewhere (*11*) the limiting residual chemical potential μ_{ar}^{∞} is given by

$$\frac{\mu_{ar}^{\infty}}{kT} = \frac{1}{kT}\int_0^1 d\xi \int \rho_b \langle u_{ab}(12)\, g_{ab}^{\infty}(12;\xi)\rangle_{\omega_1,\omega_2}\, d\boldsymbol{r}_{12} \qquad (17)$$

where $u_{ab}(12) \equiv u_{ab}(\boldsymbol{r}_{12}, \omega_1, \omega_2)$ is the pair potential energy between two molecules, one of species a, and the other of species b; $g_{ab}(12;\xi)$ is the usual angular pair correlation function for the a–b pair, in a system in which the a-species is coupled to the extent ξ with the remaining molecules; $\langle\ \rangle_{\omega_1\omega_2}$ indicates an unweighted average over all molecular orientations.

The limiting residual chemical potential may be expressed either in terms of the Henry's constant H_{ab} (*12*) or the limiting activity coefficients γ_i^{∞} (*13*) (defined within the symmetric standard-states convention), as follows

$$\frac{\mu_{ar}^{\infty}}{kT} = \ln\left(\frac{H_{ab}}{\rho_b kT}\right) \qquad (18)$$

and

$$\frac{\mu_{ar}^{\infty}}{kT} = \ln \gamma_a^{\infty} + \frac{\mu_{ar}^{*}}{kT} - \ln \frac{v_a}{v_b} \qquad (19)$$

where ρ_b and v_b are the pure solvent density and molar volume, respectively; v_a and μ_{ar}^* are the pure solute molar volume and residual chemical potential per molecule, respectively.

Equation 17 is exact and rigorous, except for the assumption of pairwise additivity; together with Equation 18 it generalizes the equation of Jonah and King (*13*) which is based on Kirkwood's expression (*14*) for the chemical potential including nonspherical molecules.

An analogous exact expression for the limiting derivatives of the activity coefficients can be readily obtained by differentiating the familiar Kirkwood–Buff expression (*15*) for γ_a with respect to composition, followed by a limiting operation. Thus the Kirkwood–Buff equation

$$\begin{aligned} \ln \gamma_a &= \int_0^{x_b} \frac{\rho \Gamma y \, dy}{1 + \rho \Gamma y(1 - y)} \\ \Gamma &\equiv G_{aa} + G_{bb} - 2G_{ab} \\ G_{ij} &\equiv \int \langle g_{ij}(12) - 1 \rangle_{\omega_1 \omega_2} \, d\mathbf{r}_{12} \end{aligned} \tag{20}$$

leads to

$$\left(\frac{\partial \ln \gamma_a}{\partial x_a} \right)_{T,P}^{\infty} = \rho_b (2G_{ab}^{\infty} - G_{bb}^{*}) \tag{21}$$

the superscripts ∞ and * denote limiting and pure quantities, respectively. Equations 20 and 21 are independent of any assumption of pairwise additivity.

Approximative Schemes. To render the exact expressions just given into useful practical forms, two approximations have been used: a perturbation about the pure solvent, and the Mansoori–Leland (*16*) approximation for the mixture radial distribution function.

PERTURBATION ABOUT PURE SOLVENT. Assuming pairwise additivity, we write the total potential energy U_N for a binary mixture consisting of N (equal to $N_a + N_b$) molecules in the form

$$U_N(\lambda) = \sum_{i>j}^{N} \sum^{N} u_{bb}(ij) + \lambda \left\{ \sum_{i}^{N_a} \sum_{j}^{N_b} [u_{ab}(ij) - u_{bb}(ij)] + \sum_{i>j}^{N_a} \sum^{N_a} [u_{aa}(ij) - u_{bb}(ij)] \right\}$$

where N_a is the number of molecules of a, N_b is the number of molecules of b, and λ is the expansion parameter. When $\lambda = 1$, we have the potential energy for the mixture, while $\lambda = 0$ gives the potential energy of pure solvent (b molecules). Expansion of the canonical ensemble partition function for the system with potential energy $U_N(\lambda)$ about $\lambda = 0$ up to first-order terms, followed by setting $\lambda = 1$, and proceeding to the limit of infinite dilution, leads at once to the following expression for the residual chemical potential,

$$\frac{\mu_{ar}^{\infty}}{kT} = \frac{\mu_{br}^{*}}{kT} - \frac{2E_{br}^{*}}{NkT} + \frac{\rho_b}{kT}\int dr_{12}\,\langle u_{ab}(12)\,g_{bb}^{*}(1,2)\rangle_{\omega_1\omega_2} \quad (22)$$

where superscripts ∞ and $*$ again refer to the infinitely dilute and pure component states, respectively. By splitting up the total potential energy as we have done above, we have succeeded in expressing an infinitely dilute solution property as the sum of two parts: the pure solvent part (i.e., the first two terms) and a mixture part arising from the a–b interactions. This is a physically transparent form, considering that in an infinitely dilute system we are looking at solute–solvent interactions against a background of solvent–solvent interactions. We further note that in the limit when a tends to b, so that $u_{aa} = u_{bb} = u_{ab}$, we have the correct expression for μ_{ar}^{∞}. That is

$$\lim_{a\to b}\frac{\mu_{ar}^{\infty}}{kT} = \frac{\mu_{br}^{*}}{kT}$$

In the special case of Lennard–Jones spherical molecules, i.e., with

$$u_{ab}(r) = 4\varepsilon_{ab}\left[\left(\frac{\sigma_{ab}}{r}\right)^{12} - \left(\frac{\sigma_{ab}}{r}\right)^{6}\right] \quad (23)$$

the integral term in Equation 22 is very readily expressed in terms of the internal energy and compressibility factor of the pure solvent, so that

$$\frac{\mu_{ar}^{\infty}}{kT} = \frac{\mu_{br}^{*}}{kT} - \frac{2E_{br}^{*}}{NkT} + \frac{\varepsilon_{ab}}{\varepsilon_{bb}}\left\{(z_b - 1)\left[\left(\frac{\sigma_{ab}}{\sigma_{bb}}\right)^{12} - \left(\frac{\sigma_{ab}}{\sigma_{bb}}\right)^{6}\right] - \frac{2E_{br}^{*}}{NkT}\left[\left(\frac{\sigma_{ab}}{\sigma_{bb}}\right)^{12} - 2\left(\frac{\sigma_{ab}}{\sigma_{bb}}\right)^{6}\right]\right\} \quad (24)$$

At liquid densities, the compressibility term is small compared with the internal energy term (enclosed in braces), so that Equation 23 may be approximated by

$$\frac{\mu_{ar}^{\infty}}{kT} = \frac{\mu_{br}^{*}}{kT} - \frac{2E_{br}^{*}}{NkT}\left\{1 + \frac{\varepsilon_{ab}}{\varepsilon_{bb}}\left[\left(\frac{\sigma_{ab}}{\sigma_{bb}}\right)^{12} - 2\left(\frac{\sigma_{ab}}{\sigma_{bb}}\right)^{6}\right]\right\} \quad (24a)$$

Equation 24 (or 24a) is also valid for weakly or moderately polar fluids with molecules interacting via an angle-averaged modified Stockmayer pseudopotential (*17*), having the same form as Equation 23 but with temperature-dependent parameters ε_{ab} and σ_{ab}. The use of such an angle-averaged modified Stockmayer model is known to be exact, provided that the Pople expansion for the thermodynamic properties can be terminated at the second-order term (*18*).

These equations have been tested against Monte Carlo results for μ_{ar}^{∞}, and found to perform well only for σ_{aa}/σ_{bb} and $\varepsilon_{aa}/\varepsilon_{bb}$ close to unity; however, they will be shown to lead to useful semiempirical correlations in the next section.

MANSOORI–LELAND APPROXIMATION. The Mansoori–Leland (*13*) approximation for the pair correlation function at infinite dilution is of the form

$$g_{ab}^{\infty}\left(\frac{r}{\sigma_{ab}},\ \omega_1,\ \omega_2;\ \frac{kT}{\varepsilon_{ab}},\ \rho\sigma_{ab};\ \left\{\frac{\sigma_{\gamma\zeta}}{\sigma_{ab}}\right\},\ \left\{\frac{\varepsilon_{\gamma\zeta}}{\varepsilon_{ab}}\right\}\right) \simeq \tag{25}$$

$$g_{bb}^{*}\left(\frac{r}{\sigma_{ab}},\ \omega_1,\ \omega_2;\ \frac{kT}{\varepsilon_{ab}},\ \rho\sigma_{bb^3}\right)$$

where $\{\sigma_{\gamma\zeta}/\sigma_{ab}\}$, etc., means all of these ratios for the mixture. Using this approximation in Equation 17 leads to the interesting relation,

$$\frac{\mu_{ar}^{\infty}}{kT} = \left(\frac{\sigma_{ab}}{\sigma_{bb}}\right)^3 \frac{\mu_{br}^{*}(T')}{kT'} \tag{26}$$

where $T' \equiv (\varepsilon_{bb}/\varepsilon_{ab})T$ and $\mu_{br}^{*}(T')$ denotes the residual chemical potential of the pure solvent b, at the scaled temperature T'.

We also consider an approximation similar to Equation 25 but differing slightly from it; in place of Equation 25, we assume that the pair correlation function for the pure solvent, b, pertains to a saturated liquid at temperature T' instead of to an unsaturated one at temperature T' and density ρ as mixture. With this assumption we have a relation slightly different from Equation 26 viz.

$$\frac{\mu_{ar}^{\infty}}{kT} = \frac{v_b(T')}{v_b(T)}\,\frac{\sigma_{ab}^3}{\sigma_{bb}^3}\,\frac{\mu_{br}^{*}(T')}{kT'} \tag{26a}$$

The exact status of the approximation leading to Equation 26a is doubtful, not having been tested against simulation data as the Mansoori–Leland approximation has been. However, Equation 26a has the desirable feature that it is exact in the dilute gas (second virial coefficient) limit.

The conformality condition (i.e., the same functional form for u_{aa}, u_{ab}, and u_{bb}) on which Equations 26 and 26a are based will usually hold only for mixtures of spherical molecules; for nonspherical molecules, the condition of conformality requires that the *aa*, *ab* and *bb* parameters differ only in the terms ε_{ab} and σ_{ab}. Other dimensionless anisotropic parameters (quadrupole moment, shape parameters, and so forth) would need to have the same values for this to be rigorously correct.

An obvious limitation to the range of applicability of Equations 26 and 26a is set by the extent to which the components differ in their well-depths, as is reflected in the difference between their critical temperatures. If $(\varepsilon_{bb}/\varepsilon_{aa})^{1/2}$ is such as to make $T' > T_b^c$, the critical temperature of the pure solvent b, it means that this component no longer exists as a liquid, and therefore the equations are inapplicable.

In the next section, we deduce a number of semiempirical linear correlations from the approximate equations so far obtained.

Semiempirical Correlations of Experimental Data. Equation 17 with Equation 18 is rather suggestive of a general correlation between the Henry's constant and the entropy of vaporization of the pure solvent, b. Indeed, a plot of $\ln(H_{ab}v_b/RT)$ against $\Delta H_b^v/RT$ for a number of dilute solutions indicates a linear correlation over a moderate range of temperatures, (*see* Figure 2) so that we are led to the relation

$$\ln\left(\frac{H_{ab}v_b}{RT}\right) = \text{constant} - C_{ab}\frac{\Delta H_b^v}{RT} \tag{27}$$

where C_{ab} is constant for a particular solute–solvent system. This equation may be shown to result from Equation 24 if the term $(\mu_{br}^*/kT - 2E_{br}^*/NkT)$ varies more slowly with temperature than the remaining term on the right hand side of this equation. Such a behavior has been confirmed by computer simulation studies based on the use of the Lennard–Jones equation of state (*19*). The last term in Equation 27 follows from the approximation

$$-E_{br} \approx \Delta H_b^v - RT \approx \Delta H_b^v$$

where we have assumed that $Pv_b^l \ll Pv_b^g$ (v_b^l and v_b^g denote the molar volumes of the liquid and vapor phases, respectively).

Some simple systems with nonpolar molecules have been found to conform well to Equation 27; Figure 2 illustrates this for the systems argon–methane, and nitrogen–methane (*20, 21*).

In other cases, however, with systems involving polar molecules,

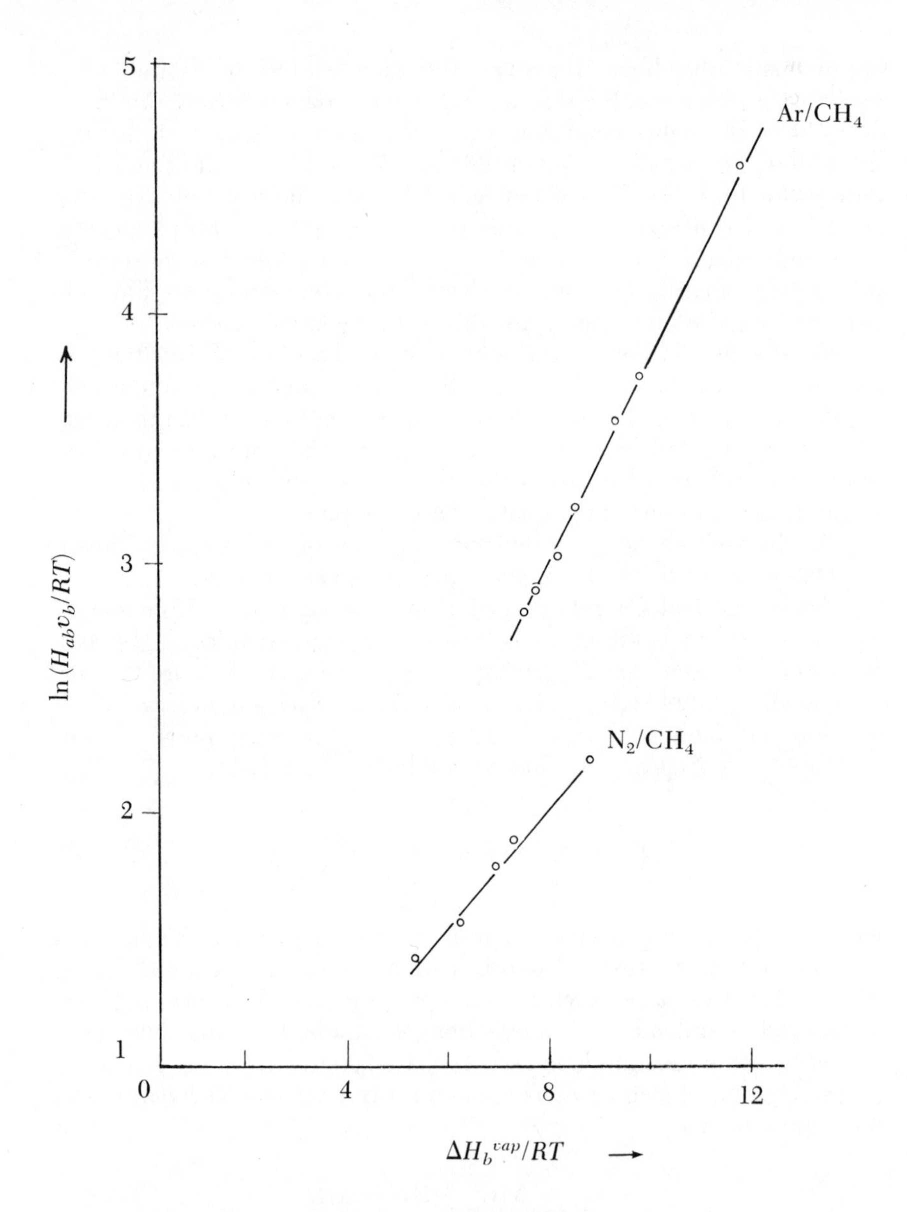

Figure 2. Test of Equation 27 for nonpolar solute (a) and solvent (b) systems argon (a) and methane (b); nitrogen (a) and methane (b) (20, 21, 38) *illustrating linear correlation of Henry's constant with the solvent entropy of vaporization. (Reproduced with permission from Ref. 4. Copyright 1981, American Society of Mechanical Engineers.)*

C_{ab} shows a marked temperature dependence. The temperature dependence of C_{ab} can be taken care of in a purely empirical fashion by fitting it to the function $a_0 + a_1T + a_2T^2$ (a_0, a_1, a_2 being constants). This is shown in Table II. Use of the angle-averaged modified Stockmayer pseudopotential, however, leads to a temperature dependence of the form

$$(a + bT)/(1 + cT)$$

so that we have

$$\ln\left(\frac{H_{ab}v_b}{RT}\right) = \text{constant} - \frac{(a + bT)}{(1 + cT)}\frac{\Delta H_b^v}{RT} \tag{28}$$

Empirically, the constant term is found to be small; thus a linear correlation can be obtained by plotting $(1 + cT)\, C'_{ab}$ versus T, where

$$C'_{ab} \equiv \ln\left(\frac{H_{ab}v_b}{RT}\right) \Big/ \frac{\Delta H_b^v}{RT} \tag{29}$$

and the constant c is chosen to give the best straight line. Figure 3 illustrates such a plot for various solutes in n-hexadecane.

Equation 26 or 26a can be used for correlating Henry's constants and limiting activity coefficients for a particular solute–solvent system with temperature by introducing an empirical temperature-independent constant

$$\frac{\mu_{ar}^{\infty}}{kT} \approx C_{ab}\frac{(V_{ac}^{1/3} + V_{bc}^{1/3})^3}{8V_{bc}} \ln\left(\frac{P_b^{sat}V_b}{RT}\right)_{T=T'} \tag{30}$$

Table II. Correlation of Henry's Constant for Several Solutes in n-Hexadecane with Entropy of Vaporization for Solvent by Equation 27

Solutes	$a_0 \times 10^3$	$a_1 \times 10^3$	$a_2 \times 10^3$
NH_3	−67.38	0.1643	0
CO	+13.20	0.1487	0
CO_2	−59.13	0.1871	0
N_2	+22.77	0.1565	0
HCl	−59.09	0.1327	0
SO_2	−162.33	0.4828	−0.0004
H_2S	−69.62	0.1089	0

Note: $C_{ab} = a_0 + a_1T + a_2T^2$. Maximum error less than 3%. Temperature range 300–475 K. From Ref. 36.

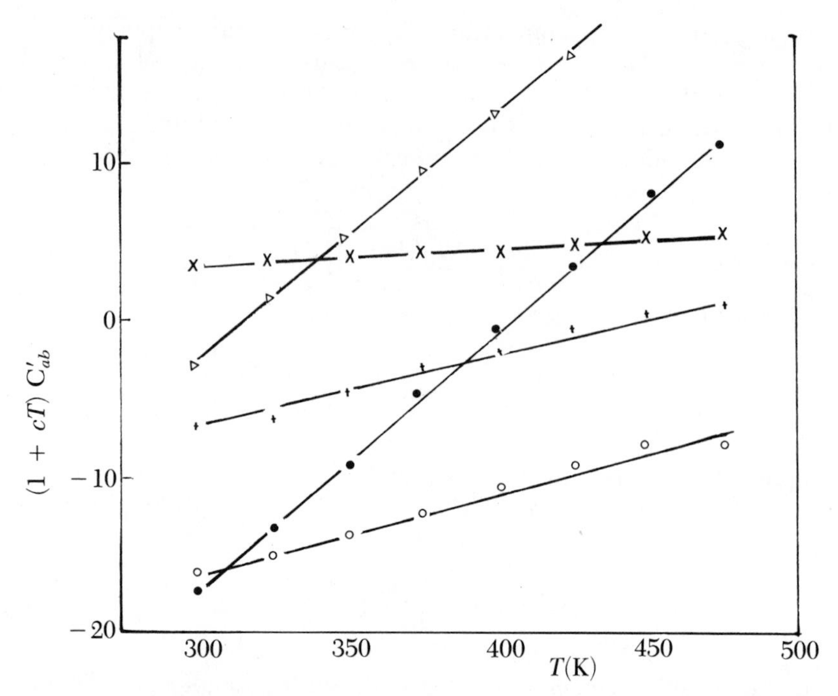

Figure 3. Test of Equation 28 for various solutes in n-*hexadecane* (36) *at 1 atm. The term* C'_{ab} *is defined by Equation 29 and the constant* c *is fitted to the data. Key: ○,* H_2S*; ●,* NH_3*; +, HCl; ×, CO; and △,* CO_2*. (Reproduced with permission from Ref. 11. Copyright 1981, American Society of Mechanical Engineers.)*

or

$$\frac{\mu^{\infty}_{ar}}{kT} \approx C'_{ab} \frac{(V^{1/3}_{ac} + V^{1/3}_{bc})^3}{8V_{bc}} \frac{V_b(T')}{V_b(T)} \ln\left(\frac{P^{sat}_b V_b}{RT}\right)_{T=T'} \tag{30a}$$

where the subscript c is used to denote critical properties and the term $(B_b P^{sat}_b/RT)$ (for a second virial coefficient vapor) has been neglected, as it is small in practice.

Table III illustrates this situation for the three systems argon–methane, nitrogen–methane, and methane–ethane. Equation 30a seems to give a linear correlation over a slightly wider temperature range than does Equation 30.

Equation 26 or 26a can also be made the basis of a corresponding states correlation for the Henry's constant and limiting activity coefficients of various solutes in fixed solvent. In each case we plot the function

$$\Phi \equiv \left(\frac{T_{bc}}{T_{ac}}\right)^{1/2} \frac{8V_{bc}}{(V^{1/3}_{ac} + V^{1/3}_{bc})^3} \frac{\mu^{\infty}_{ar}}{kT} \tag{31}$$

Table III. Experimental and Calculated Values of $\ln(H_{ab}v_b/RT)$ by Equations 30 and 30a

T (K)	Expt.	Eq. 30	Eq. 30a
	Argon–Methane[a] ($C_{ab} = 0.849$, $C'_{ab} = 0.862$)		
90.67	−4.59	−4.70	−4.63
105.00	−3.75	−3.74	−3.72
109.00	−3.56	−3.53	−3.52
111.80	−3.30	−3.39	−3.38
115.90	−3.22	−3.18	−3.19
119.60	−3.03	−3.01	−3.04
123.33	−2.88	−2.86	−2.89
126.00	−2.80	−2.76	−2.80
	Methane–Ethane[b] ($C_{ab} = 0.649$, $C'_{ab} = 585$)		
124.0	−3.79	−3.87	−3.71
140.7	−3.13	−3.17	−3.07
154.0	−2.65	−2.72	−2.65
170.4	−2.38	−2.27	−2.26
183.3	−2.06	−1.99	−2.02
194.4	−1.79	−1.76	−1.82
224.0	−1.45	−1.23	−1.43
	Nitrogen–Methane[c] ($C_{ab} = 0.657$, $C'_{ab} = 0.57$)		
113.0	−2.21	−2.38	−2.29
127.7	−1.88	−1.87	−1.88
131.5	−1.79	−1.61	−1.71
139.0	−1.56	−1.52	−1.63
150.0	−1.41	−1.07	−1.46

[a] From Refs. 20, 21, and 38.
[b] From Ref. 21.
[c] From Ref. 38.

against

$$\ln T^r_{ab} \left(T^r_{ab} \equiv \frac{T}{T^{1/2}_{ac}\, T^{1/2}_{bc}} \right)$$

using the appropriate expression μ^{∞}_{ar}/kT as given by Equations 18 and 19. The detailed arguments leading to these plots have been given (*11*).

Figure 4 illustrates such a plot of corresponding states for various solutes in two solvents, benzene and carbon tetrachloride (*22–27*), solvents with similar enough critical properties to be regarded as a single solvent for practical purposes. It can be seen that the plot is essentially

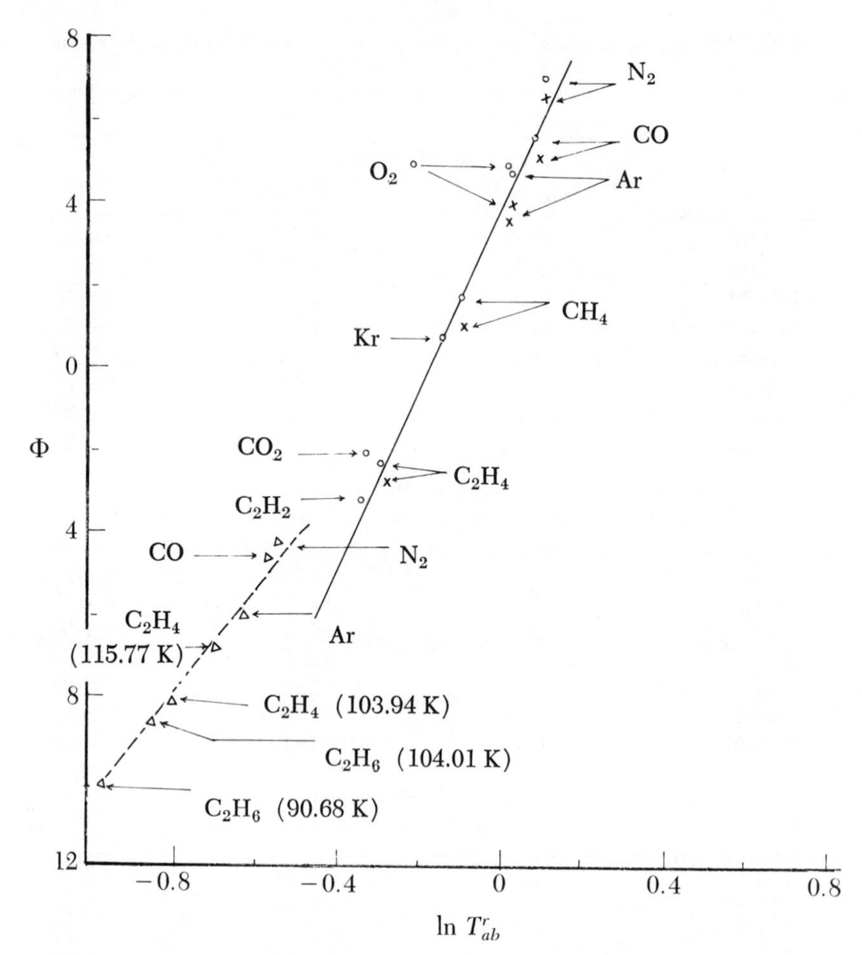

Figure 4. Corresponding states plot for Henry's constant and limiting activity coefficients for various solutes in fixed solvents; benzene, carbon tetrachloride (22–27), and methane (28–30). Key: ○, benzene; ×, carbon tetrachloride (Equations 31 and 18); and △, methane (Equations 31 and 19). (Reproduced with permission from Ref. 11. Copyright 1981, American Society of Mechanical Engineers.)

linear. Also shown is a plot for various solutes in methane (*28–30*); again, the points all fall approximately on a common line.

The conventional correlations of the Henry constant with temperature make use of basically two kinds of plots: ln H versus $1/T$ and ln H versus ln T. The linear correlations introduced in this section differ somewhat from these conventional plots, though some of the former are not completely unrelated to the latter. For example, the correlations,

$$\ln (H_{ab} v_b / RT) = C_{ab} \frac{\Delta H_b}{RT}$$

and

$$\ln(H_{ab}v_b/RT) = (a_0 + a_1T)\frac{\Delta H_b^v}{RT}$$

are very similar to the Valentiner equation (*31*), which is a combination of the two kinds of conventional correlation mentioned above

$$\log_{10}(\text{solubility}) = A + B\log_{10}T + C/T$$

where A, B, and C are temperature-independent constants. King (*32*) has shown how, under certain restrictive conditions, this equation can be obtained by purely thermodynamic considerations. In this derivation, however, one of the constants is related to the heat of absorption of the gaseous solute, rather than to the heat of vaporization of the pure solvent, as is the case with our own correlations above. Equation 28 obviously reduces to the second of the correlations above if $CT \ll 1$; otherwise the relationship with the conventional correlations or the Valentiner equation is not obvious. Equation 30 can be recast into a Valentiner form, if we ignore the temperature variation of the pure solvent molar volume (for example when the temperature range considered is not too wide) and use the Antoine expression for the logarithm of the saturated vapor pressure of the pure solvent.

As regards performance, our linear correlations seem to be a definite improvement over the conventional ln H versus $1/T$ and ln H versus ln T plots, as can be readily verified for the systems considered in Figure 3. Apart from the system hydrogen sulfide–$C_{16}H_{34}$, the conventional plots for the other systems show linearity only over a limited temperature range of between 50 and 75 K, as compared with the much wider temperature range of 175 K for our own correlations.

The importance of linear correlations of the kind we have introduced in this section cannot be overemphasized. With linear plots, extrapolations can be made with confidence, while interpolations can be made with very limited data, say two or three points. Further, such linear plots are very convenient for evaluating the temperature coefficient of the Henry constant—$\partial \ln H/\partial(1/T)$ or the partial molar entropy of solution—which we know is proportional to ln H, thus leading to linear plots for various solutes in a fixed solvent, at a fixed temperature (*13, 22*).

The potentialities of the these linear correlations are therefore quite obvious.

Solubility of Solids in Supercritical Fluids. An integro-differential relation that provides a convenient basis for correlating solubility of solids in supercritical fluids (*33*) is

$$\frac{\partial}{\partial P}\left\{Y \exp\left[-\int \frac{v_b}{kT}\,dP\right]\right\} = Z \exp\left(-\int \frac{v_b}{kT}\,dP\right) \qquad (32)$$

where

$$Y \equiv \ln\left(\frac{f_a^s v_b}{y_a RT}\right) - \frac{\mu_{br}^*}{RT}$$

$$Z \equiv \frac{-N}{(kT)^2}\int_0^1 d\xi \int dr_{12}[\overline{\langle g_{ab}^{\infty}(r,T,v;\ \xi)\, u_{ab}(r)\rangle}$$
$$- \overline{\langle g_{bb}^{*}(r,T,v;\ \xi)\, u_{bb}(r)\rangle}] \quad (32a)$$

where f_a^s is the fugacity of the solid component a, and y_a is its solubility in the gaseous component b; the bars over the integrands in Z denote averaging over the N–P–T (or constant pressure) ensemble (*see* Reference 34).

To make Equation 32 useful for practical application, we make the crucial hypothesis that the two integral terms constituting Z—one pertaining to the infinitely dilute mixture, and the other to the pure solvent b—respond similarly to the application of pressure so that Z itself is a weak function of pressure and hence is practically constant.

Under these circumstances, Equation 32 can be integrated, leading to the conclusion that the thermodynamic function

$$\Psi \equiv \frac{1}{f_b}\left\{\ln\frac{P_a^{sat} z_b}{f_b P y_a} + \frac{v_a^s(P - P_a^{sat})}{RT}\right\} \quad (33)$$

varies linearly with

$$\Phi \equiv \int_1^P \frac{1}{f_b}\, dP \quad (33a)$$

and that the function,

$$\Psi^* \equiv \frac{v_a^s - v_b^g}{RT} - \frac{v_b^g}{RT}\left\{\ln\left(\frac{P_a^{sat} z_b}{y_a P f_a}\right) + \frac{v_a^s P}{RT}\right\} \quad (34)$$

is nearly constant at maximum solubility y_a. The superscripts sat, s and g refer to saturation condition, solid, and gas, respectively.

Empirically, for a suitably chosen reference pressure P_{ref}, we can to a good approximation write,

$$\Phi \approx \text{constant} + c(P - P_{ref})$$

c being some mean value of $1/f_b$ in the pressure range of interest. To this approximation then Ψ would be expected to be a linear function of pressure over this pressure range. Figure 5 illustrates such linear cor-

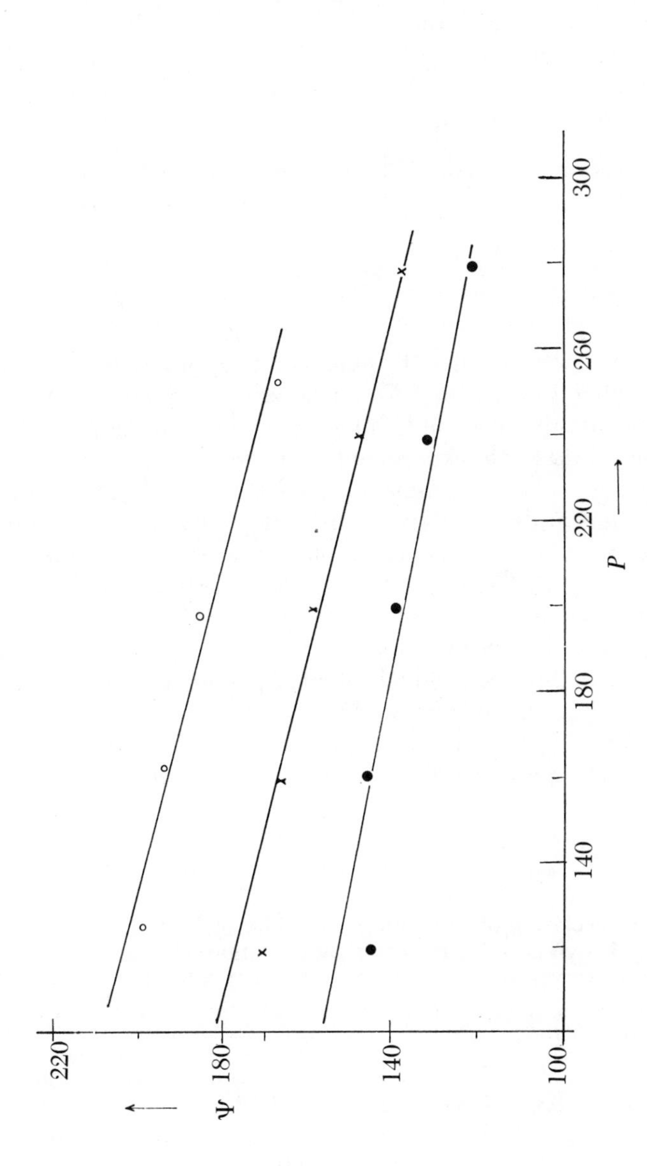

Figure 5. Linear correlation of the solubility of solids in carbon dioxide (35) with pressure; Ψ is defined by Equation 33. Key: ○, naphthalene, 328 K, 0.1 mm; ×, phenanthrene, 318 K, 0.1 mm; and ●, benzoic acid, 318 K, 0.75 mm.

relations for three different solids, benzoic acid (35), phenanthrene, and naphthalene, in carbon dioxide; in these plots, P_a^{sat}, the saturated vapor pressure of the solid, is taken as a parameter that is indicated on each graph. P_{ref} is about 150 bars.

Table IV illustrates the approximate constancy of the function Ψ^* of maximum solubility for naphthalene in supercritical ethylene over a range of temperatures and pressures. The molar volume v_a^s of naphthalene has been taken as 112 cm^3 and its variation with temperature has been neglected; P_a^{sat} is taken as 0.1 mm (Hg). Equation 34 neglects a term $\frac{\partial \ln v_b^g}{\partial P}$, which is about two orders of magnitude smaller than the other terms.

Summary

Attention has been drawn in this chapter to appropriate modifications of the familiar two-constant Van Laar and Margules equations, which admit of constants that can be identified with limiting properties and still produce close fits to experimental data.

It has been shown how to estimate these constants reliably by using a method of numerical differentiation, which depends on the measurement of intercepts rather than slopes, applied to a few data points near each end of the composition range. It is thus possible to extrapolate data from very dilute regions to the rest of the composition range.

Based on molecular thermodynamic considerations, tested linear correlations between limiting residual chemical potentials and pure component properties have been suggested.

I believe the approach in this article holds much promise in the direction of predicting mixture properties from pure component properties.

Table IV. Correlation of Maximum Solubility of Naphthalene in Ethylene with Pressure by Equation 34

T (*K*)	P_{max} (*bar*)	y_a^{max} ($\times 10^2$)	z_b	f_b (*bar*)	v_b (*cm*^{3})	$\partial \ln v/\partial P$ ($\times 10^4$)	Ψ^* ($\times 10^2$)
303.2	612	4.31	1.408	194.6	57.98	2.41	3.57
308.2	590	5.68	1.356	195.3	58.91	2.72	3.66
313.2	576	7.84	1.322	199.3	59.76	2.85	3.75
318.2	477	1.17	1.128	166.5	62.59	3.68	3.45

Note: Data from Ref. 37.

Acknowledgments

It is a pleasure to acknowledge the hospitality of the School of Chemical Engineering, Cornell University, where this reseach was carried out. I also acknowledge useful discussions and suggestions from K. E. Gubbins, W. B. Streett, K. Shing, and V. Venkatasubramanian. The willingness of J. C. G. Calado to make available his experimental data prior to publication is also gratefully acknowledged. Finally, I express gratitude to the National Science Foundation Grant INT–8016843 and to the Council for the International Exchange of Scholars for a Fulbright Award.

Literature Cited

1. Jonah, D. A. Ph.D. Thesis. University of Birmingham, England, 1961.
2. Jonah, D. A.; Ellis, S. R. M. *J. Appl. Chem.* **1965**, *15*, 151.
3. Abbott, M. M.; Van Ness, H. C. *AIChEJ* **1975**, *21*, 62.
4. Landis, F.; Nilson, E. N. "Progress in International Research on Thermodynamic and Transport Properties"; Academic Press: New York, 1962; p. 218.
5. Van Ness, H. C.; Soczek, C. A.; Kochar, N. K. *J. Chem. Engr. Data* **1967**, *12*, 346.
6. Brown, I.; Fock, W. *Aust. J. Chem.* **1945**, *4*, 417; Ibid. **1956,** *9*, 180.
7. Jonah, D. A. submitted for publication.
8. Gautreaux, M. F.; Coates, J. *AIChEJ* **1955**, *1*, 496.
9. Jonah, D. A. submitted for publication.
10. Klaus, R. L.; Van Ness, H. C. *AIChEJ* **1967**, *13*, 1132.
11. Jonah, D. A.; Shing, K. S.; Gubbins, K. E. "Proc. Eighth Symposium on Thermophysical Properties"; American Society of Mechanical Engineers: New York, 1981; in press.
12. Goldman, S. *J. Solut. Chem.* **1977**, *6*, 461.
13. Jonah, D. A.; King, M. B. *Proc. Roy. Soc.* **1971**, *323A*, 361.
14. Kirkwood, J. G. *J. Chem. Phys.* **1935**, *3*, 300.
15. Kirkwood, J. G.; Buff, F. P. *J. Chem. Phys.* **1951**, *19*, 774.
16. Mansoori, G. A.; Leland, T. W. *J. Chem. Soc., Faraday Trans. II* **1972**, *68*, 320.
17. Reed, T. M.; Gubbins, K. E. "Applied Statistical Mechanics"; McGraw Hill: New York, 1973; p. 161.
18. Egelstaff, P. A.; Gray, C. G.; Gubbins, K. E. "Molecular Structure and Properties"; Buckingham, A. D., Ed., Butterworths: London, 1975; Vol. 2, Series 2.
19. Nicolas, J. J.; Gubbins, K. E.; Streett, W. B.; Tildesley, D. J.; Brown, I.; Fock, W. *Aust. J. Chem.* **1956**, *9*, 180.
20. Perez, J. L. Ph.D. Thesis, Universidad Nacional de Mexico, Mexico City, 1977.
21. Kidnay, A. J.; Miller, R. C.; Parrish, W. R.; Hiza, M. J. *Cryogenics* **1975,** *15*, 531.
22. Jolley, J. E.; Hildebrand, J. H. *J. Am. Chem. Soc.* **1958**, *80*, 1050.
23. Horinti, J. *Sci. Pap. Inst. Phys. Chem. Res. Tokyo* **1931**, *17*, 125.
24. Vitovec, J.; Fried, V. *Collection Czech. Chem. Commun.* **1960**, *25*, 1552.
25. Gjaldbaek, J. C. *Acta Chem. Scand.* **1953**, *7*, 537.

26. Clever, L. H.; Battino, R.; Saylor, J. H.; Cross, P. M. *J. Phys. Chem.* **1957**, *61*, 1078.
27. Reeves, L. W.; Hildebrand, J. H. *J. Am. Chem. Soc.* **1957**, *79*, 1313.
28. Calado, J. C. G. unpublished data.
29. Calado, J. C. G.; Gomes De Azevedo, E. J. S.; Soares, V. A. M. *Chem. Eng. Commun.* **1980**, *5*, 149.
30. Sprow, F. B.; Prausnitz, J. M. *AIChEJ* **1966**, *12*, 780.
31. Valentiner, Z. *Z. Physik.* **4927**, *42*, 253.
32. King, M. B. "Phase Equilibrium in Mixtures"; Pergamon Press: Oxford, 1969; p. 235.
33. Jonah, D. A.; Shing, K. S.; Venkatasubramanian, V. submitted for publication.
34. Hill, T. L. "Statistical Mechanics"; McGraw Hill: New York, 1956; pp. 66–68.
35. Kurnir, R. T.; Holla, S. J.; Reid, R. C. *J. Chem. Eng. Data* **1981**, *26*, 47.
36. Tremper, K. K.; Prausnitz, J. M. *J. Chem. Eng. Data* **1976**, *21*, 295.
37. Van Welie, G. S. A.; Diepen, G. A. M. *Rec. Trav. Chim.* **1961**, *80*, 673.
38. Prausnitz, J. M.; Chueh, P. L. "Computer Calculations for High Pressure Vapour–Liquid Equilibria"; Prentice Hall: Englewood Cliffs, NJ, 1968.

RECEIVED for review January 27, 1982. ACCEPTED for publication October 5, 1982.

17

Liquid State Dynamics of Alkane Chains

GLENN T. EVANS

Oregon State University, Department of Chemistry, Corvallis, OR 97331

In this chapter, we summarize a few of the computational and analytical approaches used to understand the motion of chain molecules in a liquid. Emphasis is placed on calculating properties that are, in principle, detectable by experiments involving dielectric relaxation, NMR, depolarized light scattering, torsional photoisomerization and ring-closure rates. Comparison with experiment is made when possible.

THE DYNAMICS AND EQUILIBRIUM STRUCTURE of liquids composed of torsionally flexible chain molecules have received concentrated interest during the past few years. Interest has been aroused because machine computations can be performed to account realistically for the relevant classical degrees of freedom of a chain molecule in a polyatomic or flexible molecular solvent. The numerical work, combined with the rediscovery of the classic papers of Kramers (*1*), Kirkwood (*2*), and Eckhart (*3*, *4*), and the use of modern theories involving projection operators (*5*) and mode-coupling (*6*), has equipped this new field with the tools for a plausible approach to an understanding of flexible molecules.

In this chapter, we restrict our attention to the dynamics of the small alkanes, ranging from butane to undecane. We begin with a brief description of the flavor of the numerical approaches to simulation of chain motions. Alternatively, chain motions can be analyzed in terms of solutions to differential equations describing the ensemble-averaged trajectory of a chain, and this is also discussed.

The introduction of flexibility into a chain alters its transport and reactive properties. Dielectric relaxation, NMR, and light scattering are analyzed for evidence of overall and internal rotation. We also address a few chemical reactions, such as ring closure and torsional isomerization, because these are indicators of the isomerization process as well.

The views presented, and the selection of topics, will be biased in this short chapter. Our goal is not to write a comprehensive review but rather to summarize our findings after 4 or 5 years work.

0065-2393/83/0204-0423$06.50/0

Computer Modeling

Computer modeling is as important for chain molecules as it was for argon. Whereas real alkanes have uncontrollable and, to a certain extent, unknown intermolecular and intramolecular potentials, computer "alkanes" have the virtue of definable mechanical properties. There are several levels on which a chain molecule can be simulated using classical mechanics.

The first level is a full molecular dynamics calculation incorporating intermolecular and intramolecular forces explicitly (*6–11*). The intramolecular forces control bond lengths, bond angles, and torsion angles. The bond lengths and bond angles can be handled either with rigid constraining forces, which do not allow fluctuations in these coordinates or, more realistically, by means of differentiable potentials, which allow nonvanishing vibrational amplitudes. Torsional coordinates are always controlled by a potential consisting of a few cosine functions of the dihedral angles. Intermolecular forces are implemented via site–site or atom–atom Lennard–Jones interactions between the solvent and the chain atoms.

Without question, and without controversy as well, the full molecular dynamics approach is the best route to proceed by, although it is also the most expensive. For the small chain systems of interest to us, a correct treatment of the chemical topology is essential, and therefore there is no shortcut to simplify the intramolecular part of the problem. Whether one uses constraints or smooth forces, the algorithm must be faithful to spectroscopic evidence regarding bond lengths, bond angles, and torsional potentials in order to produce a believable simulation of a chain in solution.

An approximation can be made regarding the treatment of intermolecular forces. In the Langevin dynamics approach, one writes an approximate equation for the velocity of a backbone atom of the chain

$$\frac{mdv}{dt} = -fv(t) + F(\text{intra}) + F(\text{inter}) \qquad (1)$$

where m is the mass of a backbone atom, f is the friction coefficient, which is in turn related to the fluctuating intermolecular force F(inter) by

$$f = (^1/_6 k_B T) \int_0^\infty \langle F(\text{inter}) \cdot F(\text{inter}, t)\rangle\, dt \qquad (2)$$

and F(intra) includes the intramolecular forces as well as the potential of mean force (*12–16*). The advantage of Equation 1 is that it removes all of the degrees of freedom associated with the solvent and substitutes a random number generator as F(inter). When integrated, Equation 1 provides the momentum and position of each atom of the chain as a function of time, and hence one can determine momentum as well as position correlation functions.

In the next level of approximation, one can dispose of the momentum degrees of freedom (*13–23*). This is tantamount to arguing that the momentum of each backbone atom fluctuates on a rapid collisional time scale, and, on this rapid time scale, the positions of the atoms or backbone bonds of the chain do not move appreciably. Hence, for long times, the average of mdv/dt is zero. Thus, one can replace the left-hand side of Equation 1 by zero, solve the equation for $v(t)$ and integrate that for the position $x(t)$

$$x(t) = (1/f) \int_0^t dt \, \{F(\text{intra}, t') + F(\text{inter}, t')\} \tag{3}$$

This approach is called Brownian dynamics (BD). There are a number of complications in the BD method, especially regarding the time-dependence of the intermolecular forces (*15, 18*) but BD has the advantage of being fast enough for simulations of long chains to be a reality, and it gives a good accounting for processes that require overall and internal rotation of the chain. As will be seen, for pure torsional isomerizations Equation 3 is not adequate. Predictions of Fixman's BD algorithm (*18*) (modified by the inclusion of the metric force) will be compared with experiment later in this chapter.

Analytic Approaches

Consistent with the three levels of simulation (that is, molecular, Langevin, and Brownian dynamics) one can develop differential equations for the specific single-particle phase-space density for a chain. This density function describes the probability that a tagged chain will have a particular set of Euler angles, torsion angles, and conjugate momenta at time t, given that it has some distribution of phase-space coordinates at an earlier time. Consider butane as an example. The phase-space coordinates of butane are the three Euler angles Ω (which relate the orientation of the principal axes with respect to the laboratory), the torsional coordinate (which specifies the shape of butane), the angular momentum $\mathbf{L}$ along the principal axes, and the angular velocity L_4 along the torsional coordinate. The distribution function $f(\Omega,\phi,L,L_4,t)$ obeys a kinetic equation of the form (*24, 25*)

$$\partial_t f(\Omega,\phi,L,L_4,t) = -(iL_S + iL_C) f(\Omega,\phi,L,L_4,t) \tag{4}$$

where iL_S is the streaming (or convective) operator and iL_C is the collision operator. Streaming, in this context, describes the gas-phase, noncollisional motion of a torsionally flexible molecule, which can undergo overall and internal rotation. Although we will not write the explicit form of the Liouville operators here (because of space restrictions), qualitatively the convective term consists of operators responsible for the overall rotation of an asymmetric top, the internal rotation of the two ethyl fragments relative to one another, and the coupling terms between overall and internal rotation.

The collision operator is more complicated. Using projection operator techniques akin to those of Lebowitz and Resibois (26), one can derive a Fokker–Planck-like collision operator. It is similar to the Fokker–Planck (FP) operator in that collisions do not alter the position of the chain atoms, but do change the momenta by a small amount. The FP collision operator for butane incorporates the torsion angle dependence of the friction constant and the couplings of overall and internal rotation. The kinetic equation for f, with convective and collisional operators as described above, has been derived only for butane (22). In approximating the collision operator with a torque correlation function independent of frequency and operator, we have generated a differential equation appropriate to Langevin dynamics. To date, no analysis of the torsion angle dynamics of butane has been made with the approximate collision operator. In principle, this problem is solvable using techniques like those employed by Fixman and Rider (27) in the analysis of the rotational motion of spheres.

The FP equation reduces to a diffusion equation by integrating over the momenta of the chain and making the usual assumptions that momentum relaxation is fast compared with torsional isomerization and overall rotation. The diffusion equation for the density function $\rho(\Omega, \phi, t)$ is of the form

$$\partial_t \rho(\Omega,\phi,t) = i\boldsymbol{J} \cdot \mathbf{D} \exp[-\beta U(\phi)] \cdot i\boldsymbol{J}\{\exp[\beta U(\phi)]\rho(\Omega,\phi,t)\} \qquad (5)$$

with

$$\rho(\Omega,\phi,t) = \int d\boldsymbol{L}\, dL_4\, f(\Omega,\phi,\boldsymbol{L},L_4,t)$$

where $\mathbf{D}$ is a 4×4 diffusion tensor, $U(\phi)$ is the torsional free energy, and $i\boldsymbol{J}$ is a four-vector with elements

$$i\boldsymbol{J} = \left(iJ_x, iJ_y, iJ_z, \frac{\partial}{\partial\phi}\right)$$

The free energy $U(\phi)$ includes the gas-phase torsional component as well as contributions from the liquid. Equation 5 can be reduced further because the elements of **D**, which are reciprocals of torque correlation functions, depend only on the dihedral angle and not on the Euler angles. Consequently, one can integrate Equation 5 over the Euler angles to produce a diffusion equation for the torsional coordinate alone (*28*)

$$\partial_t \rho(\phi,t) = \partial_\phi \{D(\phi)\exp[-\beta U(\phi)]\partial_\phi \{\exp[\beta U(\phi)]\rho(\phi,t)\}\} \qquad (6)$$

with

$$\rho(\phi,t) = \int d\Omega \rho(\Omega,\phi,t)$$

The $D(\phi)$ in Equation 6 is the diffusion coefficient for the torsion angle and has an explicit and known dependence on ϕ (*28*). Equation 6 is a diffusion equation for an internal degree of freedom, derived from Newton's equations. No couplings of the overall rotational coordinates to the internal coordinate have been omitted, and hence Equation 6 is equivalent to an exact BD simulation.

Equation 5 was certainly not the first of its kind. It was predated by a generalized coordinate diffusion equation for chain molecules by the work of Kirkwood and collaborators (*2*). Despite the fact that Kirkwood's work on flexible chain systems and generalized diffusion equations took place in the late 1940s, it was only recently that the correct form of the diffusion equation could be derived (*18*). Fixman demonstrated that the distribution function $\Psi(\{Q\},t)$ for the generalized coordinates $\{Q\}$ of a chain molecule satisfied a diffusion equation

$$\partial_t \Psi(\{Q\},t) = \frac{1}{\sqrt{g}} \frac{\partial}{\partial Q^i} \sqrt{g}\, h^{ij} \left\{ \frac{\partial}{\partial Q^j} - \beta F^j \right\} \Psi(\{Q\},t) \qquad (7)$$

where h^{ij} is the contravariant component of the metric tensor, g is the metric determinant, and F^j is the generalized force on the jth coordinate. Fixman showed that F^j consists of two parts, the ordinary term arising from the derivative of the torsional free energy and a metric force that must be inserted so that dihedral angle space is evenly populated in the absence of torsional potentials.

Prior to the derivation of Equation 7, Fixman and Kovac (*29*) derived a diffusion equation in terms of the bond vectors for the N-bond chain. The advantage of bond-vector coordinates is that they provide a natural language for the constraints on bond-vector length and the nearest-neighbor bond angles. In this language, the polymer equation of motion becomes

$$\partial_t \Psi(\{\boldsymbol{b}\},t) = -\nabla^b \cdot \mathbf{S} \cdot (\nabla^b - \beta \boldsymbol{F})\Psi(\{\boldsymbol{b}\},t) \tag{8}$$

The **S** operator consists of bilinear products of the bond vectors, i.e., $\boldsymbol{b}_j\boldsymbol{b}_k$, as well as the scalar products $\boldsymbol{b}_j \cdot \boldsymbol{b}_k$, where j and k range from one to N. If bond angle and bond length constraints are used, then the action of **S** causes diffusion to take place in the hyperspace in which only the Euler angles and the torsion angles are dynamical variables.

The bond vector and the generalized coordinate form of the diffusion equation are identical, provided that one includes the metric force in the generalized coordinate version. Equations 5, 7, and 8 have been derived from different starting points but are all merely restatements of one another. The suitability of one equation versus another is determined by the types of questions posed.

Derived Quantities

The reason for providing a mechanical theory of chain molecules in solution is to construct a framework by which various time correlation functions of vectorial and tensorial combinations of bond vectors can be determined systematically. Time correlation functions and their corresponding time integrals, or correlation times, show a sensitivity to the type of probe placed on the chain, to the location of the probe, and to the potential parameters of the chain. In most cases, our interest in particular combinations of bond vectors is motivated by the fact that certain combinations correspond to an experimentally detectable phenomenon, such as a dipole moment or a light scattering spectrum.

Figure 1 shows the time correlation functions of the six-bond chain for various modes of first-rank and second-rank spherical harmonics. The correlation functions are all normalized, and hence begin at one and decay monotonically in time. Time is measured in reduced units so that real time (in units of seconds) can be obtained by multiplying the time axis by a factor of fb^2/k_BT. Using Stokes' law for the friction coefficient of each backbone atom (of radius $b/2$) yields

$$f = 3\pi\eta b \tag{9}$$

where η is the solution viscosity. In a Brownian dynamics algorithm, the solvent viscosity acts as a scale factor for the time axis and does not affect the shape of the time correlation functions in any way. The time correlation functions reported in Figure 1, and throughout this chapter, were usually determined using 100–150 trajectories of a BD calculation.

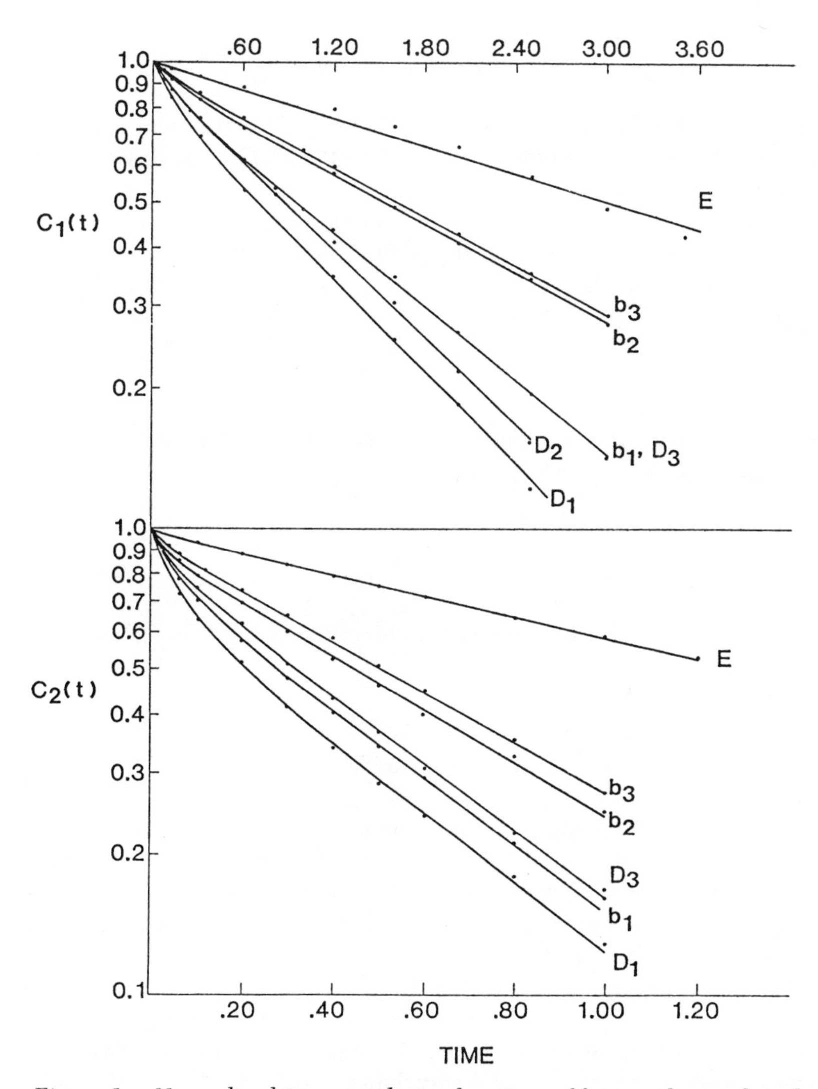

Figure 1. Normalized time correlation functions of first- and second-rank Legendre polynomials for various modes of the six-bond chain with Q = $4k_BT$. The torsional potential is defined by Equation 11. Top, l = 1; bottom, l = 2. (Reproduced with permission from Ref. 21. Copyright 1980, American Institute of Physics.)

Each trajectory consisted of 1600 time steps of 0.0025 reduced time units. A torsional potential energy $U(\phi)$, where

$$U(\phi) = (Q/2)[1 - \cos(3\phi)] \tag{10}$$

was used in the runs for Figure 1, where Q, the barrier height, was $4k_BT$. In all runs, bond angles were assumed to be 110°, in approximation of the tetrahedral value.

The labels to the right of the correlation functions in Figure 1 indicate the "mode" observed. Defined are: bond vector modes, b, whose first-rank time correlation function is $\langle \boldsymbol{b}_j \cdot \boldsymbol{b}_j(t) \rangle$; the difference modes, D,

$$\boldsymbol{D}_j = \boldsymbol{b}_j - \boldsymbol{b}_{j+1}$$

whose time correlation functions are $\langle \boldsymbol{D}_j \cdot \boldsymbol{D}_j(t) \rangle$; and the end-to-end vector modes, E, where

$$\boldsymbol{E} = \sum_{j=1}^{N} \boldsymbol{b}_j$$

The second-rank time correlation functions of $\boldsymbol{b}_j$, $\boldsymbol{D}_j$ and $\boldsymbol{E}$ are formed by the prescription

$$C_2(t) = \frac{3\langle [\boldsymbol{v} \cdot \boldsymbol{v}(t)]^2 \rangle - \langle v^2 \rangle^2}{\langle (v^2)^2 \rangle} \tag{11}$$

where $\boldsymbol{v}$ is any of the three vectors mentioned.

Figure 1 shows trends characteristic of most medium-size chains. The fastest processes in the chain involve the difference and bond modes at the chain extremities, and the correlation times increase as one approaches the chain center. Furthermore, motion of the end-to-end vector is always the slowest process. The $C_1(t)$ terms decay more slowly than the $C_2(t)$ terms, as a rule, since $C_1(t)$ angular functions have one node in the 0 to π range, whereas the $C_2(t)$ angular functions have two nodes. Typically, the ratio of C_1 to C_2 correlations times for the same mode is 2 to 3.

In addition to investigating the microcanonical ensemble time correlation functions, one can also look at the chain during one trajectory. Figure 2 shows "snapshots" of a nonane molecule described by the torsional potential of Equation 10, but with $Q = 5k_BT$. Running down the left-hand margin are the reduced time values, and on the right-hand side are the time scales appropriate to various experiments in a solution at room temperature with viscosity of 1 centipoise. The relationship of the

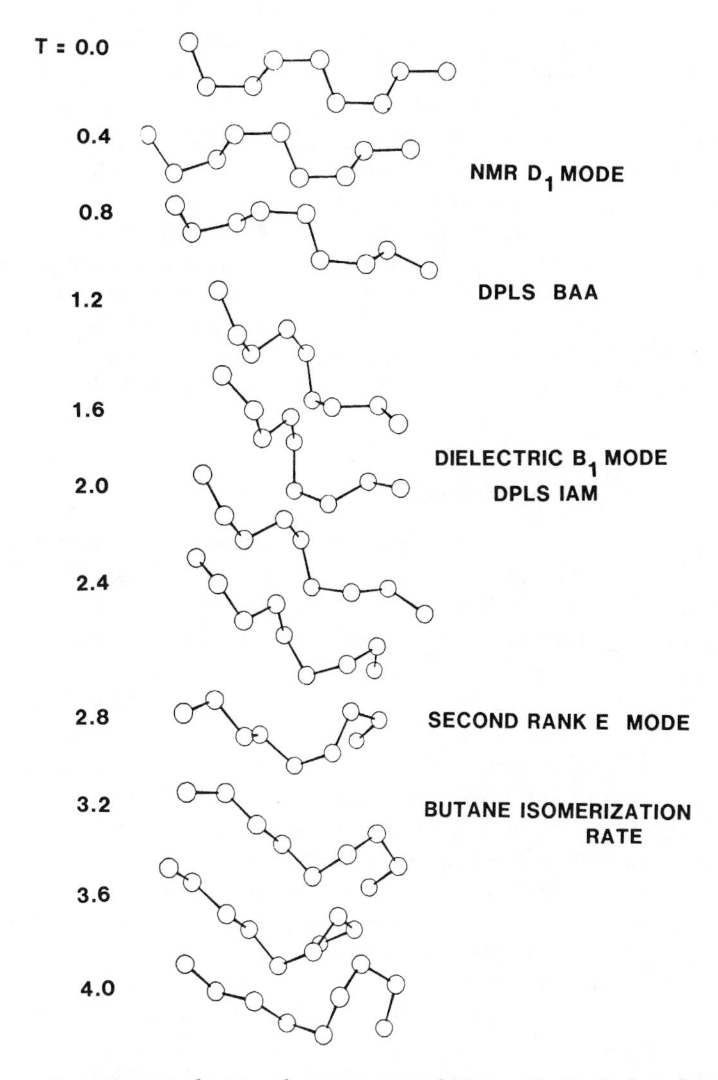

Figure 2. Time evolution of n-*nonane with* $Q = 5k_BT$. *Reduced times are labeled on the left, and on the right are the time scales for various experiments in a room-temperature solution with viscosity of 1 centipoise. To convert reduced times to real time, multiply by 8.4 ps.*

various modes to experiment will be discussed later. It is evident from Figure 2 that the terminal bonds reorient on a faster time scale than the end-to-end vector, which hardly moves at all during the length of the run. During the middle of the trajectory, a crankshaft is formed but diffuses away; hence, from this single run, we see no evidence of coherent crankshaft rotation. Other events occurring during the trajectory include the formation of structures that look as though a five- or six-membered ring is about to form. No evidence supports the formation of an eight- or nine-membered ring, in agreement with the organic chemist's adage that five- and six-membered rings form easily, whereas medium-size rings form with some difficulty. Clearly, from a single trajectory, one cannot draw sweeping general conclusions unless the trajectory is very long. Nevertheless, it is heartening that the same trends observed in the ensemble averaged time correlation functions are seen directly in a single trajectory.

Comparison with Experiment

There are several experiments that, under favorable circumstances, are capable of detecting the motion of a single chain in a simple solvent. A few candidates are dielectric relaxation, NMR relaxation, depolarized light scattering, torsional isomerization, and ring-closure rates. Before we compare BD calculations with actual experimental correlation times, we must attempt to substantiate the use of Equation 10 for the friction constant per backbone atom. Since all the high-friction results scale linearly with friction constant, it is important to fix a value for f. Toward this end, consider the translational and rotational diffusion coefficients D_T and D_R, respectively. D_T is an insensitive indicator of the subtleties of chain dynamics. Kirkwood's expression for the center-of-mass translational diffusion constant is not amended in any way by the theory presented earlier, therefore D_T can be determined using the standard Oseen formula. Mazo (*30*) has analyzed D_T for several alkanes and has found that, to fit experiment, the friction coefficient per backbone item must be chosen as in Equation 9. To test Mazo's necklace model of an alkane, wherein each backbone atom is represented by a sphere using its covalent radius, we have determined the rotational diffusion tensor for several small rigid molecules such as carbon disulfide, benzene, and chloroform (*31*). Here, too, the agreement of theory and experiment is very good, even so far as representing the anisotropy of the rotational diffusion tensor as well as its trace. Given this supportive evidence from two sources, we will select the friction constants for flexible chains by using Stokes' law expressed in terms of the carbon covalent radius.

Dielectric Relaxation. The study of dielectric relaxation in chain systems is a mature discipline, with its pioneering work dating back to

Debye (*32*). In the simplest case, the frequency-dependent dielectric constant $\varepsilon(\omega)$ is related to the one-sided Fourier transform of the normalized dipole–dipole correlation function; thus

$$\frac{\varepsilon(\omega) - \varepsilon_\infty}{\varepsilon_0 - \varepsilon_\infty} = \int_0^\infty dt \exp(-i\omega t) \frac{d}{dt} C_1(t) \tag{12}$$

where ε_0 and ε_∞ are the zero-frequency and infinite-frequency dielectric constants. Equation 12 is strictly applicable to a dilute solution of dipolar molecules in a nonpolar solvent. In this limit, dielectric relaxation can measure single particle, as opposed to collective, orientational motions (*32*). Our goal is to compare dipolar time correlation functions from BD simulations of an infinitely dilute solution of a homogeneous chain molecule with experimental correlation times and Cole–Cole plots. Figure 3 shows the experimental relaxation times of 1-bromoalkanes in cyclohexane (*33*), along with the BD calculations.

We have assumed that a bromine atom can be mimicked as a carbon atom, and that the dipole in 1-bromoalkane lies entirely along $\boldsymbol{b}_1$. Clearly, the first approximation is not entirely satisfactory, although the second is adequate. The correlation times increase with increasing barrier height, Q, and with increasing chain length. The freely rotating chain shows very little dependence on chain length. The $Q = \infty$ value is a rigid-body result, and was not obtained from BD. This *rigid body* (RB) is a composite structure consisting of an average of all possible rigid bodies consistent with the three minima in the potential given by Equation 10. Thus, we have calculated the correlation time associated with each of the 3^{N-2} conformers for an N-bond chain and have simply added them. The BD calculations approach the RB limit as they should. The experimental points lie between the cases of $Q = 4k_BT$ and $Q = 8k_BT$. Since $Q = 5k_BT$ is usually assumed to be a typical barrier, the agreement between experiment and theory in this case looks reasonable, although there is more to say about the agreement (which we believe to be somewhat fortuitous).

A more interesting test of the dynamics arises in the Cole–Cole plot displayed in Figure 4 for neat n-octyl iodide. Despite all the qualifications made earlier regarding the restrictive nature of our calculations and the applicability of Equation 12 to dilute solutions, we shall make a cavalier comparison of a neat solution with the exact results of a BD simulation. Figure 4 contains the experimental data (*34, 35*), a rigid-body simulation (actually, 3^{N-2} rigid-body arcs superimposed) and the BD results. The BD plot was obtained by a pointwise Fourier transform of the correlation function using 200 time points. Several interesting features emerge from the analysis of Figure 4.

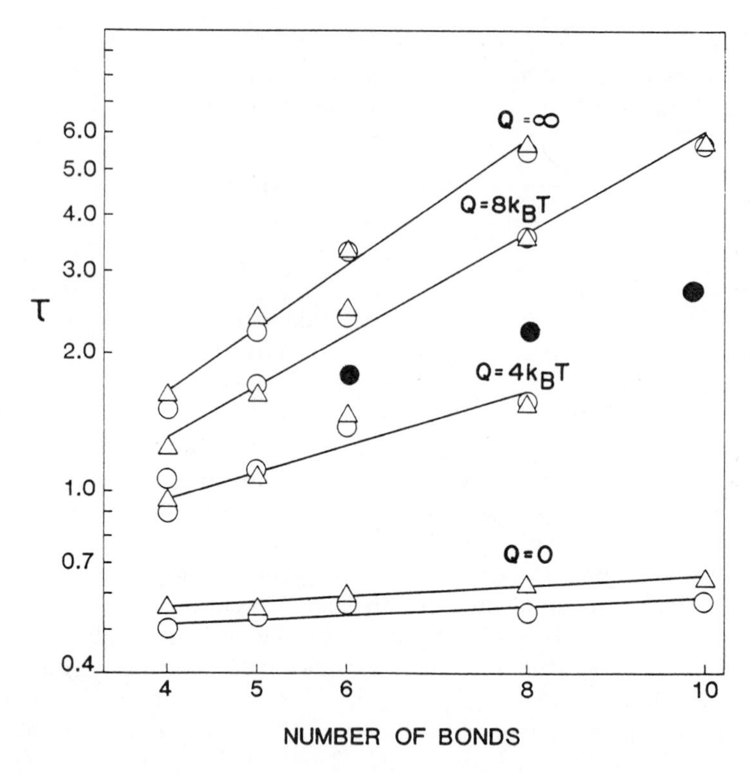

Figure 3. Variation of the first-rank terminal dipole correlation times with N *and barrier height,* Q. *The* Q = ∞ *line is a composite rigid body. Key:* △, *first rank;* ○, *second rank;* ●, *experimental values* (33). *(Reproduced with permission from Ref. 21. Copyright 1980, American Institute of Physics.)*

First, BD and experiment do not match. There is a high frequency skewness in the BD calculations, and the angle at which the BD arc approaches the (0,0) point is neither 90° nor the 60° of the experimental data. Furthermore, the BD arc nearly approaches the (0.5,0.5) point expected for the Debye semicircle. By comparison, the medium frequency regime of the RB simulation is closer to experiment than the true BD. The obvious conclusion to draw about these calculations is that experiment and theory do not match, since they do not pertain to the same system. Although we cannot disagree with that fact, it is important to point out that the BD calculation does not provide the desired skewed arc. Any theory that approximates the BD scheme, by modeling torsional isomerization in some way, will also fail to produce the desired skewed arc. Our Cole–Cole plot has some resemblance to the results of Glauber (36), which again approximate the true dynamics. It would appear that

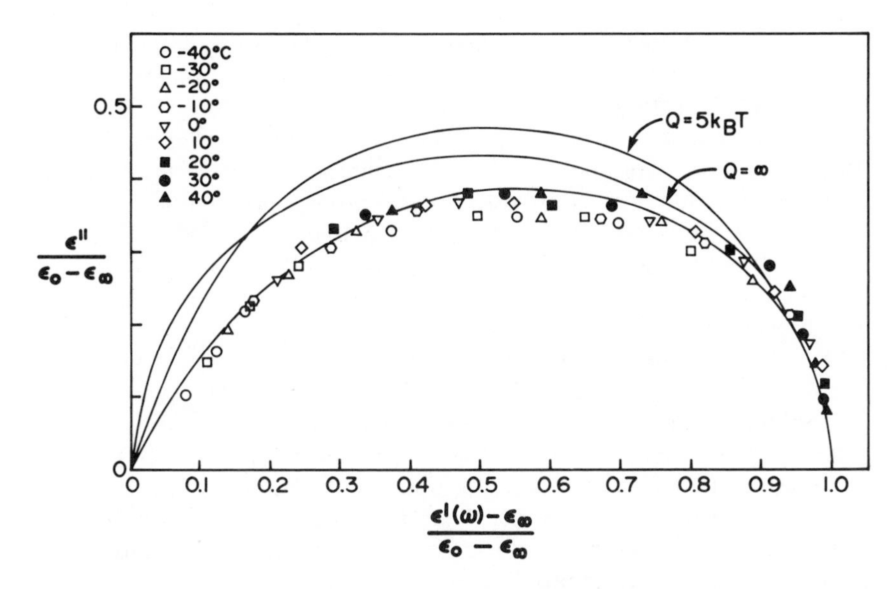

Figure 4. Calculated and experimental (35) *Cole–Cole plot for* n-*octyl iodide. The calculated results were derived from the* $\mathbf{b}_1 \cdot \mathbf{b}_1(t)$ *time correlation function with* $Q = 5k_BT$ *and for a composite rigid body.*

the nature of skewness resides in the multiparticle nature of the dipolar correlation function rather than in an exotic model for single particle reorientations. In constructing the Cole–Cole plot we encountered some pitfalls, which turned out to be very instructive. At first, we curve-fitted the $\langle b_1 \cdot b_1(t)\rangle$ time correlation function piecewise, using a power law at short times and an exponential at long time. Equipped with the curve-fitting parameters, we should execute the Fourier transform analytically and assemble the plot. In particular, for several chains we found that

$$C(t) = \begin{cases} 1 - \alpha t_\beta \quad , & \text{short } t \\ a \exp(-bt) \, , & \text{long } t \end{cases} \tag{13}$$

From chains with five bonds or more, the β values for the b_1 modes are between 0.6–0.7 for barrier heights of the order of $5k_BT$. The short time expansion worked well down to times when the correlation function had decayed to 0.5–0.6. The Cole–Cole plot achieved using Equation 13 was indeed a skewed arc and resembled the experimental data. Unfortunately, although Equation 13 fits the data in the time domain, it poorly represents the data in the frequency domain as it forces nonanalytic behavior for all small time increments. Whereas the BD algorithm generates a manifestly analytic expansion in powers of t, Equation 13 overrides this analyticity. Suffice it to say that the BD results, when properly manipulated, do not explain the skewness of the Cole–Cole plot.

NMR Relaxation. NMR relaxation in chain systems will be discussed by Levy and Karplus (*37*) in this volume, and we will not reanalyze their data. However, because NMR is such a good detector of single particle reorientation, as opposed to collective reorientation, we must mention our results. The appropriate NMR experiment involves the measurement of the T_1 for protons (or ^{13}C) down the backbone of a chain molecule (*38*). By means of these experiments, one can measure the second-rank dynamics of the hydrogen–hydrogen vector, or the ^{13}C–hydrogen vector. To a good approximation, measurements of T_1 for methylene groups are measurements of the D_j modes corresponding to the different carbons down the chain. Correlation times for these modes have been reported in tabular form (*21*). Comparison of the BD results with experimental correlation times obtained by Allerhand et al. (*38*) on heptane indicates that the theoretical times are too small by a factor of two. Theory and experiment do agree in the finding that the correlation times decrease at the chain extremities.

Light Scattering. As in the case of dielectric relaxation, depolarized light scattering (DPLS) can measure single-particle orientational correlation times provided that the chain molecule is the dilute component of the solution. Unlike dielectric relaxation, DPLS has an added complication arising from the uncertain form of the polarizability tensor for a polyatomic molecule (*39*).

The observable in DPLS is the Fourier transform of the autocorrelation function of the total molecular polarizability tensor $\boldsymbol{\alpha}$; thus

$$I(\omega) \sim \int_{-\infty}^{\infty} dt \, \exp(-i\omega t)\langle \boldsymbol{\alpha}:\boldsymbol{\alpha}(t)\rangle \tag{14}$$

where $I(\omega)$ is the spectrum of scattered light. The correlation time is derived from the time integral of the normalized polarizability correlation function. For chain molecules, problems arise regarding the form of the single-particle molecular polarizability. Qualitatively, there are two ways to model $\boldsymbol{\alpha}$. The *bond additive approximation* (BAA) assigns an axially symmetric polarizability tensor to all bonds in the chain (in our case, this means all C–H bonds as well as the C–C bonds). The total polarizability is the sum of all the individual bond polarizability tensors. Alternatively, one can view the chain as consisting of a collection of polarizable spheres interacting with all the other atoms in the chain by the dipole tensor, and this model is called the *interacting atom model* (IAM). Neither the BAA nor the IAM have strict validity for chain molecules in solution, but possibly the use of both models could enable one to bracket the behavior of a real alkane.

In Figure 5, we show the second-rank orientational correlation times (*40*) for the D_1 mode, the BAA and IAM polarizabilities, the E mode, and the experimental measurements on neat alkanes (*41*). The D_1 mode was chosen because it is the fastest second-rank mode, and its correlation time is appropriate to an NMR measurement of a relaxation time for the second carbon. As a rule, the IAM times are longer than the BAA times, since the IAM has more long-wavelength, slowly fluctuating components. The E mode is the upper limit for the single-particle times; it, too, lies below the experimental results. It is premature to read too much into our findings, but it would appear that to force theory on dilute-solution scattering to mimic scattering from concentrated solutions, one would have to implement a static orientation-pair correlation function of the order of one to two for these alkanes; there is some support for this notion (*42*). Certainly, experiments on dilute solutions of alkanes would be very valuable in guiding our insights into the nature of the molecular polarizability and the accuracy of the predictions of BD.

Torsional Isomerization. There are no real experiments on the torsional dynamics of butane, but because butane is the simplest molecule

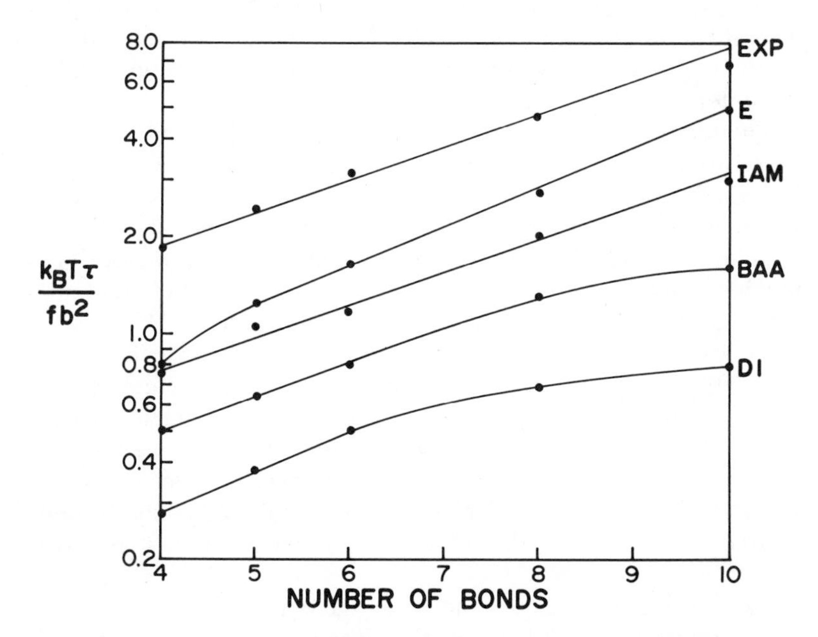

Figure 5. Second-rank orientational time correlation functions for the D_1*, BAA, IAM, and* E *modes compared with DPLS correlation times obtained from neat alkanes. Points, experimental values* (41). *(Reproduced with permission from Ref. 40. Copyright 1981, American Institute of Physics.)*

with a torsional degree of freedom, it has been the vehicle for testing theory versus theory. It is safe to say that, at this moment (December, 1981), the density dependence of the correlation time for the cosine of the torsion angle and the isomerization rate are not well understood.

Our early work on this subject used the diffusion equation for butane Equation 6, which was solved exactly using Green's function methods for the cosine correlation functions and the *trans*-to-*gauche* isomerization rate k (*43*). Comparison of the results of the diffusion equation with molecular dynamics calculations of Weber (*7*) and Ryckaert and Bellemans (*8, 9*) indicated that the differences in rates were of the order of 5%. It appeared that a BD model worked very well, even though, from a strict standpoint, the free energy function for the torsional coordinate varied too quickly to be followed accurately by a diffusion equation. It was argued by Chandler (*16, 44*), Berne (*44, 45*), Skinner and Wolynes (*45, 46*), and Hynes (*47*) that the diffusional regime was not reached in normal low-viscosity solvents, and that the momentum degrees of freedom must be included in the equations of motion for flexible systems. Furthermore, in a simulation of butane in a rigid lattice, where the viscosity is ill-defined, it was found that the isomerization rate was still in the neighborhood of the predictions of the diffusion model equipped with a shear viscosity of 1 centipoise (*11*). It is clear that the role of the medium or the shear viscosity in governing the isomerization rate is not well understood at solid densities.

Qualitatively, at nonsolid or fluid densities, the isomerization rate shows two different dependences (*44–46*). At low density, k increases with density to a maximum of the order of the transition state rate. Further increases in density cause a diminution of the rate. Physically, this is understandable because, at low friction, those molecules with enough energy to isomerize once can escape the other potential minima as well, thereby leading to no effective isomerizations. Molecules with energy less than the barrier height, on the other hand, will never isomerize. By increasing the friction, one is able to trap more effectively those molecules with energies greater than the barrier height. Increasing friction in the high-friction regime decreases the rate, since the drag on a conformer makes it harder for it to cross the barrier. Once it has crossed the barrier, the molecule is locked into torsional oscillations around the potential minimum, and therefore the rate-determining step is barrier crossing.

Recent experiments on the chair–chair interconversion in cyclohexane (*48*) provide evidence for the nonmonotonic dependence of the isomerization rate predicted by the original Kramers theory (*1*), as well as by the modern revisions (*44–46*). Cyclohexane has a barrier of 12–15 kcal/mol to internal rotation, as opposed to the 3-kcal barrier in butane. As a result of the increased barrier height, the potential in cyclohexane has a sharper curvature in the vicinity of the maxima and minima. Consequently, one would expect that the failure of a diffusion equation would be more striking than for butane. Nonetheless, the cyclohexane experiments give strong support to the contentions that one must include momentum transport in the isomerization kinetics in order to account properly for the density dependence of the rate.

Torsional Dynamics in Stilbene. The mechanism of *trans–cis* isomerization of stilbene has been the subject of much debate. Both quantum mechanical calculations and experiment support the following mechanism for photoisomerization (*49*). It is thought that stilbene can absorb a photon to form an excited S_1 singlet state. There is another state S_2, called the *phantom singlet*, which is a 90° torsional conformer with respect to the planar S_1 state. Separating the S_1 and S_2 states is a barrier whose height is roughly 2.5–4.0 kcal/mol. Crossing over the barrier from S_1 to S_2 produces the phantom singlet, which can convert radiationlessly to the *cis* isomer, thereby completing the isomerization.

The proposed mechanism for the isomerization involves a barrier-crossing event. In order to determine the frequency of barrier crossing, one must obtain the inertia component appropriate to the reaction coordinate, the coupling of the internal coordinate to the overall rotational degrees of freedom, and the molecular resistance tensor. Equipped with these quantities, one can then determine $D(\phi)$, the diffusion coefficient for the internal degree of freedom, for use in Equation 6. The differential equation for the orientational Green's function can be solved exactly; hence, one can determine the correlation time associated with the fluctuation of the number of conformers in the *trans* well. Finally, the isomerization rate is just the inverse of the number-density correlation time. The calculations indicate that the diffusive model can accommodate Hochstrasser's experimental isomerization rates (*50*) using the experimental shear viscosity, provided that the barrier heights hindering internal rotation are of the order of $3k_BT$. In light of the experimental findings on the isomerization rate of cyclohexane (*48*) one would have to anticipate effects of momentum relaxation to be operative here as well. However, since the curvature of the torsional potential is much smaller for stilbene than for cyclohexane, an unambiguous assignment of the role of momentum relaxation might not be possible.

Ring-Closure Reaction Rates. Reactive chains longer than four bonds can close to form cyclized products. The probability of a ring-closure reaction and the rate of closure will not depend on the rate of overall rotation, but rather on the torsional flexibility of the chain and the equilibrium probability of chain-end contact. There is interest in the ring-closure phenomenon for two reasons: (1) it may provide another route to understanding chain motion, and (2) it has chemical importance in understanding a class of chemical reactions.

One can learn about the dynamics of ring closure from two types of experiments, the magnetic-field dependence of biradical chemically induced dynamic nuclear polarization (CIDNP) (*51*) and the electron exchange reactions of alkane chains with terminal benzenoid substituents (*52*). Neither of these is as unequivocal in interpretation as NMR. For example, CIDNP is sensitive to magnetic parameters and a host of other noninteresting chemical reactions, as well as the ring-closure rate. Likewise, the electron transfer reactions involve doping the chain with large plate-like substituents that can alter the reorientation dynamics of a simple homogeneous chain. Nonetheless, these experiments do yield some insights into chain motion and therefore should be investigated. Szwarc et al. (*52*), have demonstrated that the rate of electron transfer, k_R, between the terminal naphthyl substituents on an alkane chain exhibits an $N^{-3/2}$ power law. One can attempt to understand this finding by applying the Wilemski and Fixman (WF) (*53*) theory of diffusion-controlled reactions. In the WF theory, the diffusion-controlled reaction rate is given by the square of the probability that the chains are within a reaction volume divided by the time integral of the autocorrelation function for the fluctuation of the number of conformers with the reaction volume. Employing the prescription of WF, together with BD simulations of the correlation function, it follows that the rate of ring closure is given approximately by (*54*)

$$k \cong \frac{Q}{3\pi\eta b r_x^2} \exp(-Q/k_B T) N^{-3/2} \tag{15}$$

for a chain with a torsional potential given by Equation 10, and where r_x is the distance at which chemical reaction takes place. Several approximations were made in calculating Equation 15, but, to within a factor of two, it will accommodate the dependence of the chemical reaction on the parameters incorporated.

Summary

In closing, there are a few points that we wish to emphasize. First, different experiments measure different properties of chain molecules. Although this statement falls somewhat short of profundity, it is a point worth stressing, as a few examples will demonstrate.

BD calculations illustrate that the DPLS correlation time (using IAM) is nearly equal to the dielectric correlation time for a terminal dipole. Normally, for small molecules when the first- and second-rank orientational correlation times are equal, one then asserts that Ivanov's (*55*) large-angle jump diffusion is operative. Here, that is completely wrong. The BD algorithm implements infinitesimal steps; moreover, from the analysis of specific trajectories, there is no evidence of cooperative motions that could resemble a large-angle jump. Dielectric relaxation of a terminal vector detects the fast motions associated with the chain, but these are observed through a first-rank function that is itself a slowly varying function. On the other hand, depolarized light scattering detects a slow variable, viz., the total polarizability, but this function is of second rank, has more nodes, and fluctuates rapidly. As a result, the dielectric and depolarized light scattering correlation times will be similar in medium-size alkane chains. Figures 2 and 5 also indicate that the DPLS and the NMR D_1 correlation times are vastly different, even though both experiments detect tensors of the same rank. In this case, NMR detects a local high frequency process, whereas DPLS is more sensitive to the longer wavelength global processes. Returning to Figure 2, it is somewhat surprising that the butane isomerization rate is slow compared to most of the correlation times already mentioned. Despite the fundamental slowness of the isomerization process (due to the barrier-hindering rotation), several shape changes occur during the nonane trajectory, presumably because the generic isomerization rate for and N-bond chain is roughly $N - 2$ times the value for butane.

Second, it is commonplace to use the Rouse modes (*56–59*) to understand chain dynamics. For long chains, the Rouse modes can predict quantitatively the correct eigenvalues of the chain Liouville operator (*19*). For short chains, it has been found that the Rouse eigenvalue associated with the lowest Rouse eigenfunction is in agreement with BD calculations; for all other Rouse eigenfunctions, BD calculations yield smaller eigenvalues than the original Rouse values (*18*, *60*). In other words, the Rouse eigenvalues get progressively worse for shorter wavelength motions. Furthermore, by means of a mode-coupling reduction of the diffusion equations, one can show that the relaxation rates for the Rouse basis functions depend on the Kramers isomerization rates for the short wavelength fluctuations (*60*).

Finally, theory most likely can account for most reorientational transport coefficients to within factors of two. The definitive experiments on dilute solutions of alkanes have yet to be performed and there is much work remaining to determine the legitimacy of the approaches summarized here.

Acknowledgments

We are grateful to the National Science Foundation and the donors of the Petroleum Research Fund, administered by the American Chemical Society, for partial support of this research. Support from the Alfred P. Sloan foundation and from the National Resource for Computation in Chemistry is also appreciated. The dielectric work reported here originated from a collaboration with R. H. Cole. Finally, this review was prepared at UCLA and has benefited from discussions with, and hospitality of, Daniel Kivelson.

Note Added in Proof

Since December 1981, a number of pertinent papers have been written. Velsko et. al. (*61*) provided experimental evidence that a frequency-dependent friction coefficient might be necessary to explain the viscosity dependence of isomerization rates in viscous media. Also, Garcia de la Torre et al. (*62*) improved our parameterization of the friction coefficient (*31*) by using Oseen hydrodynamic interactions and slip boundary conditions. Recently, we applied the FP and BGK transport equations to investigate the full density (or friction) dependence of the butane isomerization rate (*63*). Also, in collaboration with Ladanyi (*64*), we have reanalyzed the photoisomerization rates of *trans*-stilbene implementing a FP equation for the torsion angle and its canonical momentum, the potential of mean force, and the Rotne–Prager hydrodynamic interaction tensor. The operation of convective effects and hydrodynamic interaction decreased the stilbene isomerization rate by 30–40% from that predicted by a diffusion equation.

Literature Cited

1. Kramers, H. A. *Physica* **1940,** *7*, 284.
2. Erpenbeck, J. J.; Kirkwood, J. G. *J. Chem. Phys.* **1958,** *29*, 909.
3. Kirkwood, J. G. *J. Polymer. Sci.* **1954,** *12*, 1.
4. Eckhart, C. *Phys. Rev.* **1935,** *47*, 552.
5. Zwanzig, R. W. In “Lectures in Theoretical Physics”; Interscience: 1961; Vol. 3, p. 106.
6. Kawasaki, K. *Ann. Phys.* **1970,** *61*, 1.
7. Weber, T. A. *J. Chem. Phys.* **1978,** *69*, 2347.
8. Ryckaert, J. P.; Bellemans, A. *J. Chem. Soc., Faraday Disc.* **1978,** *66*, 95.
9. Ryckaert, J. P.; Bellemans, A. *Chem. Phys. Lett.* **1975,** *30*, 123.
10. Rebertus, D. W.; Berne, B. J.; Chandler, D. *J. Chem. Phys.* **1979,** *70*, 3395.
11. Rosenberg, R. O.; Berne, B. J.; Chandler, D. *Chem. Phys. Lett.* **1980,** *75*, 162.
12. Pear, M. R.; Weiner, J. H. *J. Chem. Phys.* **1979,** *71*, 212.
13. Levy, R. M.; Karplus, M.; McCammon, J. A. *Chem. Phys. Lett.* **1979,** *65*, 4.
14. Helfand, E. *J. Chem. Phys.* **1978,** *69*, 1010.
15. Helfand, E. *Bell Syst. Tech. J.* **1979,** *58*, 2289.
16. Montgomery, J. A.; Holmgren, S. L.; Chandler, D. *J. Chem. Phys.* **1980,** *73*, 3688.
17. Fixman, M. *J. Chem. Phys.* **1978,** *68*, 2983.
18. Ibid., **1978,** *69*, 1527.
19. Ibid., **1978,** *69*, 1538.

20. Helfand, E.; Wasserman, Z. R.; Wever, T. A. *J. Chem. Phys.* **1979,** *70*, 2016.
21. Evans, G. T.; Knauss, D. C. *J. Chem. Phys.* **1980,** *72*, 1504.
22. Pear, M. R.; Northrup, S. H.; McCammon, J. A. *J. Chem. Phys.* **1980,** *73*, 4703.
23. van Gunsteren, W. F.; Berendsen, H. J. C.; Rullman, J. A. C. *Mol. Phys.* **1981,** *44*, 69.
24. Evans, G. T. *J. Chem. Phys.* **1980,** *72*, 3849.
25. Evans, G. T.; Knauss, D. C. *J. Chem. Phys.* **1981,** *75*, 4647.
26. Lebowitz, J. L.; Resibois, P. *Phys. Rev.* **1965,** *139A*, 1001.
27. Fixman, M.; Rider, K. *J. Chem. Phys.* **1969,** *51*, 2425.
28. Knauss, D. C.; Evans, G. T. *J. Chem. Phys.* **1980,** *72*, 1499. Ibid., **1980,** *73*, 2017.
29. Fixman, M.; Kovac, J. *J. Chem. Phys.* **1974,** *61*, 4939.
30. Paul, E.; Mazo, R. M. *J. Chem. Phys.* **1968,** *48*, 1405.
31. Knauss, D. C.; Evans, G. T.; Grant, D. M. *Chem. Phys. Lett.* **1980,** *71*, 158.
32. McQuarrie, D. A. "Statistical Mechanics"; Harper & Row: NY, 1976.
33. Crossley, J. *Adv. Mol. Relax. Proc.* **1974,** *6*, 39.
34. Mopsik, F. I.; Cole, R. H. *J. Chem. Phys.* **1966,** *44*, 1015.
35. Mopsik, F. I., Ph.D. thesis, Brown Univ., 1964.
36. Glauber, R. J. *J. Math. Phys.* **1963,** *4*, 294.
37. Levy, R. M.; Karplus, M. Chap. 18 in this volume.
38. Doddrell, D.; Glushko, V.; Allerhand, A. *J. Chem. Phys.* **1972,** *56*, 3683.
39. Kivelson, D.; Madden, P. A. *Annu. Rev. Phys. Chem.* **1980,** *31*, 523.
40. Keyes, T.; Evans, G. T.; Ladanyi, B. M. *J. Chem. Phys.* **1981,** *74*, 3779.
41. Champion, J. V.; Dandridge, A.; Meeten, G. H. *J. Chem. Soc., Faraday Disc.* **1979,** *66*, 266.
42. Patterson, G. D.; Flory, P. J. *J. Chem. Soc., Faraday Trans. II* **1972,** *58*, 1098.
43. Knauss, D. C.; Evans, G. T. *J. Chem. Phys.* **1980,** *73*, 3423.
44. Montgomery, J. A.; Chandler, D.; Berne, B. J. *J. Chem. Phys.* **1979,** *70*, 4056.
45. Berne, B. J.; Skinner, J. L.; Wolynes, P. G. *J. Chem. Phys.* **1980,** *73*, 4314.
46. Skinner, J. L.; Wolynes, P. G. *J. Chem. Phys.* **1980,** *72*, 4913; Ibid., **1978,** *69*, 2143.
47a. Northrup, S. H.; Hynes, J. T. *J. Chem. Phys.* **198,** *73*, 2700.
47b. Groote, R. F.; Hynes, J. T. *J. Chem. Phys.* **1980,** *73*, 2715.
48. Hasha, D. L.; Eguchi, T.; Jonas, J. *J. Chem. Phys.* **1981,** *75*, 1571.
49. Knauss, D. C.; Evans, G. T.; *J. Chem. Phys.* **1981,** *74*, 4627, and references mentioned therein.
50. Greene, B. I.; Hochstrasser, R. M.; Weissman, R. B. *Chem. Phys. Lett.* **1975,** *62*, 427.
51. Kanters, F. J.; den Hollander, J. A.; Huizer, A. H. *Mol. Phys.* **1977,** *34*, 857.
52. Shimada, K.; Szwarc, M. *J. Am. Chem. Soc.* **1975,** *97*, 3313.
53. Wilemski, G.; Fixman, M. *J. Chem. Phys.* **1974,** *60*, 866.
54. James, C.; Evans, G. T. *J. Chem. Phys.* submitted for publication.
55. Ivanov, E. N. *Sov. Phys. JETP* **1964,** *18*, 1041.
56. Rouse, P. E. *J. Chem. Phys.* **1952,** *21*, 1272.
57. Stockmayer, W. H. *Pure App. Chem.* **1967,** *15*, 539.
58. Stockmayer, W. H.; Gobush, W.; Norvich, R. *Pure App. Chem.* **1971,** *26*, 537.

59. Stockmayer, W. H. *Pure App. Chem. Suppl. Macro. Chem.* **1973,** *8*, 379.
60. Evans, G. T. *J. Chem. Phys.* **1981,** *74*, 4621.
61. Velsko, S. P.; Waldeck, D. H.; Fleming, G. R. *J. Chem. Phys.* **1982,** *78*, 249.
62. Garcia de la Torre, J.; Lopez Martinez, C. *Chem. Phys. Lett.* **1982,** *88*, 564.
63. Evans, G. T. *J. Chem. Phys.* **1983,** *78*, March 1.
64. Ladanyi, B. M., Evans, G. T. *J. Chem. Phys.* **1982,** *78*, submitted.

RECEIVED for review January 27, 1982. ACCEPTED for publication September 17, 1982.

18

Trajectory Studies of NMR Relaxation in Flexible Molecules

RONALD M. LEVY
Rutgers University, Department of Chemistry, New Brunswick, NJ 08903
MARTIN KARPLUS
Harvard University, Department of Chemistry, Cambridge, MA 02138

Stochastic dynamics trajectories for alkanes in aqueous solution have been used to examine a variety of problems that arise in the interpretation of ^{13}C-NMR relaxation experiments. Exact results for spin-lattice and spin-spin relaxation times, and nuclear Overhauser enhancement values obtained from these trajectories, have been employed to analyze the relaxation behavior of small alkanes and macromolecular side chains and to test the validity of simplified relaxation models for these systems. A molecular dynamics simulation of a protein has been used to demonstrate the effects of picosecond fluctuations on observed ^{13}C spin-lattice relaxation times. It is shown how an increase in spin-lattice relaxation time can be related to order parameters for the picosecond motional averaging of the carbon–hydrogen dipolar interactions, and how the order parameters can be calculated from a protein molecular dynamics trajectory. The present work provides a firm theoretical foundation for the continuing effort to use NMR measurements for the experimental analysis of the dynamics of molecules with internal degrees of freedom.

NUCLEAR MAGNETIC RESONANCE RELAXATION MEASUREMENTS provide an important probe of the dynamics of molecules because the spin-lattice (T_1) and spin-spin (T_2) relaxation times and the nuclear Overhauser enhancement (NOE) factor (η) all depend on the thermal motions. For protonated carbons, ^{13}C NMR is particularly well suited for the study of dynamics because the relaxation is dominated by the fluctuating dipolar interactions between ^{13}C nuclei and directly bonded protons. Applications of ^{13}C NMR have been made to the dynamics of small molecules in solution (*1–6*), polymers (*7–13*), and molecules of biological interest including lipids (*14–16*) and proteins (*17–23*). Because the motions of

0065-2393/83/0204-0445$07.00/0

molecules with many internal degrees of freedom (e.g., macromolecules) are complicated, the interpretation of NMR measurements for such systems is often not unique. Empirical rules have been developed to fit the relaxation data to the molecular tumbling time combined with internal segmental motions (*6–8*). Alternatively, the experimental results have been interpreted in terms of analytically tractable descriptions of the dynamics based on continuous diffusion (*9–24*), restricted diffusion (*25–28*), and lattice jump models (*24, 26, 29, 30*). While it is usually possible to fit the experimental results in this way, the data in themselves generally are not sufficient to determine whether or not a model gives the correct description of the dynamics.

A powerful method of testing relaxation models is provided by computer-generated trajectories that make use of realistic potentials to simulate the motion of the system of interest. From such trajectories, the time-dependent correlation functions can be determined and T_1, T_2, and η can be calculated. Thus the trajectory provides simultaneously a complete knowledge of the dynamics and exact values of the NMR parameters. It is now possible to proceed by using the calculated values of T_1, T_2, and η as "experimental" quantities and fitting the various models to them. The resulting model dynamics can then be compared with the exact dynamics from the trajectory to determine how well the former corresponds to the latter. Also, the results of the trajectory studies may be compared directly with experiment without recourse to simplified theories.

In this chapter we review our work, which employs the results of computer simulations for the analysis of experimental and theoretical NMR studies of the motions of flexible molecules. The systems considered include butane and heptane tumbling in aqueous solution and the short- and long-time dynamics of side chains attached to macromolecules. The first section of this chapter reviews the procedures used to generate the trajectories and outlines the methodology employed to extract the NMR parameters from them. Then the butane and heptane trajectories are used, in the next section, to evaluate the relaxation times for these alkane chains in the motional narrowing limit, and the results are compared with experiment and with simplified models. The following section analyzes the dynamics of a side chain protruding from a macromolecule and employs the heptane trajectories to critically examine a lattice jump model for the relaxation. In the final section we make use of a full molecular dynamics simulation of a protein to demonstrate the effects of fast protein motions on observed ^{13}C-NMR relaxation times.

Methodology

In this section we outline the method used for obtaining the alkane trajectories and present the procedure for determining NMR relaxation parameters from these trajectories. We show how trajectory results can be applied to study simplified dynamical models that have been used to interpret NMR relaxation data. For the analysis of ^{13}C spin lattice relaxation in proteins we show how an increase in the relaxation time is related to order parameters for the picosecond motional averaging of the $^{13}C,H$ dipolar interactions, and how these order parameters are calculated from a molecular dynamics trajectory.

Diffusive Langevin Dynamics of Model Alkanes. The equation of motion descriptive of Brownian particles is the Langevin equation (*31*)

$$m_i \frac{d\boldsymbol{v}_i}{dt} = -\zeta_i \boldsymbol{v}_i + \boldsymbol{F}_i + \mathbf{A}_i \tag{1}$$

where $m_i \boldsymbol{v}_i$, and ζ_i are the mass, velocity, and friction constant of the *i*th particle and $\boldsymbol{F}_i$ is the systematic force acting on the *i*th particle due to the potential of mean force; $\boldsymbol{F}_i$ in general depends on the coordinates of all the particles. The stochastic term $\mathbf{A}_i(t)$ represents the randomly fluctuating force on the particle due to the solvent. In the diffusive regime, the particle momenta relax to equilibrium much more rapidly than the displacements. With the assumption that the forces are slowly varying, i.e., that it is possible to take time steps that are large compared to the momentum relaxation time $\left(\Delta t >> \frac{m_i}{\zeta_i}\right)$, Equation 1 can be integrated to obtain the equation (*31–33*) for the displacement vector, $\boldsymbol{r}_i$

$$\boldsymbol{r}_i(t + \Delta t) = \boldsymbol{r}_i(t) + [\boldsymbol{F}_i(t)/\zeta_i]\Delta t + \Delta \boldsymbol{r}_i(t) \tag{2}$$

The term $\Delta \boldsymbol{r}_i(t)$ is the random displacement due to the stochastic force; it is chosen from a Gaussian distribution with zero mean and second moment $6D_i\Delta t$, where D_i, the diffusion coefficient, is KT/ζ_i.

Equation 2 is the equation used to calculate the trajectories of the hydrocarbon chains (*34*). An extended-atom model was introduced, in which the CH_3 and CH_2 units are represented as spheres of van der Waals radius 1.85 Å; the bond lengths and angles of the alkane chains were set equal to 1.523 Å and 111.3°, respectively. Each extended-atom group along the chain acted as a point center of frictional resistance with a Stokes' law friction constant $\zeta_i = 6\pi\eta a$. To determine η, the viscosity of water at 25 °C ($\eta = 0.01$ poise) was used and a was set equal to the extended-atom van der Waals radius; with these parameters, $\zeta_i = 3.45 \times 10^{-9}$ g/s. This is to be compared with the value of $\zeta_i = 2.2 \times 10^{-9}$ g/s obtained from the experimental diffusion coefficient of methane in water at 25° C.

POTENTIAL FUNCTION. The negative gradient of the potential of mean force is used for $\boldsymbol{F}_i$ in Equations 1 and 2. This is a combination of the potential energy arising from the interactions among the particles composing the solute molecule and the effective potential due to the solvent molecules. The molecular potential was expressed as a sum of two types of terms; the first is a torsional potential energy for each of the dihedral angles, and the second consists of pairwise Lennard–Jones interactions between extended atoms separated by four or more bonds (*35, 36*). For butane there is one angular degree of freedom and only the associated torsional potential was used, but for heptane both torsional and nonbonded terms are required.

The solvent contribution to the torsional nonbonded potential was obtained from the work of Pratt and Chandler (*3*), which expresses the potential due to the solvent as $-kT \ln (y)$ where y is the cavity distribution function. The differences in energy at 25 °C between the *trans* and *gauche* butane geometry in the vacuum and in the solvent are 0.70 kcal/mol and 0.16 kcal/mol, respectively. For the solvent effect on the nonbonded interaction, the cavity distribution function for two methane molecules dissolved in water was used (*37*). The solvent-modified potential has a slightly deeper energy minimum (−0.38 versus −0.30 kcal/mol), which occurs at a smaller interparticle separation (3.0 versus 4.15 Å); at larger separations the attraction is more quickly screened in the solvent.

For all the systems considered, Equation 2 was integrated in Cartesian coordinates with the bond lengths and angles of the molecule constrained to the initial value by use of the SHAKE algorithm (*35*). In general, a correction term should be introduced when the Langevin equation is solved with constraints (*38*). This term was omitted from the present simulation, in which the effect is expected to be small (*39*).

For the butane simulation, time steps of 0.005 ps and 0.05 ps were compared; for the other molecules, time steps between 0.025 and 0.05 ps were used. Butane trajectories were run for 90 ns (25 °C) and 20 ns (50 °C) on the solvent-modified potential surface and 20 ns (25 °C) on the vacuum molecular potential surface. Heptane trajectories on the solvent-modified surface (25 °C) were recorded for 20 ns without constraints and for 10 ns with three atoms held fixed. For 10 ns of the butane trajectory (0.05-ps time step), the required central processing unit time on an IBM 370/168 computer was 15 min.

Exact Calculation of NMR Parameters. The NMR relaxation parameters probe angular correlation functions of the relaxing nucleus. These correlation functions and their Fourier transforms, the spectral densities, can be obtained directly from the alkane trajectories, which provide the complete chain dynamics in the diffusive limit. For the ^{13}C nucleus in an alkane chain, the dynamical quantities of interest are the spherical polar coordinates (with respect to a laboratory frame) of the C–H internuclear vectors. The well-known relations between the NMR relaxation parameters T_1, T_2, and η and the spectral densities for the case of dipolar relaxation involving two different spin 1/2 nuclei (*40*) are

$$\frac{1}{T_1} = \frac{N\hbar^2\gamma_C^2\gamma_H^2}{r^6}\frac{4\pi}{10}\{J_0(\omega_C - \omega_H) + 3J_1(\omega_C) + 6J_2(\omega_C + \omega_H)\} \quad (3)$$

$$\frac{1}{T_2} = \frac{N\hbar^2\gamma_C^2\gamma_H^2}{r^6}\frac{4\pi}{20}\{4J_0(0) + J_0(\omega_C - \omega_H) + 3J_1(\omega_C) + 6J_1(\omega_H) + 6J_2(\omega_C + \omega_H)\} \quad (4)$$

$$\eta = \frac{\gamma_H}{\gamma_C}\left[\frac{6J_2(\omega_C + \omega_H) - J_0(\omega_C - \omega_H)}{J_0(\omega_C - \omega_H) + 3J_1(\omega_C) + 6J_2(\omega_C + \omega_H)}\right] \quad (5)$$

where γ_C, γ_H and ω_C, ω_H are the gyromagnetic ratios and Larmor frequencies of the ^{13}C and ^1H nuclei, r is the C–H bond distance, and N is the number of protons directly bonded to the relaxing carbon nucleus. The spectral densities, $J_m(\omega)$, are the Fourier transforms of the second-order spherical harmonics, $Y_m^2(\theta, \phi)$ given by

$$J_m(\omega) = \int_0^\infty \langle Y_m^2[\theta(t)\phi(t)]\, Y_m^{*2}\,[\theta(0)\phi(0)]\rangle \cos \omega t\, dt \quad (6)$$

where the spherical-harmonic time correlation function is defined as the ensemble average

$$\langle Y_m^2[\theta(t)\phi(t)]Y_m^{*2}[\theta(0)\phi(0)]\rangle = \int d\theta d\phi \int d\theta' d\phi'\, Y_m^2(\theta, \phi)\, G(\theta, \phi, t; \theta', \phi', 0) Y_m^{*2}(\theta'\phi') \quad (7)$$

Here $G(\theta, \phi, t; \theta', \phi', 0)$ is the conditional probability that a CH vector has spherical polar coordinates $(\theta\phi)$ at times t, given that they were equal to $(\theta'\phi')$ at time 0. The spherical-harmonic time correlation functions are obtained from the alkane trajectories by replacing the ensemble average (Equation 7) with a time average (*41*). Since the orientation of the laboratory z axis is arbitrary for freely rotating molecules, the angular correlations and their spectral densities are independent of subscript m.

Once the correlation functions (Equation 7) have been evaluated from the time integral over the trajectory, the spectral densities are obtained by numerical Fourier transformation, as indicated in Equation 6. The resulting values of $J_m(\omega)$ are then introduced into Equations 3–5 to determine the NMR parameters.

Relaxation Models. The ^{13}C-NMR spectra of low molecular weight species in solution have tumbling times that are much smaller than the reciprocal of the Larmor frequencies of current spectrometers. Consequently, the resonances of such systems are observed in the motional narrowing limit $(\omega_C\tau)^2 << 1$, where the relaxation time τ is the time integral of the angular correlation function

$$\tau \equiv \int_0^\infty \langle Y_m^2[\theta(t)\phi(t)]\ Y_m^{*2}\ [\theta(0)\phi(0)]\rangle\ dt \tag{8}$$

In the motional narrowing limit, the NMR relaxation times T_1 and T_2 are equal and independent of the Larmor frequency and the ^{13}C dipolar NOE is maximal (*1*, *40*); that is, Equations 3–5 reduce to

$$\frac{1}{T_1} = (4\pi\tau)\frac{N\hbar^2\gamma_C^2\gamma_H^2}{r^6} \tag{9}$$

$$T_2 = T_1 \tag{10}$$

$$\text{NOE} = 1 + \eta = 2.99 \tag{11}$$

For rigid spherical molecules in solution, the rotational motion is described by the angular Debye diffusion equation and the second-order spherical harmonics decay as a single exponential (*41*) with relaxation time $\tau = 1/(6D)$, where D is the rotational diffusion coefficient. For flexible molecules, a distribution of relaxation times is required to describe the motion of a given C–H vector. For liquid alkanes, an effective relaxation time τ_{eff} has been extracted from experimental T_i values; τ_{eff} may be thought of as a weighted average of the distribution of relaxation times. Empirically, τ_{eff} has been separated into contributions from molecular tumbling and segmental motions (*6*, *10*, *11*)

$$\frac{1}{\tau_{\text{eff}}} = \frac{1}{\tau_0} + \frac{1}{\tau_i} \tag{12}$$

where τ_0 is the molecular tumbling time, and τ_i is the segmental motion relaxation time for the *i*th carbon. To test the validity of Equation 12 using the alkane trajectories, we have considered a local coordinate system centered on the relaxing nucleus and analyzed its motion relative to a coordinate frame embedded in the molecule (*41*). To express the angular correlation functions (Equation 7) in terms of these coordinate systems we use the Wigner rotation matrices D_{ma}^2, which transform the spherical harmonics between coordinate frames related by the Euler angles Ω. The resulting expression is

$$\langle Y_m^2[\theta(t)\phi(t)]\ Y_m^{*2}[\theta(0)\phi(0)]\rangle$$
$$= \sum_{\substack{aa' \\ bb'}} \langle[D_{ma}^{*2}(\Omega_0(t))\ D_{ma'}^2\ (\Omega_0(0))]\ [D_{ab}^{*2}(\Omega_i(t))\ D_{a'b'}^2(\Omega_i(0))]\rangle$$
$$\times\ Y_b^2(\theta_{\mathrm{mol}},\ \phi_{\mathrm{mol}})\ Y_{b'}^{*2}(\theta_{\mathrm{mol}},\ \phi_{\mathrm{mol}}) \quad (13)$$

Where $D_{ab}^2(\Omega)$ is the Wigner rotation matrix, $\Omega_0(t)$ are the Euler angles for the transformation from the laboratory to the molecular coordinate system, and $\Omega_i(t)$ are the Euler angles for the transformation from the molecular coordinate system to the local coordinates centered on the *i*th relaxing nucleus. The angles $(\theta_{\mathrm{mol}},\ \phi_{\mathrm{mol}})$ are the time-independent spherical polar coordinates of the C–H internuclear vector in the local coordinate system. An expression of the form of Equation 12 for τ_{eff} can be obtained with the assumptions that tumbling and internal motions are independent, and that the correlation functions describing the tumbling and internal motions decay as a single exponential. For this situation, the correlation functions of Equation 13 can be broken into a sum over products of correlation functions of the form:

$$\langle D_{ma}^{*2}[\Omega_0(t)D_{ma'}^2(\Omega_0(0)]\rangle \simeq e^{-t/\tau_0}\delta_{aa'} \quad (14a)$$

$$\langle D_{ab}^{*2}[\Omega_i(t)]D_{ab'}^2[\Omega_i(1)]\rangle \simeq e^{-t/\tau_i}\delta_{bb'} \quad (14b)$$

Substitution of this result into Equation 13 permits one to obtain an expression of the form of Equation 12 for τ_{eff}. In the next section of this chapter we employ the stochastic trajectory results for butane and heptane to evaluate the correlation functions (Equation 13), compare calculated NMR relaxation times with experimental trends for liquid alkanes, and analyze the validity of the approximations inherent in Equation 12 and Equations 14a and 14b.

More realistic models of NMR relaxation in polymers take into account the presence of torsional barriers and of excluded volume effects. One type of analytic model, which has been applied to hydrocarbon chains in membranes and in solution, is the so-called *jump model* (*26, 27*). In practice, there have been few attempts to interpret experimental NMR relaxation data within the context of a complete lattice jump model because the number of adjustable parameters rapidly becomes unwieldly. For the hydrocarbon relaxation problem, it has been common to employ the model with the product approximation for the relaxation; this omits excluded volume effects and the possibility of concerted motions. Each carbon–carbon bond is allowed to jump between three states, *trans*, *gauche* (+), and *gauche* (−). The rate constant for jumping from the *trans* (τ) to *gauche* (g^+, g^-) states is K_T, and the inverse rate constant is K_G. For this simplified lattice jump model the correlation functions for each of the internal degrees of freedom (Equation 13) can be written as a sum over the conditional probabilities for the allowed transitions in the form (*24, 29, 41*)

$$\langle D^{*2}_{ab}[\Omega(t)] D^{2}_{a'b'}[\Omega(0)]\rangle = d^2_{ab}(\beta) d^2_{a'b'}(\beta) \sum_{\phi=\tau,\,g^-,g^+} \sum_{\phi=\tau,g^-,g^+} e^{i[b\phi - b'\phi']} P_0(\phi') P(\phi t|\phi' 0) \quad (15)$$

The d^2_{ab} terms are real reduced Wigner matrix elements, with β the complement of the rigid CCC bond angle, $P(\phi t|\phi' 0)$ is the conditional probability that a rotational angle is ϕ at time t, given that it was ϕ' at time 0, and $P_0(\phi')$ is the equilibrium probability of a ϕ' state. Explicit expressions for the conditional probabilities are derived in terms of the eigenvalues and eigenvectors of the rate matrix (*29, 41*).

The lattice jump model is studied in some detail later in this chapter. The heptane trajectory is used to evaluate T_1, T_2, and NOE for a model amino acid side chain with four internal rotational angles. Figure 1 shows the four rotational angles and indicates the nature of the three types of coordinate systems that are used to define the configurations of the chain; that is, the laboratory system, the macromolecular system with respect to which the overall tumbling is defined, and the four local coordinate frames associated with the internal degrees of freedom. For the analysis, the coordinate frame centered on C2 is assumed to be rigidly attached to the macromolecule, which is tumbling isotropically with relaxation time τ_0. Only carbons C2 through C6 are considered, because the C7 methyl protons cannot be located from an extended-atom alkane model. The NMR parameters calculated exactly from the trajectory are compared with those obtained from the independent lattice jump model. For the latter, only the isomerization rate constants K_T and K_G for each of the rotational angles are required (*34, 43*).

Another class of NMR relaxation problems for which excluded volume effects are important deals with the contribution of fast (picosecond) motions to the relaxation of amino acid side chains in the interior of proteins (*21–29*). In the last section of this chapter we review our use of a full molecular dynamics simulation of the protein pancreatic trypsin inhibitor (PTI) to demonstrate the effects of picosecond fluctuations on observed ^{13}C T_i values. The protein trajectory has been used to evaluate the short time decay of internal correlation functions (Equation 14b), which determine the NMR relaxation (*44–45*). Because of the highly restricted nature of the motion in the protein interior, the internal correlation functions generally do not decay to zero. Instead, a plateau value

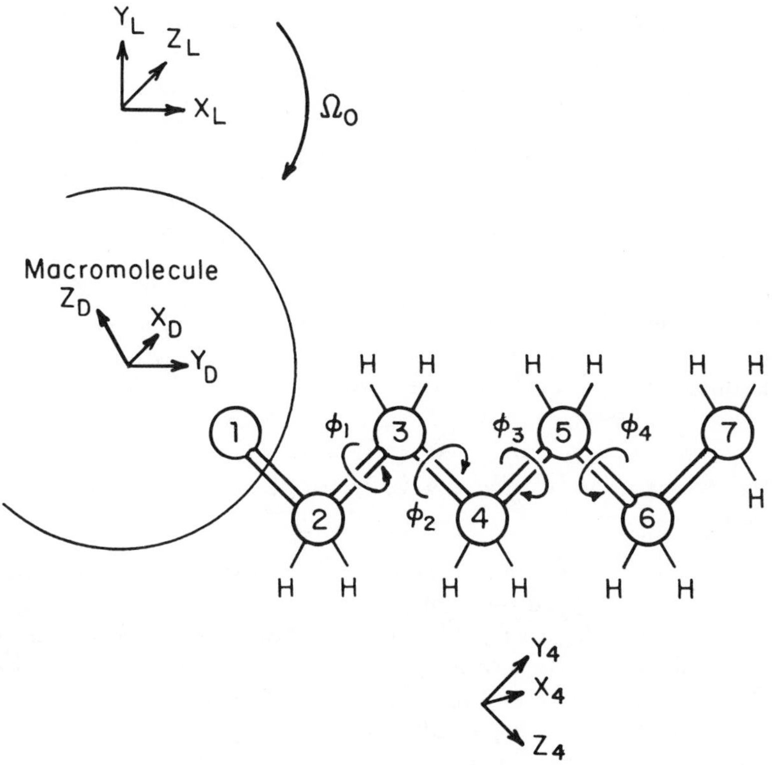

Figure 1. Schematic representation of a side chain with four internal rotational angles attached to a tumbling macromolecule. The laboratory coordinate system, molecular tumbling coordinate system, and a coordinate system attached to C4 are labeled. (Reprinted from Ref. 4. Copyright 1981, American Chemical Society.)

is often reached after t_p picoseconds, where t_p is a time short compared with the length of the trajectory. We have previously shown that, for such a plateau value, the internal correlation function is equal to the equilibrium orientation distribution obtained from the entire run (*45, 46*)

$$(\mathcal{S}_j)^2 = \frac{4\pi}{5} \sum_a |\langle Y_a^2[\theta_j(t)\phi_j(t)]\rangle|^2 \qquad (16)$$

The quantity $\mathcal{S}_j$ defined by Equation 16 is the generalized order parameter for the restricted motion of the internuclear dipole vector (*45–48*). The carbon relaxation, corrected for the picosecond motional averaging of the internuclear dipole vector, is

$$T_{1j} \simeq \mathcal{S}^{-2}\, T_{1j}^{R} \tag{17}$$

where the rigid relaxation time, T_{1j}^{R} is calculated from Equation 3 with

$$J(\omega) = \frac{1}{4\pi} \frac{\tau_0}{1 + (\omega\tau_0)^2} \tag{18}$$

In a later section we review our results concerning the evaluation of order parameters from protein trajectories and show that under suitable conditions, ^{13}C-NMR relaxation data can serve to probe the picosecond reorientation dynamics of the C–H bond vector.

NMR Relaxation of Model Alkanes in the Motional Narrowing Limit

In this section we employ the stochastic trajectory results for butane and heptane to analyze the contributions to the ^{13}C relaxation times in simple alkanes. The calculated relaxation times are compared with experimental trends. We also examine some simplified models that have been proposed for interpreting alkane NMR data and test them by comparison with the trajectory results.

Relaxation Time: Theory and Experiment. In the motional narrowing limit applicable to the small alkanes, the ^{13}C dipolar spin lattice relaxation times (T_1; see Equation 9) are inversely proportional to the relaxation times (τ) of the second-order spherical harmonics. The computed relaxation times for each of the internuclear vectors of butane and heptane, obtained from a least-squares fit of a single exponential to the decay of the calculated correlation functions, are listed in Tables I and

Table I. Relaxation Times of Butane Angular Correlation Functions

Bond	$\langle Y_0^{*1}(0) Y_0^1(t)\rangle$ $\tau(ps)$	$\langle Y_0^{*2}(0) Y_0^2(t)\rangle$ $\tau(ps)$	$\langle Y_1^{*2}(0) Y_1^2(t)\rangle$ $\tau(ps)$	$\langle Y_2^{*2}(0) Y_2^2(t)\rangle$ $\tau(ps)$
C1–C2	(18.5)[a] 16.0	5.7	5.6	5.8
C2–C3	(19.1) 16.9	6.1	5.9	6.4
C3–C4	(19.0) 15.9	6.0	5.6	5.8
C2–H	(7.6)[b] 8.1	(3.9)[b] 5.0	(4.0)[b] 4.9	(3.9)[b] 5.1
C3–H	(7.8) 8.3	(4.1) 5.2	(3.9) 4.8	(3.9) 5.1

Note: Relaxation time from least-squares fit of the relaxation to a single exponential over 10 ps, except as noted.

[a] Least-squares fit to single exponential over 20 ps.

[b] Least-squares fit to single exponential over 5 ps.

Table II. Relaxation Times of Heptane Angular Correlation Functions

Bond	$\langle Y_0^{*1}(0)Y_0^1(t)\rangle$ τ(ps)	$\langle Y_0^{*2}(0)Y_0^2(t)\rangle$ τ(ps)	$\langle Y_1^{*2}(0)Y_1^2(t)\rangle$ τ(ps)	$\langle Y_2^{*2}(0)Y_2^2(t)\rangle$ τ(ps)
C1–C2	(39.3)[a] 33.4	13.4	14.7	15.9
C2–C3	(46.7) 42.2	16.5	15.1	16.5
C3–C4	(51.9) 49.8	17.6	19.6	20.2
C4–C5	(50.3) 46.7	18.1	16.9	18.3
C5–C6	(47.3) 41.6	16.5	17.2	18.3
C6–C7	(37.3) 35.7	13.7	12.9	13.5
C2–H	(32.6)[a] 29.1	12.6	11.7	11.9
C3–H	(31.5) 29.4	13.9	11.3	13.5
C4–H	(35.1) 33.1	15.1	11.6	13.3
C5–H	(33.8) 33.4	13.4	12.3	12.5
C6–H	(25.4) 23.3	9.9	12.9	12.7

Note: Relaxation time from least-squares fit of the relaxation to a single exponential over 10 ps, except as noted.
[a] Fit to a single exponential over 20 ps.

II, respectively. Since the relaxation is not due to isotropic rigid body rotation, the single exponential fit is approximate. In Table III we compare the relaxation times obtained from a single exponential fit to the C6–H vector in heptane with the results from the time integral of the angular correlation function (Equation 8). For the complete dynamics (top row of results), the time integral of the correlation function gives a value for the relaxation time about 14% smaller than that estimated from a single exponential fit. Such a difference is not important for the analysis described below.

The relaxation times listed in Tables I and II demonstrate a number of important trends. They can be summarized as follows:

1. The relaxation times for these small alkanes in water, whose overall tumbling puts them in the motional narrowing limit, are typically between 5 and 50 ps.
2. The relaxation times increase as the chain length increases from butane to heptane.
3. The relaxation times increase from the ends of the heptane chain toward the center.
4. The relaxation of the Y_0^1 correlation functions are always slower than that of the Y_M^2 functions, but the ratio of $\tau_{l=1}/\tau_{l=2}$ is less than three.
5. The C–H internuclear vectors relax faster than the C–C vectors.

Table III. Time Integral of Angular Correlation Function of C6–H Vector and Exponential Relaxation Times

	$\int_0^\infty \langle Y_0^2[\theta(0)\phi(0)]Y_0^2[\theta(t)\phi(t)]\rangle dt$ *(ps)*	*Relaxation Time by Single Exponential Fit (ps)*
Complete dynamics	0.68	9.9[a] (0.79)[b] 10.5[c] (0.84)
Tumbling and internal motions uncoupled	0.58	9.7[a] (0.77) 11.7[c] (0.93)
Tumbling relaxation only	1.21	13.0[a] (1.03) 15.8[c] (1.26)
Internal relaxation only	—	39[c] (3.1) 40.1[d] (3.2)

[a] Least-squares fit to a single exponential over 10 ps.
[b] Values in parentheses are $1/4\pi$ times the exponential fit to correspond to the time integral (Equation 8).
[c] Least-squares fit to a single exponential over 20 ps.
[d] Least-squares fit to a single exponential over 50 ps.

The first three results are in accord with experiment; there are no data concerning the final two.

From the C–H relaxation times listed in Tables I and II, the ^{13}C-spin-lattice relaxation times (T_1) are calculated by means of Equation 9. Table IV lists the predicted ^{13}C T_1 values for each of the methylene carbons of butane and heptane; the final column gives the ratio of T_1 for carbon C2 and the *i*th internal carbon. Unfortunately, experimental measurements of these relaxation times for small alkanes in aqueous solution are not available. However, measurements of T_1 values have been reported for neat liquid alkanes (*10*, *11*), and some of these results are given in Table IV. Quantitative comparison of the heptane results shows that the T_1 values calculated from the trajectories are shorter than the experimental results by a factor of six. One source of this difference is in the larger viscosity of the aqueous solution relative to the neat liquids. The viscosities of the neat alkanes at the experimental temperature of 40 °C are three to four times smaller than the viscosity used for the aqueous solution calculation ($\eta = 0.27$ centipoise for hexane at 40 °C, $\eta = 0.34$ centipoise for heptane at 40 °C, compared with $\eta = 1$ centipoise for aqueous solution). If we assume the trajectory results scale linearly with the viscosity (as they are expected to do in the diffusive limit), we find that the calculated relaxation times for heptane are a factor of 1.5–2 shorter than the experimental values. Differences of this order may be due to the limitations of the stochastic dynamics model (*34*); these include the neglect of the inertial term of the complete Langevin equa-

Table IV. Alkane NMR Spin-Lattice Relaxation Times

	T_1 (*s*)	T_1^{C2}/T_1^{Ci}
	Trajectory Results[a]	
Butane carbons		
C2, C3	5.9 ± 0.12	
Heptane carbons		
C2, C6	1.94 ± 0.18	—
C3, C5	1.82 ± 0.14	1.07
C4	1.75 ± 0.24	1.11
	Experimental Results[b]	
Heptane carbons		
[C1, C7]	[10.9]	[0.81[c]]
C2, C6	13.2	—
C3, C5	12.8	1.03
C4	12.0	1.10
Eicosane carbons		
[C1, C20]	[3.6]	[0.43[c]]
C2, C19	2.3	—
C3, C18	1.6	1.44
C4, C17	1.1	2.09
Interior[d]	0.8	2.88

[a] 25 °C, η = 1 centipoise, aqueous solution.
[b] 39 °C, η = 0.34 centipoise for neat heptane (*10*).
[c] Normalized to the same number of directly bonded protons.
[d] Individual values not obtained.

tion, the neglect of hydrodynamic interaction, and the use of an atomic friction coefficient obtained from the translational diffusion coefficient of a monomer unit in the Langevin equation for the alkane chain. In this regard, if the covalent radius (0.77 Å) is used to obtain the monomer friction coefficient, as has been suggested (*49*), the calculated T_1 values are increased by 2.4 and are closer to experimental values. It is important to note that, while the absolute values of the predicted T_1 values are somewhat too short, the trajectory results reproduce the experimentally observed gradient in the values of T_1 along the heptane chain.

Test of Simplified Models. To explore whether relaxation in the motional narrowing limit can be divided into contributions from tumbling and segmental motions, the relaxation of the C6 carbon of heptane was analyzed. The molecular tumbling axis of the heptane molecule was defined by atoms C1–C2–C3. Both the tumbling correlation function and the correlation functions describing the internal relaxation were calculated directly from the trajectory. Figure 2 compares the decay of the second-order spherical harmonics of the C6–H internuclear vector cal-

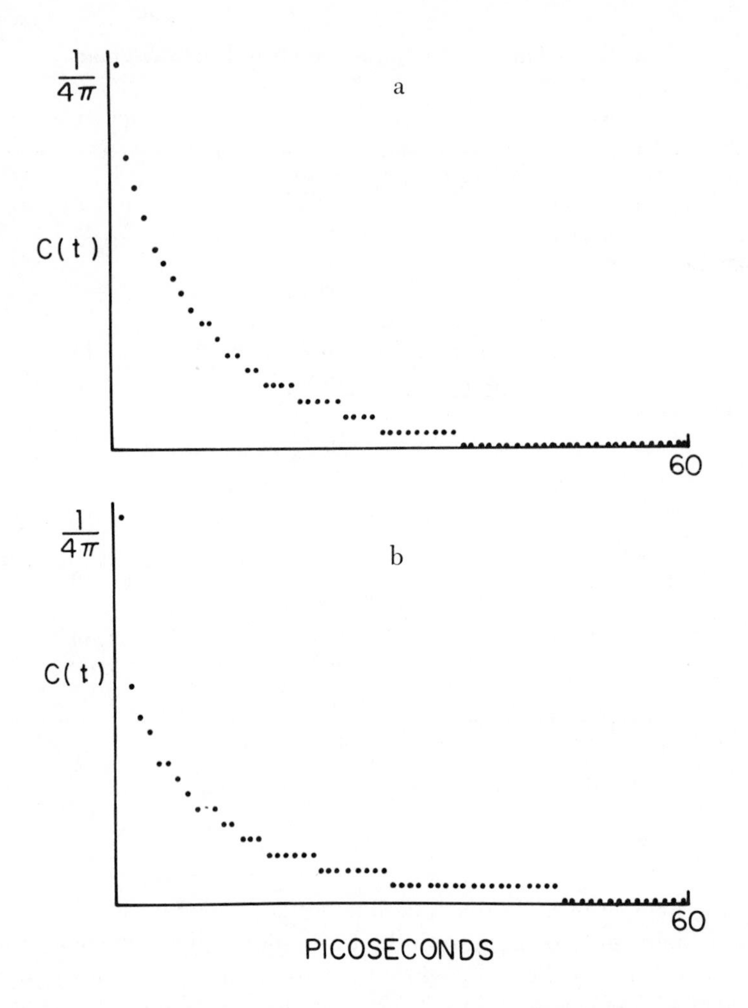

Figure 2. Relaxation of the second-order spherical harmonics $\langle Y_0^2(\theta(t)\phi(t)) \times Y_0^2(\theta(0)\phi(0))\rangle$ *of the C6–H internuclear vector calculated from the heptane trajectory. (a) Exact results from trajectory (Equation 13). (b) Results when correlations between tumbling and internal motions are broken (Equation 14a). (c) Relaxation of diagonal tumbling correlation function,* $\frac{1}{4\pi}\sum_a \langle D_{ma}^{*2}(\Omega_0(t))\, D_{ma}^2(\Omega(0))\rangle$ *in Equation 14a. (d) Relaxation of internal correlation function,* $\frac{1}{5}\sum_{bb'} \langle D_{ab}^{*2}(\Omega_i(t))\, D_{ab'}^2(\Omega_i(0))\rangle\, Y_b^2(\theta_{mol}\phi_{mol})\, Y_{b'}^{*2}(\theta_{mol}\phi_{mol})$*, in Equation 14b. (Reprinted from Ref. 41. Copyright 1981, American Chemical Society.)*

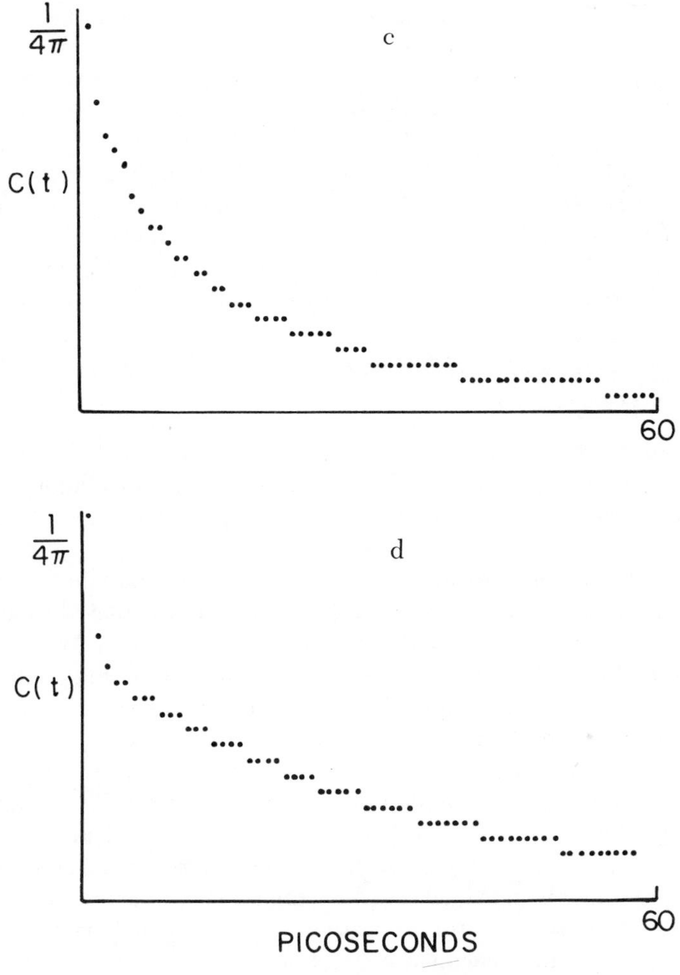

Figure 2. (continued)

culated exactly from the trajectory with the relaxation of this vector obtained when correlations between the tumbling and internal motions are broken. Also given in the figure are the separate relaxation of the tumbling and the internal motion parts of the correlation function. The first three correlation functions (Figure 2, a–c) decay to zero, but the internal motions part (Figure 2d) does not, because there is not enough freedom to sample all solid angles in this short alkane. Table IV lists the time integrals (Equation 8) of these correlation functions and the fits to them of single exponentials over several time spans. The results indicate that, for this particular carbon and choice of molecular tumbling axis, separating the dynamics into contributions from tumbling and internal relaxation is a rather good approximation. Further analysis of the applicability of Equation 12 requires individual values for τ_0 and τ_i from the stochastic trajectory. With estimates of the tumbling and internal relaxation times of $\tau_0 \simeq 16$ ps and $\tau_i \simeq 39$ ps (*41*), the empirical relation (Equation 12) yields an estimate for the total relaxation time of $\tau_{eff} = 11$ ps, as compared with the value of $\tau_{eff} = 10$ ps obtained from a fit of the exact dynamics to a single exponential. Thus, the present trajectory results support the usefulness of the empirical relation, Equation 12. A more detailed analysis for longer hydrocarbons is in progress (*50*).

NMR Relaxation of a Side Chain Attached to a Tumbling Macromolecule

In this section, the heptane trajectory is used as outlined in the section on methodology to evaluate T_1, T_2, and NOE for a model aliphatic side chain with four internal rotational angles protruding from a tumbling macromolecule.

Exact Trajectory Results. The NMR relaxation parameters T_1, T_2, and NOE for each of the carbons (C2 through C6) calculated exactly by means of Equations 3–6 and 13 from the trajectory are listed in Table V. Results obtained for three isotropic molecular tumbling times (τ = 1 ns, 10 ns, and 100 ns) and for two spectrometer frequencies (15 MHz and 68 MHz) are compared.

Of the many comparisons that can be made with the results in Table V, we consider first the calculations for 15 MHz, $\tau_0 = 1$ ns where the NMR relaxation parameters are characteristic of the motional narrowing limit; i.e., T_1 equals T_2 for all of the carbons along the chain and the NOE value is almost maximal, even for the slowest-relaxing carbon. The NMR relaxation times for this case are inversely proportional to the time integrals of the angular correlation functions (Equations 8–10). The much more rapid decay for the C6–H vector results in a 25-fold increase in the value of T_1 over that for the C2 carbon; i.e., the fast reorientation of the C6–H vector averages out the effect of the macromolecular tumbling, which is in the frequency range to produce highly efficient relaxation (as found for C2). Thus, the internal flexibility has a dramatic effect on the NMR relaxation. This is to be contrasted with the results for the heptane chain in solution. For that system, which is also in the motional narrowing limit, the much higher tumbling rate (on the same order as the internal motions) leads to longer relaxation times and a much weaker variation of T_1 with position in the chain. It is of interest that in the macromolecule the T_1 value of C6 with its large internal motional freedom approaches that obtained for the free heptane.

To examine the contributions of internal motions to the relaxation,

Table V. Side Chain NMR Parameters, Exact Result

	15 MHz			*68 MHz*		
Tumbling Time	T_1 *(ms)*	T_2 *(ms)*	*NOE* $(1+\eta)$	T_1 *(ms)*	T_2 *(ms)*	*NOE* $(1+\eta)$
τ_0 = *1 ns*						
C2 (analytic)	26.5	25.8	2.80	58.3	48.9	1.70
C3	112	110	2.86	179	163	2.34
C4	290	290	2.97	312	308	2.90
C5	452	450	2.96	493	486	2.90
C6	627	625	2.97	682	672	2.90
τ_0 = *10 ns*						
C2 (analytic)	11.9	7.3	1.30	133	11	1.16
C3	60	40	1.63	230	57	2.44
C4	243	223	2.74	292	247	2.91
C5	368	313	2.50	492	358	2.95
C6	519	454	2.60	665	515	2.92
τ_0 = *100 ns*						
C2 (analytic)	62.4	1.2	1.16	1,270	1.2	1.15
C3	171	7.1	2.18	302	7.2	2.87
C4	274	101	2.93	293	103	2.93
C5	466	99	2.86	496	100	2.94
C6	617	163	2.89	667	165	2.94

we compare the high field (68 MHz) results at the two slower tumbling times (10 ns and 100 ns). For the rigid carbon C2, the macromolecule is tumbling too slowly for efficient relaxation; consequently, T_2, T_1 and the NOE value are minimal. Side chain flexibility can enhance relaxation rates because of the presence of additional frequency components closer to the Larmor frequency. This does not occur for the side chain tumbling at τ = 10 ns, where the increase in T_1 along the chain is monotonic. However, the micelle tumbling (100 ns) is so slow with respect to the Larmor frequency that internal motions do increase the relaxation rates. The rigid C2 carbon T_1 is longer than 1 s while C3, which can reorient about one side chain rotation axis, has a T_1 of only 302 ms. The high-field T_1 and T_2 values for C6, which is farthest out along the chain, demonstrate that T_1 is sensitive to the high frequency molecular motions, whereas T_2 probes primarily the low frequency motions. For C6, the T_1 values at τ = 10 ns and at 100 ns are almost identical (665 and 667 ms, respectively). By contrast, T_2 is much shorter for the longer tumbling time (for C6, T_2 = 515 ms at τ = 10 ns and T_2 = 165 ms at τ = 100 ns). This result demonstrates the importance of measuring both T_1 and T_2 in order to obtain information about molecular dynamics from NMR.

Simplified Models. We now evaluate the NMR parameters obtained from simplified models for the side chain dynamics and compare the results obtained from these models with the exact results. In the product approximation, motions about the side chain rotational axes are assumed to be uncorrelated. We have found, for most of the cases studied, that the NMR parameters calculated from the product approximation are close to the exact results. For example, at τ_0 = 1 ns or 10 ns, the results for C4 are calculated in the product approximation to be within 5% of the exact values (*41*). Even for C6 the approximate NMR parameters are within 50% of the exact results. That the product approximation generally gives good results suggests that correlations among neighboring dihedral angles do not play a significant role in the chain isomerization dynamics.

The independent lattice jump model idealizes the chain dynamics to instantaneous jumps among stable side chain configurations; between jumps the side chain is assumed to be tumbling rigidly with the macromolecule. NMR parameters calculated from the independent lattice jump model using isomerization rate constants calculated from the heptane trajectory (*34*) are listed in Table VI. In contrast to the product

Table VI. Side Chain NMR Parameters, Lattice Jump Model

	15 MHz			*68 MHz*		
Tumbling Time	T_1 (*ms*)	T_2 (*ms*)	*NOE* $(1+\eta)$	T_1 (*ms*)	T_2 (*ms*)	*NOE* $(1+\eta)$
τ_0 = *1 ns*						
C2 (analytic)	26.5	25.8	2.80	58.3	48.9	1.70
C3	99	97	2.86	162	147	2.31
C4	226	225	2.96	252	247	2.85
C5	339	339	2.98	358	355	2.92
C6	511	510	2.98	520	518	2.97
τ_0 = *10 ns*						
C2 (analytic)	11.9	7.3	1.30	133	11	1.16
C3	52	35	1.59	216	50	2.40
C4	179	158	2.63	235	183	2.85
C5	296	278	2.81	338	301	2.92
C6	477	469	2.94	496	481	2.96
τ_0 = *100 ns*						
C2 (analytic)	62.4	1.2	1.7	1270	1.2	1.15
C3	159	6.0	2.11	295	6.1	2.88
C4	219	60	2.90	241	61	2.91
C5	324	151	2.94	339	153	2.95
C6	437	377	2.98	496	380	2.98

approximation, which yields T_1 and T_2 values that are uniformly longer than the exact results, those calculated from the jump model are uniformly shorter than the exact results. The jump model relaxation times are generally within 30% of the exact values, although the high field T_2 values for micellar tumbling (τ_0 = 100 ns) differ by 50% for C6. An important source of error introduced by the jump model is readily apparent when plots of the angular correlation functions are examined (Figure 3). The plots of the product approximation and the exact relaxation for the C6–H vector are very similar. The jump model, however, exhibits a considerably slower initial decay as compared with the two other functions. The reason for the absence of the fast initial decay in the jump model is that it does not include the high frequency, small-amplitude oscillations that occur within a given potential well. The good results obtained from the jump model in the present case are due to a cancellation of errors. Uncoupling the correlation in the motions along the chain leads to a more rapid decay (less efficient NMR relaxation), while ignoring the short time oscillations of the chain leads to a slower decay (more efficient NMR relaxation). Although these corrections are not very large for the present system, there are cases (e.g., in the interior of proteins) where jumps are so rare that the oscillations within a well make a more important contribution to the relaxation. In the following section we briefly review the results of our trajectory studies of the picosecond motional averaging of NMR probes in the protein interior.

Picosecond Motional Averaging of NMR Probes of Protein Dynamics

In spite of the close-packed structure of native proteins, molecular dynamics simulations (*51*) have shown that significant atomic fluctuations occur on a picosecond time scale. We have used a molecular dynamics simulation of the pancreatic trypsin inhibitor (PTI) to demonstrate the effects of picosecond fluctuations on the observed ^{13}C T_i values. The NMR relaxation rates are determined by time correlation functions, and these correlation functions decay on several different time scales. On the shortest time scale, there is a rapid initial loss of correlation in the first few picoseconds (*44*). The fast decay results from the combined effect of the vibrational potential of the residue containing the nucleus and of collisions between the atoms of the residue and those of the surrounding cage in the protein. For residues that are involved in larger, more complex fluctuations, the initial decay is followed by a much slower loss of correlation over the next tens to hundreds of picoseconds. If the internal motions are restricted, the correlation functions will decay to a plateau value that is equal to the inverse of the square of the order parameter

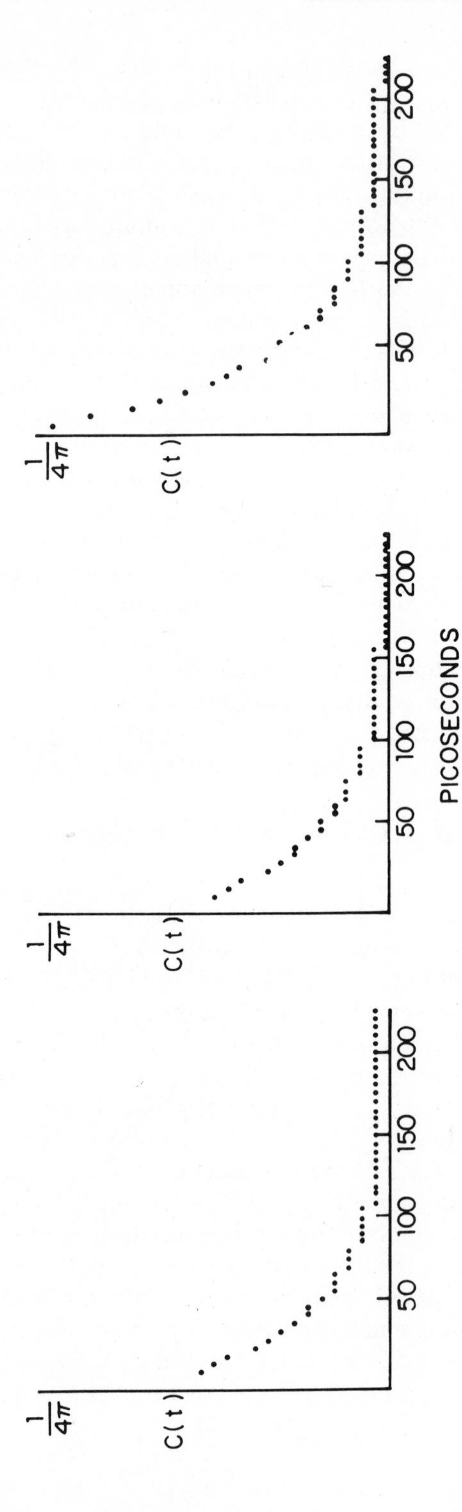

Figure 3. The C6–H angular correlation function: left, calculated exactly from the trajectory (Equation 13); center, calculated from the product approximation; and right, calculated from the independent lattice jump model (Equation 15). (Reprinted from Ref. 41. Copyright 1981, American Chemical Society.)

that describes the distribution of NMR probe orientations (Equation 16). Finally, for experiments in solution, the correlation functions relax to zero due to the overall tumbling of the proteins.

For an examination of the effect of fast motional averaging on T_1 values, we have evaluated the order parameters (Equation 16) and the increase in T_1 values (Equation 17) for 62 protonated (*44*) and 4 non-protonated (*45*) carbons in PTI. In 13 out of 14 α-carbons studied, motional averaging increased the T_1 values by less than 20% ($1.05 \leq T_i/T_i^R \leq 1.11$). An examination of the NMR correlation functions for these atoms demonstrated that they do approach a plateau value within 2 ps. We therefore concluded that it is unlikely that lower frequency internal fluctuations contribute significantly to the relaxation of these atoms. For the 12 residues studied that have side chains, all of the β carbons exhibit more motional averaging than the α carbons to which they are attached. An approximate plateau value appears within 2 ps ($1.11 \leq T_i/T_i^R \leq 1.23$). Of all the carbons studied, the aliphatic γ carbons had the greatest increase in T_1. However, for these carbons the motional averaging does not plateau within 2 ps and lower frequency fluctuations contribute to the NMR relaxation.

The motions of aromatic side chains in proteins have been the subject of several recent experimental and theoretical investigations (*45, 52–55*). A large number of experimental techniques can be used to study the ring motions, including ^{13}C NMR relaxation, ^{1}H and ^{2}H lineshape, and fluorescence depolorization measurements. Levy et al. (*43–46*) have used molecular dynamics simulations of pancreatic trypsin inhibitor (PTI) to evaluate the picosecond motional averaging of order parameters for NMR and fluorescence probes of ring motions. In the course of this work it was observed that the order parameters were anomalously small for certain probe orientations on the two rings in PTI that flipped during the trajectory. Levy and Sheridan (*56*) subsequently developed a simple analytical model that can be used to evaluate order parameters for probes of ring motions when the rings are undergoing both restricted diffusion and 180° jumps about the side chain axes. It was demonstrated that, within the model, the restricted high frequency ring motions can have a large effect on the probe order parameters. Extension of this work to analyze the effect of restricted ring motions on NMR line-shapes in the solid state is underway.

Conclusions

Stochastic dynamics trajectories for alkanes in aqueous solution have been used to examine a variety of problems that arise in the interpretation of ^{13}C-NMR relaxation experiments. Exact results for T_1, T_2, and NOE values obtained from these trajectories have been employed to analyze the relaxation behavior of small alkanes and macromolecular side chains and to test the validity of simplified relaxation models for these systems. Results were obtained for the spin lattice relaxation times of butane and heptane in aqueous solution. A gradient in relaxation times along the heptane chain was found that is in agreement with the measured values for the neat liquid. The empirical separation of the relaxation contributions in these molecules into a tumbling term plus an internal motion term was shown to yield useful results. We have also been able to analyze models for side chain relaxation in macromolecules. For macromolecules with short alkanelike side chains moving freely in aqueous solvent, the lattice jump model was shown to provide a satisfactory description of the NMR relaxation.

A molecular dynamics simulation of a protein has been used to demonstrate the effects of picosecond fluctuations on observed ^{13}C T_1 values. It is shown how an increase in T_1 can be related to order parameters for the picosecond motional averaging of the C, H dipolar interactions, and how these order parameters can be calculated from a protein molecular dynamics trajectory.

In summary, the present work provides a firm theoretical foundation for the continuing effort to use NMR measurements for the experimental analysis of the dynamics of molecules with internal degrees of freedom.

Acknowledgments

We thank C. M. Dobson for helpful discussion. We thank G. Lipari and A. Szabo for many discussions concerning generalized order parameters and for making preprints of their work available to us. This work has been supported by grants from the National Institutes of Health, the National Science Foundation, the Petroleum Research Fund, and the Rutgers University Research Council and a Biochemical Research Support Grant. R. M. Levy is an Alfred P. Sloan Fellow and the recipient of a National Institutes of Health Research Career Award.

Literature Cited

1. Kuhlman, K.; Grant, D.; and Harris, R. *J. Chem. Phys.* **1970,** *52*, 3439.
2. Lyerla, J. R., Jr.; Grant, D. *J. Phys. Chem.* **1972,** *76*, 3213.
3. Bauer, D.; Alms, G.; Brauman, J.; Pecora, R. *J. Chem. Phys.* **1974,** *61*, 2255.
4. Lyerla, J. R.; Levy, G. C. *Top. Carbon–13 NMR Spect.* **1974,** *1*, 79.
5. Jonas, J. *Proc. Nato ASI High Press. Chem.* **1978,** Reidel, Dordrecht.
6. Doddrell, D.; Allerhand, A. *J. Am. Chem. Soc.* **1971,** *93*, 1558.
7. Connor, T. M. *J. Chem. Soc., Trans. Faraday* **1963,** *60*, 1579.
8. Schaefer, J. *Macromolecules* **1973,** *6*, 882.
9. Levine, Y. K.; Birdsall, N.; Lee, A. G.; Metcalfe, J. C.; Partington, P.; Roberts, G. C. K. *J. Chem. Phys.* **1974,** *60*, 2890.

10. Lyerla, J. R., Jr.; McIntyre, H. M.; Torchia, D. A. *Macromolecules* **1974,** *7*, 11.
11. Lyerla, J. R., Jr.; Horikawa, T. T. *J. Phys. Chem.* **1976,** *80*, 1106.
12. Levy, G. C.; Axelson, D. E.; Schwartz, R.; Hochmann, J. *J. Am. Chem. Soc.* **1978,** *100*, 410.
13. Canet, D.; Brondeau, J.; Nery, H.; Marchol, J. P. *Chem. Phys. Lett.* **1980,** *72*, 184.
14. Lee, A. G.; Birdsall, N. J. M.; Metcalfe, J. E.; Warren, G. B.; Roberts, G. C. K. *Proc. Roy. Soc. Lond.* **1976,** *193B*, 253.
15. Godini, P.; Landsberger, F. *Biochemistry* **1975,** *14*, 3927.
16. Bocian, D.; Chan, S. *Annu. Rev. Phys. Chem.* **1978,** *29*, 307.
17. Oldfield, E.; Norton, R.; Allerhand, A. *Biochemistry* **1975,** *250*, 6338.
18. *Ibid.*, **1975,** *250*, 6381.
19. Visscher, R.; Gurd, F. R. N. *J. Biol. Chem.* **1975,** *250*, 2238.
20. Jones, W. C., Jr.; Rothgeb, T. M.; Gurd, F. R. N. *J. Biol. Chem.* **1976,** *251*, 7452.
21. Wittebort, R. J.; Rothgeb, T. M.; Szabo, A.; Gurd, F. R. N. *Proc. Nat. Acad. Sci. USA* **1979,** *76*, 1059.
22. Jelinski, L. W.; Torchia, D. *J. Mol. Biol.* **1979,** *133*, 45.
23. Jelinski, L. W.; Torchia, D. *J. Mol. Biol.* **1980,** *138*, 255.
24. Wallach, D. *J. Chem. Phys.* **1967,** *47*, 5258.
25. Wittebort, R. J.; Szabo, A.; Gurd, F. R. N. *J. Am. Chem. Soc.* **1980,** *102*, 5723.
26. Wittebort, R. J.; Szabo, A. *J. Chem. Phys.* **1978,** *69*, 1722.
27. London, R. E.; Avitable, J. *J. Am. Chem. Soc.* **1978,** *100*, 7159.
28. Howarth, O. W. *J. Chem. Soc., Faraday Discuss. II*, **1979,** *75*, 863.
29. London, R. E.; Avitable, J. *J. Am. Chem. Soc.* **1977,** *99*, 7765.
30. Tsutsumi, A.; Chachaty, C. *Macromolecules* **1979,** *12*, 429.
31. Chandrasekhar, S. *Rev. Mod. Phys.* **1943,** *15*, 1.
32. Ermak, D. L.; McCammon, J. A. *J. Chem. Phys.* **1978,** *69*, 1352.
33. Ermak, D. L.; *J. Chem. Phys.* **1975,** *62*, 4189.
34. Levy, R. M.; Karplus, M.; McCammon, J. A. *Chem. Phys. Lett.* **1979,** *65*, 4.
35. Ryckaert, J. P.; Cicotti, G.; Berendsen, H. J. C. *J. Comp. Phys.* **1977,** *23*, 327.
36. Ryckaert, J. P.; Bellemans, A. *Chem. Phys. Lett.* **1975,** *30*, 123.
37. Pratt, L. R.; Chandler, D. *J. Chem. Phys.* **1977,** *67*, 3683.
38. Fixman, M. *J. Chem. Phys.* **1978,** *69*, 1527.
39. Levy, R. M.; In "Stochastic Molecular Dynamics"; National Resources for Computation in Chemistry; Lawrence Berkeley Laboratory, Univ. CA, Berkeley; 1979.
40. Abragam, A. "The Principles of Nuclear Magnetism"; Oxford Press: London, 1978.
41. Levy, R. M.; Karplus, M.; Wolynes, P. G. *J. Am. Chem. Soc.* **1981,** *103*, 5998.
42. Berne, B.; Pecora, R. "Dynamic Light Scattering"; Wiley: New York, 1976; Chap. 7.
43. Levy, R. M. In "Diffusive Dynamics of Alkane Chains"; CECAM Workshop Report, Universite de Paris, 1978.
44. Levy, R. M.; Karplus, M.; McCammon, J. A. *J. Am. Chem. Soc.* **1981,** *103*, 994.
45. Levy, R. M.; Dobson, C. M.; Karplus, M. *Biophys. J.* **1982,** *39*, 107.
46. Levy, R. M.; Szabo, A. *J. Am. Chem. Soc.*, **1982,** *104*, 2073.

47. Bocian, D. F.; Chan, S. I. *Ann. Rev. Phys. Chem.* **1979,** *29*, 307.
48. Lipari, G.; Szabo, A. *J. Am. Chem. Soc.* in press.
49. Knauss, D. C.; Evans, G. T.; Grant, D. M. *Chem. Phys. Lett.* **1980,** *71*, 158.
50. Pastor, R.; Karplus, M.; Levy, R. M. to be submitted for publication.
51. Karplus, M.; McCammon, J. A. *CRC Crit. Rev. Biochem.* **1981,** *9*, 293.
52. Campbell, I. D.; Dobson, C. M.; Moore, G. R.; Perkins, S. J.; Williams, R. J. P. *FEBS Lett.* **1976,** *70*, 91.
53. Lakowicz, J. R.; Weber, G. *Biophys. J.* **1980,** *32*, 591.
54. Gelin, B. R.; Karplus, M. *Proc. Nat. Acad. Sci.* **1975,** *81*, 801.
55. McCammon, J. A.; Wolynes, P. G.; Karplus, M. *Biochemistry* **1979,** *18*, 927.
56. Levy, R. M.; Sheridan, R. P. *Biophys. J.* in press.

RECEIVED for review March 15, 1982. ACCEPTED for publication September 13, 1982.

19

Influence of Flexibility on the Properties of Chain Molecules

R. SZCZEPANSKI and G. C. MAITLAND

Imperial College, Department of Chemical Engineering, London, S.W.7., England

Simulation results are reported for conformational properties and diffusion coefficients for a system of n-*octane-like molecules in the liquid phase at about 400 K. The effects of chain flexibility have been studied by changing the nature of intramolecular interactions, keeping the same intermolecular potential. The presence of an excluded volume potential acting between atoms on the same chain is found to be of crucial importance in determining the size and shape of molecules in a liquid. The influence of a torsional potential is of secondary importance for the properties we have examined. In the case of a realistic* n-*alkane model, the balance between intermolecular and intramolecular forces is such that chain dimensions in the liquid are close to their values at infinite dilution.*

CHAIN FLEXIBILITY INFLUENCES both the static and dynamic properties of large molecules, but for experimental systems, it is not possible to identify the precise effects of flexibility changes because they are usually accompanied by modifications to the intermolecular interactions. It is, however, possible to vary the details of molecular interactions in a well-controlled way by computer simulation. By using this technique, the effects of intermolecular and intramolecular potentials on molecular and bulk properties may be separately assessed.

The flexibility of chain molecules depends on the possibility of rotation about the bonds that connect the backbone atoms. In our model, there is a potential associated with torsional motion, an excluded volume potential that (except at the most extreme and unrealistic temperatures) prevents chains from passing through themselves, and an intermolecular potential. We have examined the effects on chain shape and dynamic properties of changing the two intramolecular interactions while keeping the same intermolecular potential. By varying the potentials we have been able to simulate a range of molecular types from the very flexible

0065-2393/83/0204-0469$06.00/0

(freely rotating chains) to the very rigid (the type of molecules that form liquid crystals). The real n-alkane forms an intermediate case. A considerable amount of work has been done on the torsional and orientational dynamics of chain molecules using both analytical techniques and approximate numerical simulation (*1*). However, the work described in this chapter has a somewhat different scope because it is concerned with the effects of rather large changes in the nature of intramolecular potentials in the presence of intermolecular interactions, rather than small variations solely in the torsional potential.

In this chapter we present the results of molecular dynamics simulations of n-octane-like molecules in the liquid phase, using an alkane model that is based on those used in previous simulations of chain molecules (*2–7*). This work forms part of a more general study of the effects of the interactions within and between polymer molecules on their bulk properties. In the first section, we describe in more detail the molecular model and potentials used in the simulations. Brief details of the molecular dynamics method and specifications of the simulation experiments are then given. The following section of this chapter describes results for those static properties that characterize the size and shape of individual chains and the structure of the liquid. The dynamic properties of molecules that are of interest in this study include correlations between individual bonds, the rate of change of chain length and shape, and both rotational and translational diffusion of molecules as a whole. Results for translational diffusion are included in the final section. An analysis of the other dynamic properties will be presented elsewhere (*8*).

The n-*Alkane Model*

The model used in this work is a combination of those adopted by Weber (*2*) and Ryckaert and Bellemans (*6*). The n-alkane is represented by a carbon-atom backbone, and hydrogen atoms do not appear explicitly. The total mass of the molecule is distributed equally between backbone atoms, with no distinction between CH_3 and CH_2 groups. Interactions between and within molecules are modeled by a set of potentials as described below.

Intermolecular Lennard–Jones Potential. Molecules interact by a site–site potential between backbone atoms. The form of this potential is

$$\begin{aligned} U(r) &= u_{\mathrm{LJ}}(r) + \left(\frac{du_{\mathrm{LJ}}}{dr}\right)_{r_c}\left(\frac{r_c^2 - r^2}{2r_c}\right) - u_{\mathrm{LJ}}(r_c) \qquad r \leq r_c \\ U(r) &= 0 \qquad r > r_c \end{aligned} \tag{1}$$

where

$$u_{LJ}(r) = 4\varepsilon\left[\left(\frac{\sigma}{r}\right)^{12} - \left(\frac{\sigma}{r}\right)^{6}\right]$$

and r is the distance between atom–atom centers. The form of Equation 1 ensures that both the potential and force go smoothly to zero at the cut-off distance r_c. Values for the potential parameters in Equation 1 are taken from Ryckaert and Bellemans (7)

$$\sigma = 3.923 \times 10^{-10}\ m$$

$$\varepsilon/k = 72\ \mathrm{K}$$

$$r_c = 2.5\ \sigma$$

Torsional and Bond-Bending Potentials. Part of the n-alkane chain is shown in Figure 1. The bond vector $\boldsymbol{b}_i$ runs from atom i to $i+1$, and all bonds have a fixed length b_0. The potential associated with deformation of the bond angle θ_i is defined as

$$V_\theta = \frac{1}{2}\gamma_\theta\left(\cos\theta_0 + \frac{\boldsymbol{b}_{i-1}\cdot\boldsymbol{b}_i}{b_0^2}\right)^2 \tag{2}$$

where γ_θ is a force constant and θ_0 is the equilibrium bond angle. The torsional potential for rotation of adjacent parts of the chain about bond $\boldsymbol{b}_i$ is defined as

$$V_\phi = \gamma_\phi(1.116 + 1.462\cos\phi - 1.578\cos^2\phi - 0.368\cos^3\phi + 3.156\cos^4\phi - 3.788\cos^5\phi) \tag{3}$$

where

$$\cos\phi = \frac{\xi_i - \cos^2\theta_0}{\sin^2\theta_0} \tag{4}$$

$$\xi_i = \frac{\boldsymbol{b}_{i-1}\cdot\boldsymbol{b}_{i+1}}{b_0^2} \tag{5}$$

and γ_ϕ is a constant.

The rotational potential and molecular conformations corresponding to the three minima are shown in Figure 1. At fixed bond length b_0, the angle ϕ corresponds to the torsional angle when all bond angles are at the equilibrium value θ_0.

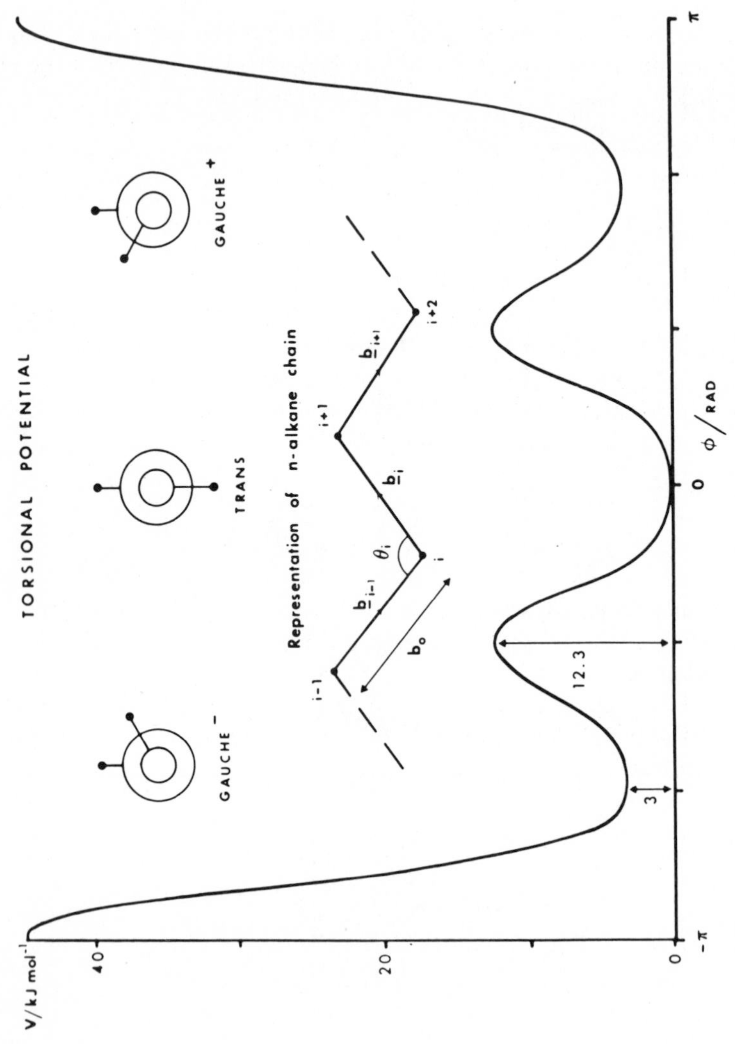

Figure 1. Geometry of alkane chain and torsional potential calculated by Equation 3.

Molecular flexibility may be adjusted by varying the parameters in the bond-bending and torsional potentials. We have chosen to fix the chain geometry by using the same bond angle potential in all calculations. The parameter γ_ϕ in the torsional potential is varied by writing it as

$$\gamma_\phi = \alpha(\gamma_\phi)_0 \tag{6}$$

where $(\gamma_\phi)_0$ is the value of γ_ϕ corresponding to a torsional potential that is thought to be approximately the same as that for a real n-alkane. The adjustable parameter α ranges from 0, for a molecule with no barriers to rotation, to values exceeding unity for more rigid molecules.

It is our primary aim to compare differences between molecular models with varying flexibilities rather than to find optimum potential parameters for an n-alkane. For this reason, we have been content to adopt the following values for parameters, which have been used in previous n-alkane models (2–7), in Equations 2–6

$$\gamma_\theta = 1.3 \times 10^5 \text{ J/mol}$$

$$b_0 = 1.53 \times 10^{-10} \text{ m}$$

$$\theta_0 = 109.47°$$

$$(\gamma_\phi)_0 = 8.314 \times 10^3 \text{ J/mol}$$

Intramolecular Lennard–Jones Potential. The potentials described above take no account of the so-called pentane effect, whereby the simultaneous occurrence of adjacent *gauche*$^+$ and *gauche*$^-$ conformations along a chain has a low probability because of repulsive exclusion forces acting between the two CH_2 groups brought into close proximity.

We include this effect by allowing Lennard–Jones interactions between atoms on the same chain, provided they are separated by at least three other atoms. The following potential is used

$$V_{LJ}(r) = \beta U(r) \tag{7}$$

where $U(r)$ is defined in Equation 1, and β is an adjustable parameter that has a role similar to α; in our realistic alkane model, $\beta = 1$. Any value of $\beta > 0$ introduces an excluded volume effect, making it impossible for a chain to pass through itself (except, as noted earlier, at extreme temperatures).

Molecular Dynamics Simulations

All of the simulations described here were carried out using 45 molecules, each of 8 atoms, enclosed in a cubic box. The usual periodic boundary conditions were applied with nearest image interactions (9) for the intermolecular potential. The bulk density of the fluid was 564 kg/m^3 (molar mass = 0.114 kg), which corresponds to a saturated liquid at 440 K for real *n*-octane.

The equations of motion were integrated in Cartesian coordinates using the Verlet algorithm and the SHAKE procedure (*10*) for introducing bond length constraints. Although it has been shown that rigid constraints on both bond length and bond angle do not always produce correct physical behavior (*11, 12*), there is evidence (*13*) that using bond length constraints alone, as is the case here, does not introduce any significant error.

Details of the simulations runs are given in Table I, where the potentials *A–E* are defined in terms of the parameters α and β. Potential *A* is the reference potential, which is thought to be a reasonable model of a real *n*-alkane. Potential *C* describes a freely rotating chain, and *B* corresponds to a freely rotating chain with excluded volume. Potentials *D* and *E* both represent very rigid molecules.

The starting configuration for each run was derived from a previous simulation, using a potential similar to *A*, which represented 250 ps of real time. Model parameters were then changed to those shown in Table I, and the simulations were continued. The velocities were rescaled at 5-ps intervals until a temperature of about 400 K was achieved; then a further period of equilibration (typically 50 ps) was allowed.

Static Properties

Chain Conformation. The size and shape of a chain molecule is most simply characterized by its end-to-end distance and radius of gyration about the center of mass. It is interesting to compare our results with those for some idealized models of isolated chains. For a freely

Table I. Specifications of Simulation Runs

Potential	α	β	T(*K*)	$\frac{\Delta U^a}{kT}$	*Integration*[b] *Time Step* *(10^{-3} ps)*	*Total Simulation Time (ps)*
A	1	1	396	3.8	5	80
B	0	1	393	0.	5	41
C	0	0	405	0.	5	35
D	5	5	354	25.7	1	77
E	10	10	444	33.6	1	52

Note: Chain length was eight units; volume was 5.56 σ^3/molecule.
[a] ΔU is the potential barrier between *trans* and *gauche* states.
[b] Execution time per step on a CDC 7600 computer was 0.24 s.

joined chain of n bonds, each of length b_0, it may be shown that (*14*)

$$\begin{aligned} \langle R^2/nb_0^2\rangle &= 1 \\ \langle S^2/nb_0^2\rangle &= \frac{1}{6}\frac{(n+2)}{(n+1)} \end{aligned} \tag{8}$$

and for a chain with fixed bond angles but no rotational barriers

$$\begin{aligned} \langle R^2/nb_0^2\rangle &= \left(\frac{1+\alpha}{1-\alpha}\right) - \frac{2\alpha}{n}\frac{(1-\alpha^n)}{(1-\alpha)^2} \\ \langle S^2/nb_0^2\rangle &\approx \frac{(n+2)(1+\alpha)}{6(n+1)(1-\alpha)} - \frac{\alpha}{(n+1)(1-\alpha)^2} \end{aligned} \tag{9}$$

where $\alpha = \cos\theta_0$.

In Table II, conformational averages for our simulations are compared with Equations 8 and 9 and with Flory's calculations (*15*) by using the rotational isomeric state approximation with parameters appropriate to a polymethylene chain.

Even though there is no rotational barrier for potential *B*, the sizes and shapes of the molecules are surprisingly close to those for *A*. The effect of intermolecular and excluded volume potentials for *A* and *B* is to cause chain dimensions to be expanded relative to the predictions of Equations 8 and 9. In fact, the two potentials show close similarities in nearly all the properties we have examined, indicating the dominance of long-range intramolecular interactions along the chain as compared with near-neighbor interactions.

Table II. Conformational Averages and Diffusion Coefficients for Octane Simulations

Potential	T(*K*)	⟨R⟩	⟨R²⟩	⟨S⟩	⟨S²⟩	⟨R²⟩/7	7/⟨S²⟩	⟨R²⟩/⟨S²⟩	D_{tr} (*10⁻⁸ m²/s*)
A	396	4.73	22.64	1.69	2.88	3.23	2.43	7.86	0.94[a]
B	393	4.50	20.52	1.64	2.69	2.93	2.60	7.62	1.10
C	405	2.61	8.54	1.27	1.67	1.22	4.19	5.11	1.20
D	354	4.57	21.25	1.65	2.75	3.04	2.54	7.23	0.63
E	444	4.51	20.73	1.64	2.72	2.96	2.57	7.62	1.22
Freely jointed chain, Equation 8						1	5.33	5.33	
Fixed bond angles, Equation 9						1.79	3.42	6.13	
Rotational isomeric state approximation (*14*), 413 K						3.1	2.5	7.75	

Note: All distances are in units of bond length, b_0.

[a] Estimate of experimental value based on an extrapolation of high-temperature data of Douglass and McCall (*16*) is $0.7 - 1.1 \times 10^{-8}$ m²/s.

Results for potentials *A* and *B* compare well with the rotational isomeric state approximation for isolated polymethylene chains. This agreement between chain dimensions at infinite dilution and in a dense liquid shows that there is a balancing out of intermolecular and intramolecular forces in the dense phase. Even for the relatively small chain length studied in this work, the details of the torsional potential are of secondary importance compared with the excluded volume potential.

Potential *C* does not have rigidly fixed bond angles, but otherwise this model should conform to Equation 9 in the limit of isolated chains. The results in Table II show that the infinite dilution limit (in terms of chain size) is not obtained for potential *C*, demonstrating large effects due to intermolecular forces.

In the cases of potentials *D* and *E*, both torsional and excluded volume terms are very large. Because a small time step is required (*see* Table I) in order to maintain energy conservation, it is not practical to carry out calculations covering very long real times. It is evident from the results for dynamic properties that the simulations for *D* and *E* were not long enough to allow a sufficient number of configurations to be sampled. As a result, reliable averages cannot be estimated, and too much significance should not be attached to the precise values in Table II for potentials *D* and *E*.

Distributions of Intramolecular Distances. Probability distributions for end-to-end distances are compared in Figure 2. The similar distributions for *A* and *B* are as expected in view of the similarities in conformational averages. The peak for *B* is shifted relative to *A* because conformations are not restricted by a torsional potential. About 40% of molecules with potential *C* have an end-to-end distance of two bond lengths or less. The absence of any intrachain interactions allows a molecule to curl up on itself in a ring conformation. The peak in $P(r)$ at 1.6 b_0 is consistent with such a ring structure. Figure 2 (bottom) shows the very similar distributions for potentials *D* and *E*. Both simulations were started from the same configuration, which was itself the result of a previous simulation with a lower torsional potential. It is clear that no major conformational changes have taken place for either *D* or *E*.

Further information about chain conformation is provided by the *intramolecular* pair distribution function, $g'(r)$, which is proportional to the probability of finding two atoms on the same molecule with centers separated by a distance between r and $r+dr$. The actual values of $g'(r)$ shown in Figure 3 were obtained using the same normalization as used for $g(r)$, the *intermolecular* pair distribution function (*see* the section on Bulk Fluid Structure).

For a dense fluid, potential *C* clearly leads to unrealistic behavior, with different atoms on a molecule occupying the same position in space. By assuming a compact shape, molecules minimize the energy of inter-

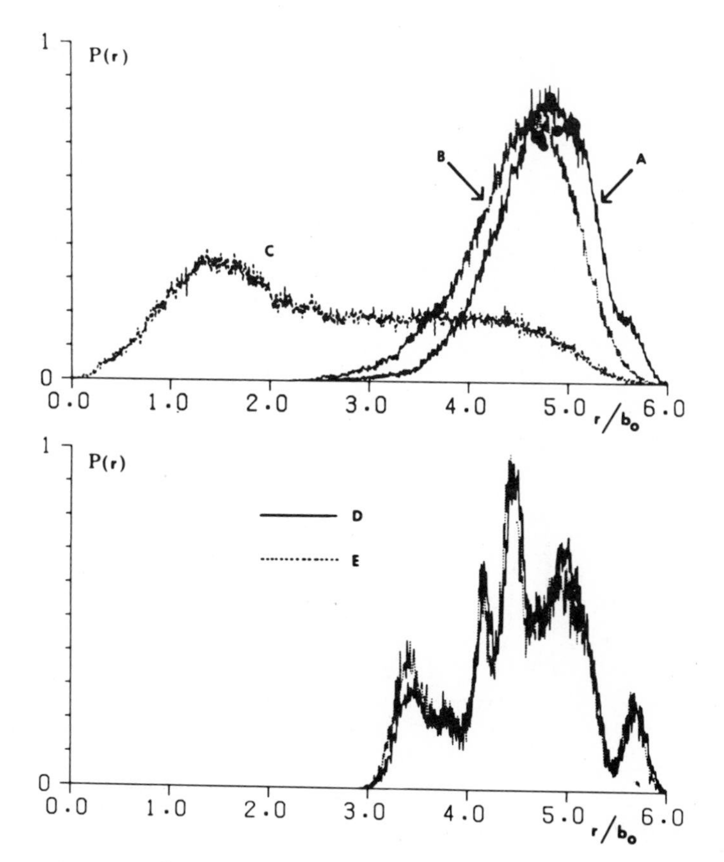

Figure 2. Distribution of end-to-end distances for octane molecules. Top, Potentials A, B, C. Bottom, Potentials D and E.

molecular interactions—the ring structure is energetically favored in the absence of intramolecular forces.

The pair distribution function for potential A clearly shows peaks that may be associated with different combinations of *trans* and *gauche* torsional angles along the chain. The excluded volume term in potential B tends to stretch out the chain, and hence favors the *trans* state without particularly favoring intermediate rotational states. The general shape of $g'(r)$ is thus similar to that for A, but the only detailed feature is the *trans* peak.

Distribution of Rotational States. The octane molecule has 5 torsional angles, corresponding to rotations about internal bonds, which we denote as $\phi_1, \ldots, \phi_5$. Because of molecular symmetry, ϕ_1 and ϕ_5 are equivalent for the purposes of calculating the distribution of torsional angles. The probability distributions for potential A, shown in Figure 4

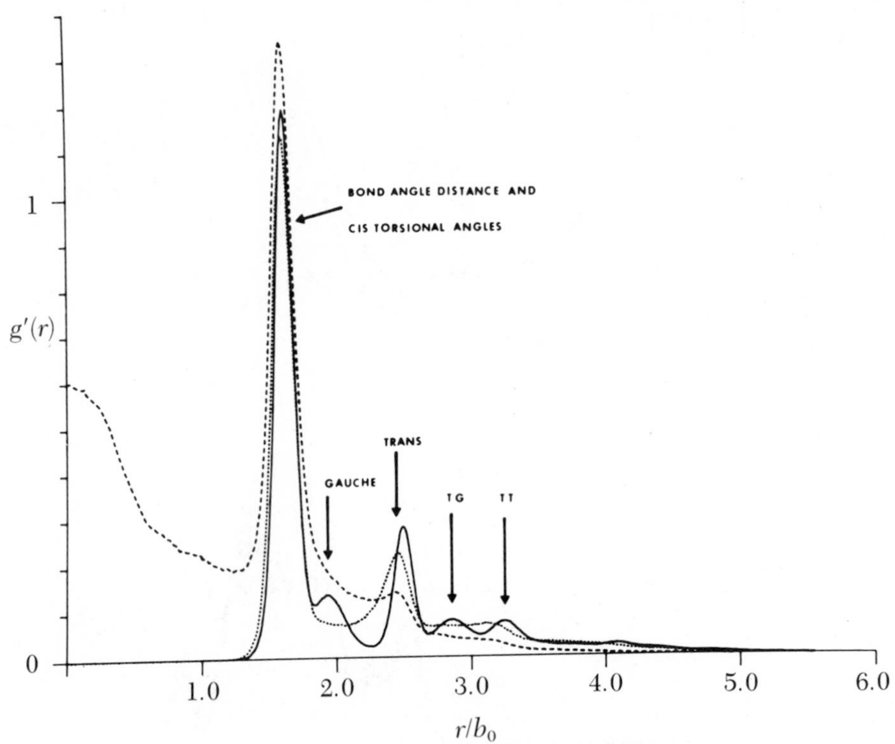

Figure 3. Octane intramolecular pair distribution function. Potential A(—). Potential B (····). Potential C (----).

(top), are highly symmetric, indicating that equilibrium between *gauche*$^{\pm}$ and *trans* states has been established. Distributions for inner angles ϕ_2 and ϕ_3 are virtually identical, but there is a pronounced shift from *trans* to *gauche* conformations for the outer angle, ϕ_1 ($\cos \phi > 0.5$ defines a *trans* state).

Removing the torsional potential, Figure 4 (center), results in a much more uniform distribution, but as mentioned above, the excluded volume potential favors a stretched-out conformation with most angles in the *trans* state. There is a well-defined variation in the proportion of near-*trans* conformers for different angles along the chain that must be a result of the balance between intramolecular and intermolecular Lennard–Jones potentials. The distribution for ϕ_1 is again the most uniform because rotations of the end bond vectors do not involve cooperative movements of the rest of the molecule.

Removing the excluded volume potential (potential C) makes angles near $\pm\pi$ the preferred conformation; this is shown in Figure 4 (bottom). The large number of *cis* angles indicates a compact globular structure in

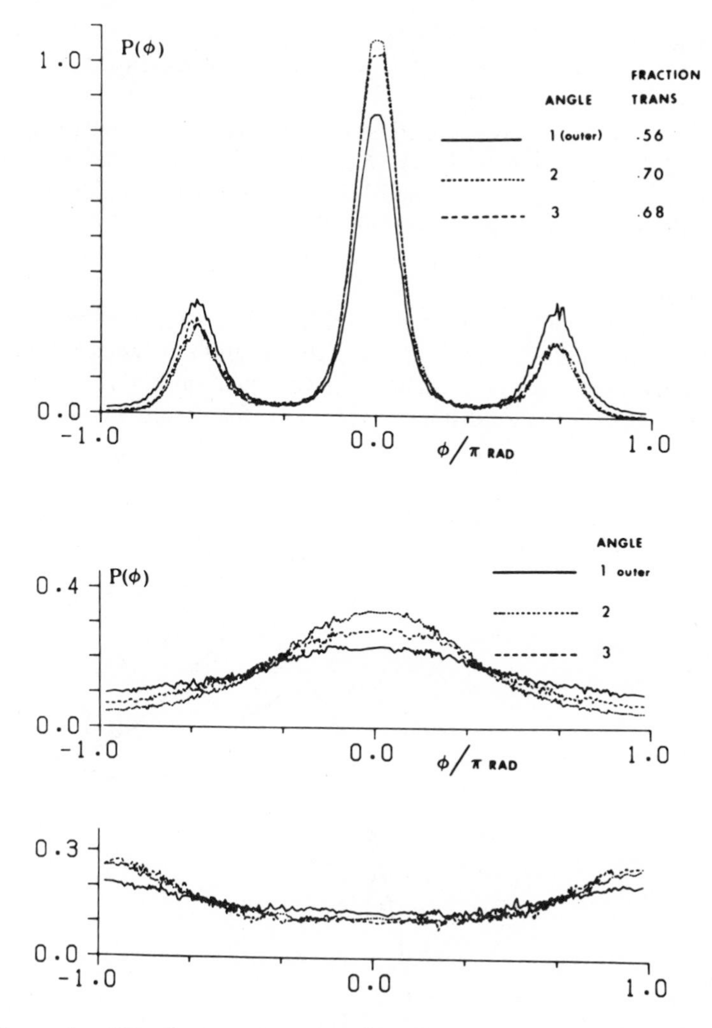

Figure 4. Distribution of torsional angles in octane. Top, Potential A. Center, Potential B. Bottom, Potential C.

accordance with other results already described. Differences between the three distributions are small, but once again, the distribution of ϕ_1 is the most uniform.

Distributions for potentials *D* and *E* are not shown because they simply consist of spikes as $\phi = 0$ and $\pm 2\pi/3$. There were no transitions between *gauche* and *trans* states during the simulations, for the reasons given earlier.

Bulk Fluid Structure. The intermolecular pair distribution function, $g(r)$, provides a measure of structure in the fluid as a whole. The value of $g(r)$ gives the probability of finding atoms on different molecules separated by distances in the range r to $r+dr$ relative to the same probability for a uniform fluid.

The distributions for potentials *A* and *B* (which are virtually identical), and *C* are shown in Figure 5 out to a distance of 2.5σ, which is the cut-off distance for the Lennard–Jones potential. Oscillations in $g(r)$ die out rapidly, and no peaks beyond the second were observed in any simulation. For $r > 2.5\sigma$, $g(r)$ was evaluated with lower precision and is not shown.

The distribution for potentials *A* and *B* is rather featureless compared with $g(r)$ for monatomic Lennard–Jones fluids, and it is typical of previous calculations on alkanes (2–7). The large first peak in $g(r)$ for potential *C* is shifted out to greater distances than that for *A* and *B*, showing the

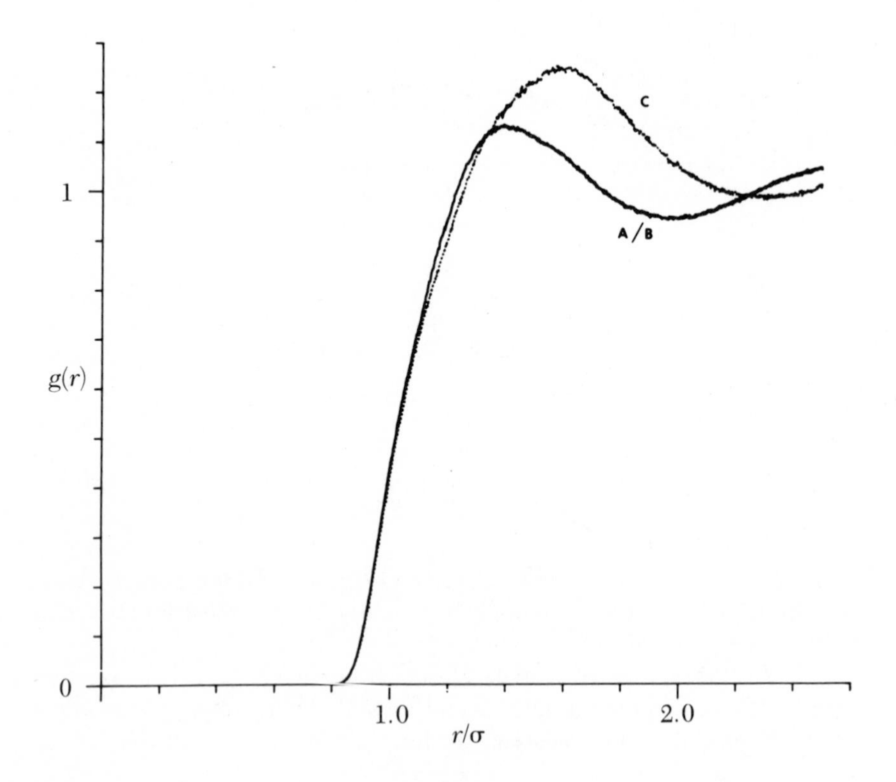

Figure 5. Octane intermolecular pair distribution function.

effect of intramolecular forces on intermolecular spacings through changes in molecular shape. The compact symmetric structure of molecules with potential C enhances the effective intermolecular potential by concentrating several Lennard–Jones atoms in a small volume. Molecules pack in a more regular way but at larger separations.

Translational Diffusion

Values of the self-diffusion coefficient were computed from the mean squared displacement of the centers of mass of the octane molecules, according to the equation (*10*)

$$D_{tr} = \frac{1}{6t} \langle (\mathbf{r}(0) - \mathbf{r}(t))^2 \rangle \qquad (10)$$

A typical plot of $\langle (\Delta r)^2 \rangle$ is shown in Figure 6 (top). As observed by Weber (*3*), there is a considerable delay before the center of mass of a molecule moves any distance from its initial position. Discarding this initial start-up period, the diffusion coefficients were estimated from the subsequent linear portions of $\langle \Delta r^2 \rangle$ plotted against t. Values of D_{tr} are listed in Table II.

Weber's calculations give a value of $D_{tr} \simeq 1.15 \times 10^{-8}$ m^2/s at 400 K, compared with 0.94×10^{-8} m^2/s for this work using potential A. Even though these simulations are not intended to correspond exactly to real liquid octane, it is worthy of note that these two values lie within the range of uncertainty for the extrapolated experimental measurements of Douglass and McCall (*16*). Because the results reported here are mostly isothermal, it is not possible to say how well the temperature dependence of D_{tr} is reproduced.

Comparing potentials A, B, and C, there is a trend to more rapid translational diffusion as the molecule becomes more flexible. With potentials B and C it is easier for molecules to change their shapes in order to move past each other. Molecules with potential C are also smaller on average than are those with A or B, hence diffusion is easier. In contrast, potentials D and E allow very little change in molecular shape, and diffusion is thus hindered (note that simulation of D and E are at temperatures significantly different from those in other runs).

We have also evaluated the velocity autocorrelation function, defined as

$$C_V(t) = \frac{\langle \mathbf{V}(0) \cdot \mathbf{V}(t) \rangle}{\langle \mathbf{V}^2(0) \rangle} \qquad (11)$$

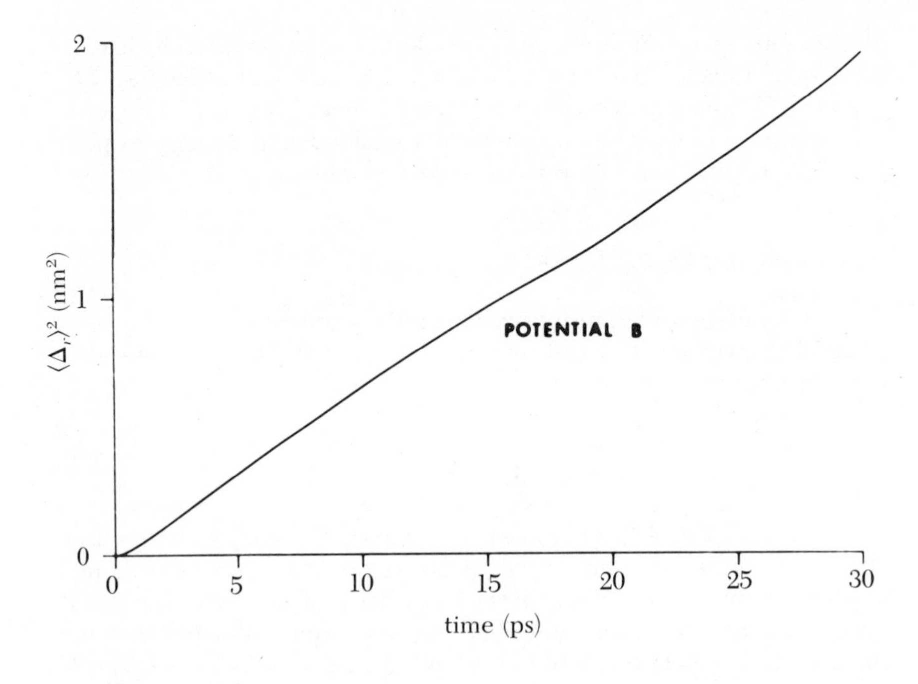

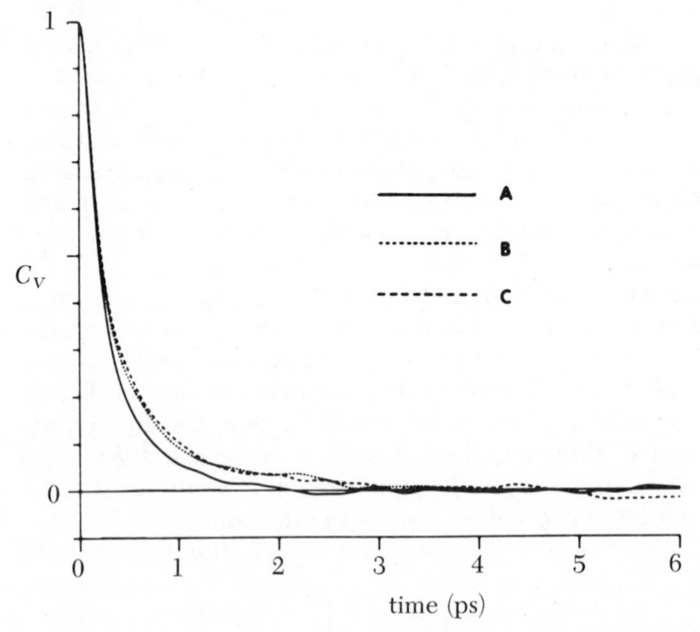

Figure 6. Functions related to translational diffusion. Top, Mean squared displacement of centers of mass. Bottom, Velocity autocorrelation functions.

where $\mathbf{V}(t)$ is the center of mass velocity of a molecule at time t. Diffusion coefficients may be obtained by integrating Equation 11 (*17*)

$$D_{tr} = \frac{1}{3} \langle \mathbf{V}^2(0) \rangle \int_0^\infty C_V(t)\, dt \tag{12}$$

We find that diffusion coefficients calculated from Equation 12 are consistent with those derived from the mean squared displacement of centers of mass (Equation 10).

For potentials *A–C*, $C_V(t)$ is shown in Figure 6 (bottom). The absence of any large negative region in C_V has been noted in previous simulations of chain molecules (*5, 6*).

Discussion

In this study we have carried out simulations with molecular models that represent a wide range of chain flexibilities. Our results characterize in detail the properties of a realistic *n*-alkane model (potential *A*) and two models of flexible molecules (potentials *B* and *C*). Results on very rigid molecules (potentials *D* and *E*) are less complete because of the limitations of the molecular dynamics technique.

We find that both static and dynamic properties are sensitive to the balance between intermolecular and intramolecular forces. For conformational properties, this is illustrated by the fact that although our simulations are of the liquid phase, the values of $\langle R^2 \rangle$ and $\langle S^2 \rangle$ are in good agreement with calculations using the rotational isomeric state approximation for isolated chains. Similar agreement has also been noted by Dettenmaier in neutron scattering experiments (*18*) on liquid *n*-alkanes. We are currently investigating the detailed reasons for this agreement by performing simulations for a system as a function of density.

In our model, intramolecular forces are divided into short-range forces, (i.e., those associated with torsional rotation) and long-range forces that provide excluded volume effects. For all of the properties we have considered, it is the excluded volume potential that is most important, because it is responsible for the size and shape of the molecules. Conformational properties for the realistic model and the model with only excluded volume interactions (*B*) are very similar. Removing rotational barriers and excluded volume effects (*C*) leads to a very compact molecular shape with the chain tightly folded into a ring structure. The behavior of this model is interesting but unrealistic for a liquid phase. We have carried out some preliminary gas-phase calculations with potential *C* and obtain reasonable agreement with the freely rotating chain model (Equation 9). Our conclusions concerning the importance of intramolecular interactions in determining the conformations of longer alkane chains are in broad agreement with the findings of van Gunsteren et al. (*19*) in their recent study of Brownian dynamics simulations for these systems.

Large rotational barriers for potentials D and E make the molecules very rigid. We have not calculated the number of *trans–gauche* transitions for any of the simulation runs, but it is evident from the distribution of end-to-end distances (Figure 2) and from the very sharp spikes of the distribution of torsional angles (not shown) that no large-scale conformational changes took place. In contrast, the runs with potentials A–C have reached equilibrium—further results (*8*) show that correlations between bonds decay with a characteristic time of about 1 ps, and van Gunsteren et al. (*19*) have found an average residence time of 3 ps for *trans* bonds in liquid *n*-decane at 481 K.

The standard molecular dynamics technique is not an efficient way of examining the high-rigidity limit for chain molecules. For static properties, the Monte Carlo method described by Bigot and Jorgensen (*20*) should offer a better approach. Molecular dynamics calculations for somewhat smaller increases in the height of the torsional barrier from that of potential A are now being carried out.

The translational diffusion coefficient is not greatly affected by changes in the intramolecular potentials. The relatively small difference between D_{tr} for potentials A and C is particularly surprising, in view of the large difference in both chain length and overall molecular shape between these two cases. The effect of changes in intramolecular forces is much greater for other dynamic properties and longer chains (*8*) although, again, not nearly as marked as the changes in molecular shape and size would lead one to expect. This illustrates the importance of specific intermolecular interactions rather than shape factors (such as hydrodynamic volume) in determining the dynamic properties of chain molecules. The present work represents a first step towards interpreting phenomena generally attributed to such concepts as "chain entanglements" in terms of more specific interactions.

Acknowledgments

We should like to thank Dr. G. Saville for advice on computational techniques and the Imperial College Computer Centre for help with many practical problems. The provision of a travel grant by the Royal Society is gratefully acknowledged.

Literature Cited

1. Evans, G. T. *J. Chem. Phys.* **1981,** *74*, 4621.
2. Weber, T. A. *J. Chem. Phys.* **1978,** *69*, 2347.
3. Ibid. **1979,** *70*, 4277.

4. Weber, T. A.; Helfand, E. *J. Chem. Phys.* **1979,** *71*, 4760.
5. Ibid. **1980,** *72*, 4014.
6. Ryckaert, J. P.; Bellemans, A. *Chem. Phys. Lett.* **1975,** *30*, 23.
7. Ryckaert, J. P.; Bellemans, A. *J. Chem. Soc., Faraday Discuss.* **1978,** *66*, 95.
8. Szczepanski, R.; Maitland, G. C., submitted for publication.
9. Hansen, J. P.; McDonald, I. R. "Theory of Simple Liquids"; Academic Press: New York, 1976; Chap. 3.
10. Ryckaert, J. P.; Ciccotti, G.; Berendsen, H. J. C. *J. Comp. Phys.* **1977,** *23*, 327.
11. Rallison, J. M. *J. Fluid Mech.* **1979,** *93*, 251.
12. Helfand, E. *J. Chem. Phys.* **1979,** *71*, 5000.
13. van Gunsteren, W. F. *Mol. Phys.* **1980,** *40*, 1015.
14. Flory, P. J. "Statistical Mechanics of Chain Molecules"; Wiley: New York, 1969; Chap. 1, pp. 11–18.
15. Ibid. 1969; Chap. 5, pp. 144–149.
16. Douglass, D. C.; McCall, D. W. *J. Phys. Chem.* **1958,** *62*, 1102.
17. Zwanzig, R. *Annu. Rev. Phys. Chem.* **1965,** *16*, 67.
18. Dettenmaier, M. *J. Chem. Phys.* **1978,** *68*, 2319.
19. van Gunsteren, W. F.; Berendsen, H. J. C.; Rullmann, J. A. C. *Mol. Phys.* **1981,** *44*, 69.
20. Bigot, B.; Jorgensen, W. L. *J. Chem. Phys.* **1981,** *75*, 1944.

RECEIVED for review January 27, 1982. ACCEPTED for publication September 10, 1982.

20

Simulation of Polyethylene

THOMAS A. WEBER, EUGENE HELFAND, and ZELDA R. WASSERMAN[1]

Bell Laboratories, Murray Hill, NJ 07974

Both Brownian and molecular dynamics simulations have been performed for polyethylene fluids. The polymer is represented as a backbone skeleton of carbon atoms with flexible bonds and angles in addition to a torsional rotational potential. Conformational transitions g $\rightleftarrows$ t *are observed to occur with an activation energy of one barrier height, although some second-neighbor cooperativity is observed. Reaction rates have been measured from first passage times using hazard analysis and also from orientational and conformational autocorrelation functions. The dynamics of vectors along and perpendicular to the polymer chain have been studied. The time dependence of the correlation functions can be reasonably fitted on the basis of a model with three types of processes. The two fastest correspond to independent conformational transitions and to two-bond correlated transitions. A much slower process, probably related to concerted motion of large segments of the chain, is also observed.*

A RECENT PROGRAM OF STUDY involving computer simulation of polymers has yielded a wealth of both quantitative and qualitative information about these systems (*1–3*). These studies have been of two types. For one, we have simulated the Brownian motion of a single polymer molecule in a viscous medium. From this one learns about polymer internal dynamics, particularly the kinetics and mechanisms of conformational transitions. We have also performed molecular dynamics simulations of a collection of polymer molecules at normal polymer melt densities. In addition to being able to analyze the internal dynamics, we are then able to study equilibrium properties, such as correlations arising from the packing of the molecules.

This report will be devoted exclusively to discussing some of the latest findings from the Brownian dynamics simulations. First, let us summarize several of the results of the earlier research (more detail will

[1] Current address: DuPont Experimental Station, Wilmington, DE 19898

0065-2393/83/0204-0487$06.00/0

be presented later). From the trajectories it has been possible to extract the total transition rate for the *trans*(t) ⇄ *gauche* plus or minus ($g^{\pm}$). For this purpose the method of hazard analysis, borrowed from reliability theory, has proved to be a valuable technique for sharpening quantitative analysis. An Arrhenius plot of the rates (3) reveals that the activation energy is approximately the height of the potential barrier for the rotation of a single bond. Hazard analysis and inspection of the trajectories (at a level of detail unavailable to experimentalists) reveals that many of the transitions occur as cooperative pairs, involving two second-neighbor bonds separated by a *trans*. These two bonds undergo counterrotation in a crank-like fashion (cf. Equations 7b and 7c). Such a process helps to localize the overall motion involved in the transition process.

In the latest series of very long runs (trajectories corresponding to the order of 10 ns) reported here, we have been able to calculate several different orientational autocorrelation functions similar to those measured in dielectric relaxation, NMR, and ESR experiments. Using a diffusional model to describe the conformational transitions, it is possible to compare directly the three different approaches: orientational autocorrelation functions, conformational autocorrelation functions, and hazard analysis (3). We thus aim to establish a closer connection between what the experimentalist can measure and what the simulator can observe. In this chapter we will restrict our focus to only one of the four different vector-orientational autocorrelation functions we have calculated—that vector which bisects two carbon–carbon bonds. Analysis is more complicated for those vectors that retain significant correlation after $g \rightleftarrows t$ relaxation.

In the next section, the model potential used to specify the polyethylene is outlined. Both the bisector-orientational and the bond-conformation autocorrelation functions are defined. Then, in the following section, the three-process model used to approximate the time dependence of the autocorrelation function is defined. In the final section a comparison is made between the rates determined using the different methods. In all instances the results on which we will focus were obtained using the Brownian dynamics method. This allows the calculation of long trajectories, which are necessary for accurate determination of the autocorrelation functions.

Potential and Autocorrelation Functions

The Brownian dynamics simulation of an infinite chain of polyethylene uses interactions along the backbone to maintain bond lengths and bond angles and to model the rotational potential around the torsional angle. The carbon–carbon bonds are maintained near the equilibrium bond length, b_0, by a quadratic potential in bond length (the force constants and parameters chosen correspond to potential A of Reference 3).

The bond angles are maintained near the tetrahedral configuration by a quadratic potential in the cosine of the bond angle. The rotational potential chosen is the Scott and Scheraga (*4*) potential originally used in the Ryckaert and Bellemans (*5*) simulation of butane. The potential is a fifth-order polynomial in the cosine of the torsional angle and is shown in Figure 1. No other explicit interactions between carbon centers along the chain have been included. This potential is thus a type of phantom chain model since the polymer may pass through itself. Excluded volume interactions such as the pentane effect, which prohibits $g^{\pm}g^{\mp}$ pairs, are neglected. Also, specific interactions involving hydrogen atoms are not

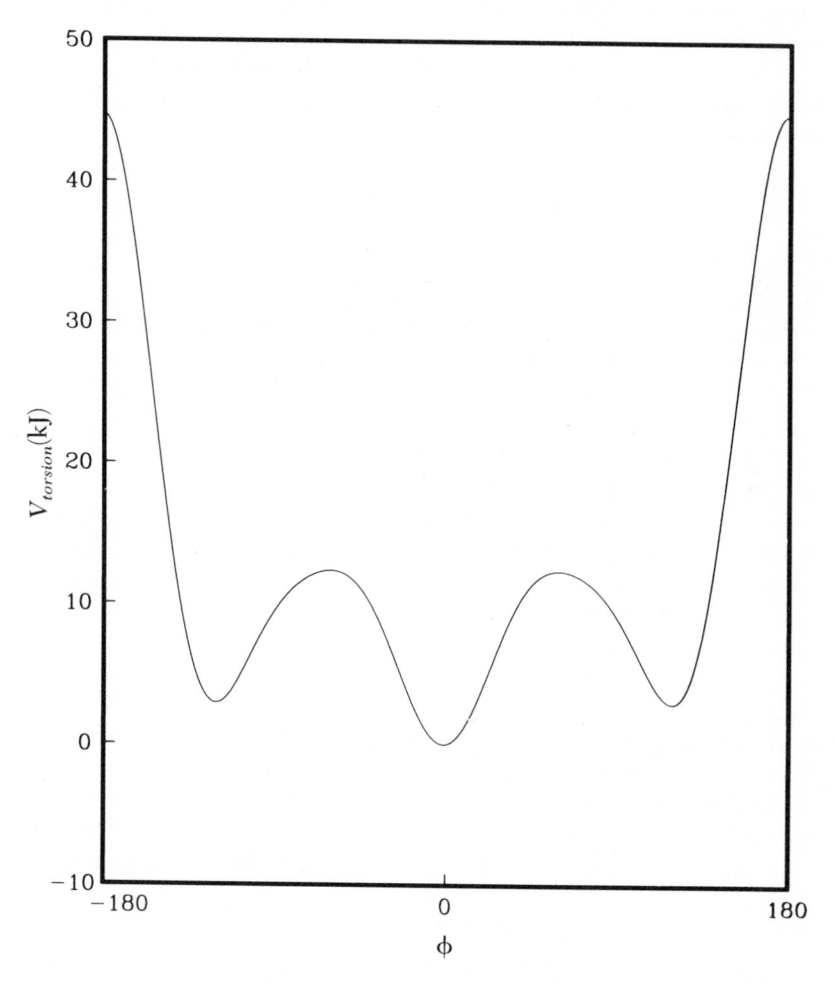

Figure 1. Potential for bond rotation as a function of torsional angle. Trans *corresponds to* $\phi = 0$.

included, although the mass of the CH_2 group is used for the mass of the "carbon" center.

The simulations were performed on a system of 200 carbon centers, which have been periodically repeated to form an infinite chain. This additional boundary condition makes all carbon centers equivalent, improving transition statistics. The position of the 201st carbon center is given as

$$\boldsymbol{r}_{201} = \boldsymbol{r}_1 + \langle R^2 \rangle^{1/2} \boldsymbol{e}_x \tag{1}$$

where $\boldsymbol{e}_x$ is a unit vector along the x direction and $\langle R^2 \rangle$ is the average mean-squared end-to-end distance for a chain of 200 units with the model potential.

Once an initial starting configuration for the polymer chain is chosen, the equations of motion for a Brownian particle, in the high-friction limit, are solved; that is

$$\beta \frac{dr_i}{dt} = -\left(\frac{1}{m}\right) \nabla_i V + \mathbf{A}_i(t) \tag{2}$$

In Equation 2, β is the friction constant divided by the mass, V is the potential for the polyethylene chain and $\mathbf{A}_i(t)$ is a random force representing the incoherent part of the solvent interactions on the ith particle. A representative value for β of 10^5 ns^{-1} has been chosen; since the time scales with β, rates for other values are easily determined.

A representative trajectory is generated using the extension of the Runge–Kutta method to stochastic differential equations recently proposed by Helfand (6). The integration time step is $\Delta t = 5 \times 10^{-6}$ ns. For each of the two temperatures reported here, over 40,000 transitions were observed. Between 2 and 3 million integration time steps were necessary, and the total calculation required in excess of 12 hours on the CRAY-1 computer.

The primary motivation for extending the calculations reported previously (3) by a factor of 10 was to enable the accurate determination of autocorrelation functions to be made. Several different autocorrelation functions were investigated; however, here we will focus on the bisector-orientational autocorrelation function and the *trans*-conformational autocorrelation function.

The vector bisector of the two carbon–carbon bonds is given by

$$f_i = (\hat{\boldsymbol{b}}_i - \hat{\boldsymbol{b}}_{i+1})/\sqrt{2(1-\cos\theta_i)} \tag{3}$$

where $\hat{\boldsymbol{b}}_i(t)$ is the unit vector along the ith bond at time t and $\theta_i(t)$ is the angle between the ith and $i+1$st bond vector. The manner in which this vector relaxes as a function of time is given by the autocorrelation function

$$\langle f_i(t) f_i(0) \rangle = \frac{1}{N} \sum_{i=1}^{N} \int f_i(t'+t) f_i(t')\, dt' \tag{4}$$

where the summation is over all vertices.

The *trans*-conformational autocorrelation function measures the relaxation of a *trans*-conformational state of a bond, and provides a means of estimating the transition rate. This autocorrelation function is given by the function

$$\langle T(t)\, T(0) \rangle = \frac{1}{N} \sum_{i=1}^{N} \int T_i(t'+t)\, T_i(t')\, dt' \tag{5}$$

where $T_i(t)$ is defined as

$$T_i(t) = \begin{cases} 1 & \text{if bond } i \text{ is } trans \text{ at time } t \\ 0 & \text{otherwise} \end{cases} \tag{6}$$

We first postulate that the important mechanisms for $g \rightleftarrows t$ transitions are

$$t \rightleftarrows g^{\pm} \tag{7a}$$

$$ttt \rightleftarrows g^{\pm} t g^{\mp} \tag{7b}$$

$$g^{\pm} tt \rightleftarrows tt g^{\pm} \tag{7c}$$

with forward rates λ_0, λ_1, and λ_2, and reverse rates λ_0', λ_1', λ_2', respectively. Equation 7a represents the rate for single independent transitions, whereas the other two reactions are the most important cooperative transitions observed in our previous study (3) of the transition data. If we define the probabilities that a bond is g^+, t, or g^- as P_+, P_t, and P_-, respectively, then using simple first-order rate theory and a mean-field approximation for the state of neighboring bonds

$$\frac{d}{dt} \begin{bmatrix} P_+ \\ P_t \\ P_- \end{bmatrix} = -\boldsymbol{M} \begin{bmatrix} P_+ \\ P_t \\ P_- \end{bmatrix} \tag{8}$$

The overall rate of transitions is

$$\lambda = \overline{P}_+ M_{11} + \overline{P}_t M_{22} + \overline{P}_- M_{33} \tag{9}$$

where $\overline{P}_k$ is the equilibrium fraction of state k. Using a detailed balance it is easy to show that

$$\lambda_0' = \frac{2\overline{T}}{1 - \overline{T}}\lambda_0 \tag{10a}$$

$$\lambda_1' = \left[\frac{2\overline{T}}{1 - \overline{T}}\right]^2 \lambda_1 \tag{10b}$$

where $\overline{T}$ $(=\overline{P}_t)$ is the average fraction of *trans* conformer. The transition rate (7) is then given as

$$\lambda = 4\overline{T}\lambda_0 + 8\overline{T}^3\lambda_1 + 4\overline{T}^2(1-\overline{T})\lambda_2 \tag{11}$$

The autocorrelation function defined by Equation 5 is

$$\langle T(t)\,T(0)\rangle = \overline{T}P_t(t) \tag{12}$$

where $P_t(t)$ is the solution of Equation 8 subject to the boundary condition that $P_t(0) = 1$. If we focus on only the decay of the *trans*-autocorrelation function it is easy to show that

$$\frac{\langle T(t)\,T(0)\rangle - \overline{T}^2}{\langle T^2\rangle - \overline{T}^2} = \exp(-L_0 t) \tag{13a}$$

where

$$L_0 = \frac{2}{1 - \overline{T}}\lambda_0 + \frac{4\overline{T}^2}{1 - \overline{T}}\lambda_1 + 2\overline{T}\,\lambda_2 \tag{13b}$$

Another method that we have successfully employed to measure the overall rate of transition is hazard analysis. If $h(t)\,dt$ is the probability that a bond that has not undergone a transition in time t will undergo a transition in time $t+dt$, then $h(t)$ is the hazard rate. The cumulative hazard is defined as the integral of the hazard

$$H(t) = \int_0^t h(t')\,dt' \tag{14}$$

In the simulation, a transition is defined as follows: Initially, the state of each bond is specified by the region in which its torsional angle falls. The three regions, g^-, t, and g^+, are delimited by the maxima of the torsional potential curve, Figure 1. A transition to a new state is regarded as having occurred when the torsional angle reaches the bottom of the potential well of some other region. From the collection of times between transitions a hazard plot, such as Figure 2, is prepared. The asymptotic slope is the total transition rate, λ. A comparison will be made between the transition rates as measured by hazard analysis and as determined using the autocorrelation function approach.

Time Dependence of Autocorrelation Functions

The bisector-orientational and *trans*-conformational autocorrelation functions described in the previous section were measured for two very

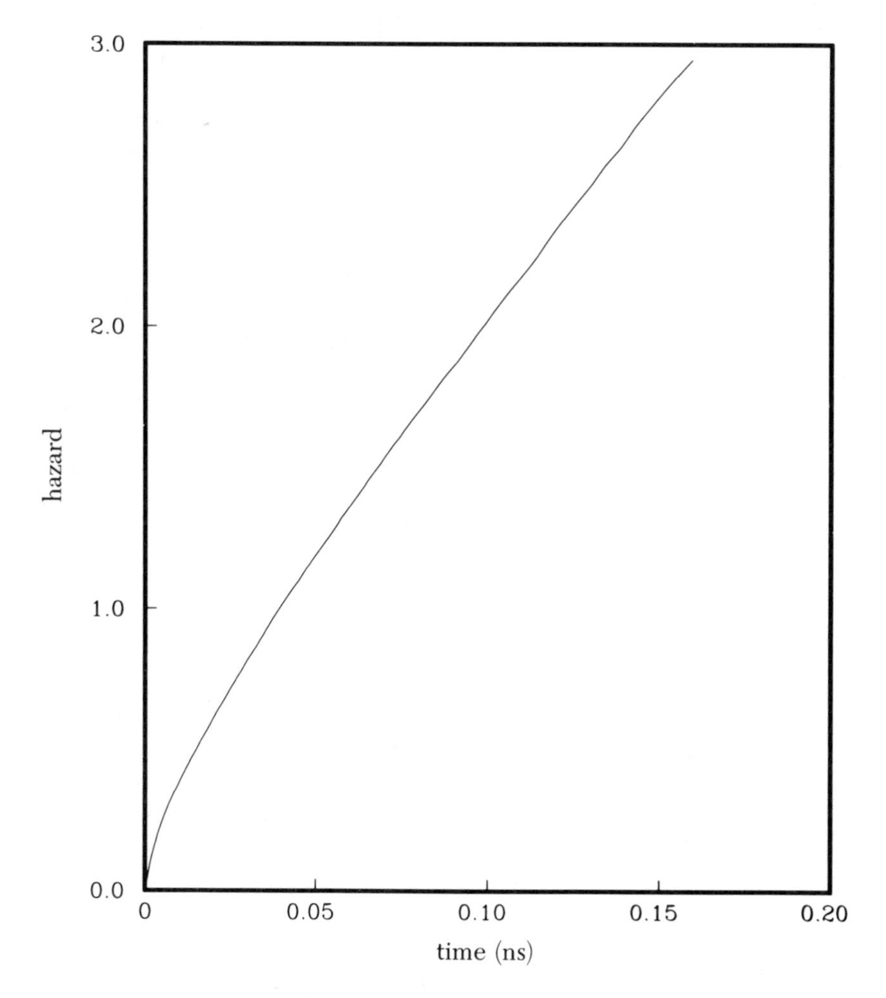

Figure 2. Hazard plot for the 425-K simulation of the polymer.

long runs at 372 K and 425 K. The nonlinearity of a log plot of these autocorrelation functions shows that more complicated processes than a single Poisson transition process occur. This reconfirms our earlier speculations deduced from studying the data via hazard analysis (3). In fact, we postulate that at least three processes are responsible for the loss of correlation. On a fast time scale, at least two processes are occurring—single isolated conformational transitions, and correlated transitions involving at least two bonds. Also, for those vectors that do not completely lose correlation when a bond undergoes $g \rightleftarrows t$ equilibration, it may be necessary to move large segments of the chain to destroy the correlation completely. This third process is so slow that accurate measurement of its rate is not possible even with the very long trajectories that we have accumulated. The bisector-orientational function, however, completely loses correlation by conformational transitions, which greatly simplifies the data analysis.

A general form for the phenomenological equation describing the time dependence of this three-process autocorrelation function is

$$\Phi(t) = (1-\alpha)e^{-t/\theta}F(t/\rho) + \alpha e^{-t/\tau} \tag{15}$$

The single exponential in θ is from simple diffusional loss of correlation due to single bond conformational transitions. The exponential in τ describes the rather slow loss of correlation due to rotational diffusion of larger segments of the polymer backbone. The parameter α partitions the correlation loss between the mechanisms, and it may in some sense be viewed as the amount of correlation remaining after many conformational transitions prior to significant loss through this third mechanism. For the bisector-orientational autocorrelation function, α is identically zero. The function $F(t/\rho)$ describes the loss of correlation due to the two-bond correlated transitions.

In our previous studies using hazard analysis (2, 3), we found that shortly after a transition there was an enhanced probability that the second neighbor would undergo a transition. This second-neighbor cooperativity was especially enhanced when the intervening bond was a *trans* conformer. In addition, the type of transition observed corresponds to counterrotation of bonds, which had the effect of causing only minor translational movement of the polymer tails. The favored types of correlated transitions we observed were those described by Equations 7b and 7c.

This two-bond cooperativity may be modeled by a diffusional process (8)

$$F(t/\rho) = \sum_{n=0}^{\infty} Q_{+n}Q_{-n} \tag{16}$$

where Q_{+n} is the probability of making n transitions along the chain and Q_{-n} is the probability of making n transitions in the reverse direction. Therefore, $Q_{+n}Q_{-n}$ is the probability of returning to the origin after an excursion of n transitions. If the transition rate for a step in one direction is $1/(2\rho)$ then according to Poisson statistics

$$F(t/\rho) = \sum_{n=0}^{\infty} \left[\frac{(t/2\rho)^n \exp(-t/2\rho)}{n!} \right]^2$$
$$F(t/\rho) = \exp(-t/\rho) I_0(t/\rho) \tag{17}$$

where I_0 is the zeroth-order Bessel function.

The *trans*-conformational and bisector-orientational autocorrelation function have been fitted by using

$$\Phi(t) = e^{-t/\theta} e^{-t/\rho} I_0(t/\rho) \tag{18}$$

Since θ represents the independent process and ρ models the correlated transitions, Equations 13 and 18 may be combined to give

$$\frac{1}{\theta} = \frac{2}{1 - \overline{T}} \lambda_0 \tag{19a}$$

$$\frac{1}{\rho} = \frac{4\overline{T}^2}{1 - \overline{T}} \lambda_1 + 2\overline{T}\lambda_2 \tag{19b}$$

A plot of the bisector-orientational autocorrelation function for the run at 425 K is shown in Figure 3. The fit between the data and the function defined by Equation 18 is exceedingly good. Attempts were made to fit the data using other functional forms, such as a sum of two exponentials and the exponential and error-function compliment form as suggested by the work of Monnerie et al. (9) and of Stockmayer and co-workers (*10*). These alternate fitting functions, however, were unable to reproduce the curvature of the data in the region where $\lambda_1 t$ was of the order of 0.1. The θ and ρ parameters, as extracted from a least-squares fit to the data at the two different temperatures, are given in Table I.

A plot of the *trans*-conformational autocorrelation function as defined by Equation 13 is shown in Figure 4 for the run at 425 K. Here again, the fit using the form given by Equation 18 is exceedingly good. The two parameters, θ and ρ, extracted from a least-squares fit to the data are given in Table I.

Figure 2 shows the hazard plot for the data at 425 K. The slope of the long time portion of the curve is proportional to the total rate. We

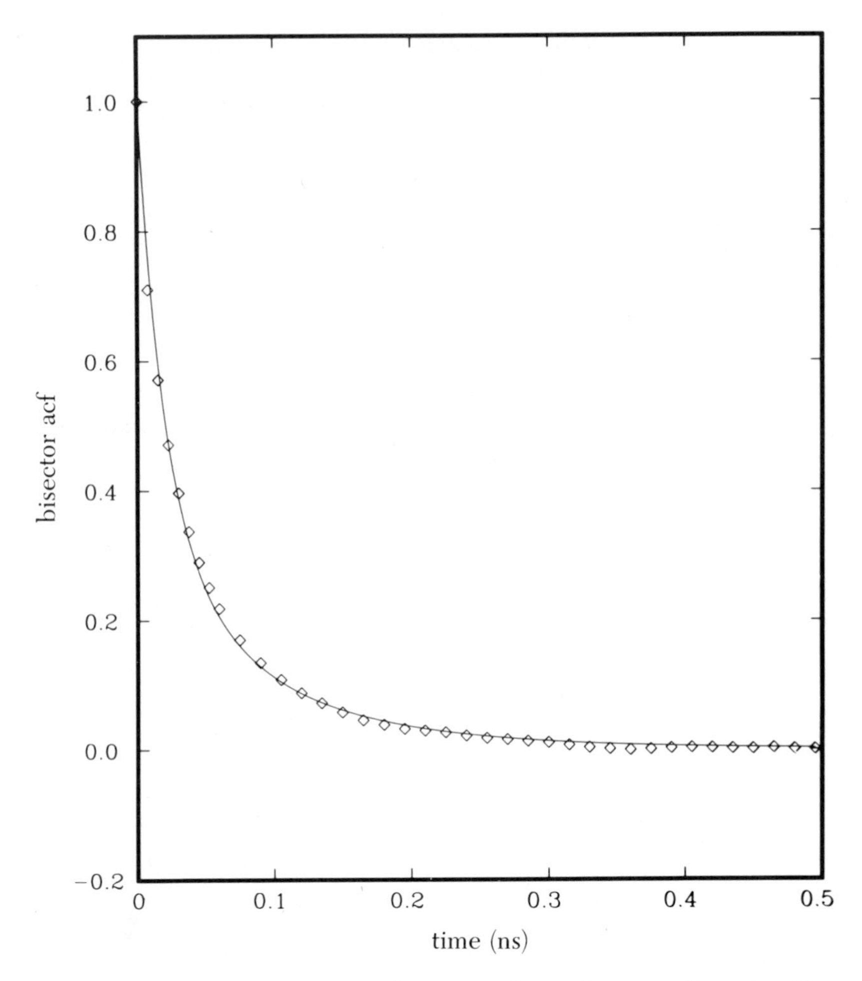

Figure 3. Bisector-orientational autocorrelation function, Equation 13, vs. time (ns) for the run at T = *425 K. The points represent calculated data, the solid line is a fit to the function of Equation 18.*

have determined the rate by a maximum likelihood estimation of the parameters of a functional fit to the hazard (*11*). The total transition rates for the runs at the two different temperatures are given in the table. This is related to λ_{tg}, the *trans* to *gauche* transition rate, which was previously (3) defined by the equation $\lambda = 4\overline{T}\lambda_{tg}$.

Comparison of Rates

The values for the θ parameter as determined by the different autocorrelation function techniques, either bisector-orientational or *trans*-

Table I. Results of Long Runs

Property	*Run 1*	*Run 2*
T (K)	372	425
Number of transitions	40,069	40,052
Total time (ns)	14.912	9.204
E^*/k_BT	3.5	4.0
% t actual	60.081	55.931
% t theoretical	59.745	56.881
Bisector acf [a]		
θ	0.244	0.133
ρ	0.0472	0.0324
Trans acf [a]		
θ	0.253	0.108
ρ	0.0555	0.0368
λ_0	0.7940	1.9927
λ_1	1.7008	3.2150
λ_2	10.0170	15.4305
λ	10.539	17.960
Hazard λ	10.740	17.528

[a] Autocorrelation function

conformational relaxation, are exceedingly close, as are values for the ρ parameter. This similarity in rates indicates that both types of relaxation studies are monitoring the same fundamental processes. The relaxation of the *trans*-conformational autocorrelation function is the easiest to understand in terms of the simple single transition and correlated transition model. We will focus on comparing the total transition rates as determined from conformational relaxation studies with those determined directly from hazard analysis.

To determine the overall transition rate λ from the relaxation data it is necessary to specify the rate λ_2 (the rate for the reaction $g^{\pm}tt \rightarrow ttg^{\pm}$) in terms of the rate λ_1 (the rate for the reaction $ttt \rightarrow g^{\pm}tg^{\mp}$). One approach that might be taken is to analyze the transition data for the relative frequency of the two types of reactions. This ratio is calculated to be

$$\chi = \frac{\overline{T}^3\lambda_1 + \left[\frac{1-\overline{T}}{2}\right]^2 \overline{T}\lambda_1'}{\overline{T}^2(1-\overline{T})\lambda_2} \qquad (20)$$

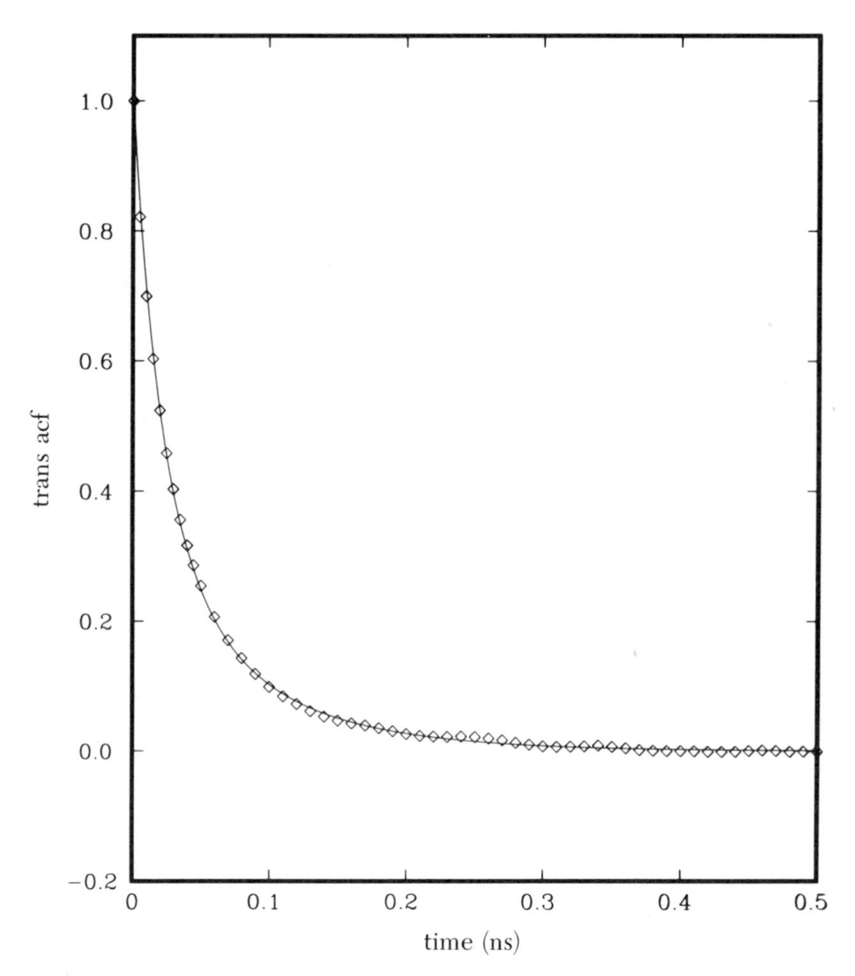

Figure 4. The trans-*conformational autocorrelation function, Equation 4, vs. time (ns) for the run at* T = *425* K. *The points represent the calculated data, the solid line is a fit using Equation 18.*

and was determined experimentally for the run (3) at 372 K to be χ = 407/523 = 0.778.

Another approach is to estimate λ_1 and λ_2 using reaction rate theory. In this case, the rate is given as the rate of crossing the highest barrier separating the initial and final states. If there are two reaction channels the rates are added. In this way we write

$$\lambda_1 = 2r \exp[-(E^* + E_g)/k_B T] \tag{21}$$

where r is a prefactor representing the rate of barrier crossing and entropy differences between initial and activated states; the number 2 arises from the two equivalent channels (which *trans* transforms first); and the final factor is the activation energy term. Similarly, we write

$$\lambda_2 = r\{\exp[-(E^* - E_g)/k_B T] + \exp[-E^*/k_B T]\} \tag{22}$$

where the sum in braces arises from the two channels (the $g^\pm \rightarrow t$ first, or the $t \rightarrow g^\pm$ first). Using these estimates we find

$$\lambda_2 = \frac{(1+\overline{T})\overline{T}}{(1-\overline{T})^2}\lambda_1 \tag{23}$$

The value for χ using Equations 20 and 23 at 372 K is found to be 0.504. This ratio is lower than is experimentally observed and amounts to an overestimate of the rate λ_2.

The rates for the single isolated transitions and the cooperative transitions may be found by using Equations 19 and 23. The values for λ_0, λ_1, and λ_2 as determined from the *trans*-conformational autocorrelation function relaxation data for the two temperatures are listed in Table I. The total transition rate for all mechanisms is then calculated from Equation 11, and these rates are listed in Table I. The agreement between the rates as determined from hazard analysis and those determined from the relaxation rates is striking. This is especially so, considering that cooperative transitions were modeled in a mean-field approximation, and that the estimate used to relate λ_1 and λ_2 is rather crude. Furthermore, only the most important two bond transition mechanisms, Equations 7b and 7c, have been included, so that approximately 20% of the correlated transitions have been neglected (*see* Table VII of Reference 3).

The study of the relaxation of the *trans*-conformational autocorrelation function provides a means of separating the rates of isolated transitions from the rates for cooperative transitions. More importantly, the temperature dependence of the two competing processes may now be studied.

From hazard analysis it is possible to determine total rates of transitions with relatively short trajectories, of the order of 4000 transitions. These rates have not changed substantially when the trajectory was increased by a factor of 10 (*3*). It is not possible, however, to determine accurate autocorrelation functions for such short runs and still extract meaningful relaxation rates through fits to the data, using Equation 18. It is thus consoling to note that the rates determined by the two methods are in fundamental agreement.

One of the most interesting aspects of the present work is the close agreement between the parameters determined by the *trans*-conformational autocorrelation function and those determined from the bisector-orientational autocorrelation function. This provides a strong verification of the conjecture that relaxation within the polymer, at least on the time scales we are considering, is due primarily to the single and cooperative conformational transitions. Work is continuing to extract rates from the vector autocorrelation functions, because experimental relaxation techniques such as NMR, ESR, and depolarized light scattering measure the manner in which various orientational vectors or tags within the polymer relax.

Literature Cited

1. Weber, T. A.; Helfand, E. *J. Chem. Phys.* **1979,** *71*, 4760.
2. Helfand, E.; Wasserman, Z. R.; Weber, T. A. *J. Chem. Phys.* **1979,** *70*, 2016.
3. Helfand, E.; Wasserman, Z. R.; Weber, T. A. *Macromolecules* **1980,** *13*, 526.
4. Scott, R. A.; Scheraga, H. A. *J. Chem. Phys.* **1966,** *44*, 3054.
5. Ryckaert, J.-P.; Bellemans, A. *Chem. Phys. Lett.* **1975,** *30*, 123.
6. Helfand, E. *Bell Syst. Tech. J.* **1979,** *58*, 2289.
7. Weber, T. A.; Helfand, E.; in press.
8. Hall, C.; Helfand, E. *J. Chem. Phys.* **1982,** *77*, 3275.
9. Valeur, B.; Jarry, J.-P.; Geny, F.; Monnerie, L. *J. Polym. Sci.* **1975,** *13*, 667.
10. Jones, A. A.; Stockmayer, W. H. *J. Polym. Sci.* **1977,** *15*, 847.
11. Mann, N. R.; Schafer, R. E.; Singpurwalla, N. D. "Methods for Statistical Analysis of Reliability and Life Data"; Wiley: New York, 1974;

RECEIVED for review January 27, 1982. ACCEPTED for publication September 10, 1982.

INDEX

INDEX

C

E

F

G

H

J

K

L

M

N

X

Y

Z

Production Editors: Florence H. Edwards and Anne G. Bigler
Indexer: Florence H. Edwards
Jacket Designer: Anne G. Bigler

Typesetting by FotoTypesetters, Inc., Baltimore, Md
Printing and binding by Maple Press, Inc., Manchester, Pa

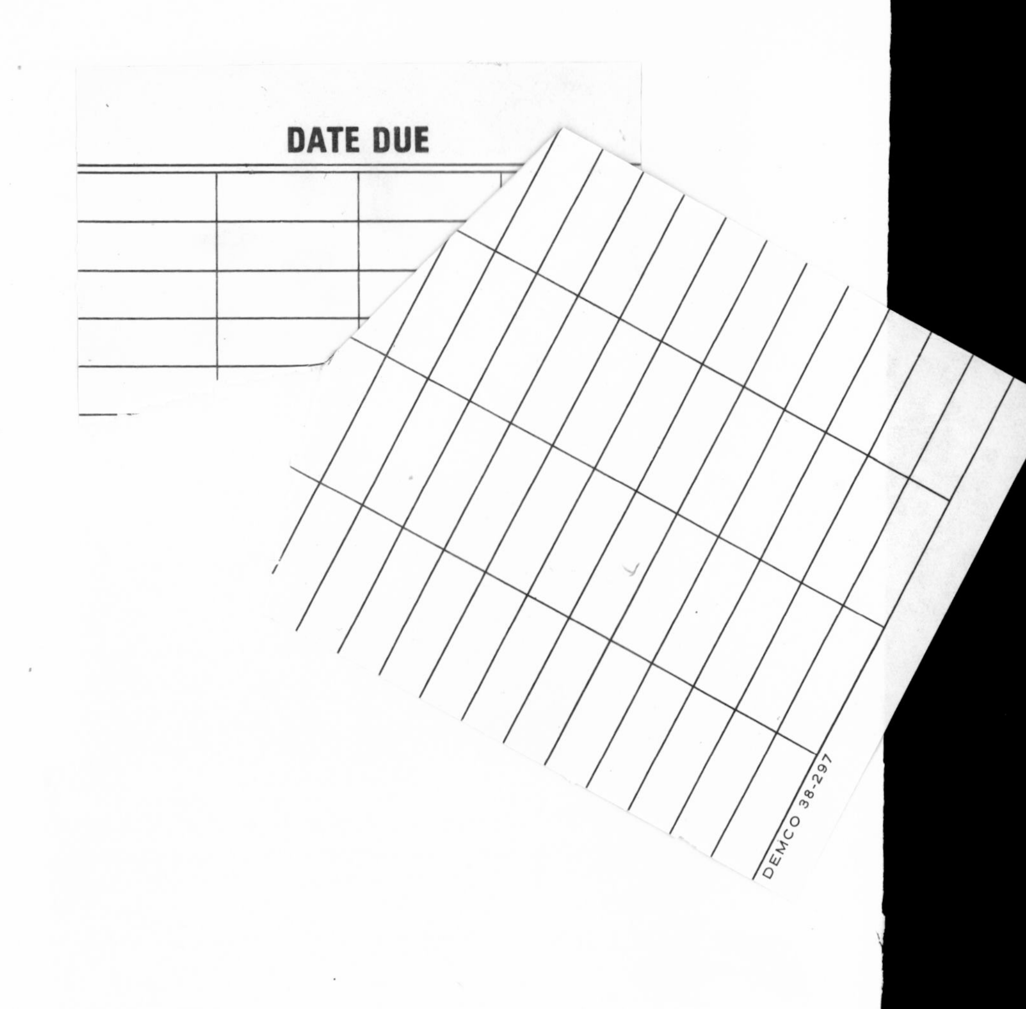
DATE DUE
DEMCO 38-297